BROADBAND MOBILE MULTIMEDIA

WIRELESS NETWORKS AND MOBILE COMMUNICATIONS

Dr. Yan Zhang, Series Editor
Simula Research Laboratory, Norway
E-mail: yanzhang@ieee.org

Unlicensed Mobile Access Technology: Protocols, Architectures, Security, Standards and Applications
Yan Zhang, Laurence T. Yang and Jianhua Ma
ISBN: 1-4200-5537-2

Wireless Quality-of-Service: Techniques, Standards and Applications
Maode Ma, Mieso K. Denko and Yan Zhang
ISBN: 1-4200-5130-X

Broadband Mobile Multimedia: Techniques and Applications
Yan Zhang, Shiwen Mao, Laurence T. Yang and Thomas M Chen
ISBN: 1-4200-5184-9

The Internet of Things: From RFID to the Next-Generation Pervasive Networked Systems
Lu Yan, Yan Zhang, Laurence T. Yang and Huansheng Ning
ISBN: 1-4200-5281-0

Millimeter Wave Technology in Wireless PAN, LAN, and MAN
Shao-Qiu Xiao, Ming-Tuo Zhou and Yan Zhang
ISBN: 0-8493-8227-0

Security in Wireless Mesh Networks
Yan Zhang, Jun Zheng and Honglin Hu
ISBN: 0-8493-8250-5

Resource, Mobility and Security Management in Wireless Networks and Mobile Communications
Yan Zhang, Honglin Hu, and Masayuki Fujise
ISBN: 0-8493-8036-7

Wireless Mesh Networking: Architectures, Protocols and Standards
Yan Zhang, Jijun Luo and Honglin Hu
ISBN: 0-8493-7399-9

Mobile WIMAX: Toward Broadband Wireless Metropolitan Area Networks
Yan Zhang and Hsiao-Hwa Chen
ISBN: 0-8493-2624-9

Distributed Antenna Systems: Open Architecture for Future Wireless Communications
Honglin Hu, Yan Zhang and Jijun Luo
ISBN: 1-4200-4288-2

AUERBACH PUBLICATIONS
www.auerbach-publications.com
To Order Call: 1-800-272-7737 • Fax: 1-800-374-3401
E-mail: orders@crcpress.com

BROADBAND MOBILE MULTIMEDIA

Techniques and Applications

Edited by
Yan Zhang • Shiwen Mao
Laurence T. Yang • Thomas M. Chen

CRC Press is an imprint of the
Taylor & Francis Group, an **informa** business
AN AUERBACH BOOK

Auerbach Publications
Taylor & Francis Group
6000 Broken Sound Parkway NW, Suite 300
Boca Raton, FL 33487-2742

Printed in the United States of America on acid-free paper
10 9 8 7 6 5 4 3 2 1

International Standard Book Number-13: 978-1-4200-5184-1 (Hardcover)

Library of Congress Cataloging-in-Publication Data

Broadband mobile multimedia : techniques and applications / edited by Yan Zhang ... [et al.].
p. cm. -- (Wireless networks and mobile communications ; v, 9)
Includes bibliographical references and index.
ISBN-13: 978-1-4200-5184-1
ISBN-10: 1-4200-5184-9
1. Mobile computing. 2. Multimedia systems. 3. Broadband communication systems. I. Zhang, Yan-Qing.

QA76.59.B76 2008
004.6'6--dc22 2007050624

Visit the Taylor & Francis Web site at
http://www.taylorandfrancis.com

and the Auerbach Web site at
http://www.auerbach-publications.com

Table of Contents

Preface .... **vii**

Editors .... **xi**

List of Contributors .... **xv**

SECTION I: MULTIMEDIA SYSTEMS

1 **Design Challenges for Wireless Multimedia Sensor Networks** **3**
Tommaso Melodia and Kaushik R. Chowdhury

2 **Performance Analysis of Multimedia Traffic over HSDPA** **47**
Irene de Bruin, Frank Brouwer, Neill Whillans, Yusun Fu, and Youqian Xiao

3 **Interactive Mobile TV Technologies: An Overview** **87**
Ramakrishna Vedantham and Igor D.D. Curcio

4 **Multiparty Audioconferencing on Wireless Networks** **119**
R. Venkatesha Prasad, Vijay S. Rao, H.N. Shankar, and R. Muralishankar

SECTION II: MULTIMEDIA OVER *AD HOC* AND SENSOR NETWORKS

5 **Routing for Video Communications over Wireless *Ad Hoc* Networks** **157**
Shiwen Mao, Y. Thomas Hou, Hanif D. Sherali, and Scott F. Midkiff

6 Multipath Unicast and Multicast Video Communication over Wireless *Ad Hoc* Networks ... 193
Wei Wei and Avideh Zakhor

7 Video Communications over Wireless Sensor Networks ... 235
Min Chen, Shiwen Mao, Yong Yuan, and Victor C.M. Leung

SECTION III: MULTIMEDIA OVER WIRELESS LOCAL AREA NETWORKS

8 Multimedia Quality-of-Service Support in IEEE 802.11 Standards ... 261
Zhifeng Tao, Thanasis Korakis, Shivendra Panwar, and Leandros Tassiulas

9 Peer-Assisted Video Streaming over WLANs ... 305
Danjue Li, Chen-Nee Chuah, Gene Cheung, and S.J. Ben Yoo

10 Multimedia Services over Broadband Wireless LAN ... 333
Jianhua He, Zuoyin Tang, Yan Zhang, and Zongkai Yang

11 Improving User-Perceived Quality for Video Streaming over WLAN ... 361
Nikki Cranley and Gabriel-Miro Muntean

SECTION IV: QUALITY OF SERVICE AND ENABLING TECHNOLOGIES

12 End-to-End QoS Support for Video Delivery over Wireless Internet ... 409
Qian Zhang, Wenwu Zhu, and Ya-Qin Zhang

13 Handoff Management of Wireless Multimedia Services: A Middleware Approach ... 435
P. Bellavista, A. Corradi, and L. Foschini

14 Packet Scheduling in Broadband Wireless Multimedia Networks ... 473
Rong Yu, Yan Zhang, Zhi Sun, and Shunliang Mei

15 The Peak-to-Average Power Ratio in Orthogonal Frequency Division Multiplexing Wireless Communication Systems ... 505
Tao Jiang, Laurence T. Yang, and Yan Zhang

Index ... 531

Preface

"Follow the money" was W. Mark Felt's (a.k.a. *Deep Throat*) advice to Bob Woodward and Carl Bernstein for unraveling the Watergate scandal. After years of research and development, there have been significant advances in signal processing, networking and delivery technologies, network infrastructure and deployment, as well as successful business models. Multimedia is now ready to hit the market. Users are not satisfied with simple forms of communications anymore. A wide range of multimedia applications are emerging, such as Voice-over-IP, online chatting, video on demand, Internet Protocol Television (IPTV) or cellvision/mobile TV, and interactive gaming, among others. Service providers are making great efforts to move toward "triple play" or "quad play." This trend is highlighted by the recent multi-billion-dollar eBay/Skype deal and the Google/YouTube deal.

An equally important advice for engineers, researchers, and for all of us is to "find the bottleneck." Thus, where is the performance bottleneck in the big picture as multimedia is becoming widely available in the Internet? The answer, we believe, is wireless access networks, such as third-generation (3G) and beyond wireless networks, Wi-Fi, WiMAX, and Bluetooth wireless local area networks (WLAN), *ad hoc*/mesh networks, and wireless sensor networks. Despite considerable advances, current wireless networks are not able to offer comparable data rates as do their wired counterparts. Although it frees users from a socket and cable, mobility brings about a new level of challenge, including time-varying wireless channels and dynamic topology and connectivity. The situation is even more serious in the case of multihop wireless networks, where end-to-end throughput quickly decreases as hop count increases, largely due to carrier sensing and spatial reuse issues. As the increasing demand for multimedia communications continues to drive the expansion of consumer and enterprise markets as well as the evolution of wireless technologies, multimedia service provisioning is believed to be one of the prerequisites to guarantee the success of the next-generation wireless networks.

This book aims to meet this compelling need by providing a collection of the latest advances in this important problem area. Given the considerable research effort being made and the vast literature that exists, it is not possible to provide a complete coverage of all the related issues. However, we aim to provide a big picture of state-of-the-art research, a representative sampling of important research outcomes, and in-depth treatments of selected topics in the area of broadband wireless multimedia communications. Overall, this book is a useful technical guide covering introductory concepts, fundamental techniques, latest advances, and open issues in broadband wireless multimedia. A large number of illustrative figures, cross-references, as well as comprehensive references for readers interested in more details are provided.

This book consists of 15 chapters, which are organized into four parts as follows:

- Multimedia systems
- Multimedia over *ad hoc* and sensor networks
- Multimedia over wireless local area networks
- QoS and enabling technologies

Part I introduces various broadband wireless multimedia systems and surveys related work. Part II focuses on the routing and cross-layer design issue of multimedia communication over multihop wireless *ad hoc*/sensor networks, where video is used as a reference application. Part III explores various issues related to multimedia communications over WLANs, which constitute a dominant part of today's broadband wireless access networks. Part IV presents latest advances in QoS provisioning mechanisms and other enabling technologies, including end-to-end QoS provisioning, middleware, mobility management, scheduling, and power control.

The salient features of the book are as follows:

- Identifies the basic concepts, key technologies, and cutting-edge research outcomes, as well as open problems and future research directions in the important problem area of broadband mobile multimedia
- Provides comprehensive references on state-of-the-art technologies for broadband wireless multimedia
- Contains a sufficient number of illustrative figures for easy reading and understanding of the materials
- Allows complete cross-referencing through a broad coverage on layers of the protocol architecture
- In-depth treatment of selected problems/technologies for enabling wireless multimedia service

The book represents a useful reference for techniques and applications of broadband wireless multimedia. Target readers include students, educators, telecommunication service providers, research strategies, scientists, researchers, and engineers working in the areas of wireless communications, wireless networking, and multimedia communications. It can also be used as a textbook for an advanced selected topic course on broadband wireless multimedia for graduate students.

This book would not have been possible without the efforts and the time invested by all the contributors. They were extremely professional and cooperative, and did a great job in the production of this book. Our reviewers provided valuable comments/feedbacks, which, we believe, greatly helped improve the quality of this book. Special thanks go to Rich O'Hanley, Jessica Vakili, and Karen Schober of Taylor & Francis Group for their continuous support, patience, and professionalism from the beginning to the final stage. Last but not least, we thank our families and friends for their constant encouragement, patience, and understanding throughout this project, which was a pleasant and rewarding experience.

Yan Zhang
Shiwen Mao
Laurence T. Yang
and **Thomas M. Chen**

Editors

Yan Zhang holds a PhD from the School of Electrical and Electronic Engineering, Nanyang Technological University, Singapore. Since August 2006, he has been working with the Simula Research Laboratory, Norway (http://www.simula.no/). He is an associate editor of *Security and Communication Networks* (Wiley) and is also on the editorial board of three international journals. He has served as coeditor for a number of special journal issues and a score of books. Dr. Zhang has been active as a chair, cochair, or organizer for many workshops since 2006. He received the Best Paper Award and Outstanding Service Award in the IEEE 21st International Conference on Advanced Information Networking and Applications (AINA-07). His research interests include resource, mobility, spectrum, energy, and security management in wireless networks and mobile computing. He is also a member of IEEE and IEEE ComSoc.

Shiwen Mao is an assistant professor in the Department of Electrical and Computer Engineering, Auburn University, Auburn, Alabama. He received his BS and MS in electrical engineering from Tsinghua University, Beijing, People's Republic of China in 1994 and 1997, respectively. He also received his BS in business management from Tsinghua University in 1994; an MS in system engineering and a PhD in electrical and computer engineering from Polytechnic University, Brooklyn, New York, in 2000 and 2004, respectively. He was a research member at IBM China Research Lab, Beijing from 1997 to 1998, and a research intern at Avaya Labs-Research,

Holmdel, New Jersey, in the summer of 2001. He was a research scientist in the Department of Electrical and Computer Engineering, Virginia Tech, Blacksburg, Virginia, from December 2003 to April 2006. He is an associate editor of two respected journals, and he served as cochair of the Wireless Communications and Networking Symposium of CHINACOM 2008. Dr. Mao coauthored two textbooks: *TCP/IP Essentials: A Lab-Based Approach* (Cambridge University Press, 2004) and *Principles of Network Performance Analysis* (Cambridge University Press, 2008). He is a member of the IEEE and received a certificate of appreciation from the IEEE Computer Society in 2007. Dr. Mao's research interests include cross-layer design and optimization in multihop wireless networks, cognitive networks, and multimedia communications.

Laurence T. Yang is a professor at St Francis Xavier University, Antigonish, Nova Scotia, Canada. His research focuses on high-performance computing and networking, embedded systems, ubiquitous/pervasive computing, and intelligence. He has published approximately 280 papers in refereed journals, conference proceedings, and book chapters in these areas. He has been involved in more than 100 conferences and workshops as a program/general conference chair, and in more than 200 conference and workshops as a program committee member. He has served on numerous IEEE Technical Committees and on the IFIP Working Group 10.2 on Embedded Systems. In addition, he is the editor-in-chief of nine international journals and a few book series. He serves as an editor for international journals. He has contributed to, edited, or coedited 30 books. He has won numerous awards for his research, including several IEEE Best Paper Awards.

Thomas M. Chen is an associate professor in the Department of Electrical Engineering at Southern Methodist University (SMU). Before joining SMU, he was a senior member of the technical staff at GTE Laboratories (now Verizon) in Waltham, Massachusetts. He is currently the editor-in-chief of *IEEE Communications Magazine*. He also serves as a senior technical editor for *IEEE Network*, and was the founding editor of *IEEE Communications Surveys*. He serves as a member-at-large on the IEEE Communications Society Board of Governors. He is a coauthor of *ATM Switching Systems* (Artech House 1995). He received the IEEE Communications Society's Fred W. Ellersick best paper award in 1996.

List of Contributors

P. Bellavista
University of Bologna
Bologna, Italy

Frank Brouwer
Twente Institute for Wireless and Mobile Communications
Enschede, The Netherlands

Irene de Bruin
Twente Institute for Wireless and Mobile Communications
Enschede, The Netherlands

Min Chen
University of British Columbia
Vancouver, British Columbia, Canada

Gene Cheung
HP Research Laboratories
Tokyo, Japan

Kaushik R. Chowdhury
Georgia Institute of Technology
Atlanta, Georgia

Chen-Nee Chuah
University of California
Davis, California

A. Corradi
University of Bologna
Bologna, Italy

Nikki Cranley
Dublin Institute of Technology
Dublin, Ireland

Igor D.D. Curcio
Nokia Research Center
Tampere, Finland

L. Foschini
University of Bologna
Bologna, Italy

Yusun Fu
Huawei Technologies Co., Ltd
Shanghai, China

Jianhua He
Swansea University
Swansea, United Kingdom

Y. Thomas Hou
Virginia Polytechnic Institute and State University
Blacksburg, Virginia

Tao Jiang
Huazhong University of Science and Technology
Wuhan, China

Thanasis Korakis
Polytechnic University
Brooklyn, New York

Victor C.M. Leung
University of British Columbia
Vancouver, British Columbia, Canada

Danjue Li
Cisco System Ltd
San Jose, California

Shiwen Mao
Auburn University
Auburn, Alabama

Shunliang Mei
Tsinghua University
Beijing, China

Tommaso Melodia
State University of New York at Buffalo
Buffalo, New York

Scott F. Midkiff
Virginia Polytechnic Institute and State University
Blacksburg, Virginia
and
National Science Foundation
Arlington, Virginia

Gabriel-Miro Muntean
Dublin City University
Dublin, Ireland

R. Muralishankar
PES Institute of Technology
Bangalore, India

Shivendra Panwar
Polytechnic University
Brooklyn, New York

R. Venkatesha Prasad
Delft University of Technology
Delft, The Netherlands

Vijay S. Rao
Delft University of Technology
Delft, The Netherlands

H.N. Shankar
PES Institute of Technology
Bangalore, India

Hanif D. Sherali
Virginia Polytechnic Institute and State University
Blacksburg, Virginia

Zhi Sun
Georgia Institute of Technology
Atlanta, Georgia

Zuoyin Tang
Swansea University
Swansea, United Kingdom

Zhifeng Tao
Mitsubishi Electric Research Laboratories
Cambridge, Massachusetts

Leandros Tassiulas
University of Thessaly
Volos, Greece

Ramakrishna Vedantham
Nokia Research Center
Palo Alto, California

Wei Wei
University of California
Berkeley, California

Neill Whillans
Twente Institute for Wireless and Mobile Communications
Enschede, The Netherlands

Youqian Xiao
Huawei Technologies Co., Ltd
Shanghai, China

Zongkai Yang
Huazhong University of Science and Technology
Wuhan, China

Laurence T. Yang
St. Francis Xavier University
Antigonish, Nova Scotia, Canada

S.J. Ben Yoo
University of California
Davis, California

Rong Yu
South China University of Technology
Guangzhou, China

Yong Yuan
Huazhong University of Science and Technology
Wuhan, China

Avideh Zakhor
University of California
Berkeley, California

Qian Zhang
Hong Kong University of Science and Technology
Hong Kong, China

Yan Zhang
Simula Research Laboratory
Lysaker, Norway

Ya-Qin Zhang
Microsoft Corporation
Beijing, China

Wenwu Zhu
Microsoft Corporation
Beijing, China

MULTIMEDIA SYSTEMS

Chapter 1

Design Challenges for Wireless Multimedia Sensor Networks

Tommaso Melodia and Kaushik R. Chowdhury

Contents

1.1 Introduction 4
1.2 Applications of Wireless Multimedia Sensor Networks 6
1.3 Design and Characteristics of Wireless Multimedia Sensor Networks 7
1.4 Network Architecture 10
1.4.1 Reference Architecture 10
1.4.2 Single- versus Multitier Sensor Deployment 12
1.4.3 Coverage 13
1.4.4 Internal Organization of a Multimedia Sensor 13
1.5 Application Layer 14
1.5.1 Traffic Classes 15
1.5.2 Multimedia Encoding Techniques 16
1.5.2.1 Pixel-Domain Wyner–Ziv Encoder 18
1.5.2.2 Transform-Domain Wyner–Ziv Encoder 20
1.5.3 System Software and Middleware 20
1.5.4 Collaborative In-Network Processing 22
1.5.5 Open Research Issues 23

1.6 Protocols 23
1.6.1 Transport-Layer Protocols 24
1.6.1.1 TCP/UDP and TCP-Friendly Schemes for WMSNs 24
1.6.1.2 Application-Specific and Nonstandard Protocols 26
1.6.2 Network Layer 27
1.6.2.1 QoS Routing Based on Network Conditions 27
1.6.2.2 QoS Routing Based on Traffic Classes 28
1.6.2.3 Routing Protocols with Support for Streaming 28
1.6.3 MAC Layer 29
1.6.3.1 Channel Access Policies 29
1.6.3.2 Contention-Based Protocols 29
1.6.3.3 Contention-Free Single-Channel Protocols 30
1.6.3.4 Contention-Free Multi-Channel Protocols 32
1.6.3.5 Scheduling 33
1.6.3.6 Link-Layer Error Control 34
1.6.4 Cross-Layer Protocols 35
1.7 Physical Layer 36
1.7.1 Ultra Wideband Communications 37
1.7.1.1 Ranging Capabilities of UWB 38
1.7.1.2 Standards Based on UWB 39
1.7.1.3 Open Research Issues 39
1.8 Conclusions 40
Acknowledgment 40
References 40

Low-cost CMOS cameras, microphones, and sensors have recently become ubiquitously available. This has fostered the development of the so-called wireless multimedia sensor networks (WMSNs), that is, distributed systems composed of wirelessly interconnected devices that can ubiquitously retrieve multimedia content such as video and audio streams, still images, and scalar sensor data from the environment. This chapter discusses the state-of-the-art in algorithms, protocols, and hardware for WMSNs; open research issues are discussed in detail. Architectures for WMSNs are explored, along with their advantages and drawbacks. Existing solutions and open research issues at the application, transport, network, link, and physical layers of the communication stack are investigated, along with possible cross-layer synergies and optimizations.

1.1 Introduction

Wireless sensor networks (WSNs) [1] have significantly attracted the attention of the research community in the past few years. Significant results in this area have resulted in the increased interest in potential civil and military applications. In fact, sensor networks can consistently enhance our ability to gather physical information from the environment. However, as of today, most deployed WSNs measure scalar physical phenomena such as temperature, pressure, humidity, or location of objects. Sensor network communication protocols are usually designed for applications that have low bandwidth demands, and are delay-tolerant.

The recent availability of inexpensive hardware such as CMOS cameras and microphones, able to ubiquitously capture multimedia content from the environment, is paving the way for the development of the so-called wireless multimedia sensor networks (WMSN) [2,3], that is, networks of wirelessly interconnected devices that allow retrieving video and audio streams, still images, and scalar sensor data. With rapid improvements and miniaturization in hardware, a single device can be equipped with audio and visual information-collection modules. As an example, the Cyclops image capturing and inference module [4] is designed for extremely light-weight imaging and can be interfaced with a host mote such as Crossbow's MICA2 [5] or MICAz [6]. In addition, WMSNs will be able to store, process in real-time, correlate, and fuse multimedia data originating from heterogeneous sources.

The challenges associated with the delivery and in-network processing of multimedia content require the sensor network paradigm to be rethought to provide applications with a predefined level of quality of service (QoS). Because the goal of minimizing the energy consumption has driven most of the research in sensor networks so far, mechanisms to efficiently deliver application-level QoS, and to map these requirements to network-layer metrics such as latency and jitter, have not been primary concerns for sensor network researchers.

Conversely, algorithms, protocols, and techniques to deliver multimedia content over large-scale networks have been the focus of intensive research in the past 20 years, especially in Asynchronous Transfer Mode (ATM) wired and wireless networks. Later, many of the results derived for ATM networks have been readapted, and architectures such as Diffserv and Intserv for Internet QoS delivery have been developed. However, there are several peculiarities that make the QoS delivery of multimedia content in sensor networks an even more challenging, and largely unexplored, task.

This chapter provides a detailed description of the main research challenges in algorithms and protocols for the development of WMSNs. In particular, Section 1.2 describes the applications of WMSNs, whereas Section 1.3 describes the main characteristics of WMSNs, including the major factors influencing their design, reference architecture, and internal organization. Section 1.4 suggests possible architectures for WMSNs and describes their characterizing features. Section 1.5 discusses the functionalities handled at the application layer of a WMSN, including possible advantages and challenges of multimedia in-network processing. Section 1.6 discusses the existing solutions and open research issues at the transport, network, link, and physical layers of the communication stack, respectively. Finally, Section 1.8 concludes the chapter.

1.2 Applications of Wireless Multimedia Sensor Networks

WMSNs will enhance existing sensor network applications such as tracking, home automation, and environmental monitoring. Moreover, new applications will be enabled by the possibility of retrieving multimedia content such as the following:

- *Multimedia surveillance sensor networks.* Wireless video sensor networks will be composed of interconnected, battery-powered miniature video cameras, each packaged with a low-power wireless transceiver that is capable of processing, sending, and receiving data. Video and audio sensors will be used to enhance and complement existing surveillance systems against crime and terrorist attacks. Large-scale networks of video sensors can extend the ability of law enforcement agencies to monitor areas, public events, private properties, and borders.
- *Storage of potentially relevant activities.* Multimedia sensors could infer and record potentially relevant activities (thefts, car accidents, traffic violations), and make video/audio streams or reports available for future query.
- *Traffic avoidance, enforcement, and control systems.* It will be possible to monitor car traffic in big cities or highways and deploy services that offer traffic routing advice to avoid congestion. In addition, smart parking advice systems based on WMSNs [7] will allow monitoring available parking spaces and provide drivers with automated parking advice, thus improving mobility in urban areas. Moreover, multimedia sensors may monitor the flow of vehicular traffic on highways and retrieve aggregate information such as average speed and number of cars. Sensors could also detect violations and transmit video streams to law enforcement agencies to identify the violator, or buffer images and streams in case of accidents for subsequent accident scene analysis.
- *Advanced healthcare delivery.* Telemedicine sensor networks [8] can be integrated with third-generation (3G) multimedia networks to provide ubiquitous healthcare services. Patients will carry medical sensors to monitor parameters such as body temperature, blood pressure, pulse oximetry, electrocardiogram (ECG), and breathing activity. Furthermore, remote medical centers will perform advanced remote

monitoring of their patients via video and audio sensors, location sensors, and motion or activity sensors, which can also be embedded in wrist devices [8].

- *Automated assistance for the elderly and family monitors*. Multimedia sensor networks can be used to monitor and study the behavior of elderly people as a means to identify the causes of illnesses that affect them such as dementia [9]. Networks of wearable or video and audio sensors can infer emergency situations and immediately connect elderly patients with remote assistance services or with relatives.
- *Environmental monitoring*. Several projects on habitat monitoring that use acoustic and video feeds are being envisaged, in which information has to be conveyed in a time-critical fashion. For example, arrays of video sensors are already being used by oceanographers to determine the evolution of sandbars via image-processing techniques [10].
- *Structural health monitoring*. Multimedia streams can be used to monitor the structural health of bridges [11] or other structures.
- *Person locator services*. Multimedia content such as video streams and still images, along with advanced signal-processing techniques, can be used to locate missing persons, or identify criminals or terrorists.
- *Industrial process control*. Multimedia content such as imaging, temperature, or pressure, among others, may be used for time-critical industrial process control. "Machine vision" is the application of computer vision techniques to industry and manufacturing, where information can be extracted and analyzed by WMSNs to support a manufacturing process such as those used in semiconductor chips, automobiles, food, or pharmaceutical products. For example, in the quality control of manufacturing processes, details or final products are automatically inspected to find defects. In addition, machine vision systems can detect the position and orientation of parts of the product to be picked up by a robotic arm. The integration of machine vision systems with WMSNs can simplify and add flexibility to systems for visual inspections and automated actions that require high-speed, high-magnification, and continuous operation.

1.3 Design and Characteristics of Wireless Multimedia Sensor Networks

There are several factors that mainly influence the design of a WMSN, which are outlined in this section.

- *Application-specific QoS requirements*. The wide variety of applications envisaged on WMSNs will have different requirements. In addition to data delivery modes typical of scalar sensor networks, multimedia data includes "snapshot" and "streaming multimedia" content. Snapshot-type multimedia data contains event-triggered observations obtained in a short time period. Streaming multimedia content is generated over longer time periods and requires sustained information delivery. Hence, a strong foundation is needed in terms of hardware and supporting high-level algorithms to deliver QoS and consider application-specific requirements. These requirements may pertain to multiple domains and can be expressed, among others, in terms of a combination of bounds on energy consumption, delay, reliability, distortion, or network lifetime.

- *High bandwidth demand.* Multimedia content, especially video streams, require transmission bandwidth that is orders of magnitude higher than that supported by currently available sensors. For example, the nominal transmission rate of state-of-the-art IEEE 802.15.4 compliant components such as Crossbow's [6] MICAz or TelosB [12] motes is 250 kbit/s. Data rates at least one order of magnitude higher may be required for high-end multimedia sensors, with comparable power consumption. Hence, high data rate and low power consumption transmission techniques need to be leveraged. In this respect, the ultra wideband (UWB) transmission technique seems particularly promising for WMSNs, and its applicability is discussed in Section 1.7.
- *Cross-layer coupling of functionalities.* In multihop wireless networks, there is a strict interdependence among functions handled at all layers of the communication stack. Functionalities handled at different layers are inherently and strictly coupled due to the shared nature of the wireless communication channel. Hence, the various functionalities aimed at QoS provisioning should not be treated separately when efficient solutions are sought.
- *Multimedia source coding techniques.* Uncompressed raw video streams require excessive bandwidth for a multihop wireless environment. For example, a single monochrome frame in the NTSC-based Quarter Common Intermediate Format (QCIF; 176 × 120) requires approximately 21 kB, and at 30 frames per second (fps), a video stream requires over 5 Mbit/s. Hence, it is apparent that efficient processing techniques for lossy compression are necessary for multimedia sensor networks. Traditional video coding techniques used for wireline and wireless communications are based on the idea of reducing the bit rate generated by the source encoder by exploiting source statistics. To this aim, encoders rely on "intraframe" compression techniques to reduce redundancy within one frame, whereas they leverage "interframe" compression (also known as "predictive encoding" or "motion estimation") to exploit redundancy among

subsequent frames to reduce the amount of data to be transmitted and stored, thus achieving good rate-distortion performance. Because predictive encoding requires complex encoders, powerful processing algorithms, and entails high energy consumption, it may not be suited for low-cost multimedia sensors. However, it has recently been shown [13] that the traditional balance of complex encoder and simple decoder can be reversed within the framework of the so-called distributed source coding, which exploits the source statistics at the decoder, and by shifting the complexity at this end allows the use of simple encoders. Clearly, such algorithms are very promising for WMSNs and especially for networks of video sensors, where it may not be feasible to use existing video encoders at the source node due to processing and energy constraints.

- *Multimedia in-network processing.* Processing of multimedia content has mostly been approached as a problem isolated from the network-design problem, with a few exceptions such as joint source-channel coding [14] and channel-adaptive streaming [15]. Hence, research that addressed the content delivery aspects has typically not considered the characteristics of the source content and has primarily studied cross-layer interactions among lower layers of the protocol stack. However, the processing and delivery of multimedia content are not independent and their interaction has a major impact on the levels of QoS that can be delivered. WMSNs will allow performing multimedia in-network processing algorithms on the raw data. Hence, the QoS required at the application level will be delivered by means of a combination of both cross-layer optimization of the communication process and in-network processing of raw data streams that describe the phenomenon of interest from multiple views, with different media, and on multiple resolutions. Hence, it is necessary to develop application-independent and self-organizing architectures to flexibly perform in-network processing of multimedia contents.
- *Power consumption.* This is a fundamental concern in WMSNs, even more than in traditional WSNs. In fact, sensors are battery-constrained devices, whereas multimedia applications produce high volumes of data, which require high transmission rates and extensive processing. The energy consumption of traditional sensor nodes is known to be dominated by the communication functionalities, whereas this may not necessarily be true in WMSNs. Therefore, protocols, algorithms, and architectures to maximize the network lifetime while providing the QoS required by the application are a critical issue.
- *Flexible architecture to support heterogeneous applications.* WMSN architectures will support several heterogeneous and independent applications with different requirements. It is necessary to develop

flexible, hierarchical architectures that can accommodate the requirements of all these applications in the same infrastructure.

- *Multimedia coverage.* Some multimedia sensors, in particular video sensors, have larger sensing radii and are sensitive to direction of acquisition (directivity). Furthermore, video sensors can capture images only when there is unobstructed line of sight between the event and the sensor. Hence, coverage models developed for traditional WSNs are not sufficient for predeployment planning of a multimedia sensor network.
- *Integration with Internet Protocol (IP) architecture.* It is of fundamental importance for the commercial development of sensor networks to provide services that allow querying the network to retrieve useful information from anywhere and at any time. For this reason, future WMSNs will be remotely accessible from the Internet, and will therefore need to be integrated with the IP architecture. The characteristics of WSNs rule out the possibility of all IP sensor networks and recommend the use of application-level gateways or overlay IP networks as the best approach for integration between WSNs and the Internet [16].
- *Integration with other wireless technologies.* Large-scale sensor networks may be created by interconnecting local "islands" of sensors through other wireless technologies. This needs to be achieved without sacrificing on the efficiency of the operation within each individual technology.

1.4 Network Architecture

The problem of designing a "scalable network architecture" is of primary importance. Most proposals for WSNs are based on a flat, homogeneous architecture in which every sensor has the same physical capabilities and can only interact with neighboring sensors. Traditionally, the research on algorithms and protocols for sensor networks has focused on "scalability," that is, how to design solutions whose applicability would not be limited by the growing size of the network. Flat topologies may not always be suited to handle the amount of traffic generated by multimedia applications including audio and video. Similarly, the processing power required for data processing and communications, and the power required to operate it, may not be available on each node.

1.4.1 Reference Architecture

In Figure 1.1, we introduce a reference architecture for WMSNs, where three sensor networks with different characteristics are shown, possibly deployed

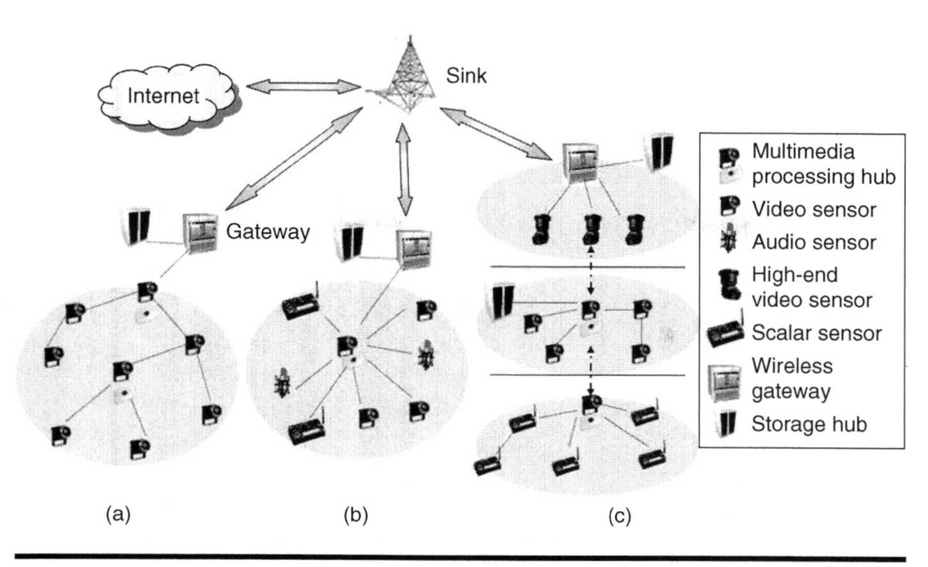

Figure 1.1 Reference architecture of a WMSN. (a) Single-tier flat, homogeneous sensors, distributed processing, centralized storage; (b) single-tier clustered, heterogeneous sensors, centralized processing, centralized storage; (c) multitier, heterogeneous sensors, distributed processing, distributed storage.

in different physical locations. The first cloud on the left shows a single-tier network of homogeneous video sensors. A subset of the deployed sensors have higher processing capabilities, and are thus referred to as "processing hubs." The union of the processing hubs constitutes a distributed processing architecture. The multimedia content gathered is relayed to a "wireless gateway" through a multihop path. The gateway is interconnected to a "storage hub," which is in charge of storing multimedia content locally for subsequent retrieval. Clearly, more complex architectures for distributed storage can be implemented when allowed by the environment and the application needs, which may result in energy savings because by storing it locally the multimedia content does not need to be wirelessly relayed to remote locations. The wireless gateway is also connected to a central "sink," which implements the software front end for network querying and tasking. The second cloud represents a single-tiered clustered architecture of heterogeneous sensors (only one cluster is depicted). Video, audio, and scalar sensors relay data to a central clusterhead (CH), which is also in charge of performing intensive multimedia processing on the data (processing hub). The CH relays the gathered content to the wireless gateway and storage hub. The last cloud on the right represents a multitiered network, with heterogeneous sensors. Each tier is in charge of a subset of the functionalities. Resource-constrained, low-power scalar sensors are in charge of performing simpler tasks, such as detecting scalar physical measurements, whereas resource-rich, high-power devices are responsible for more complex tasks. Data processing and storage can be performed in a distributed fashion at each different tier.

1.4.2 Single- versus Multitier Sensor Deployment

One possible approach for designing a multimedia sensor application is to deploy homogeneous sensors, and program each sensor to perform all possible application tasks. Such an approach yields a flat, single-tier network of homogeneous sensor nodes. An alternative, multitier approach is to use heterogeneous elements (Figure 1.2) [17]. In this approach, resource-constrained, low-power elements are in charge of performing simpler tasks, such as detecting scalar physical measurements, whereas resource-rich, high-power devices take on more complex tasks. For instance, a surveillance application can rely on low-fidelity cameras or scalar acoustic sensors to perform motion or intrusion detection, whereas high-fidelity cameras can be woken up on demand for object recognition and tracking. In Ref. 18, a multitier architecture is advocated for video sensor networks for surveillance applications. The architecture is based on multiple tiers of cameras with different functionalities, with the lower tier composed of

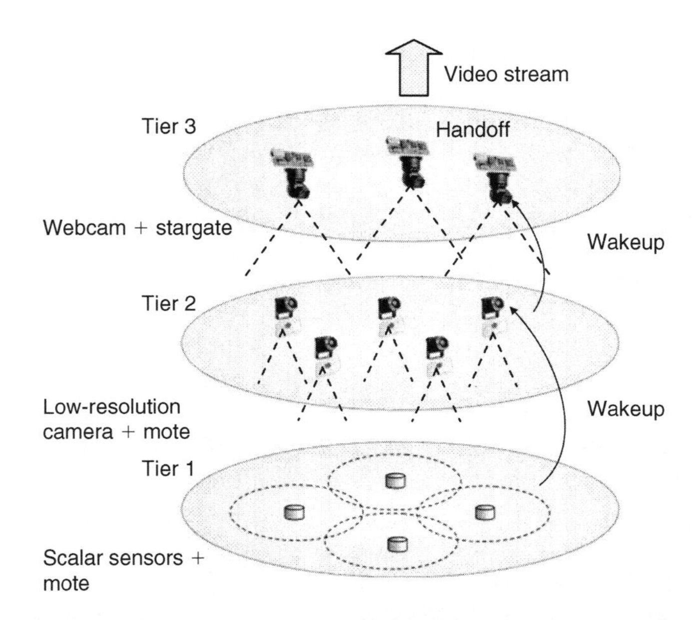

Figure 1.2 The multitier architecture of SensEye. (From Kulkarni, P., Ganesan, D., Shenoy, P., and Lu, Q. SensEye: A Multi-Tier Camera Sensor Network. *Proc. of ACM Multimedia*. Singapore, November 2005.)

low-resolution imaging sensors, and the higher tier composed of high-end pan-tilt-zoom cameras. It is argued, and shown by means of experiments, that such an architecture offers considerable advantages with respect to a single-tier architecture in terms of scalability, lower cost, better coverage, higher functionality, and better reliability.

1.4.3 Coverage

In traditional WSNs, sensor nodes collect information from the environment within a predefined "sensing range," that is, a roughly circular area defined by the type of sensor being used.

Multimedia sensors generally have larger sensing radii and are also sensitive to the direction of data acquisition. In particular, cameras can capture images of objects or parts of regions that are not necessarily close to the camera itself. However, the image can obviously be captured only when there is an unobstructed line of sight between the event and the sensor. Furthermore, each multimedia sensor/camera perceives the environment or the observed object from a different and unique viewpoint, given the different orientations and positions of the cameras relative to the observed event or region. In Ref. 19, a preliminary investigation of the coverage problem for video sensor networks is conducted. The concept of sensing range is replaced with the camera's "field of view," that is, the maximum volume visible from the camera. It is also shown how an algorithm designed for traditional sensor networks does not perform well with video sensors in terms of coverage preservation of the monitored area.

1.4.4 Internal Organization of a Multimedia Sensor

A sensor node is composed of several basic components, as shown in Figure 1.3: a sensing unit, a processing unit (central processing unit [CPU]), a communication subsystem, a coordination subsystem, a storage unit (memory), and an optional mobility/actuation unit. Sensing units are usually composed of two subunits: sensors (cameras, audio, or scalar sensors) and analog to digital converters (ADCs). The analog signals produced by the sensors based on the observed phenomenon are converted to digital signals by the ADC, and then fed into the processing unit. The processing unit, which is generally associated with a storage unit (memory), manages the procedures that make the sensor node collaborate with the other nodes to carry out the assigned sensing tasks. A communication subsystem connects the node to the network, and is composed of a transceiver unit and of communication software. The latter includes a communication protocol stack and system software such as middleware, operating systems,

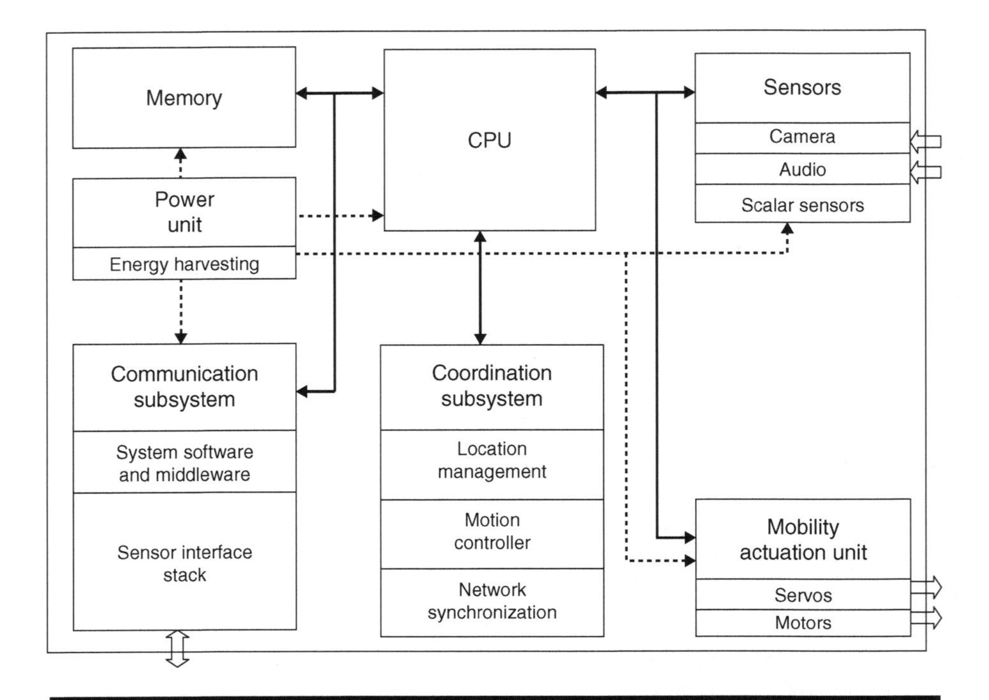

Figure 1.3 Internal organization of a multimedia sensor.

and virtual machines. A coordination subsystem is in charge of coordinating the operation of different network nodes by performing operations such as network synchronization and location management. An optional mobility/actuation unit can also be controlled by the coordination subsystem. Finally, the whole system is powered by a power unit that may be supported by an energy scavenging unit such as solar cells.

1.5 Application Layer

The functionalities handled at the application layer of a WMSN are characterized by high heterogeneity, and encompass traditional communication problems as well as more general system challenges. The services offered by the application layer include (i) providing traffic management and admission control functionalities, that is, prevent applications from establishing data flows when the network resources needed are not available; (ii) performing source coding according to application requirements and hardware constraints by leveraging advanced multimedia encoding techniques; (iii) providing flexible and efficient system software, that is, operating systems and middleware to export services for higher-layer applications to build upon; and (iv) providing primitives for applications to leverage collaborative, advanced in-network multimedia processing techniques. This section provides an overview of these challenges.

1.5.1 Traffic Classes

Admission control has to be based on QoS requirements of the overlying application. We envision that WMSNs will need to provide support and differentiated service for several different classes of applications. In particular, they will need to provide differentiated service between real-time and delay-tolerant applications, and loss-tolerant and loss-intolerant applications. Moreover, some applications may require a continuous stream of multimedia data for a prolonged period of time (multimedia streaming), whereas some other applications may require event-triggered observations obtained in a short time period (snapshot multimedia content). The main traffic classes that need to be supported are

- *Real-time, loss-tolerant, multimedia streams.* This class includes video and audio streams, or multilevel streams composed of video/audio and other scalar data (e.g., temperature readings), as well as metadata associated with the stream, that need to reach a human or automated operator in real-time, that is, within strict delay bounds, and that are however relatively loss-tolerant (e.g., video streams can be within a certain level of distortion). Traffic in this class usually has a high bandwidth demand.
- *Delay-tolerant, loss-tolerant, multimedia streams.* This class includes multimedia streams that, being intended for storage or subsequent offline processing, does not need to be delivered within strict delay bounds. However, due to the typically high bandwidth demand of multimedia streams and limited buffers of multimedia sensors, data in this traffic class needs to be transmitted almost in real-time to avoid excessive losses.
- *Real-time, loss-tolerant, data.* This class may include monitoring data from densely deployed scalar sensors such as light sensors whose monitored phenomenon is characterized by spatial correlation or loss-tolerant snapshot multimedia data (e.g., images of a phenomenon taken from several multiple viewpoints at the same time). Hence, sensor data has to be received timely, but the application is moderately loss-tolerant. The bandwidth demand is usually between low and moderate.
- *Real-time, loss-intolerant, data.* This may include data from time-critical monitoring processes such as distributed control applications. The bandwidth demand varies between low and moderate.

- *Delay-tolerant, loss-intolerant, data.* This may include data from critical monitoring processes, with low or moderate bandwidth demand, that require some form of offline postprocessing.
- *Delay-tolerant, loss-tolerant, data.* This may include environmental data from scalar sensor networks, or non-time-critical snapshot multimedia content, with low or moderate bandwidth demand.

Table 1.1 presents a possible mapping of the applications, presented in Section 1.5, into the traffic classes described earlier.

1.5.2 Multimedia Encoding Techniques

There exists a vast literature on multimedia encoding techniques. The captured multimedia content should ideally be represented in such a way as to allow reliable transmission over lossy channels (error-resilient coding) using algorithms that minimize processing power and the amount of information to be transmitted. The main design objectives of a coder for multimedia sensor networks are thus

- *High compression efficiency.* Uncompressed raw video streams require high data rates and thus consume excessive bandwidth and energy. It is necessary to achieve a high ratio of compression to effectively limit bandwidth and energy consumption.
- *Low complexity.* Multimedia encoders are embedded in sensor devices. Hence, they need to be of low complexity to reduce cost and form factors, and of low power to prolong the lifetime of sensor nodes.
- *Error resiliency.* The source coder should provide robust and error-resilient coding of source data.

To achieve a high compression efficiency, the traditional broadcasting paradigm for wireline and wireless communications, where video is compressed once at the encoder and decoded several times, has been dominated by predictive encoding techniques. These, used in the widely spread International Standards Organization (ISO) Motion Picture Experts Group (MPEG) schemes, or the International Telecommunication Union – Telecommunication Standardization Sector (ITU-T) recommendations H.263 [20] and H.264 [21] (also known as AVC or MPEG-4 part 10), are based on the idea of reducing the bit rate generated by the source encoder by exploiting source statistics. Hence, intraframe compression techniques are used to reduce redundancy within one frame, whereas interframe compression exploits the correlation among subsequent frames to reduce the amount of data to be transmitted and stored, thus achieving good rate-distortion performance. Because the computational complexity is

Table 1.1 Classification of WMSN Applications and QoS Requirements

Application	*Classification (Data Type)*	*Level QoS*	*Mapped Bandwidth*	*Description*
Surveillance networks, monitoring elderly patients, person locator services	Multimedia	Real-time, loss-tolerant	High	Multilevel streams composed of video/audio and scalar data, as well as stream metadata, that need to reach the user in real time
Storage of potent relevant activities	Multimedia	Delay-tolerant, loss-tolerant	High	Streams intended for the storage of subsequent offline processing, but need to be delivered quickly owing to the limited buffers of sensors
Traffic avoidance, enforcement, and control systems	Data	Real-time, loss-tolerant	Moderate	Monitoring data from densely deployed spatially correlated scalar sensors or snapshot multimedia data (e.g., images of phenomena taken from simultaneous multiple viewpoints)
Industrial process control	Data	Real-time, loss-intolerant	Moderate	Data from time-critical monitoring processes such as distributed control applications
Structural health monitoring	Data	Delay-tolerant, loss-tolerant	Moderate	Data from monitoring processes that require some form of offline post-processing
Environmental monitoring	Data	Delay-tolerant, loss-tolerant	Low	Environmental data from scalar sensor networks, or non-time-critical snapshot multimedia content

dominated by the motion estimation functionality, these techniques require complex encoders, powerful processing algorithms, and entail high energy consumption, whereas decoders are simpler and loaded with lower processing burden. For typical implementations of state-of-the-art video compression standards, such as MPEG or H.263 and H.264, the encoder is five to ten times more complex than the decoder [13]. Hence, to realize low-cost, low-energy-consumption multimedia sensors it is necessary to develop simpler encoders and still retain the advantages of high compression efficiency.

However, it is known from information-theoretic bounds established by Slepian and Wolf for lossless coding [22] and by Wyner and Ziv [23] for lossy coding with decoder side information that efficient compression can be achieved by leveraging knowledge of the source statistics at the decoder only. This way, the traditional balance of complex encoder and simple decoder can be reversed [13]. Techniques that build upon these results are usually referred to as "distributed source coding." Distributed source coding refers to the compression of multiple correlated sensor outputs that do not communicate with one another [24]. Joint decoding is performed by a central entity that receives data independently compressed by different sensors. However, practical solutions have not been developed until recently. Clearly, such techniques are very promising for WMSNs and especially for networks of video sensors. The encoder can be simple and of low power, whereas the decoder at the sink will be complex and loaded with most of the processing and energy burden. For excellent surveys on the state-of-the-art of distributed source coding in sensor networks and in distributed video coding, see Refs 24 and 13, respectively. Other encoding and compression schemes that may be considered for source coding of multimedia streams, including JPEG with differential encoding, distributed coding of images taken by cameras having overlapping fields of view, or multilayer coding with wavelet compression, are discussed in Ref. 3. Here, we focus on recent advances on low-complexity encoders based on Wyner–Ziv coding [23], which are promising solutions for distributed networks of video sensors that are likely to have a major impact in future design of protocols for WMSNs.

The objective of a Wyner–Ziv video coder is to achieve lossy compression of video streams and achieve performance comparable to that of interframe encoding (e.g., MPEG), with complexity at the encoder comparable to that of intraframe coders (e.g., motion-JPEG).

1.5.2.1 Pixel-Domain Wyner–Ziv Encoder

In Refs 25 and 26, a practical Wyner–Ziv encoder is proposed as a combination of a pixel-domain intraframe encoder and interframe decoder system for video compression. A block diagram of the system is reported in Figure 1.4. A regularly spaced subset of frames are coded using a conventional intraframe coding technique, such as JPEG, as shown at the bottom

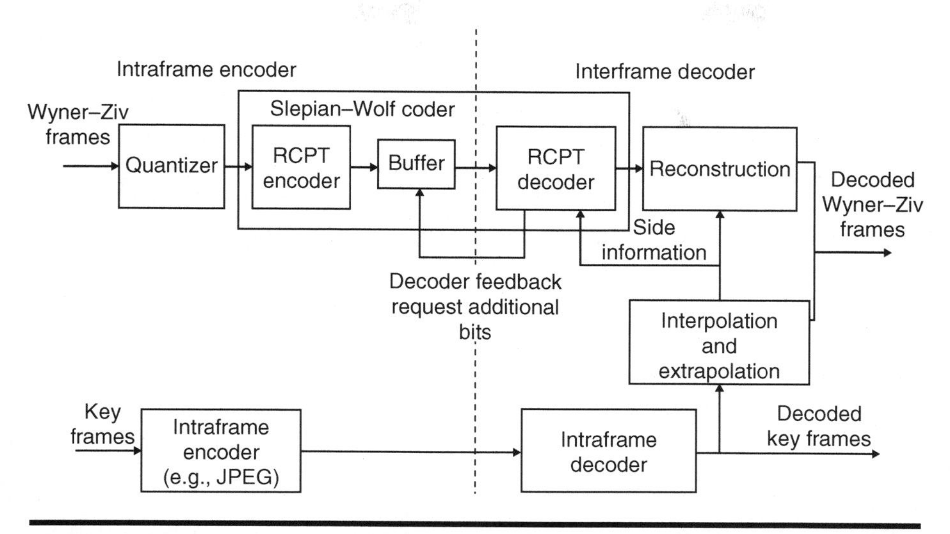

Figure 1.4 Block diagram of a pixel-domain Wyner–Ziv encoder. (From Aaron, A., Rane, S., Zhang, R., and Girod, B. Wyner–Ziv Coding for Video: Applications to Compression and Error Resilience. *Proc. of IEEE Data Compression Conf. (DCC)*, pp. 93–102. Snowbird, UT, March 2003. With permission.)

of the Figure 1.4. These are referred to as "key frames." All frames between the key frames are referred to as "Wyner–Ziv frames" and are intraframe encoded but interframe decoded. The intraframe encoder for Wyner–Ziv frames (shown on top) is composed of a quantizer followed by a Slepian–Wolf coder. Each Wyner–Ziv frame is quantized and blocks of symbols are sent to the Slepian–Wolf coder, which is implemented through rate-compatible punctured turbo codes (RCPT). The parity bits generated by the RCPT coder are stored in a buffer. A subset of these bits is then transmitted on request from the decoder. This allows adapting the rate based on the temporally varying statistics between the Wyner–Ziv frame and the side information. The parity bits generated by the RCPT coder are in fact used to "correct" the frame interpolated at the decoder. For each Wyner–Ziv frame, the decoder generates the side information frame by interpolation or extrapolation of previously decoded key frames and Wyner–Ziv frames. The side information is leveraged by assuming a Laplacian distribution of the difference between the individual pixels of the original frame and the side information. The parameter defining the Laplacian distribution is estimated online. The turbo decoder combines the side information and the parity bits to reconstruct the original sequence of symbols. If reliable decoding of the original symbols is impossible, the turbo decoder requests additional parity bits from the encoder buffer.

Compared to predictive coding such as MPEG or H.26X, pixel-domain Wyner–Ziv encoding is much simpler. The Slepian–Wolf encoder only requires two feedback shift registers and an interleaver. Its performance, in terms of peak signal-to-noise ratio (PSNR), is 2–5 dB better than the conventional motion-JPEG intraframe coding. The main drawback of this scheme is that it relies on online feedback from the receiver. Hence it may not be suitable for applications where video is encoded and stored for subsequent use. Moreover, the feedback may introduce excessive latency for video decoding in a multihop network.

1.5.2.2 Transform-Domain Wyner–Ziv Encoder

In conventional source coding, a source vector is typically decomposed into spectral coefficients by using orthonormal transforms such as the discrete cosine transform (DCT). These coefficients are then individually coded with scalar quantizers and entropy coders. In Ref. 27, a transform-domain Wyner–Ziv encoder is proposed. A blockwise DCT of each Wyner–Ziv frame is performed. The transform coefficients are independently quantized, grouped into coefficient bands, and then compressed by a Slepian–Wolf turbo coder. As in the pixel-domain encoder described in Section 1.5.2.1, the decoder generates a side information frame based on previously reconstructed frames. Based on the side information, a bank of turbo decoders reconstructs the quantized coefficient bands independently. The rate-distortion performance is between conventional intraframe transform coding and conventional motion-compensated transform coding.

A different approach consists of allowing some simple temporal dependence estimation at the encoder to perform rate control without the need for feedback from the receiver. In the PRISM scheme [28], the encoder selects the coding mode based on the frame difference energy between the current frame and a previous frame. If the energy of the difference is very small, the block is not encoded. If the block difference is large, the block is intracoded. Between these two situations, one of the different encoding modes with different rates is selected. The rate estimation does not involve motion compensation and hence is necessarily inaccurate if motion compensation is used at the decoder. Further, the flexibility of the decoder is restricted.

1.5.3 System Software and Middleware

The development of efficient and flexible system software to make functional abstractions and information gathered by scalar and multimedia sensors available to higher-layer applications is one of the most important challenges faced by researchers to manage complexity and heterogeneity of sensor systems. As in Ref. 29, the term “system software” is used here to refer to operating systems, virtual machines, and middleware, which export services to higher-layer applications. Different multimedia sensor network applications are extremely diverse in their requirements and in the way they interact with the components of a sensor system. Hence, the main desired characteristics of a system software for WMSNs can be identified as follows:

- Provides a high-level interface to specify the behavior of the sensor system. This includes semantically rich querying languages that allow specifying what kind of data is requested from the sensor network, the quality of the required data, and how it should be presented to the user.
- Allows the user to specify application-specific algorithms to perform in-network processing on the multimedia content [30]. For example, the user should be able to specify particular image-processing algorithms or multimedia coding format.
- Long-lived, that is, needs to smoothly support evolutions of the underlying hardware and software.
- Shared among multiple heterogeneous applications.
- Shared among heterogeneous sensors and platforms. Scalar and multimedia sensor networks should coexist in the same architecture, without compromising on performance.
- Scalable.

There is an inherent trade-off between degrees of flexibility and network performance. Platform independence is usually achieved through layers of abstraction, which usually introduces redundancy and prevents the developer from accessing low-level details and functionalities. However, WMSNs are characterized by the contrasting objectives of optimizing the use of the scarce network resources and not compromising on performance. The principal design objective of existing operating systems for sensor networks such as TinyOS is high performance. However, their flexibility, interoperability, and reprogrammability are very limited. There is a need for research on systems that allow for this integration.

We believe that it is of paramount importance to develop efficient, high-level abstractions that will enable easy and fast development of sensor network applications. An abstraction similar to the famous Berkeley Transmission Control Protocol (TCP) sockets, that fostered the development of Internet applications, is needed for sensor systems. However, different from the Berkeley sockets, it is necessary to retain control on the efficiency of the low-level operations performed on battery-limited and resource-constrained sensor nodes.

As a first step toward this direction, Chu et al. [31] recently proposed the Sensor data Library (Sdlib), a sensor network data and communications library built upon the nesC language [32] for applications that require best-effort collection of large-size data such as video monitoring applications. The objective of the effort is to identify common functionalities shared by several sensor network applications and to develop a library of thoroughly tested, reusable, and efficient nesC components that abstract high-level operations common to most applications, although leaving differences among them to adjustable parameters. The library is called Sdlib as an analogy to the traditional C++ Standard Template Library. Sdlib provides an abstraction for common operations in sensor networks although the developer is still able to access low-level operations, which are implemented as a collection of nesC components, when desired. Moreover, to retain the efficiency of operations that are so critical for sensor networks battery lifetime and resource constraints, Sdlib exposes policy decisions such as resource allocation and rate of operation to the developer, although hiding the mechanisms of policy enforcement.

1.5.4 Collaborative In-Network Processing

As discussed earlier, collaborative in-network multimedia processing techniques are of great interest in the context of a WMSN. It is necessary to develop architectures and algorithms to flexibly perform these functionalities in-network with minimum energy consumption and limited execution time. The objective is usually to avoid transmitting large amounts of raw streams to the sink by processing the data in the network to reduce the communication volume.

Given a source of data (e.g., a video stream), different applications may require diverse information (e.g., raw video stream versus simple scalar or binary information inferred by processing the video stream). This is referred to as application-specific querying and processing. Hence, it is necessary to develop expressive and efficient querying languages, and to develop distributed filtering and in-network processing architectures, to allow real-time retrieval of useful information.

Similarly, it is necessary to develop architectures that efficiently allow performing data fusion or other complex processing operations in-network. Algorithms for both inter- and intramedia data aggregation and fusion need to be developed, as simple distributed processing schemes developed for existing scalar sensors are not suitable for computation-intensive processing required by multimedia contents. Multimedia sensor networks may require computation-intensive processing algorithms (e.g., to detect the presence of suspicious activity from a video stream). This may require considerable processing to extract meaningful information and to perform compression. A fundamental question to be answered is whether this processing can be done on sensor nodes (i.e., a flat architecture of multifunctional sensors that can perform any task), or if the need for specialized devices, for example, "computation hubs," arises.

1.5.5 Open Research Issues

- Although theoretical results on Slepian–Wolf and Wyner–Ziv codings exist since 30 years, there is still a lack of practical solutions. The net benefits and the practicality of these techniques still need to be demonstrated.
- It is necessary to completely explore the trade-offs between the achieved fidelity in the description of the phenomenon observed and the resulting energy consumption. As an example, the video distortion perceived by the final user depends on source coding (frame rate, quantization) and channel coding strength. For example, in a surveillance application, the objective of maximizing the event detection probability is in contrast to the objective of minimizing the power consumption.
- As discussed earlier, there is a need for higher-layer abstractions that will allow the fast development of sensor applications. However, due to the resource-constrained nature of sensor systems, it is necessary to control the efficiency of the low-level operations performed on battery-limited and resource-constrained sensor nodes.
- There is a need for simple, yet expressive high-level primitives for applications to leverage collaborative, advanced in-network multimedia processing techniques.

1.6 Protocols

In applications involving high-rate data, the transport, network, and link layers assume special importance by providing congestion control, delay-bounded routing, fair and efficient scheduling, among other functionalities. Although a large body of work exists in this area, there is a growing trend for newer design approaches in the context of WMSNs. As an example, the User Datagram Protocol (UDP), which has been traditionally used for transporting multimedia content may not be suited for WMSNs, given its blind packet drop policy. There is a need to decouple network-layer reliability and congestion control, often weighting one over the other, to meet performance requirements of real-time traffic. Similarly, at the link layer, protocols allow higher bandwidth utilization through multiple-transceiver sensors, multiple input multiple output (MIMO) antennas may be devised.

Recent research efforts have stressed on cross-layer design principles, wherein information gathered at a given layer can be used for making better decisions in the protocols that operate at entirely different layers. As an example, high packet delays and low bandwidth can force the routing layer to change its route decisions. Different routing decisions alter the set of links to be scheduled, thereby influencing the performance of the Media

Access Control (MAC) layer. Furthermore, congestion control and power control are also inherently coupled [33], as the capacity available on each link depends on the transmission power. Moreover, specifically to multimedia transmissions, the application layer does not require full insulation from lower layers, but instead needs to perform source coding based on information from the lower layers to maximize the multimedia performance.

This section discusses the considerations in optimizing the performance of the protocols operating at each of the layers, as well as exploring their interdependencies.

1.6.1 Transport-Layer Protocols

To explore the functionalities and support provided by the transport layer, the following discussion is classified into (1) TCP/UDP and TCP-friendly schemes for WMSNs and (2) application-specific and nonstandardized protocols. Figure 1.5 summarizes the discussion in this section.

1.6.1.1 TCP/UDP and TCP-Friendly Schemes for WMSNs

For real-time applications such as streaming media, UDP is preferred over TCP as "timeliness" is of greater concern than "reliability." However, in WMSNs, it is expected that packets are significantly compressed at the source, and redundancy is reduced as far as possible owing to the high transmission overhead in the energy-constrained nodes. Simply dropping packets during congestion conditions, as undertaken in UDP, may introduce discernable disruptions if they contain important original content not captured by interframe interpolation, such as the region of interest (ROI) feature used in JPEG2000 [34] or the I-frame used in the MPEG family.

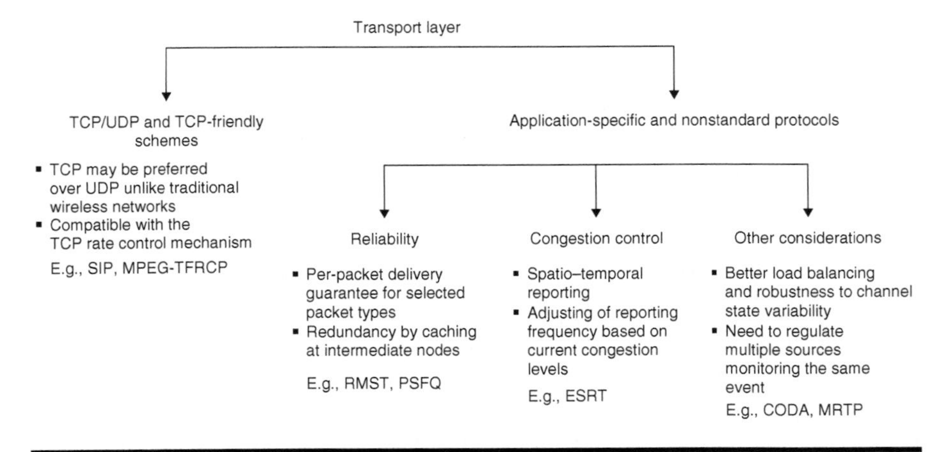

Figure 1.5 Classification of existing transport-layer protocols.

Unlike UDP, the TCP header can be modified to carry data-specific information including minimum resource requirements of different service classes in case of differentiated services.

We thus believe that TCP with appropriate modifications is preferable over UDP for WMSNs, if standardized protocols are to be used. With respect to sensor networks, several problems and their likely solutions such as large TCP header size, data versus address-centric routing, energy efficiency, among others, are identified and solutions are proposed in Ref. 35. We next indicate the recent work in this direction that evaluates the case for using TCP in WMSNs.

- *Effect of jitter induced by TCP.* A key factor that limits multimedia transport based on TCP and TCP-like rate control schemes is the jitter introduced by the congestion control mechanism. This can, however, be mitigated to a large extent by playout buffers at the sink, which is typically assumed to be rich in resources. As an example, the MPEG-TFRCP (TCP-Friendly Rate Control Protocol for MPEG-2 Video Transfer) [36] is an equation-based rate control scheme designed for transporting MPEG video in a TCP-friendly manner.
- *Overhead of the reliability mechanism in TCP.* The reliability mechanism provided by TCP introduces an end-to-end message passing overhead. Distributed TCP caching (DTC) [37] overcomes these problems by caching TCP segments inside the sensor network and by local retransmission of TCP segments. The nodes closest to the sink are the last-hop forwarders on most of the high-rate data paths and thus run out of energy first. DTC shifts the burden of the energy consumption from nodes close to the sink to the network, apart from reducing networkwide retransmissions.
- *Regulating streaming through multiple TCP connections.* The availability of multiple paths between the source and sink can be exploited by opening multiple TCP connections for multimedia traffic [38]. Here, the desired streaming rate and the allowed throughput reduction in presence of bursty data, such as video, is communicated to the receiver by the sender. This information is used by the receiver, which then measures the actual throughput and controls the rate within the allowed bounds by using multiple TCP connections and dynamically changing its TCP window size for each connection.

Although the recent TCP-based protocols for sensor networks, such as the Sensor Internet Protocol (SIP) [39] and the open source uIP [35] look promising, the inability to distinguish between bad channel conditions and network congestion is a major problem in TCP. This has motivated a new family of specialized transport-layer protocols where the design practices followed are entirely opposite to that of TCP [40], or stress on a particular functionality of the transport layer such as reliability or congestion control.

1.6.1.2 Application-Specific and Nonstandard Protocols

Depending on the application, both reliability and congestion control may be equally important functionalities or one may be preferred over the other. As an example, a video capture of a moving target enjoys a permissible level of loss tolerance and should be prevented from consuming a large proportion of network resources. The presence or absence of an intruder, however, may require a single data field but needs to be communicated without any loss of fidelity or information. We next list the important characteristics of such TCP-incompatible protocols in the context of WMSNs.

- *Reliability*. As discussed earlier in this section, certain packets in a flow may contain important information that is not replicated elsewhere, or cannot be regenerated through interpolation. Thus, if a prior recorded video is being sent to the sink, all the I-frames could be separated and the transport protocol should guarantee that each of these reach the sink. Reliable Multisegment Transport (RMST) [41] or the Pump Slowly Fetch Quickly (PSFQ) protocol [42] can be used for this purpose as they buffer packets at intermediate nodes, allowing for faster retransmission in case of packet loss. However, there is an overhead of using the limited buffer space at a given sensor node for caching packets destined for other nodes, as well as performing timely storage and flushing operations on the buffer.
- *Congestion control*. Multimedia data rates are typically in the order of 64 kbit/s for constant bit rate voice traffic, whereas video traffic may be bursty and approximately 500 kbit/s [43]. Although these data generation rates are high for a single node, multiple sensors in overlapped regions may inject similar traffic on sensing the same phenomenon, leading to network congestion. The Event-to-Sink Reliable Transport (ESRT) protocol [44] leverages the fact that spatial and temporal correlation exists among the individual sensor readings [45]. The ESRT protocol regulates the frequency of event reporting in a remote neighborhood to avoid congestions in the network. However, this approach may not be viable for all sensor applications as nodes transmit data only when they detect an event, which may be a short duration burst, as in the case of a video monitoring application. The feedback from the base station may hence not reach in time to prevent a sudden congestion due to this burst.
- *Other considerations*. As in TCP implementations, the use of multiple paths improves the end-to-end data transfer rate. The COngestion

Detection and Avoidance (CODA) protocol [46] allows a sink to regulate multiple sources associated with a single event in case of persistent network congestion. However, as the congestion inference in CODA is based on queue length at intermediate nodes, any action taken by the source occurs only after a considerable time delay. Other solutions include the Multiflow Real-time Transport Protocol (MRTP) [47] that is suited for real-time streaming of multimedia content by splitting packets over different flows, but does not consider energy efficiency or retransmissions for scalar data, both important considerations for a heterogeneous WMSN.

The success in energy-efficient and reliable delivery of multimedia information extracted from the phenomenon directly depends on selecting the appropriate coding rate, number of sensor nodes, and data rate for a given event [45]. For this purpose, new reliability metrics coupled with the application-layer coding techniques should be investigated.

1.6.2 Network Layer

Data collected by the sensor nodes may consist of different traffic classes, each having its own QoS requirement. We focus our discussion on the primary network-layer functionality of routing, although stressing on the applicability of the existing protocols for delay-sensitive and high bandwidth needs.

The concerns of routing, in general, differ significantly from the specialized service requirements of multimedia streaming applications. As an example, multiple routes may be necessary to satisfy the desired data rate at the destination node. Also, different paths exhibiting varying channel conditions may be preferred depending on the type of traffic and its resilience to packet loss. Primarily, routing schemes can be classified based on (i) network conditions that leverage channel and link statistics, (ii) traffic classes that decide paths based on packet priorities, and (iii) specialized protocols for real-time streaming that use spatio–temporal forwarding. Figure 1.6 provides a classification of the existing routing protocols and summarizes the discussion in this section.

1.6.2.1 QoS Routing Based on Network Conditions

Network conditions include interference seen at intermediate hops, the number of backlogged flows along a path, residual energy of the nodes, among others. A routing decision based on these metrics can avoid paths that may not support high bandwidth applications or introduce retransmission owing to bad channel conditions.

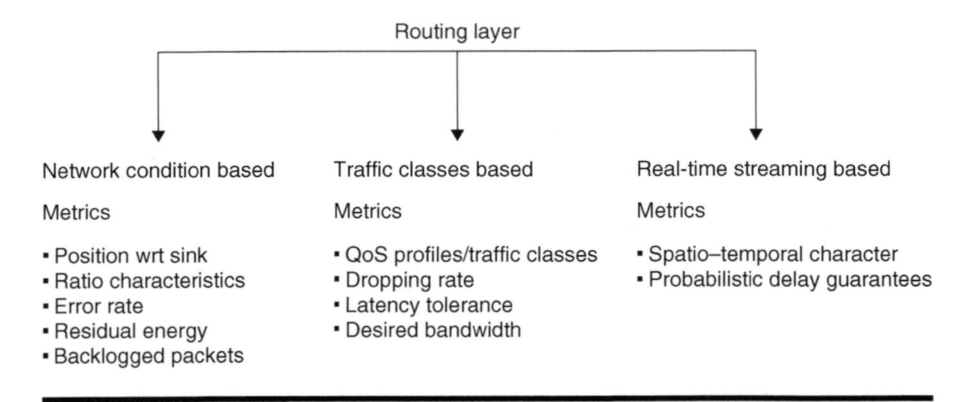

Figure 1.6 Classification of existing routing protocols.

The use of image sensors for gathering topology information [48] and generating QoS metrics [49] have been recently explored. Both construct a weighted cost function that takes into account different factors such as position, backlogged packets in the queue, and residual energy and communication parameters such as error rate to derive energy-efficient paths bounded by maximum allowed delays.

1.6.2.2 QoS Routing Based on Traffic Classes

As an example that highlights the need for network-level QoS, consider the task of bandwidth assignment for multimedia mobile medical calls, which include patients' sensing data, voice, pictures, and video data [8]. Unlike the typical source-to-sink multihop communication used by classical sensor networks, the proposed architecture uses a 3G cellular system in which individual nodes forward the sensed data to a cellular phone or a specialized information-collecting entity taking into account handoff effects. Different priorities are assigned to video data originating from sensors on ambulances, audio traffic from elderly people, and images returned by sensors placed on the body. It is important that some flows be preferred over the others given their time-critical nature and the level of importance to the patient's treatment.

1.6.2.3 Routing Protocols with Support for Streaming

The SPEED protocol [50] provides three types of real-time communication services, namely, real-time unicast, real-time area-multicast, and real-time area-anycast. It uses geographical location for routing and a key difference with other schemes of this genre is its spatio–temporal character, that is, it takes into account the timely delivery of the packets. As it works satisfactorily under scarce resource conditions and can provide service differentiation, SPEED takes the first step in addressing the concerns of real-time routing in WMSNs.

A significant extension over SPEED, the MMSPEED protocol [51] can efficiently differentiate between flows with different delay and reliability requirements. MMSPEED is based on a cross-layer approach between the network and the MAC layers in which a judicious choice is made over the reliability and timeliness of packet arrival. It is argued that the differentiation in reliability is an effective way of channeling resources from flows with relaxed requirements to flows with tighter requirements. Although current research directions make an effort to provide real-time streaming, they are still best-effort services.

Giving firm delay guarantees and addressing QoS concerns in a dynamically changing network is a difficult problem and yet is important for seamless viewing of the multimedia frames. Although probabilistic approaches such as MMSPEED take the first step toward this end, clearly, further work is needed in the networking layer.

1.6.3 MAC Layer

Research efforts to provide MAC-layer QoS can be mainly classified into (i) channel access policies, (ii) scheduling and buffer management, and (iii) error control. We next provide a brief description of each and highlight their support to multimedia traffic. The scope of this chapter is limited to the challenges posed by multimedia traffic in sensor networks and the efforts at the MAC layer to address them. A detailed survey of MAC protocols for classical sensor networks using scalar data can be found in Ref. 52. Figure 1.7 provides a classification of relevant MAC-layer functionalities and summarizes the discussion in this section.

1.6.3.1 Channel Access Policies

The main causes of energy loss in sensor networks are attributed to "packet collisions" and subsequent retransmissions, "overhearing packets" destined for other nodes, and "idle listening," a state in which the transceiver circuits remain active even in the absence of data transfer. Thus, regulating access to the channel assumes primary importance and several solutions have been proposed in the literature.

1.6.3.2 Contention-Based Protocols

Most contention-based protocols such as S-MAC [53], and protocols inspired by it [54], have a single-radio architecture. They alternate between sleep cycles (low power modes with transceiver switched off) and listen cycles (for channel contention and data transmission). However, we believe

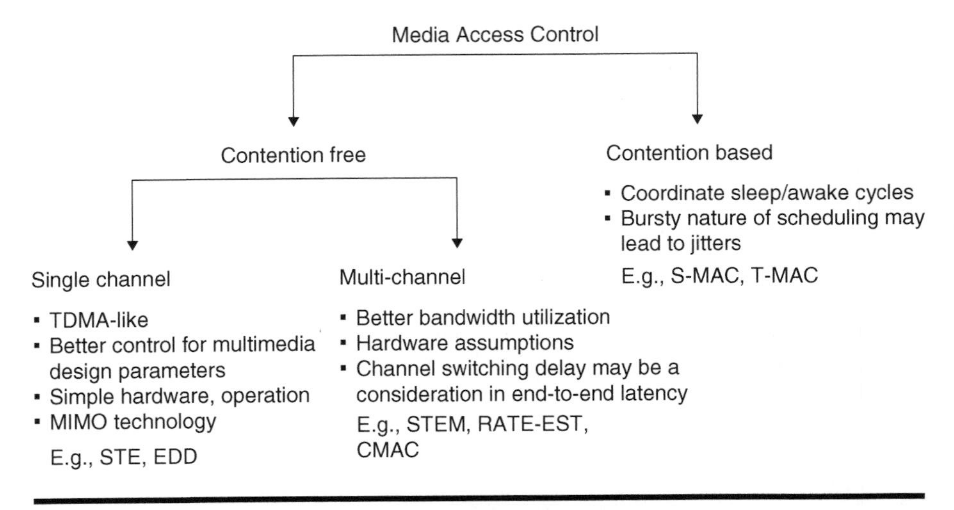

Figure 1.7 Classification of protocols at the data link layer.

that their applicability to multimedia transmission is limited owing to the following reasons:

- The primary concern in the protocols of this class is saving energy, and this is accomplished at the cost of latency and by allowing throughput degradation. A sophisticated duty cycle calculation based on permissible end-to-end delay needs to be implemented and coordinating overlapping "listen" period with neighbors based on this calculation is a difficult research challenge.
- Video traffic exhibits an inherent bursty nature and can lead to sudden buffer overflow at the receiver. This problem is further aggravated by the transmission policy adopted in T-MAC [54]. By choosing to send a burst of data during the listen cycle, T-MAC shows performance improvement over S-MAC, but at the cost of monopolizing a bottleneck node. Such an operation could well lead to strong jitters and result in discontinuous real-time playback.

1.6.3.3 Contention-Free Single-Channel Protocols

Time division multiple access (TDMA) is a representative protocol of this class in which the CH or sink helps in slot assignment, querying particular sensors, and maintaining time schedules. We believe that such protocols can be easily adapted for multimedia transmission and highlight the likely design considerations.

- TDMA schemes designed exclusively for sensor networks [55; and references therein] have a small reservation period (RP) that is generally contention-based, followed by a contention-free period that spans the rest of the frame. This RP could occur in each frame or at predecided intervals to assign slots to active nodes taking into consideration the QoS requirement of their data streams. The length of the TDMA frames and the frequency of the RP interval are some of the design parameters that can be exploited when designing a multimedia system. However, clock drift and synchronization issues must be accounted for they have a pronounced effect on small TDMA slot sizes.

- For real-time streaming video, packets are time-constrained and scheduling policies such as shortest time to extinction (STE) [56] or earliest due date (EDD) [57] can be adopted. Both of these are similar in principle as packets are sent in the increasing order of their respective delay tolerance, but differ in respect that EDD may still forward a packet that has crossed its allowed delay bound. Based on the allowed packet loss of the multimedia stream, the dependencies between packet dropping rate, arrival rate, and delay tolerance [56] can be used to decide the TDMA frame structure and thus ensure the smooth replay of data. This allows greater design choices as against Ref. 57, where the frame lengths and slot duration are considered constant.
- As sensor nodes are often limited by their maximum data transmission rate, depending on their multimedia traffic class, the duration of transmission could be made variable. Thus, variable TDMA (V-TDMA) schemes should be preferred when heterogeneous traffic is present in the network. Tools for calculating the minimum worst-case per-hop delay provided in Ref. 58 could be used in conjunction with the allowed end-to-end delay when V-TDMA schemes are used.
- The high data rate required by multimedia applications can be addressed by spatial multiplexing in MIMO systems that use a single channel but employ interference cancellation techniques. Recently, "virtual" MIMO schemes have been proposed for sensor networks [59], where nodes in proximity form a cluster. Each sensor functions as a single antenna element, sharing information and thus simulating the operation of a multiple antenna array. A distributed compression scheme for correlated sensor data, which specially addresses multimedia requirements, is integrated with the MIMO framework in Ref. 60. However, a key consideration in the MIMO-based systems is the number of sensor transmissions and the required signal energy per transmission. As the complexity is shifted from hardware to sensor coordination, further research is needed at the MAC layer to ensure that the required MIMO parameters such as channel state and desired diversity/processing gain are known to both the sender and receiver at an acceptable energy cost.

1.6.3.4 Contention-Free Multi-Channel Protocols

As discussed earlier in this section, existing bandwidth can be efficiently utilized by using multiple channels in a spatially overlapped manner. We observe that Berkeley's 3G MICA2 Mote has an 868/916 GHz multi-channel transceiver [5]. In Rockwell's WINS nodes, the radio operates on one of the 40 channels in the Industrial, Scientific and Medical (ISM) frequency band, selectable by the controller [61]. We next outline the design parameters that could influence MAC design in multi-channel WMSNs.

- Recent research has focused on a two-transceiver paradigm in which the main radio (MR) is supplemented by the presence of a wake-up radio (LR) having similar characteristics [62–64] or a simple low-energy design [65] that emits a series of short pulses or a busy tone. The LR is used for exchanging control messages and is assigned a dedicated channel. In high bandwidth applications, such as streaming video, the use of a separate channel for channel arbitration alone does not allow best utilization of the network resources.
- We propose that WMSNs use in-band signaling, where the same channel is used for both data and channel arbitration [66]. Although such protocols undoubtedly improve bandwidth efficiency, they introduce the problem of distinct channel assignment and need to account for the delay to switch to a different channel [67], as its cumulative nature at each hop affects real-time media. This work leaves an open question on whether switching delay can be successfully hidden with only one interface per node. If this is possible, it may greatly simplify sensor design, while performing, as well as a multi-channel, multi-interface solution.
- Multi-channel protocols are not completely collision-free as seen in the case of control packet collision [62,63,66], and the available channels cannot be assumed to be perfectly nonoverlapping. This may necessitate dynamic channel assignment, taking into account the effect of adjacent channel interference, to maintain the network QoS.

The use of multi-channel communication paradigm can also be used to highlight the need for cross-layer design strategies, described in detail in Section 1.6.4. Consider the scenario in Figure 1.8 in which a multiradio architecture is assumed. Existing data rates of approximately 40 and 250 kbit/s supported by the MICA2 and MICAz motes are not geared to support multimedia traffic and hence the use of multiple channels for the same flow greatly increases the data rate. The network layer must preferentially route packets to nodes that can support reception on more than one channel. The link layer undertakes sender–receiver negotiations and decides on the particular channels to be used for the impending data transfer.

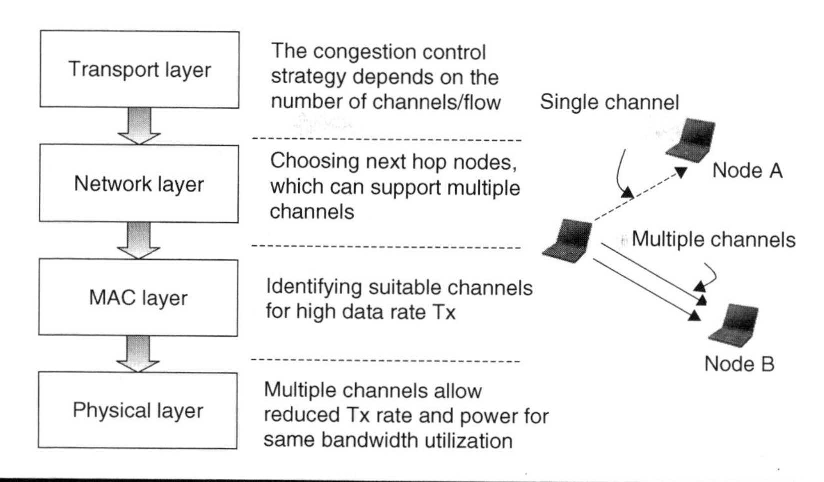

Figure 1.8 Interdependencies in the protocol stack leading to a cross-layer architecture in WMSNs.

The number of channels to be used may decide the congestion control algorithm adopted at the transport layer. Thus, although these complex interdependencies help in realizing high data rates on the resource-constrained sensor nodes, they also introduce significant challenges in protocol design and must be investigated further.

1.6.3.5 Scheduling

MAC-layer scheduling in the context of WMSNs differs from the traditional networking model in the sense that apart from choosing the queueing discipline that accounts for latency bounds, rate/power control, and consideration of high channel error conditions needs to be incorporated.

To generate optimal schedules that minimize both power consumption and the probability of missing deadlines for real-time messages, PARM [68] integrates the EDD metric described in Section 1.6.3 into an energy consumption function. Although significant performance improvements are demonstrated, this work needs to be extended for large-scale networks that are typically envisaged for WMSNs.

Queueing at the MAC layer has been extensively researched and several schemes with varying levels of complexity exist. Of interest to multimedia applications is the development of schemes that allow a delay bound and thus assure smooth streaming of multimedia content. E^2WFQ [69], a variant of the established weighted fair queueing (WFQ) discipline, allows adjustments to be made to the energy–latency–fidelity trade-off space. Extending WFQ, the wireless packet scheduling (WPS) presented in Ref. 70, addresses the concerns of delay and rate-sensitive packet flows, thus making it suitable for multimedia traffic. WPS, however, assumes that the channel error is completely predictable at any time and its practical implementation shows marked deviations from the idealized case in terms of worst-case complexity. This work is suitable for single-hop sensor–sink communication and multihop forwarding issues are not explored.

Network calculus [71,72] is a theory for deterministic queueing systems that allows the assignment of service guarantees by traffic regulation and deterministic scheduling. Through tools provided by network calculus, bounds on various performance measures, such as delay and queue length, at each element of the network can be derived and thus QoS of a flow can be specified. Arrival, Departure, and Service curves reflect the constraints that flows are subjected to within a network. The calculus relies on Min-plus algebra, in which addition and multiplication are replaced by minimum and addition, respectively, to operate on these curves. Current network calculus results have been mostly derived for wired networks, and assume static topologies and fixed link capacity, indicating that further work needs to be undertaken in dynamic scenarios typically envisaged for WMSNs.

1.6.3.6 Link-Layer Error Control

The inherent unreliability of the wireless channel, coupled with a low-frame loss rate requirement of the order of 10^{-2} for good-quality video, poses a challenge in WMSNs. Two main classes of mechanisms are traditionally employed to combat the unreliability of the wireless channel at the physical (PHY) and data link layer, namely forward error correction (FEC) and automatic repeat request (ARQ), along with hybrid schemes. ARQ mechanisms use bandwidth efficiently at the cost of additional latency. Hence, although carefully designed selective repeat schemes may be of some interest, naive use of ARQ techniques is clearly infeasible for applications requiring real-time delivery of multimedia content.

An important characteristic of multimedia content is "unequal importance," that is, not all packets have the same importance, as seen in the case of the I-frames in the MPEG family. Applying different degrees of FEC to different parts of the video stream, depending on their relative importance (unequal protection) allows a varying overhead on the transmitted packets. For example, this idea can be applied to layered coded streams to provide graceful degradation in the observed image quality in the presence of error losses, thus avoiding the so-called "cliff" effects [2].

In general, delivering error-resilient multimedia content and minimizing energy consumption are contradicting objectives. For this reason, and due to the time-varying characteristics of the wireless channel, several joint source and channel coding schemes have been developed (e.g., Ref. 14), which try to reduce the energy consumption of the whole process. Some recent papers [73,74] even try to jointly reduce the energy consumption of the whole process of multimedia content delivery, that is, jointly optimize source coding, channel coding, and transmission power control. However, most of these efforts have originated from the multimedia or coding communities, and thus do not jointly consider other important networking aspects of content delivery over a multihop wireless networks of memory-, processing-, and battery-constrained devices.

In Ref. 75, a cross-layer analysis of error control schemes for WSNs is presented. The effects of multihop routing and of the broadcast nature of wireless communications are investigated to model the energy consumption, latency, and packet error rate performance of error control schemes. As a result, error control schemes are studied through a cross-layer analysis that considers the effects of routing, medium access, and physical layer. This analysis enables a comprehensive comparison of FEC and ARQ schemes in WSNs. FEC schemes are shown to improve the error resiliency compared to ARQ. In a multihop network, this improvement can be exploited by reducing the transmit power (transmit power control) or by constructing longer hops (hop length extension) through channel-aware routing protocols. The analysis reveals that, for certain FEC codes, hop length extension decreases both the energy consumption and the end-to-end latency subject to a target packet error rate compared to ARQ. Thus, FEC codes are an important candidate for delay-sensitive traffic in WSNs. However, transmit power control results in significant savings in energy consumption at the cost of increased latency. There is also a need to integrate source and channel coding schemes in existing cross-layer optimization frameworks. The existing schemes mostly consider point-to-point wireless links, and neglect interference from neighboring devices and multihop routes.

1.6.4 Cross-Layer Protocols

Although a consistent amount of recent papers have focused on cross-layer design and improvement of protocols for WSNs, a systematic methodology to accurately model and leverage cross-layer interactions is still largely missing. Most of the existing studies decompose the resource allocation problem at different layers, and consider allocation of the resources at each layer separately. In most cases, resource allocation problems are treated either heuristically, or without considering cross-layer interdependencies, or by considering pairwise interactions between isolated pairs of layers.

In Ref. 76, the cross-layer transmission of multimedia content over wireless networks is formalized as an optimization problem. Several different approaches for cross-layer design of multimedia communications are discussed, including "bottom–up approach," where the lower layers try to insulate the higher layers from losses and channel capacity variations, and "top–down," where the higher layer protocols optimize their parameters at the next lower layer. However, only single-hop networks are considered.

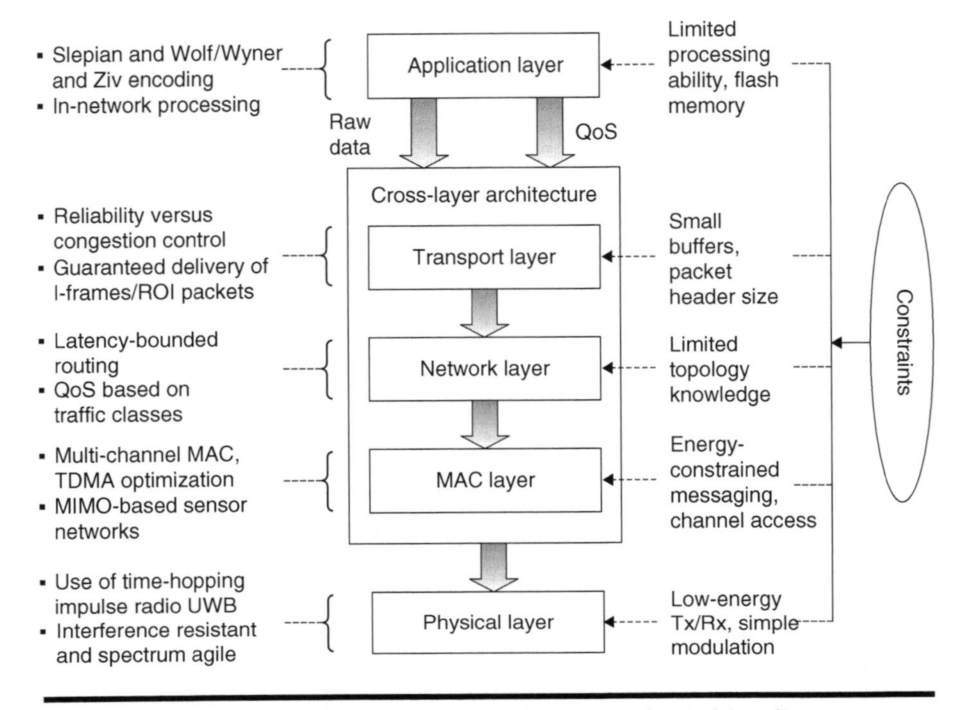

Figure 1.9 Cross-layer communication architecture of a multimedia sensor.

In Ref. 77, several techniques that provide significant performance gains through cross-layer optimizations are surveyed. In particular, the improvements of adaptive link-layer techniques such as adaptive modulation and packet-size optimization, joint allocation of capacity and flows (i.e., MAC and routing), joint scheduling, and rate allocation are discussed. Although still maintaining a strict layered architecture, it is shown how these cross-layer optimizations help to improve the spectral efficiency at the physical layer, and the PSNR of the video stream perceived by the user. Clearly, energy-constrained multimedia sensors may need to leverage cross-layer interactions one step further. At the same time, optimization metrics in the energy domain need to be considered as well.

Figure 1.9 summarizes our discussion on protocol design showing how application-layer QoS requirements and design constraints of WMSN can be addressed by carefully designing a cross-layer communication architecture.

1.7 Physical Layer

This section discusses the applicability of the UWB transmission technique, which we advocate as a potential technology for multimedia sensors.

1.7.1 Ultra Wideband Communications

The UWB* technology has the potential to enable low power consumption and high data rate communications within tens of meters, characteristics that make it an ideal choice for WMSNs.

UWB signals have been used for several decades in the radar community. Recently, the U.S. Federal Communications Commission (FCC) Notice of Inquiry in 1998 and the First Report and Order in 2002 [79] inspired a renewed flourish of research and development efforts in both academy and industry due to the characteristics of UWB that make it a viable candidate for wireless communications in dense multipath environments.

Although UWB signals, as per the specifications of the FCC, use the spectrum from 3.1 to 10.6 GHz, with appropriate interference limitation, UWB devices can operate using spectrum occupied by existing radio services without causing interference, thereby permitting scarce spectrum resources to be used more efficiently. Instead of dividing the spectrum into distinct bands that are then allocated to specific services, UWB devices are allowed to operate overlaid and thus interfere with existing services, at a low enough power level that existing services would not experience performance degradation. The First Report and Order by the FCC includes standards designed to ensure that existing and planned radio services, particularly safety services, are adequately protected.

There exist two main variants of UWB. The first, known as time-hopping impulse radio UWB (TH-IR-UWB) [78], and mainly developed by Win and Scholtz [80], is based on sending very short duration pulses (in the order of hundreds of picoseconds) to convey information. Time is divided into frames, each of which is composed of several chips of very short duration. Each sender transmits one pulse in a chip per frame only, and multi-user access is provided by pseudorandom time-hopping sequences (THS) that determine in which chip each user should transmit. A different approach, known as multicarrier UWB (MC-UWB), uses multiple simultaneous carriers, and is usually based on Orthogonal Frequency Division Multiplexing (OFDM) [81].

MC-UWB is particularly well suited for avoiding interference because its carrier frequencies can be precisely chosen to avoid narrowband interference to or from narrowband systems. However, implementing an MC-UWB front-end power amplifier can be challenging due to the continuous variations in power over a very wide bandwidth. Moreover, when OFDM is

* The U.S. Federal Communications Commission (FCC) defines UWB as a signal with either a fractional bandwidth of 20 percent of the center frequency or 500 MHz (when the center frequency is above 6 GHz). The FCC calculates the fractional bandwidth as $2(f_H - f_L)/(f_H + f_L)$, where f_H represents the upper frequency of the -10 dB emission limit and f_L represents the lower frequency limit of the -10 dB emission limit [78].

used, high-speed FFT processing is necessary, which requires significant processing power and leads to complex transceivers.

TH-IR-UWB signals require fast switching times for the transmitter and receiver and highly precise synchronization. Transient properties become important in the design of the radio and antenna. The high instantaneous power during the brief interval of the pulse helps to overcome interference to UWB systems, but increases the possibility of interference from UWB to narrowband systems. The RF front end of a TH-IR-UWB system may resemble a digital circuit, thus circumventing many of the problems associated with mixed-signal integrated circuits. Simple TH-IR-UWB systems can be very inexpensive to construct.

Although no sound analytical or experimental comparison between the two technologies is available to our knowledge, we believe that TH-IR-UWB is particularly appealing for WMSNs for the following reasons:

- It enables high data rate, very low-power wireless communications, on simple-design, low-cost radios (carrierless, baseband communications) [80].
- Its fine delay resolution properties are appropriate for wireless communications in dense multipath environment, by exploiting more resolvable paths [82].
- It provides large processing gain in the presence of interference.
- It provides flexibility, as data rate can be traded for power spectral density and multipath performance.
- Finding suitable codes for THS is trivial (as opposed to CDMA codes), and no assignment protocol is necessary.
- It naturally allows for integrated MAC/PHY solutions [83]. Moreover, interference mitigation techniques [83] allow realizing MAC protocols that do not require mutual temporal exclusion between different transmitters. In other words, simultaneous communications of neighboring devices are feasible without complex receivers as required by CDMA.
- Its large instantaneous bandwidth enables fine time resolution for accurate position estimation [84] and for network time distribution (synchronization).
- UWB signals have extremely low-power spectral density, with low probability of intercept/detection (LPI/D), which is particularly appealing for military covert operations.

1.7.1.1 Ranging Capabilities of UWB

Particularly appealing for WMSNs are UWB high data rate with low power consumption, and its positioning capabilities. Positioning capabilities are needed in sensor networks to associate physical meaning to the information gathered by sensors. Moreover, knowledge of the position of each network device allows for scalable routing solutions [85]. Angle-of-arrival techniques and signal strength-based techniques do not provide considerable advantages with respect to other transmission techniques, whereas time-based approaches in UWB allow ranging accuracy in the order of centimeters [84]. This can be intuitively explained by expression 1.1, which gives a lower bound on the best achievable accuracy of a distance estimate $\hat{d}$ [84]:

$$\sqrt{\mathrm{Var}(\hat{d})} \geq \frac{c}{2\sqrt{2}\pi\sqrt{\mathrm{SNR}}\beta'} \tag{1.1}$$

where c is the speed of light, SNR the signal-to-noise ratio, and β the effective signal bandwidth. As can be seen, the accuracy of the time-based localization technique can be improved by increasing either the effective bandwidth or the SNR. For this reason, the large bandwidth of UWB systems allows extremely accurate location estimations, for example, within 1 in. at SNR = 0 dB and with a pulse of 1.5 GHz bandwidth. Excellent comprehensive surveys of the UWB transmission technique, and of the localization techniques for UWB systems, are provided in Refs 86 and 84, respectively.

1.7.1.2 Standards Based on UWB

The IEEE 802.15.3a task group has been discussing for three years an alternate physical layer for its high data rate wireless personal area networks (WPAN) standard. However, in early 2005, the group has been disbanded after not being able to reach a consensus on a single UWB-based standard between two competing proposals from two leading industry groups—the UWB Forum and the WiMedia Alliance. The UWB Forum proposal was based on a direct sequence (DS)-UWB technology, whereas the WiMedia alliance was proposing a multiband orthogonal frequency division multiplexing (MB-OFDM). The IEEE 802.15.4a task group is developing an alternate physical layer for low data rate, very low power consumption sensors, based on impulse radio UWB.

1.7.1.3 Open Research Issues

- Although the UWB transmission technology is advancing rapidly, many challenges need to be solved to enable multihop networks of UWB devices. In particular, although some recent efforts have been undertaken in this direction [83,87], how to efficiently share the medium in UWB multihop networks is still an open issue.
- As a step ahead, research is needed aimed at designing a cross-layer communication architecture based on UWB with the objective of

reliably and flexibly delivering QoS to heterogeneous applications in WMSNs, by carefully leveraging and controlling interactions among layers according to the application's requirements.

- It is necessary to determine how to provide provable latency and throughput bounds to multimedia flows in an UWB environment.
- It is needed to develop analytical models to quantitatively compare different variants of UWB to determine trade-offs in their applicability to high data rate and low power consumption devices such as multimedia sensors.
- A promising research direction may also be to integrate UWB with advanced cognitive radio [88] techniques to increase the spectrum utilization. For example, UWB pulses could be adaptively shaped to occupy portions of the spectrum that are subject to lower interference.

1.8 Conclusions

We discussed the state-of-the-art of research on WMSNs and outlined the main research challenges. Algorithms and protocols for the development of WMSNs were surveyed, and open research issues were discussed in detail. We discussed existing solutions and open research issues at the application, transport, network, link, and physical layers of the communication stack, along with possible cross-layer synergies and optimizations. We pointed out how recent work undertaken in Wyner–Ziv coding at the application layer, specialized spatio–temporal transport-layer solutions, delay-bounded routing, multi-channel MAC protocols, and UWB technology, among others, seem to be the most promising research directions in developing practical WMSNs. We believe that this research area will attract the attention of many researchers and that it will push one step further our ability to observe the physical environment and interact with it.

Acknowledgment

The authors would like to thank Prof. Ian F. Akyildiz for his valuable comments and continuous guidance and encouragement.

References

1. Akyildiz, I. F., W. Su, Y. Sankarasubramaniam, and E. Cayirci. Wireless sensor networks: A survey. *Computer Networks (Elsevier)*, 38(4), 393–422, 2002.
2. Gurses, E. and O. B. Akan. Multimedia communication in wireless sensor networks. *Annals of Telecommunications*, 60(7–8), 799–827, 2005.

3. Misra, S., M. Reisslein, and G. Xue. A survey of multimedia streaming in wireless sensor networks. *IEEE Communications Surveys and Tutorials* (accepted for publication).
4. Rahimi, M., R. Baer, O. Iroezi, J. Garcia, J. Warrior, D. Estrin, and M. Srivastava. Cyclops: *In Situ* Image Sensing and Interpretation in Wireless Sensor Networks. *Proc. of the ACM Conf. on Embedded Networked Sensor Systems (SenSys).* San Diego, CA, November 2005.
5. Crossbow MICA2 Mote Specifications. http://www.xbow.com.
6. Crossbow MICAz Mote Specifications. http://www.xbow.com.
7. Campbell, J., P. B. Gibbons, S. Nath, P. Pillai, S. Seshan, and R. Sukthankar. IrisNet: An Internet-Scale Architecture for Multimedia Sensors. *Proc. of the ACM Multimedia Conference.* Singapore, November 2005.
8. Hu, F. and S. Kumar. Multimedia Query with QoS Considerations for Wireless Sensor Networks in Telemedicine. *Proc. of Society of Photo-Optical Instrumentation Engineers—Intl. Conf. on Internet Multimedia Management Systems.* Orlando, FL, September 2003.
9. Reeves, A. A. Remote Monitoring of Patients Suffering from Early Symptoms of Dementia. *Intl. Workshop on Wearable and Implantable Body Sensor Networks.* London, UK, April 2005.
10. Holman, R., J. Stanley, and T. Ozkan-Haller. Applying video sensor networks to nearshore environment monitoring. *IEEE Pervasive Computing*, 2(4), 14–21, 2003.
11. Basharat, A., N. Catbas, and M. Shah. A framework for intelligent sensor network with video camera for structural health monitoring of bridges. *Proc. of the First IEEE Intl. Workshop on Sensor Networks and Systems for Pervasive Computing (PerSeNS 2005).* Kauai Island, Hawaii, March 2005.
12. Crossbow TelosB Mote Specifications. http://www.xbow.com.
13. Girod, B., A. Aaron, S. Rane, and D. Rebollo-Monedero. Distributed video coding. *Proceedings of the IEEE*, 93(1), 71–83, 2005.
14. Eisenberg, Y., C. E. Luna, T. N. Pappas, R. Berry, and A. K. Katsaggelos. Joint source coding and transmission power management for energy efficient wireless video communications. *IEEE Transactions on Circuits and Systems for Video Technology*, 12(6), 411–424, 2002.
15. Girod, B., M. Kalman, Y. Liang, and R. Zhang. Advances in channel-adaptive video streaming. *Wireless Communications and Mobile Computing*, 2(6), 549–552, 2002.
16. Zuniga, M. and B. Krishnamachari. Integrating Future Large Scale Sensor Networks with the Internet. USC Computer Science Technical Report CS 03-792, 2003.
17. Kulkarni, P., D. Ganesan, P. Shenoy, and Q. Lu. SensEye: A Multi-Tier Camera Sensor Network. *Proc. of ACM Multimedia.* Singapore, November 2005.
18. Kulkarni, P., D. Ganesan, and P. Shenoy. The Case for Multi-Tier Camera Sensor Network. *Proc. of the ACM Workshop on Network and Operating System Support for Digital Audio and Video (NOSSDAV).* Stevenson, WA, June 2005.
19. Soro, S. and W. B. Heinzelman. On the Coverage Problem in Video-Based Wireless Sensor Networks. *Proc. of the IEEE Intl. Conf. on Broadband Communications, Networks, and Systems (BroadNets).* Boston, MA, October 2005.
20. Video Coding for Low Bit Rate Communication. ITU-T Recommendation H.263.

21. Advanced Video Coding for Generic Audiovisual Services. ITU-T Recommendation H.264.
22. Slepian, D. and J. Wolf. Noiseless coding of correlated information sources. *IEEE Transactions on Information Theory*, 19(4), 471–480, 1973.
23. Wyner, A. and J. Ziv. The rate-distortion function for source coding with side information at the decoder. *IEEE Transactions on Information Theory*, 22, 1–10, 1976.
24. Xiong, Z., A. D. Liveris, and S. Cheng. Distributed source coding for sensor networks. *IEEE Signal Processing Magazine*, 21, 80–94, 2004.
25. Aaron, A., S. Rane, R. Zhang, and B. Girod. Wyner–Ziv Coding for Video: Applications to Compression and Error Resilience. *Proc. of IEEE Data Compression Conf. (DCC)*, pp. 93–102. Snowbird, UT, March 2003.
26. Aaron, A., E. Setton, and B. Girod. Towards Practical Wyner–Ziv Coding of Video. *Proc. of IEEE Intl. Conf. on Image Processing (ICIP)*. Barcelona, Spain, September 2003.
27. Aaron, A., S. Rane, E. Setton, and B. Girod. Transform-Domain Wyner–Ziv Codec for Video. *Proc. of Society of Photo-Optical Instrumentation Engineers—Visual Communications and Image Processing*. San Jose, CA, January 2004.
28. Puri, R. and K. Ramchandran. PRISM: A New Robust Video Coding Architecture Based on Distributed Compression Principles. *Proc. of the Allerton Conf. on Communication, Control, and Computing*. Allerton, IL, October 2002.
29. Koshy, J. and R. Pandey. VM*: Synthesizing Scalable Runtime Environments for Sensor Networks. *Proc. of the ACM Conf. on Embedded Networked Sensor Systems (SenSys)*. San Diego, CA, November 2005.
30. Feng, W. and N. B. W. Feng. Dissecting the Video Sensing Landscape. *Proc. of the ACM Workshop on Network and Operating System Support for Digital Audio and Video (NOSSDAV)*. Stevenson, WA, June 2005.
31. Chu, D., K. Lin, A. Linares, G. Nguyen, and J. M. Hellerstein. Sdlib: A Sensor Network Data and Communications Library for Rapid and Robust Application Development. *Proc. of Information Processing in Sensor Networks (IPSN)*. Nashville, TN, April 2006.
32. Gay, D., P. Levis, R. von Behren, M. Welsh, E. Brewer, and D. Culler. The nesC Language: A Holistic Approach to Network Embedded Systems. *Proc. of the ACM SIGPLAN 2003 Conf. on Programming Language Design and Implementation (PLDI)*. San Diego, CA, June 2003.
33. Chiang, M. Balancing transport and physical layers in wireless multihop networks: Jointly optimal congestion control and power control. *IEEE Journal on Selected Areas in Communications*, 23(1), 104–116, 2005.
34. JPEG2000 Requirements and Profiles. ISO/IEC JTC1/SC29/WG1 N1271, March 1999.
35. Dunkels, A., T. Voigt, and J. Alonso. Making TCP/IP Viable for Wireless Sensor Networks. *Proc. of European Workshop on Wireless Sensor Networks (EWSN)*. Berlin, Germany, January 2004.
36. Miyabayashi, M., N. Wakamiya, M. Murata, and H. Miyahara. MPEG-TFRCP: Video Transfer with TCP-friendly Rate Control Protocol. *Proc. of IEEE Intl. Conf. on Communications (ICC)*. Helsinki, June 2001.
37. Dunkels, A., T. Voigt, J. Alonso, and H. Ritter. Distributed TCP Caching for Wireless Sensor Networks. *Proc. of the Mediterranean Ad Hoc Networking Workshop (MedHocNet)*. Bodrum, Turkey, June 2004.

38. Nguyen, T. and S. Cheung. Multimedia Streaming with Multiple TCP Connections. *Proc. of the IEEE Intl. Performance Computing and Communications Conference (IPCCC)*. Arizona, April 2005.
39. Luo, X., K. Zheng, Y. Pan, and Z. Wu. A TCP/IP Implementation for Wireless Sensor Networks. *Proc. of IEEE Conf. on Systems, Man and Cybernetics*, pp. 6081–6086. October 2004.
40. Sundaresan, K., V. Anantharaman, H.-Y. Hsieh, and R. Sivakumar. ATP: A reliable transport protocol for ad hoc networks. *IEEE Transactions on Mobile Computing*, 4(6), 588–603, 2005.
41. Stann, F. and J. Heidemann. RMST: Reliable Data Transport in Sensor Networks. *Proc. of IEEE Sensor Network Protocols and Applications (SNPA)*, pp. 102–112. Anchorage, AK April 2003.
42. Wan, C. Y., A. T. Campbell, and L. Krishnamurthy. PSFQ: A Reliable Transport Protocol for Wireless Sensor Networks. *Proc. of ACM Workshop on Wireless Sensor Networks and Applications (WSNA)*. Atlanta, GA, September 2002.
43. Zhu, X. and B. Girod. Distributed Rate Allocation for Multi-Stream Video Transmission Over *Ad Hoc* Networks. *Proc. of IEEE Intl. Conf. on Image Processing (ICIP)*, pp. 157–160. Genoa, Italy, September 2005.
44. Akan, O. and I. F. Akyildiz. Event-to-sink reliable transport in wireless sensor networks. *IEEE/ACM Transactions on Networking*, 13(5), 1003–1017, 2005.
45. Vuran, M. C., O. B. Akan, and I. F. Akyildiz. Spatio-temporal correlation: Theory and applications for wireless sensor networks. *Computer Networks (Elsevier)*, 45(3), 245–259, 2004.
46. Wan, C. Y., S. B. Eisenman, and A. T. Campbell. CODA: Congestion Detection and Avoidance in Sensor Networks. *Proc. of the ACM Conf. on Embedded Networked Sensor Systems (SenSys)*. Los Angeles, CA, November 2003.
47. Mao, S., D. Bushmitch, S. Narayanan, and S. S. Panwar. MRTP: A Multiflow Real-time Transport Protocol for *ad hoc* networks. *IEEE Transactions on Multimedia*, 8(2), 356–369, 2006.
48. Savidge, L., H. Lee, H. Aghajan, and A. Goldsmith. QoS-Based Geographic Routing for Event-Driven Image Sensor Networks. *Proc. of IEEE/CreateNet Intl. Workshop on Broadband Advanced Sensor Networks (BaseNets)*. Boston, MA, October 2005.
49. Akkaya, K. and M. Younis. An Energy-Aware QoS Routing Protocol for Wireless Sensor Networks. *Proc. of Intl. Conf. on Distributed Computing Systems Workshops (ICSDSW)*. Washington, 2003.
50. He, T., J. A. Stankovic, C. Lu, and T. F. Abdelzaher. A spatiotemporal communication protocol for wireless sensor networks. *IEEE Transactions on Parallel and Distributed Systems*, 16(10), 995–1006, 2005.
51. Felemban, E., C.-G. Lee, and E. Ekici. MMSPEED: Multipath Multi-SPEED protocol for QoS guarantee of reliability and timeliness in wireless sensor networks. *IEEE Transactions on Mobile Computing*, 5(6), 738–754, 2006.
52. Kredo, K. and P. Mohapatra. Medium Access Control in wireless sensor networks. *Computer Networks (Elsevier)*, 51(4), 961–964, 2007.
53. Ye, W., J. Heidemann, and D. Estrin. Medium Access Control with coordinated, adaptive sleeping for wireless sensor networks. *IEEE Transactions on Networking*, 12(3), 493–506, 2004.

54. Dam, T. V. and K. Langendoen. An Adaptive Energy-Efficient MAC Protocol for Wireless Sensor Networks. *Proc. of the ACM Conf. on Embedded Networked Sensor Systems (SenSys)*. Los Angeles, CA, November 2003.
55. Kulkarni, S. S. and M. Arumugam. TDMA Service for Sensor Networks. *Proc. of the Intl. Conf. on Distributed Computing Systems Workshops (ICDCSW)*, pp. 604–609. Washington, 2004.
56. Santivanez, C. and I. Stavrakakis. Study of various TDMA schemes for wireless networks in the presence of deadlines and overhead. *IEEE Journal on Selected Areas in Communications*, 17(7), 1284–1304, 1999.
57. Capone, J. and I. Stavrakakis. Delivering QoS requirements to traffic with diverse delay tolerances in a TDMA environment. *IEEE/ACM Transactions on Networking*, 7(1), 75–87, 1999.
58. Cui, S., R. Madan, A. Goldsmith, and S. Lall. Energy-Delay Tradeoffs for Data Collection in TDMA-based Sensor Networks. *Proc. of IEEE Intl. Conf. on Communications (ICC)*, pp. 3278–3284. Seoul, Korea, May 2005.
59. Jayaweera, S. K. An Energy-Efficient Virtual MIMO Communications Architecture Based on V-BLAST Processing for Distributed Wireless Sensor Networks. *Proc. of IEEE Intl. Conf. on Sensor and Ad-hoc Communications and Networks (SECON)*. Santa Clara, CA, October 2004.
60. Jayaweera, S. K. and M. L. Chebolu. Virtual MIMO and Distributed Signal Processing for Sensor Networks—An Integrated Approach. *Proc. of IEEE Intl. Conf. on Communications (ICC)*. Seoul, Korea, May 2005.
61. Agre, R., L. P. Clare, G. J. Pottie, and N. P. Romanov. Development Platform for Self-Organizing Wireless Sensor Networks. *Proc. of Society of Photo-Optical Instrumentation Engineers—Aerosense*. Orlando, FL, March 1999.
62. Miller, M. J. and N. H. Vaidya. A MAC protocol to reduce sensor network energy consumption using a wakeup radio. *IEEE Transactions on Mobile Computing*, 4(3), 228–242, 2005.
63. Schurgers, C., V. Tsiatsis, S. Ganeriwal, and M. Srivastava. Optimizing sensor networks in the energy-latency-density design space. *IEEE Transactions on Mobile Computing*, 1(1), 70–80, 2002.
64. Schurgers, C., V. Tsiatsis, S. Ganeriwal, and M. Srivastava. Topology Management for Sensor Networks: Exploiting Latency and Density. *Proc. of ACM Intl. Conf. on Mobile Computing and Networking (MobiCom)*. Atlanta, GA, September 2002.
65. Guo, C., L. C. Zhong, and J. M. Rabaey. Low Power Distributed MAC for *Ad Hoc* Sensor Radio Networks. *Proc. of the IEEE Global Communications Conference (GLOBECOM)*. San Antonio, TX, November 2001.
66. Chowdhury, K. R., N. Nandiraju, D. Cavalcanti, and D. P. Agrawal. CMAC—A Multi-Channel Energy Efficient MAC for Wireless Sensor Networks. *Proc. of IEEE Wireless Communications and Networking Conference (WCNC)*. Las Vegas, NV, April 2006.
67. Kyasanur, P. and N. H. Vaidya. Capacity of Multi-Channel Wireless Networks: Impact of Number of Channels and Interfaces. *Proc. of ACM Intl. Conf. on Mobile Computing and Networking (MobiCom)*. Cologne, Germany, August 2005.
68. Alghamdi, M. I., T. Xie, and X. Qin. PARM: A Power-Aware Message Scheduling Algorithm for Real-time Wireless Networks. *Proc. of ACM Workshop on Wireless Multimedia Networking and Performance Modeling (WMuNeP)*, pp. 86–92. Montreal, Quebec, Canada, 2005.

69. Raghunathan, V., S. Ganeriwal, M. Srivastava, and C. Schurgers. Energy efficient wireless packet scheduling and fair queuing. *ACM Transactions on Embedded Computing Systems*, 3(1), 3–23, 2004.
70. Lu, S., V. Bharghavan, and R. Srikant. Fair scheduling in wireless packet networks. *IEEE/ACM Transactions on Networking*, 7(4), 473–489, 1999.
71. Boudec, J.-Y. L. and P. Thiran. *Network Calculus*. Springer, Berlin, LNCS 2050, 2001.
72. Cruz, R. A calculus for network delay. I. Network elements in isolation. *IEEE Transactions on Information Theory*, 37(1), 114–131, 1991.
73. Lu, X., E. Erkip, Y. Wang, and D. Goodman. Power efficient multimedia communication over wireless channels. *IEEE Journal on Selected Areas of Communications*, 21(10), 1738–1751, 2003.
74. Yu, W., Z. Sahinoglu, and A. Vetro. Energy-Efficient JPEG 2000 Image Transmission over Wireless Sensor Networks. *Proc. of IEEE Global Communications Conference (GLOBECOM)*, pp. 2738–2743. Dallas, TX, January 2004.
75. Vuran, M. C. and I. F. Akyildiz. Cross-Layer Analysis of Error Control in Wireless Sensor Networks. *Proc. of IEEE Intl. Conf. on Sensor and Ad-hoc Communications and Networks (SECON)*. Reston, VA, September 2006.
76. Schaar, M. V. D. and S. Shankar. Cross-layer wireless multimedia transmission: Challenges, principles and new paradigms. *IEEE Wireless Communications*, 12(4), 50–58, 2005.
77. Setton, E., T. Yoo, X. Zhu, A. Goldsmith, and B. Girod. Cross-layer design of *ad hoc* networks for real-time video streaming. *IEEE Wireless Communications*, 12(4), 59–65, 2005.
78. Reed, J. *Introduction to Ultra Wideband Communication Systems*. Prentice-Hall, Englewood Cliffs, NJ, 2005.
79. Revision of Part 15 of the Commission's Rules Regarding Ultra-Wideband Transmission Systems. First note and Order, Federal Communications Commission, ET-Docket 98–153, Adopted February 14, 2002, released April 22, 2002.
80. Win, M. Z. and R. A. Scholtz. Ultra-wide bandwidth time-hopping spread-spectrum impulse radio for wireless multiple-access communication. *IEEE Transactions on Communications*, 48(4), 679–689, 2000.
81. Batra, A., J. Balakrishnan, A. Dabak. Multi-Band OFDM Physical Layer Proposal for IEEE 802.15 Task Group 3a. IEEE P802.15 Working Group for Wireless Personal Area Networks (WPANs), March 2004.
82. Scholtz, R. A. and N. Z. Win. Impulse radio: How it works. *IEEE Communications Letters*, 2(2), 36–38, 1998.
83. Merz, R., J. Widmer, J.-Y. L. Boudec, and B. Radunovic. A joint PHY/MAC architecture for low-radiated power TH-UWB wireless ad-hoc networks. *Wireless Communications and Mobile Computing Journal*, 5(5), 567–580, 2005.
84. Gezici, S., Z. Tian, G. B. Giannakis, H. Kobayashi, A. F. Molisch, H. V. Poor, and Z. Sahinoglu. Localization via ultra-wideband radios. *IEEE Signal Processing Magazine*, 22(4), 70–84, 2005.
85. Melodia, T., D. Pompili, and I. F. Akyildiz. On the interdependence of distributed topology control and geographical routing in ad hoc and sensor networks. *IEEE Journal on Selected Areas in Communications*, 23(3), 520–532, 2005.
86. Yang, L. and G. B. Giannakis. Ultra-WideBand communications: An idea whose time has come. *IEEE Signal Processing Magazine*, 3(6), 26–54, 2004.

87. Cuomo, F., C. Martello, A. Baiocchi, and F. Capriotti. Radio resource sharing for ad-hoc networking with UWB. *IEEE Journal on Selected Areas in Communications*, 20(9), 1722–1732, 2002.
88. Akyildiz, I. F., W.-Y. Lee, M. C. Vuran, and S. Mohanty. NeXt generation/dynamic spectrum access/cognitive radio wireless networks: A survey. *Computer Networks (Elsevier)*, 50(13), 2127–2159, 2006.

Chapter 2

Performance Analysis of Multimedia Traffic over HSDPA

Irene de Bruin, Frank Brouwer, Neill Whillans, Yusun Fu, and Youqian Xiao

Contents

2.1 Introduction 48
2.2 Scenario and Model Description 50
2.2.1 Multicell Propagation Model 51
2.2.2 System and User Characteristics 53
2.3 Scheduling........ 54
2.3.1 Round-Robin Scheduler 55
2.3.2 Max C/I Scheduler 55
2.3.3 Proportional Fair Scheduler........ 56
2.3.4 Time-To-Live Scheduler 57
2.4 Analysis of the Default Scenario........ 57
2.4.1 Validation of Cell Selection 57
2.4.2 Mixed Traffic Scenario 59
2.5 Mobility: Optimization of Handover Parameters........ 62
2.5.1 Handover Threshold........ 62
2.5.2 Handover Delay 64

2.5.3 Discard Timer 67
2.6 Performance of Different Scheduler Algorithms 68
2.6.1 Specific Scheduler Properties 68
2.6.2 Priority Scheduling 75
2.6.3 Comparison of Scheduling Algorithms 78
2.6.4 Consumption of System Resources 80
2.7 Conclusion 82
2.8 Open Issues 84
Acknowledgments 85
References 85

2.1 Introduction

The rapid advances in wireless technologies have brought about a worldwide demand for high quality multimedia applications and services. These will particularly appear in the Internet, as well as in wireless scenarios [1,2]. Real-time services over high speed downlink packet access (HSDPA) is an issue under discussion in 3rd Generation Partnership Project (3GPP) [3,4]. A key improvement of Universal Mobile Telecommunications System (UMTS) has been realized with HSDPA, which has been defined for release 5 of the UMTS standard within 3GPP. HSDPA aims at increasing the systems capacity and the users' peak throughput from 2 to over 10 Mbps. Its transmission time feedback loop is smaller than that of regular UMTS and the scheduling functionality is no longer in the radio network controller (RNC) but in the Node-B (base station). As a result, the fluctuations in the radio channel fading characteristics can be better tracked. Scheduling of data flows is a key mechanism to provide quality of service (QoS) and optimize resource efficiency. Whereas the former objective is clearly the most important to the user, the latter is essential for the operator.

Although the initial scope of HSDPA was on best effort services, the interest in also using HSDPA for real-time applications is growing. An important reason for this interest is the increasing role of Voice-over-IP (VoIP) in both fixed and wireless networks. A general consensus in the market is that VoIP is the new and better way of communicating voice traffic.

The introduction of VoIP also creates new opportunities. As the VoIP connection is a session between IP addresses, enhancing the speech connection with video or data connections is rather easy. In addition, VoIP is better linked to the concept of the upcoming IP multimedia subsystem (IMS). VoIP, being a real-time service, has significantly different requirements on the connection than best effort services have. Delay is the most critical issue for VoIP, whereas loss of some packets can be tolerated. At the same time, best effort services require guaranteed delivery of the data, but do accept additional delay. The increasing demand for multimedia communication drives the development of consumer and enterprise markets, as well as the evolution of wireless technologies.

To be able to carefully investigate multimedia traffic over HSDPA, the two traffic types mentioned earlier form the basis of this chapter, and have complementary constraints on delay and error behavior:

- VoIP traffic
  - Delay sensitive
  - A low error rate is tolerable
- Best effort traffic
  - Less constraints on delay
  - Packets should not be dropped

To be able to compete with the other upcoming wireless technologies like WiMAX, EDGE, and others, the traditional wireless networks, with their infrastructure-based architecture, are challenged to smoothly transmit and deliver multimedia services. This requires the development of techniques, protocols, and algorithms to achieve the broadband high-speed, high-capacity, and high-reliability requirements. HSDPA enhances both the capacity and the peak rate of Wideband Code Division Multiple Access (WCDMA)-based systems. Although HSDPA will be used to serve users on a shared channel, the total capacity of the corresponding channel definitely competes with other broadband alternatives. A network simulator has been used to study the end-to-end behavior of enhanced UMTS (enhanced UMTS radio access networks extensions [EURANE] [5]). This simulator focuses on Media Access Control (MAC) and Radio Link Control (RLC). High-speed versions of these protocols are implemented for HSDPA, according to 3GPP standards. For earlier studies performed with the EURANE simulator, see Refs 6 and 7.

To investigate the feasibility of real-time services over HSDPA, the handover process is one of the issues that should be investigated. Therefore, we assume users moving at 120 km/h. Shorter transmission time interval (TTI) of HSDPA, as an extension to the WCDMA standard, enables fast scheduling. As a result, it is important to investigate different scheduling algorithms. The performance of these algorithms should be considered for the total system as well as user performance. The algorithms considered in this chapter are Round-Robin (RR), Max C/I, Proportional Fair (PF), and a Time-To-Live (TTL) scheduler.

The main idea underlying the RR scheduler is fairness between all users. The downside of the overall better system throughput of Max C/I scheduling is the relatively larger variance of the user throughput. The periodic character in the generation of VoIP packets in first hand fits best to the RR way of ordering the scheduling queues. However, the bursty traffic of best effort services is consuming the system's resources in an efficient way when the Max C/I scheduler is employed. Between these two extremes, we have the PF and TTL schedulers. These two try to mediate the extremes of the RR and Max C/I schedulers, so that the variance of the user throughput becomes less, but without compromising system throughput too much. Another way to differentiate in the scheduling of users is the assignment of priorities to different services. The impact of prioritizing the VoIP traffic is also investigated.

Section 2.2 discusses the simulation model used in this analysis, and also includes the description of the scenario, and the propagation modeling. This is followed by a description of the scheduling algorithms and other related aspects in Section 2.3. Section 2.4 presents results from the default simulation scenario. Several parameters will next be varied. Investigations on the handover process are discussed in Section 2.5. The impact of different scheduling algorithms is described in Section 2.6. Finally, Section 2.7 concludes this chapter, and the open issues that remain are presented in Section 2.8.

2.2 Scenario and Model Description

The main focus of this chapter is a network layer study of multimedia traffic over HSDPA. This section describes assumptions of the network-level simulator, propagation model, traffic mix, and finally some system and user characteristics. All results described are gathered with the ns-2-based network-level simulator EURANE [5].

The EURANE network-level simulator in ns-2 implements a complete end-to-end connection from the external IP network through the Core Network and UMTS Terrestrial Radio Access Network (UTRAN) to the user equipment (UE). The simulator focuses on MAC and RLC, and employs protocols implemented for HSDPA according to the 3GPP standard (release 5). The network simulator also requires a model that mimics the main characteristics of the radio channel based on physical layer simulation results. This is done by means of a link-level simulator that implements all physical layer aspects of enhanced UMTS (release 5) as specified by 3GPP. The enhancements include HSDPA (with both QPSK and 16-QAM modulations [8]). Results of this link-level simulator are used in the network-level simulator by means of input trace files resulting in signal-to-noise ratio (SNR) values fluctuating over time.

For the High-Speed Downlink Shared Channel (HS-DSCH; the wideband equivalent of the downlink shared channel in UMTS), a fast link adaptation method is applied in HSDPA. This involves a selection of the modulation and coding scheme. On its turn, these define several transport block sizes (TBS), alongside different UE types [8]. The UE reports the observed quality to the Node-B by means of the channel quality indicator (CQI). The Node-B decides, based on the CQI and additional information, such as the amount of available data to send and what TBS to use in its transmission. Large TBSs require several channelization codes of a cell to be allocated to the HS-DSCH, for example, a ratio of 15/16 of all codes [8]. Thus, a bad channel condition does not result in a higher transmit power (as is the case for the downlink-shared channel in UMTS), but instead prescribes another coding and modulation scheme.

The path loss model for the EURANE network-level simulator consists of three parts: distance loss, shadowing, and multipath. The model treats these three components independently. The simulator includes many details, in particular the MAC-hs and RLC functionality, as well as a more abstract modeling of the PHY, MAC-d, and Iub flow control, of which high-speed versions of these protocols are implemented for HSDPA, according to 3GPP standards [9,10]. The Hybrid-Automatic Repeat reQuest (H-ARQ) model assumes chase combining, which utilizes retransmissions to obtain a higher likelihood of packet reception. For the analysis described in this chapter, it is assumed that the UEs are of categories 1–6. The simulator selects the TBS from Table 7A of 3GPP specification 25.214 [9].

To estimate the performance of each single TBS, a link-level simulator has been used. This simulator considers radio communication between the Node-B and the UE using the HS-DSCH, based on the 3GPP specifications (for a detailed description, see Ref. 11). The link-level results provide an Eb/No versus block-error-rate (BLER) curve for all possible CQIs, that is, the interval [0, 30]. It should be mentioned that consecutive CQIs have a near constant offset of 1 dB at a BLER of 10 percent [11,12]. Inside the network-level simulator the UE indicates the CQI to the Node-B. The CQI represents the largest TBS resulting in a BLER of less than 0.1. The relation between CQI and SNR for a BLER of 0.1 is approximated through a linear function, based on the 3GPP standard [8]:

$$\mathrm{CQI} = \begin{cases} 0 & \mathrm{SNR} \leq -16 \\ \left\lfloor \frac{\mathrm{SNR}}{1.02} + 16.62 \right\rfloor & -16 < \mathrm{SNR} < 14 \\ 30 & 14 \leq \mathrm{SNR} \end{cases} \tag{2.1}$$

The root mean square error (RMSE) of this approximation is less than 0.05 dB, based on integer CQI values [11]. Note that a CQI equal to zero indicates out of range and the maximum CQI equals 30. Thus the function truncates at these limits [8].

2.2.1 Multicell Propagation Model

Under high mobility situations, handovers become an important issue for VoIP and other real-time services, when making use of HSDPA. As the soft and softer handover do not apply to the HS-DSCH (the data channel of HSDPA), the handover is a hard handover, with "break before make." This

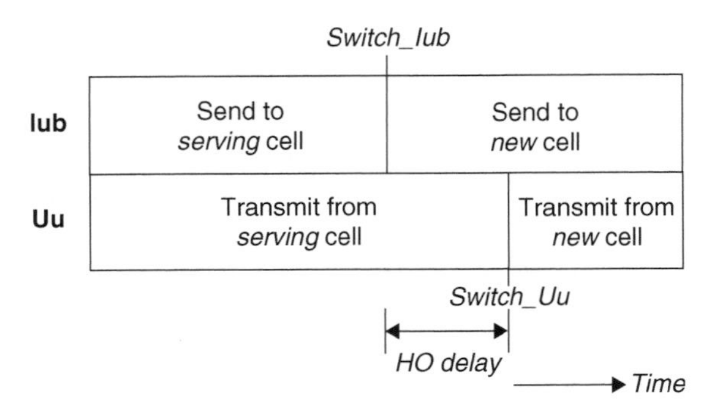

Figure 2.1 Time evolvement of the handover process, including the two switch moments that have HO delay in between them.

process is depicted in Figure 2.1; see also Ref. 13. To empty the buffer of the old cell, a handover delay is introduced. After sending the last packet over the Iub to the serving cell, some time needs to be included to empty the buffer of the serving cell (i.e., from the time moment marked by switch_Iub to the time moment marked by switch_Uu). In the meantime, no transmission can start from the new cell yet. The delay between these two switch moments is referred to as the "handover delay." A handover threshold level is used to prevent the unstable handover behavior of a user unnecessarily hopping back and forth between two cells.

A multicell propagation model is used to create a realistic pattern of handovers. However, this should be done with as little cells as possible to keep the simulation time of the complex network simulator within reasonable time limits. The cell layout has a hexagonal structure, where each Node-B covers three cells. A wraparound technique creates an infinite world in which the multicell character of UMTS has been modeled by means of three 3-sector sites, adding up to nine cells. Through the small number of cells and base stations used for the base layout, the simulations run considerably fast, creating more results in the same time. It has been verified that this model is large enough to create a realistic pattern of handovers.

We consider a macro cell with a COST 231-Hata distance loss model, which is a radio propagation model that extends the Hata and Okumura models to cover a more elaborated range of frequencies. The site-to-site distance is 1000 m, and the users are moving at 120 km/h in a Vehicular A environment. In the radio base station (RBS), the maximum power is set to 43 dBm per cell. The antenna pattern has a front-to-back ratio of 30 dB, a 3 dB opening angle of 66°, a gain of 17 dBi and a 3 dB feeder loss. It should be stressed that the interpretation of some results may change for different scenario settings. Also a different overall load or traffic mix may give different conclusions, in particular, for example, the optimization of the discard timer settings.

Position updates take place every 90 ms. With a typical handover execution time of round 100 ms, setting the update interval much smaller makes little sense. On a position update, the model uses the new position to calculate the distance loss. The precalculated map contains distance loss values for the nine cells, including an antenna gain. Next, the shadowing adds to the distance loss. The model assumes a single shadowing value per 3-sector site. The shadowing is modeled as a distance correlated stochastic process with a lognormal distribution, assuming a standard deviation of 5 dB and a correlation distance of 40 m.

Apart from the distance loss and shadowing, a multipath component is also fed into the actual physical layer model. The multipath fading is in line with standard UMTS models. The model is a correlated stochastic process for which the correlation depends on the UE speed. The multipath component uses a tapped delay line model, including intersymbol interference. The number of calculations in the model prevents the in-line calculation of the multipath model. Instead, the value is obtained from a trace. The trace is sufficiently long: at least 2000 s, corresponding to 1 million samples that are correlated in time based on the user velocity, which is assumed to be 120 km/h throughout this chapter. Each UE cell combination has a different starting point in the trace. These locations can be random with sufficient distance between the various locations. By generating the trace using an fast Fourier transform, the trace-file can be wrapped around. The simulator uses the path loss results without multipath fading to trigger the handover events. The model assumes that the handover measurement functionality, including time to trigger and threshold, suppresses the effect of the multipath fading.

The success or failure of a transmission is modeled based on previous work carried out in this area [11,14]. Link-level simulator results are used to analytically map a given SNR and CQI to a block error probability in an additive white Gaussian noise channel [12]. Overall, a delay of three TTIs (2 ms) is assumed between the CQI measurement at the user and the use of this control information at the Node-B. At the high user speed considered here, this control information will be very outdated: in the 6 ms that have passed, the user meanwhile has traveled 20 cm, which is more than a UMTS wavelength. As a result of this, the BLER will definitely exceed the target value of 10 percent.

2.2.2 System and User Characteristics

Randomly divided over the nine cells, both voice and data traffic are roughly one-half of the downlink traffic expressed in offered data rate. The system assumes 300 VoIP users and 20 best effort users. Each best effort user carries a User Datagram Protocol (UDP) stream of 118 kbps, which is 15 times the traffic load of a VoIP user. To investigate the system utilization, results from

Table 2.1 System Parameter Settings

Parameter	*Value*
Fixed delay from transmitter to RNC	70 ms
Maximum number of H-ARQ (re)transmissions	3
Number of HS-SCCHs	4
Number of parallel H-ARQ processes	8
Number of HS-PDSCHs (codes)	8
Flow control interval	10 ms
Discard time of PDUs in MAC-hs	200 ms
MAC-d PDU size	296 bits (37 bytes)

simulations with other traffic loads are also included for the RR scheduler. The modeled voice traffic is 12.2 kbps adaptive multi rate (AMR). Taking robust header compressing into account, silence detection, an activity factor of 50 percent and a packet interarrival time of 20 ms, the VoIP traffic is modeled with a packet data unit (PDU) size of 37 bytes. This is done by means of a continuous UDP stream. The BLER requirement for AMR is set to 7×10^{-3}. As the VoIP packets are time-critical and the AMR codec allows some packet loss, the RLC transmits the packets in unacknowledged mode (UM). This also avoids specific choices in the Transmission Control Protocol (TCP) that may bias the analysis.

Some remaining parameter settings are collected in Table 2.1. The fixed 70 ms delay from transmitter to the RNC assumes a voice connection with a UE at the far end, where the far-end UE uses a circuit-switched voice connection. Scheduling at the Node-B depends on many factors including the scheduling algorithm, the channel condition, the traffic load, and HS-DSCH power. The Iub flow control takes place at intervals of 10 ms. When many VoIP packets are waiting at the Node-B, the availability of the High-Speed Shared Control Channels (HS-SCCH) also becomes important. To prevent delayed packets from consuming capacity, a discard timer is set to remove outdated packets from the Node-B queue.

2.3 Scheduling

The shorter TTI of HSDPA, as an extension to the WCDMA standard, enables fast scheduling. As a result, it is important to investigate different scheduling algorithms. The performance of these algorithms should be considered for the total system as well as for user performance. Apart from the specific algorithm, another way to differentiate in the scheduling of users is the assignment of different priorities to different services. Finally, the impact of prioritizing the VoIP traffic is also investigated. Simulation results from these three issues are presented in Section 2.6.

To make efficient use of the system resources, multiple users might be served in a single TTI by using multiple HS-SCCHs. Also the transmit power of a MAC-hs PDU should be decreased as much as possible. This implies that, once a user is selected for transmission, every scheduler maximizes the amount of MAC-d PDUs that are transmitted, although the amount of power to do that successfully should be minimized. The schedulers are located at the Node-B and collect all queues that have packets waiting. For the case of VoIP, at most scheduling moments a user only has a single MAC-d PDU waiting, and will transmit with the lowest TBS that fits just one MAC-d PDU. Rest of this section describes the four scheduling algorithms considered in this chapter: RR, Max C/I, PF, and a TTL schedulers. Assume that users 1 to M have data pending in the Node-B and have a CQI that can support the transport of at least one MAC-d PDU.

2.3.1 Round-Robin Scheduler

Each active queue, that is, each queue that has packets waiting, is marked by the time moment that the packet, which is currently at the head of the queue, was put at the head of that queue. The amount of time this packet has waited at the head of the queue m is $Q_m(t)$, with t being the current TTI index. Note that for traffic generated in bursts, this time stamp is presumably not the time that such a packet entered that queue, but is the time that this queue was last served. For packets arriving to an empty queue, the time stamp equals the arrival time of that packet.

The packet with the earliest arrival time, that is, the longest waiting time at the head of the queue, is selected for transmission. In formula notation, the RR scheduler picks a queue index, $m_{\mathrm{RR}}(t)$, as follows:

$$m_{\mathrm{RR}}(t) = \max_{m}\{Q_m(t)\}, \quad m = 1, \ldots, M \tag{2.2}$$

where $m_{\mathrm{RR}}(t)$ denotes the queue index of the user(s) that has been selected for transmission. Furthermore, it should be noted that the notion of fairness between users for this scheduler is achieved with respect to the assignment of a TTI as such, and not with respect to the amount of data that is sent during such a TTI. As a result, for services that generate traffic in bursts, a user experiencing good channel conditions needs a relatively short waiting time for the transmission of a burst because a large TBS can be used.

2.3.2 Max C/I Scheduler

The Max C/I scheduler ranks the active users based on the channel conditions. The user that has reported the highest C/I value is scheduled first. In formula notation, the Max C/I scheduler picks a queue index, $m_{\mathrm{C/I}}(t)$, based on the following procedure:

$$m_{\mathrm{C/I}}(t) = \max_{m}\{C_m(t)\}, \quad m = 1, \ldots, M \tag{2.3}$$

where C_m is the maximum number of MAC-d PDUs that can be served at the current channel condition. In the results, a high BLER is reported for this scheduler. This scheduler is designed such that it schedules users that have reported a very good channel condition. However, this information is outdated for users moving at a high velocity because it is used in a downlink data transmission three TTIs later. For example, a user moving at 120 km/h has already moved 0.2 m, and as such the channel quality has also changed. The C/I scheduler will generally schedule users that were experiencing a peak in the fast fading, and therefore have a rather large probability to have a lower channel quality at the actual moment of scheduling. To overcome this problem, we have also considered this scheduler in combination with the so-called conservative TBS selection algorithm, described in Section 2.6.1.

2.3.3 Proportional Fair Scheduler

The PF scheduler exists in many variants. The current implementation is based on the algorithm described in Ref. 15. It schedules the active user that has the highest relative CQI, which is the supported throughput, normalized with the average delivered throughput for that user. In formula notation, the PF scheduler picks a queue index, $m_{\mathrm{PF}}(t)$, based on the following procedure:

$$m_{\mathrm{PF}}(t) = \max_{m}\left\{\frac{R_m(t)}{T_m(t)}\right\} \quad m = 1, \ldots, M \tag{2.4}$$

where $R_m(t)$ is the minimum of the following two values:

1. Maximum number of MAC-d PDUs that can be served at the current channel condition
2. Number of MAC-d PDUs waiting in the queue

Moreover,

$$T_m(t) = (1 - \alpha) \cdot T_m(t - 1) + \alpha \cdot R'_m(t - 1) \tag{2.5}$$

where $R'_m(t-1)$ is equal to the number of MAC-d PDUs that were scheduled to user m in the previous TTI. Note that T_m is updated at every TTI that the Node-B has packets waiting for the user m. So, if a user has packets waiting, but is not picked by the scheduler, $R'_m = 0$ and the resulting

$T_m(t)$ is reduced, which increases expression 2.4 and thus increases the chance that this user is scheduled in the next time moment. Finally, note that $m_{\mathrm{PF}}(t)$ is initialized to 1 if calculated for the first time.

2.3.4 Time-To-Live Scheduler

The TTL scheduler ranks the active users based on a measure of the C/I ratio, normalized with a fraction of the so-called TTL [16]. The latter term becomes smaller in time, and as such increases the chance of this user to be scheduled. The idea of the target TTL is introduced to ensure that packets for users with bad channel conditions, which thus might pass this time limit, are not removed from the Node-B queue immediately but instead have a second possibility to be scheduled. In formula notation, the TTL scheduler picks a queue index, $m_{\mathrm{TTL}}(t)$, as follows:

$$m_{\mathrm{PF}}(t) = \max_{m} \left\{ \frac{C_m(t)}{\mathrm{TTL}_X} \right\} \qquad m = 1, \ldots, M \tag{2.6}$$

with

$$\mathrm{TTL}_X = \begin{cases} \mathrm{DT}_{\mathrm{target}} - q_m & \text{if } q_m < \mathrm{DT}_{\mathrm{target}} \\ \epsilon \cdot (\mathrm{DT} - q_m) & \text{if } \mathrm{DT}_{\mathrm{target}} \leq q_m \leq \mathrm{DT} \end{cases} \tag{2.7}$$

where DT denotes the discard time, $\mathrm{DT}_{\mathrm{target}}$ the targeted fraction of DT, and $q_m(t)$ the time the packet at the head of queue m has already spent in that queue. Note that a packet that is not yet scheduled after $\mathrm{DT}_{\mathrm{target}}$ will be treated with "high priority" because of the multiplication with ϵ, which is taken equal to 0.001. A packet that is still not scheduled after the discard timer has expired ($q_m > \mathrm{DT}$) and is removed from that Node-B queue. Similarly, as in the C/I scheduler, the TTL scheduler also uses the conservative TBS selection mechanism.

2.4 Analysis of the Default Scenario

This section investigates the cell selection process in the multicell environment. Furthermore, a realistic system load is determined in terms of VoIP or best effort users. All this is done based on results from the EURANE simulator.

2.4.1 Validation of Cell Selection

Figure 2.2 shows the distance loss in combination with the hexagonal cell structure. In this figure, we have omitted the effect of shadowing, and

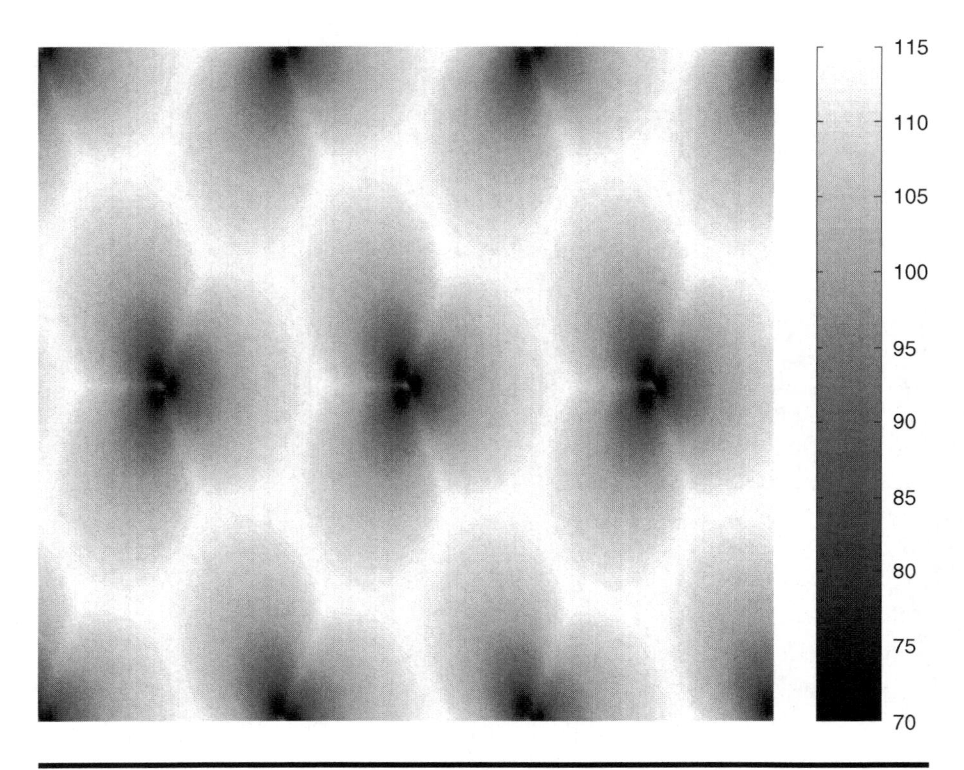

Figure 2.2 Minimum loss (in dB) for all Node-Bs and sectors.

the resulting picture perfectly displays the hexagonals and the antenna patterns. The left half is a shifted copy of the right half, and the distance loss is symmetrical in the vertical plane.

As a next step, users have been added to the system, and their paths of movement have been tracked with shadowing being included also. Figure 2.3 shows a snapshot of the locations of which the Node-B in the middle of the figure was the best cell, without applying any handover thresholds. The figure shows the selection of the three sectors using three different symbols. The snapshot shows that blurring due to shadowing clearly exists. The largest distance location a cell still selected is roughly 0.8 times the site-to-site distance. The thick arrow in Figure 2.3 indicates such a sample.

An issue to consider when using wraparound techniques to model an infinite area, is the exceptional case that when in the real world a user undergoes a handover by hopping over a neighbor cell, in the modeled setting the new cell may coincide with the old cell. In the case of HSDPA selection of the "related cells," a best cell should be geographically disconnected. That is, the distance between leaving an old cell until entering a

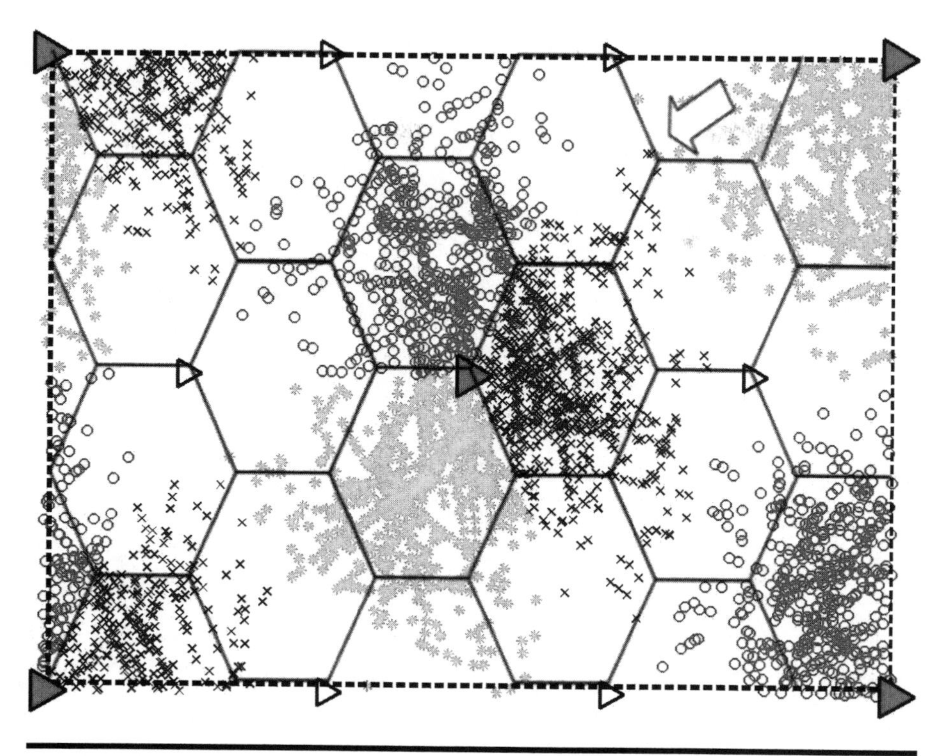

Figure 2.3 Snapshot of best cells selection. The three symbols (*, +, ∘) represent the three cells that are served by one Node-B.

new related cell should be large enough so that the system cannot keep connected, thereby missing a handover event.

2.4.2 Mixed Traffic Scenario

To determine a suitable load, we compare two traffic scenarios:

- *VoIP only.* 600 VoIP users are active in the nine cells of the system, resulting in an average load of 66.7 sources per cell.
- *Mixed traffic.* Half of the traffic is still VoIP traffic, although the other half is replaced by 20 users carrying best effort traffic.

The total offered data is the same for both scenarios, and we focus on the mixed traffic scenario. Each best effort user carries 15 times the traffic load of a VoIP user: 15×7.85 kbps $= 117.75$ kbps. The air-interface scheduler of this default scenario uses the RR principle. The scheduler sorts the transmission queues in order, according to the time the scheduler served each

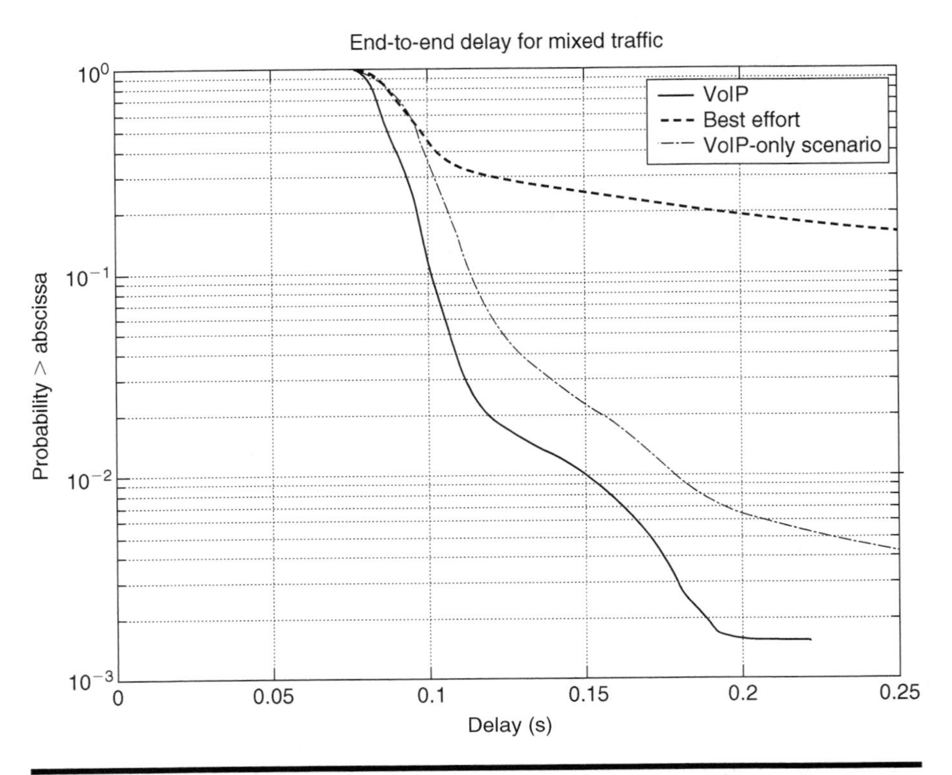

Figure 2.4 Complementary CDF of the end-to-end delay for mixed traffic scenario. The delay for the VoIP-only scenario is included as reference.

queue last. Figure 2.4 shows the resulting end-to-end delay, which consists of a static component that models the fixed network delay from the transmitter to the RNC (70 ms, see Table 2.1), and a scheduling delay, that is, the time spent in the Node-B queue and also at the user (due to the reordering procedure). Figure 2.4 also shows the delay curve for the VoIP-only scenario. For the mixed traffic situation, the delay for the VoIP users has clearly improved at the cost of the delay for the best effort users. This is due to the RR principle as the scheduler aims at providing all users an equal time-share, and the best effort users require a larger time-share to handle their traffic. As the RR scheduler does not consider this, the delay increases. The end-to-end delay for best effort traffic is allowed to be somewhat larger than VoIP traffic, but the current delay is too large. The 99th percentile can be in the order of 250 ms. Also note that the system throughput is larger for the mixed scenario because the users carrying best effort traffic, once scheduled, allow for more data to be scheduled, which implies a more efficient use of the system resources, and therefore the VoIP users in the mixed scenario have a lower delay compared to the VoIP-only scenario.

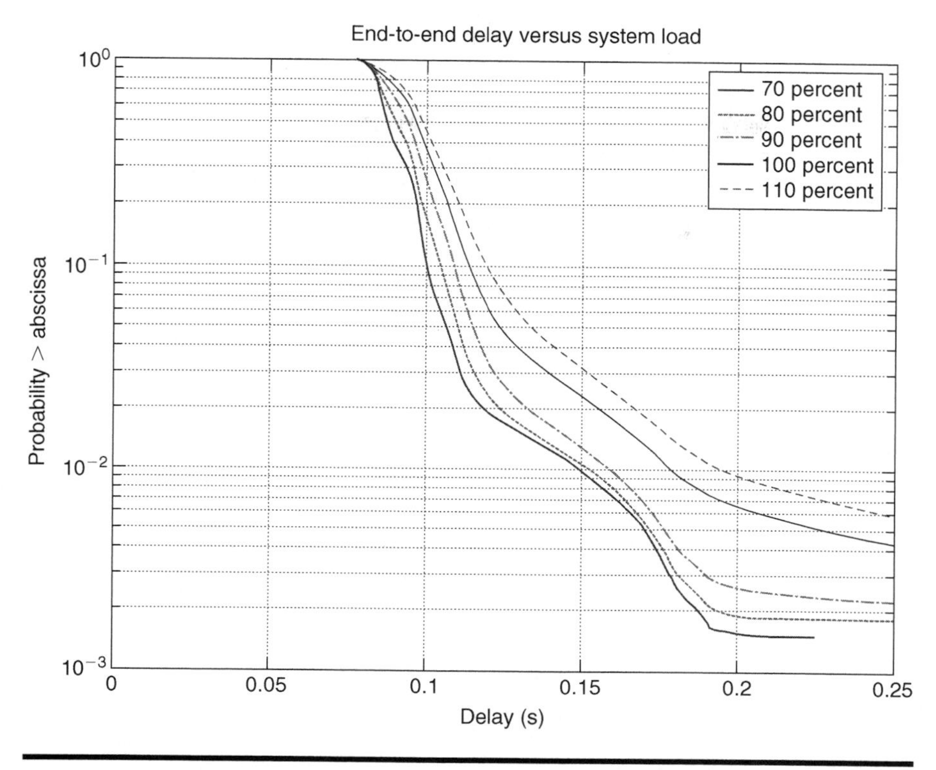

Figure 2.5 Complementary CDF for the end-to-end delay for a varying system load in the VoIP-only scenario.

Finally, we discuss the relation between system load and delay. Therefore, we investigate the effect of load in the case of the VoIP-only scenario. The load in the default scenario is used as a reference, and is indicated as 100 percent. It corresponds to an average load of 66.7 VoIP users per cell. The load is varied relative to this reference load, to observe the effect of load-variation on the end-to-end delay, and Figure 2.5 shows the relation between the delay and the system load. Figure 2.5 plots the relative capacity compared to the reference, as the absolute capacity from this analysis depends strongly on a large number of assumptions on both environmental conditions as well as system implementation. It shows that an increase in load stretches the delay curve. The shortest delays remain constant, but the longer delays increase. In addition, the different curves show a break point. The difference between the curves up to 90 percent load is small. The curves starting at 100 percent load show a larger difference. When load becomes an issue, 10 percent more load results in roughly 20 ms additional end-to-end delay.

2.5 Mobility: Optimization of Handover Parameters

Under high mobility situations, handovers become an important issue for VoIP and other real-time services, when making use of HSDPA, where the handover is a hard handover, with a "break before make." As the VoIP packets are time-critical and the AMR codec allows some packet loss, the RLC transmits the packets in UM. This requires that the time instances of stopping and starting should be such that the serving (old) cell has sufficient time to complete transmission of packets already sent over the Iub, although avoiding as much as possible, a time gap in the transmission. Too short a handover delay results in packets that are still waiting in the Node-B queue at the time of switching. These packets are, by definition, lost. At the same time, any additional waiting time causes extra delay to packets waiting in the new cell. This section investigates this trade-off, in particular with respect to handover aspects of an HSDPA system carrying VoIP services. The effect of the handover threshold, the handover delay, and the discard timer will be explored. To come closer to a realistic HSDPA scenario, best effort traffic is also included. In the remainder of this section, we will analyze the effect of the handover threshold, the handover delay, and the discard timer; see Table 2.2. First we start with a definition of the handover process, which is already depicted in Figure 2.1.

DEFINITION (Handover Process)
If $signal_{\text{target}}(\text{t}) > signal_{\text{source}}(\text{t}) + \text{HOT}$, *then trigger the handover procedure:*
Release the old Iub link, and create the new Iub link at time t
Release the old Uu link, and create the new Uu link at time t + HOD

2.5.1 Handover Threshold

The handover algorithm measures the SNR of the common pilot channel (CPICH) of each cell. When the difference between the SNR of the current cell and the best cell exceeds the threshold, the RNC initiates the handover of the HSDPA connection [10]. In 3GPP, this is referred to as repointing. During the handover procedure, the RNC stops transmission of packets over the Iub to the old cell and starts transmission to the new cell.

Table 2.2 Handover Parameters That Are Varied during This Chapter

Parameter	*Acronym*	*Default Value*	*Variation*
Handover threshold	HOT	6 dB	5, 6, 7 dB
Handover delay	HOD	100 ms	70, 100, 130, 160 ms
Discard timer	DT	200 ms	75, 100, 150, 200 ms

The packets that arrive to the RNC during a handover are already forwarded to the new cell and therefore experience extra delay while waiting in the new cell. Higher handover intensities increase the chance of a user being in handover, and therefore create higher probabilities for packets to arrive during the handover. The analysis varies the handover threshold between 5 and 7 dB.

Table 2.3 shows the resulting handover intensity expressed as handovers per UE per hour. Table 2.3 splits the handovers into inter- and intra-Node-B. The intra-Node-B cases allow for redirection of the scheduling queue at the Node-B to the new cell, if the Node-B implementation can handle it. As a reference, the bottom line shows the intensity of the border crossing of geographical hexagonal cells. This is the intensity when no fading would exist. As the model assumes that the shadowing for all cells of a single Node-B is the same, the intensity for the intra-Node-B is close to the theoretical value. The reason that it is even lower is that some intra-Node-B handovers are replaced by a combination of two inter-Node-B handovers. Furthermore, Table 2.3 shows that for realistic handover thresholds the fraction of packets that arrive during the hand-over delay of 100 ms is around 2.5 percent. This percentage is calculated by multiplying the total number of handovers (fourth column in Table 2.3) with the 0.100/3600 fraction of the time of one handover. The VoIP packets are time-critical and the AMR codec allows a packet loss of a maximum of 0.7 percent. So the delay of the packets arriving during the hand-over plays a measurable role in the overall performance.

Table 2.4 shows the transmission quality for the different handover thresholds. Owing to the choice for the RR scheduler, the quality for VoIP is significantly different from the performance of the best effort traffic. Any other scheduler will give a different balance. For both types of traffic the non-scheduled fraction of the MAC-d PDUs increases with the handover threshold, whereas the residual BLER of the H-ARQ remains the same. The difference in residual BLER between both types of traffic is due to the

Table 2.3 Handover Intensity and Handover Probability (Assuming a Handover Delay of 100 ms); Related to Handover Threshold

Handover Threshold (dB)	*Intensity (Number of Handovers/UE/Hour)*			*Time Ratio of Being in a Handover (Percent)*
	Inter-Node-B	*Intra-Node-B*	*Total*	
5	891	71	962	2.7
6	741	70	811	2.3
7	626	71	697	1.9
Geographic border crossings	176	88	265	0.7

Table 2.4 Transmission Quality Related to Handover Threshold

	VoIP			*Best Effort*		
Handover Threshold (dB)	*PDUs Not Scheduled*	*H-ARQ Residual BLER*	*End-to-End PER*	*PDUs Not Scheduled (Percent)*	*H-ARQ Residual BLER*	*End-to-End PER (Percent)*
5	1×10^{-5}	1.5×10^{-3}	1.5×10^{-3}	6.4	3.9×10^{-3}	6.8
6	3×10^{-5}	1.5×10^{-3}	1.5×10^{-3}	7.5	3.7×10^{-3}	7.9
7	12×10^{-5}	1.6×10^{-5}	1.7×10^{-3}	9.1	3.8×10^{-5}	9.4

difference in TBS that is used. As the best effort traffic produces more traffic per time instance, the traffic is packed into larger TBSs. Larger TBSs have a larger residual BLER after N transmissions. Although for VoIP the residual BLER is the dominant contributor to the end-to-end packet error rate (PER), the number of nonscheduled PDUs dominates for the best effort traffic.

For VoIP traffic, the delay decreases for an increasing handover threshold, whereas the opposite is true for the best effort traffic. This shows the balance between two forces in the system.

1. An increase in handover threshold reduces the chance of an unnecessary handover-pair triggered by a temporally extreme shadow fading contribution (the geographic border crossing has not been reached yet).
2. An increase in the value for the handover threshold causes the power to be used less efficiently, resulting in less power being available for other transmissions.

In the system under study, the first force dominates for VoIP, whereas the second force dominates for the best effort traffic. Overall, executing a handover to save power tends to prevail over avoiding handovers to limit the probability that a packet arrives during a handover.

2.5.2 Handover Delay

The handover delay, as discussed here, is the time between the moment the RNC stops transmission over the Iub to the old cell, until the first PDU can be scheduled from the new cell, which is a model of the handover process described in 3GPP (e.g., Ref. 17) and illustrated in Figure 2.1. This delay does not necessarily start directly at the time the RNC decides to execute the handover. The discussions on the handover process in 3GPP (e.g., Ref. 17) indicate that the delay for establishing a connection is at least well over 100 ms.

There are two opposite effects in the handover delay. A longer delay allows for more PDUs to be scheduled, emptying the queues of the old cell.

Table 2.5 Probability That Packets Arrive during the Handover Process

Handover delay	70 ms	100 ms	130 ms	160 ms
Packets arriving during the handover process	1.6 percent	2.3 percent	3.0 percent	3.7 percent

A shorter delay decreases the probability that a packet arrives during this time (see Table 2.5). In addition, a shorter delay will decrease the average time such that a packet has to wait until the new cell can start scheduling it.

Throughout the rest of this chapter, several statistics are presented. Depending on the variable, either a cumulative distribution function (CDF) or a probability density function (PDF) will be shown. By definition, a CDF is increasing from 0 to 1, and CDF(x) w.r.t. X denotes the chance $P(X \leq x)$. For delay statistics, one is mostly interested in the percentage of packets that is received successfully within a certain time. Thus, the end-to-end delay will be shown by means of the corresponding CDF. This also allows for a clear indication whether the AMR codec (which only for an error rate threshold of 0.7 percent) will be able to successfully perform the decoding. A PDF is the derivative of the corresponding CDF. PDF(x) w.r.t. X denotes the chance $P(X = x)$.

Figure 2.6 shows the complementary CDF of the end-to-end delay for a varying handover delay. The VoIP curves for all values of the handover delay display distinct kinks, indicating a significant increase in end-to-end delay. This percentage level is strongly correlated to the amount of VoIP packets that encounter a handover process, as shown in Table 2.4. It should also be noted that the horizontal distances between all four VoIP curves at the lower right-hand side of Figure 2.6 are separated by 30 ms, corresponding to the difference in subsequent values of the handover delay, that is, the delay of these packets is large, purely due to the handover process. The bending of all curves for best effort traffic takes place at an end-to-end delay of around 120 ms, which is mainly due to the fact that the scheduler sorts the transmissions queues, according to the time the scheduler served each queue last, whereas each best effort user carries 15 times the traffic of a VoIP user. Finally, it can be noted that the minimum of all the end-to-end delay curves is indeed the expected fixed delay of 70 ms; see Table 2.1.

Figure 2.7 shows the complementary CDF of the scheduling delay, that is, the time a packet has been waiting at the Node-B (for the case of a 70 ms handover delay). Figure 2.7 makes a distinction between PDUs arriving at the RNC during a handover, and packets arriving when no handover takes place. The packets arriving during the handover delay have to wait at least until the handover process is complete. The packets that, during the handover (HO) process, have been redirected to the new cell, and experience an extra delay because the Uu connection has not been set up

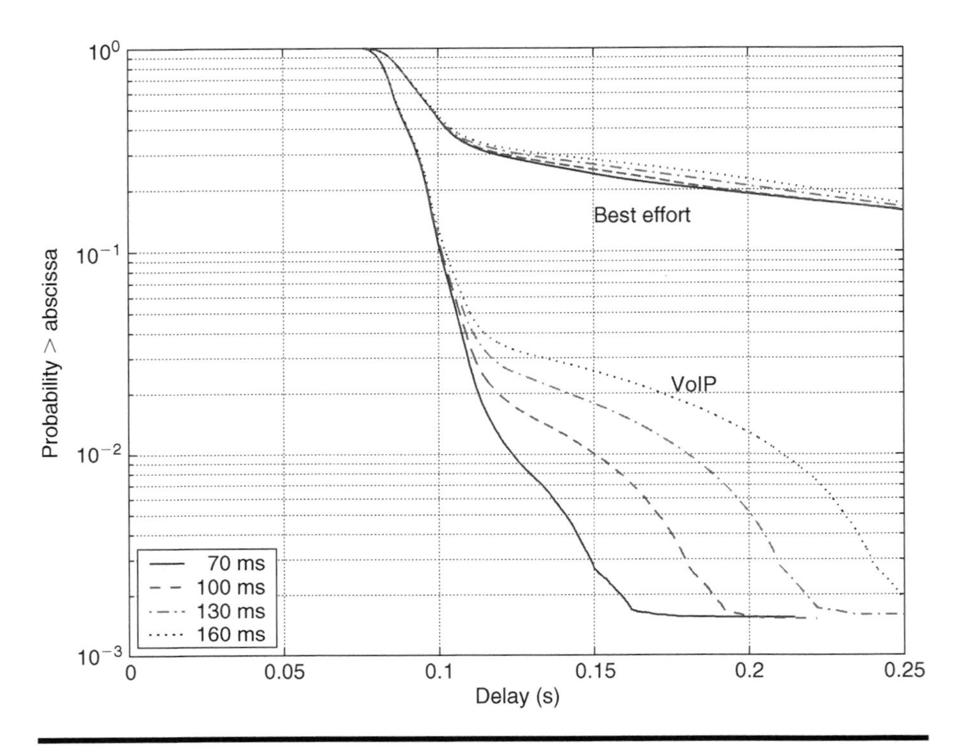

Figure 2.6 Complementary CDF of the end-to-end delay for mixed traffic scenario for varying handover delays.

yet, are marked with "in HO," whereas all others are marked with "out of HO." The current setting of the handover delay (70 ms) results in an average extra delay of at least 35 ms. Directly after the handover, the packets are sent as soon as possible, resulting in a steep curve, in particular for VoIP.

Figure 2.7 shows an additional effect for the best effort traffic. Directly after a handover, the probability of a large delay is significantly lower than during the average situation. During the average situation, the queues at the Node-B can fill-up due to lack of resources. The amount of packets that are generated during a handover process is rather small, so the queue build-up at the new cell is also limited and much smaller than the queues of a cell that has already been up and running for a while. Considering the intersection between the two curves in Figure 2.7b, about 25 percent of all best effort packets experience a scheduling delay of more than 70 ms. Thus, this value of the handover delay (70 ms) is considered too short to empty the source cell, and will therefore result in a significant loss of best effort packets. As the RLC will typically transmit best effort packets in AM (acknowledged mode), packets arriving at the UE via the new cell will trigger the RLC to retransmit the packets discarded from the Node-B,

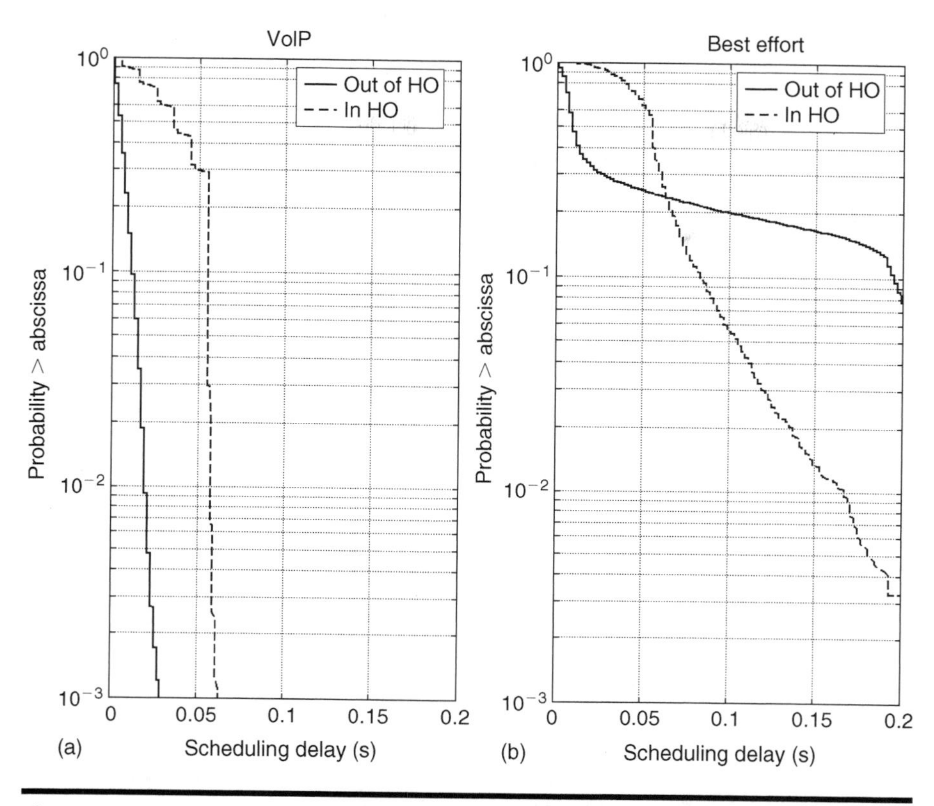

Figure 2.7 Complementary CDF of the scheduling delay for mixed traffic scenario for a handover delay of 70 ms. (a) VoIP traffic and (b) best effort traffic.

causing a limited additional delay. Figure 2.6 shows that the shorter handover delay, similarly as for VoIP, is also better for the best effort traffic with its long delays. The handover delay is preferably set in the order of 50–100 ms.

2.5.3 Discard Timer

The Node-B contains a discard timer that discards packets waiting for too long. The residual BLER of the H-ARQ process is likely to be in the order of 1×10^{-3} to 2×10^{-3}. Moreover, AMR requires a PER of less than 7×10^{-3} for class A bits. Combining these two values, approximately 5×10^{-3} of the VoIP packets could be discarded in the Node-B without violating this requirement.

For best effort traffic, the situation is different, as it is persistent traffic. The RLC will retransmit packets that have been discarded by the Node-B. In general, discarding packets does not change the traffic load. It however

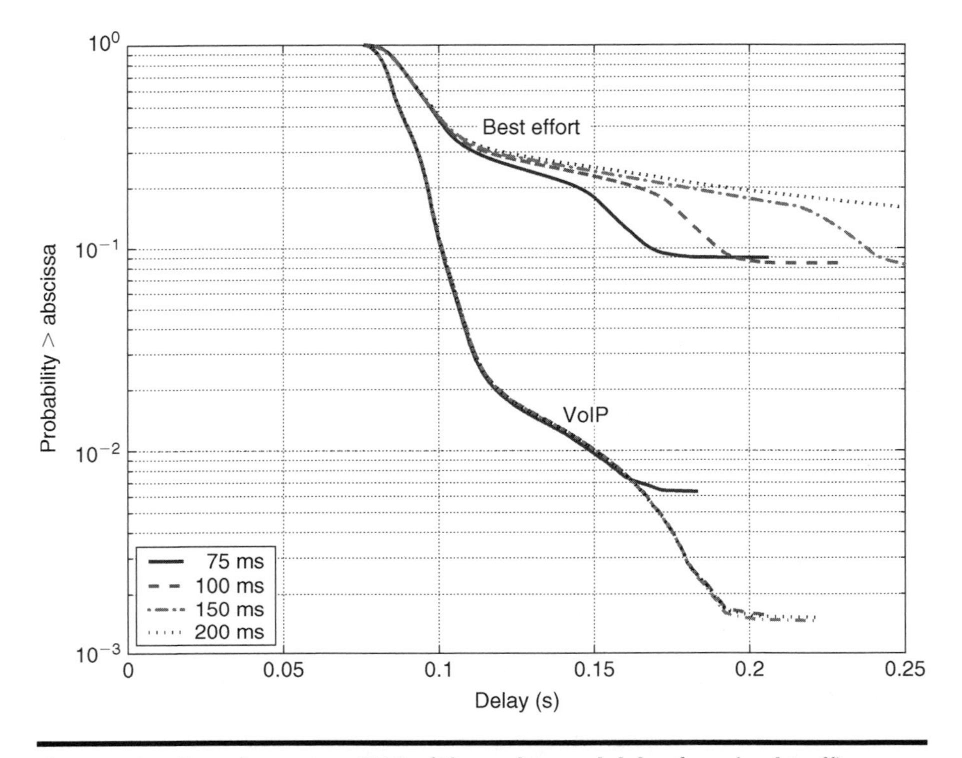

Figure 2.8 Complementary CDF of the end-to-end delay for mixed traffic scenario for a varying discard timer.

removes the peaks, making room for other traffic. Figure 2.8 shows the complementary CDF for the end-to-end delay for a varying discard timer. Considering the curves for VoIP, the only effect of decreasing the discard timer is an increase of the error floor. Discarding late packets does not provide a gain, only a risk when setting it too tight. Decreasing the discard timer also increases the error floor for the best effort traffic. However, it also decreases the delay for the majority of the packets. The discard timer relates to the handover delay, and the typical scheduling delay. Here the handover delay is set to the default value of 100 ms. The majority of the packets have a scheduling delay of a maximum of 100 ms; thus a reasonable setting for the discard timer would be 200 ms.

2.6 Performance of Different Scheduler Algorithms

2.6.1 Specific Scheduler Properties

This section starts with the BLER statistics for all schedulers. The TBS selection is based on the channel quality reported from the UE to the Node-B. Theoretically, the BLER statistics should not depend too much on the type of scheduler. However, in reality this control information is three TTIs old; thus, for users that move at high velocity, it is presumably outdated. The UEs move at a speed of 120 km/h = 33.3 m/s. As the CQI is three TTIs old at the time of the transmission, the UE moves $33.3 \times 3 \times 2$ ms = 0.2 m, where the wavelength equals 0.15 m. Therefore, the multipath fading at the time of transmission is uncorrelated from the measurement. Because the impact of this effect is different for each scheduler, it is addressed in this section. For the RR scheduler, this causes an inaccuracy where the channel condition could have improved, as well as deteriorated between the two time moments. The other two schedulers prefer UEs with high CQI. As a result, the channel condition typically deteriorates between the CQI measurement and the time of transmission.

For some schedulers a high conditional and residual BLER is experienced. For this reason, we have also included the so-called conservative TBS selection mechanism. In this algorithm, the Node-B decreases the received CQI by 1 before entering the selection and allocation functionality. After this CQI adjustment, less data is scheduled while keeping the transmit power at the proper level. As a result, the BLER decreases. Table 2.6 collects the BLER results for the RR, PF, the TTL, and the Regular and Conservative C/I schedulers. The main difference in the results involves the use of the conservative allocation of channel resources. Another result that becomes clear from the table is the drastic decrease of the conditional BLER after the first H-ARQ transmission. It is present for all schedulers and can be explained by the high UE velocity. H-ARQ retransmissions are performed 12 ms, or even 24 ms, after the original transmission. Within this time the channel condition has presumably changed a lot, for users moving at 120 km/h. Because the residual BLER for the regular C/I and TTL algorithms approaches the 0.7 percent requirement for end-to-end VoIP traffic, from now on the results of the C/I and the TTL schedulers will be considered with the CQI adjustment.

Next, some statistics of the CQI are considered for some extreme schedulers. Figure 2.9 collects the statistics of the channel conditions of all users, as well as the scheduler-specific statistics of the CQI values of the users

Table 2.6 BLER Values for Different Scheduling Algorithms

BLER	*Conditional BLER at the First, Second, and Third Transmissions (Percent)*			*Residual BLER (Percent)*
Regular C/I	34.0	9.7	8.6	0.3
C/I	21.9	5.7	5.4	0.1
Round-Robin	29.0	7.9	7.3	0.2
Time-To-Live	23.5	6.3	6.0	0.1
Proportional Fair	30.5	8.7	7.8	0.2

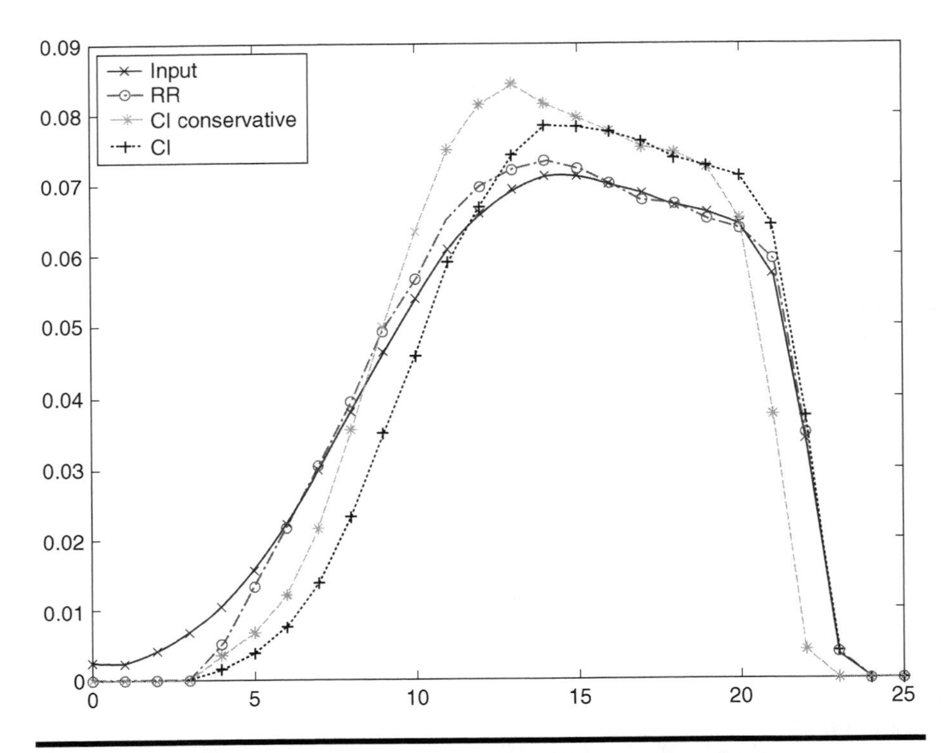

Figure 2.9 Probability density function of distribution of used CQIs.

that are actually scheduled. The "input" shows the distribution of CQI as the UEs report to the Node-B; thus this is before scheduling. The other curves show the CQIs actually assigned by the scheduler. Note that for most scheduling events, the CQI value representing the channel condition is reduced to save Node-B power, because only a small TBS is needed for the one or few VoIP MAC-d PDUs that are waiting in the queue.

For the C/I scheduler, users that experience good channel conditions presumably have only a single packet waiting in the Node-B queue, and therefore only need the smallest TBS. The RR scheduler might be scheduling a user with a high channel quality because the packet at the head of the queue has already been waiting for the longest time. This also implies that more packets are waiting for this user, or it is scheduled because there are still some resources left at the Node-B for transmission in that TTI. At a high channel condition, very little power is needed to add a packet to the transmission block of a TTI.

It is clear from Figure 2.9 that the RR results are close to the input statistics. The first CQI value that allows for a transmission is CQI = 4, corresponding to a TBS of 39 bytes. This explains the difference between the RR statistics and input statistics. Users with a bad channel condition have a long

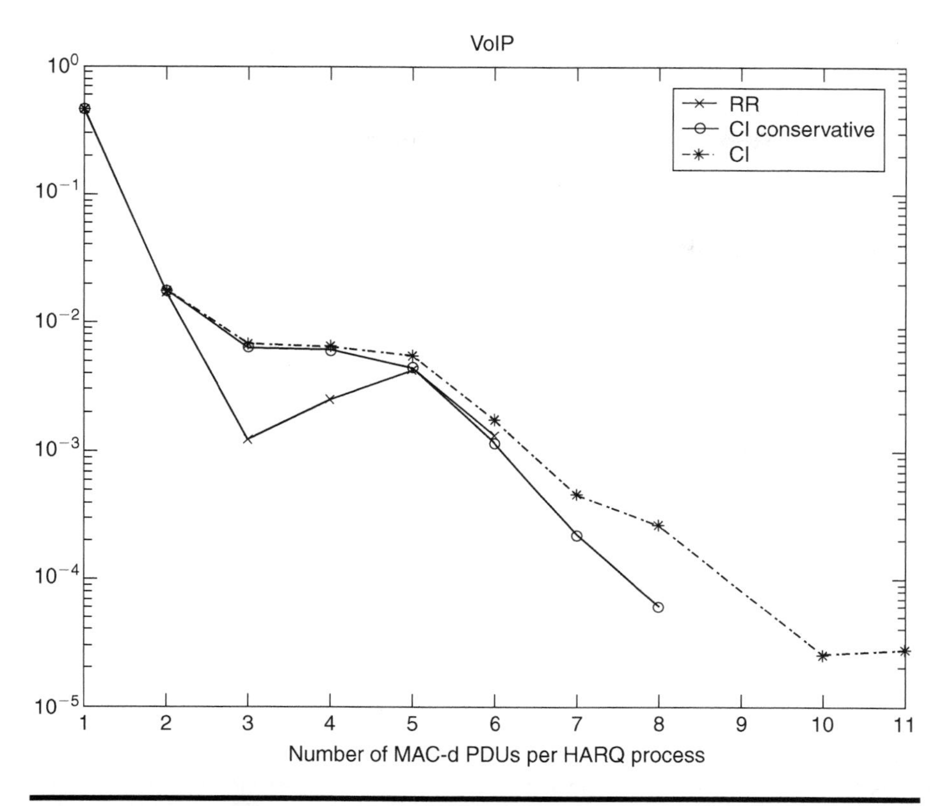

Figure 2.10 Probability density function of the number of MAC-d PDUs for the VoIP traffic.

waiting time for the packet at the head of the queue. As a result, this queue is scheduled immediately once the channel condition has become good enough for a transmission. These reasons explain why the distribution of the channel condition of users picked by the RR scheduler does not match the overall statistics of the channel condition of the users. From the CQI statistics of the regular C/I scheduler it can be concluded that it selects users of higher than normal channel conditions. It does not select more users than the RR scheduler at very good channel conditions. This is presumably because the very good channel conditions are preceded by suboptimal channel conditions at which the C/I scheduler already selects these users. Finally, the conservative scheduler shows the result after the mentioned CQI adjustment of -1, which mainly implies a transition of unit 1 to the left.

This is why we next consider the number of MAC-d PDUs that are actually scheduled during each H-ARQ process. The results are split into VoIP and best effort users and collected in Figures 2.10 and 2.11. The VoIP packets usually come one at a time. Thus, most of the MAC-hs PDUs contain only a single MAC-d PDU. Therefore, the plot for the VoIP figure

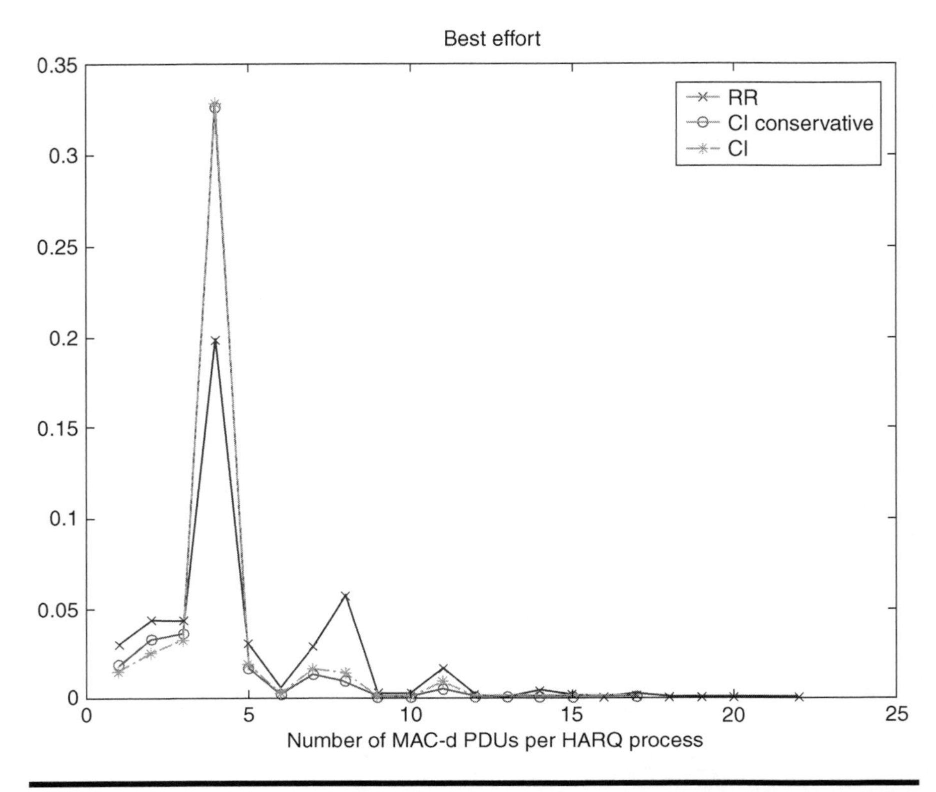

Figure 2.11 Probability density function of the number of MAC-d PDUs for the best effort traffic.

(Figure 2.10) has been shown in logarithmic format. Both plots are normalized with the total number of generated packets for VoIP and best effort traffic. For the best effort traffic of Figure 2.11, the total sum remains below 1 because a lot of these packets have already been discarded from the Node-B queue beforehand.

There are two things that distinguish the different scheduler results in Figure 2.10. First of all, there is a clear dip at 3 MAC-d PDUs, followed by an increase toward 4 and 5 MAC-d PDUs. This is a result of packets waiting at the new Node-B due to the handover timer, which is set to 100 ms. During this time, five VoIP packets are generated and put in the queue at the new Node-B. In particular, the RR scheduler will immediately assign this queue to be scheduled once the connection to the UE has been made. The tail of Figure 2.10 also behaves differently for the three schedulers. The number of MAC-d PDUs per process is never larger than six for the RR result, and is due to two reasons. Because each user is scheduled regularly, the queues do not grow too long. Moreover, the channel quality is not a criterion for scheduling, and as such usually does not allow too many MAC-d PDUs to fit into a single transport block.

Owing to the higher number of erroneous transmissions, more packets keep waiting in the Node-B queues for the C/I scheduler compared to the conservative C/I alternative, which results in more packets scheduled in a single MAC-hs PDU. The underlying idea of a fair allocation between all users is displayed by the short tail of the RR scheduler. The maximum number of MAC-d PDUs that are scheduled together equals 6 for the RR scheduler. The maximum for the C/I scheduler equals 11, which corresponds to the fact that when generating 1 packet per 18.8 ms, at most 11 packets can be waiting before the discard timer expires after 200 ms.

The equivalent plot for the best effort traffic is dominated by a preference for a certain number of MAC-d PDUs. The first peak is at 4 MAC-d PDUs. This can easily be explained with the traffic, which generates 4 MAC-d PDUs per flow control cycle. As a result, a user that is scheduled during that flow control period typically has four packets waiting, or a multiple of four packets. Another effect that is also important is the number of MAC-d PDUs that fit in a MAC-hs PDU when it is completely utilized. For the larger transport blocks, there is not always a TBS available for every number of MAC-d PDUs waiting. This is the case for 6, 9, 10, 12, 13, 16, 18, 20, and 21 MAC-d PDUs. These numbers of MAC-d PDUs form only a MAC-hs PDU in case there are no more packets waiting to fill the MAC-d PDU up to its maximum size. This explains the rather low number of times that these specific numbers of MAC-d PDUs are scheduled together. Next, the three schedulers from the figure are compared. There is only one value at which the C/I contributions are far higher than RR and that is at 4 MAC-d PDUs. This can be explained by stating that the best effort users are not scheduled immediately after the burst of packets from the RNC have arrived at the Node-B flow but have to wait for some time. Thus, once the best effort users are scheduled for transmission, they have already received a new burst of four packets, which explains the high value for the RR scheduler at 8, 12, etc., MAC-d PDUs.

Finally, we discuss some results of the two remaining schedulers that are intended to combine the benefits of RR and Max C/I. The main idea underlying the RR scheduler is fairness between all users. The downside of the overall better system throughput of Max C/I scheduling is the relatively larger variance of the user throughput. In between these two extremes, we have the PF and TTL schedulers. We will first consider different parameter settings for the PF scheduler.

The parameter $(1-\alpha)$ (see Equation 2.5) determines how much history is taken into account in the filter for the average delivered throughput. A large value of α corresponds to a more dynamic result for the average throughput, and results in a larger variety in the users that are scheduled.

Table 2.7 Results for Different Parameter Settings of α in the PF Scheduler

	Discarded (Percent)	
α (PF Scheduler)	*VoIP*	*Best Effort*
0.025	0.020	1.9
0.050	0.014	2.0
0.100	0.013	2.2
0.200	0.011	2.4
RR	0.001	3.8
C/I	0.67	4.3

Note: The Round-Robin and C/I results are also included.

Results for this scheduler have been collected in Table 2.7. The results have been split for the VoIP and the best effort users. When α is increased, less VoIP packets get discarded at the cost of more best effort packets being discarded. However, the percentages of VoIP packets being discarded (0.011–0.02 percent) are far below the target value of 0.7 percent. For all values of α, less packets are discarded compared to the RR and C/I schedulers. The resulting end-to-end delay does not depend very much on the specific setting of the parameter. In the remainder of this chapter, the case of $\alpha = 0.100$ will be considered as the default value for PF scheduling.

One of the main conclusions already found by Kolding [15] was that the scheduler performs best at UE speeds around 3–10 km/h. Because our results were obtained for UE speeds of 120 km/h, we will discuss the results briefly. Moreover, because the scheduling algorithm is based on the averaged throughput, it only performs well in situations where the channel condition is changing slowly in time. That is not the case at the very high UE speeds, which is the focus of this chapter. Furthermore, it should be stressed that at every handover that takes place, the initialization of the filter in the averaging will have a large effect. Certainly because, during a handover, packets have been waiting at the new Node-B before the transmission to the UE can be made. This will create an unusually high initial throughput that will take a long time to converge to the actual long-term throughput.

Finally, we consider the TTL scheduler, which, just like the PF scheduler, tries to combine the benefits of the RR and C/I schedulers. The TTL tries to make sure that packets are scheduled before they are discarded from the Node-B queue. The TTL target time equals 20, 40, and 60 ms for the considered results (10, 20, and 30 percent of the discard timer of 200 ms.) A lower TTL target makes the scheduler attempt to schedule a packet, which has been waiting for a while, at an earlier stage. As expected, this scheduler has the lowest rate of discarded packets; see Table 2.8. It is also clear, from

Table 2.8 Results for Different Parameter Settings of the Target Percentage in the TTL Scheduler

TTL Target Ratio (Percent)	*Discarded (Percent)*	
	VoIP	*Best Effort*
10	0.44	3.0
20	0.43	3.0
30	0.46	3.1
RR	0.003	7.5
C/I	1.35	8.6

Note: The Round-Robin and C/I results are included also.

this table, that the results of the TTL scheduler are not very sensitive, with respect to the exact value of the TTL target.

Because the TTL target is very much related to the setting of the discard timer, we have also performed some simulations in which the discard timer has been varied, although keeping the TTL target time constant at 20 ms. In this case, the behavior of the longer delay is changing because packets are discarded from the Node-B queue differently; see Figure 2.12. There is a large difference between the results for different settings of the discard timer for the VoIP as well as the best effort users. Also note that the convergence level at the right-hand side of Figure 2.12 should be considered as the percentage of packets that is lost. From now on, we will only consider the TTL results for the case of a 20 percent target, in combination with a 200 ms discard timer.

2.6.2 Priority Scheduling

In this section, we consider the three most important schedulers for the case where all VoIP packets have a scheduling priority over the best effort packets. Figure 2.13 collects the delay statistics for these schedulers. The C/I results for the nonpriority equivalent are also included. When priority is included, as expected, the VoIP users are far better off at the cost of the longer delay that is experienced by the best effort users. When comparing the priority results for the three schedulers, it is clear that the impact of the specific scheduler becomes less important. The priority as such creates a new scheduler with two classes. Inside these classes the packets are still scheduled according to the specific scheduling algorithm, but this has a smaller effect now. This is also clear from Figure 2.13: the three priority VoIP results are very close. Only the tail of the distribution is different. The choice of the scheduler does have an impact on the delay statistics of the best effort users. This is probably the result of the fact that the TTL scheduler generally on average consumes more power. This is also the case

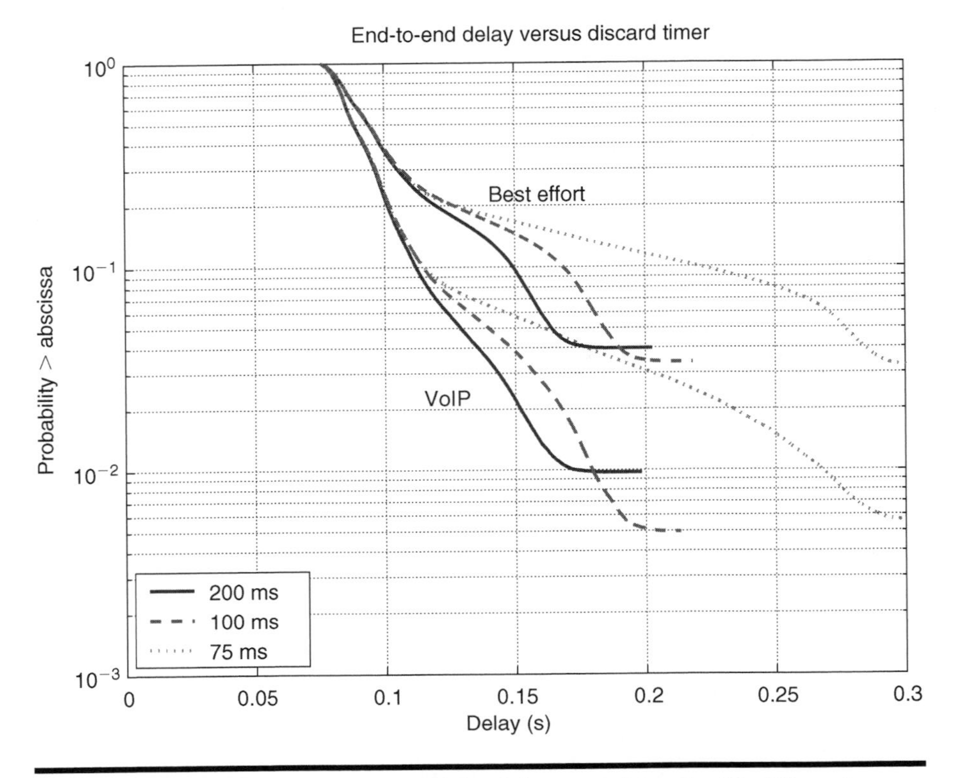

Figure 2.12 Complementary CDF of the delay experienced with different discard times of the TTL scheduler, although keeping the TTL target time constant at 20 ms.

for the VoIP users. Because the best effort packets can only be scheduled when there are no longer VoIP packets waiting, the scheduling of best effort traffic is influenced by the resources that are already in use by the VoIP traffic. This is particularly the case for the TTL scheduler.

A difference in priority between VoIP and best effort traffic can also be realized implicitly with the TTL scheduler by setting different discard timers in combination with a different scheduling target. This is particularly interesting as the best effort traffic can cope with longer delays. One simulation is repeated with the same settings for the VoIP settings (200 ms), although the discard time for the best effort traffic was extended from 200 to 1000 ms. The underlying idea is to improve the performance of the VoIP results at the cost of a longer delay in the best effort results. Figure 2.14 collects the results that indeed prove that the performance improves significantly for VoIP. With the original settings, 0.6 percent of the VoIP packets was not being received at all, most of it due to packets being discarded from the Node-B queue. The VoIP performance improves due to a larger setting of

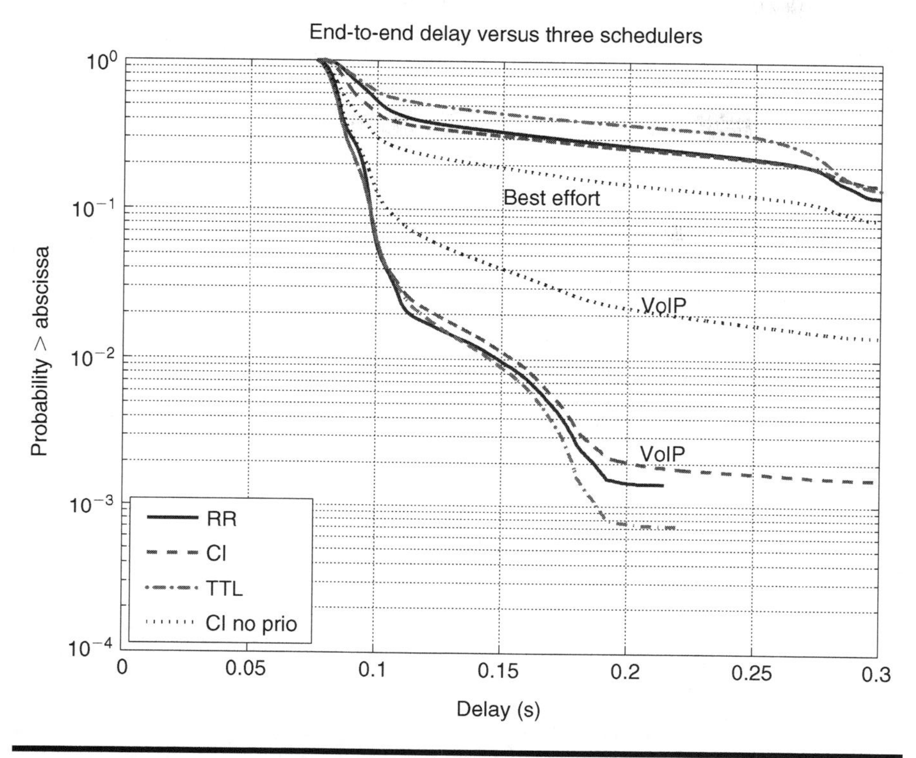

Figure 2.13 Complementary CDF of the delay experienced for different schedulers with priority on the VoIP packets. For reference, the C/I result for the case without priority is also included.

the discard timer for the best effort traffic: the number of VoIP packets that do not arrive at all reduces to less than 0.14 percent, also the delays, as such, are much lower. This improvement in the VoIP results is at the cost of a reduction in the performance of the best effort results.

The number of best effort packets that do not arrive increases from over 3 to over 5 percent. Owing to the higher discard timer setting for the best effort traffic, the packets also stay much longer in the Node-B queue, thereby causing packets to be also put on hold at the RNC. For this simulation, the best effort traffic source generates approximately 400 MAC-d PDUs per second, whereas the Node-B buffer size is set to 250 MAC-d PDUs. This implies that when no packet is scheduled or discarded, the Node-B buffer is full after 600 ms of packet generation. As a result, packets do arrive successfully with an end-to-end delay of up to 1.3 s, which includes a waiting time at the RNC queue of up to 250 ms. Finally, it should be stressed that the high number of packets that are not received are mainly lost during a handover. These packets will be retransmitted by the RLC.

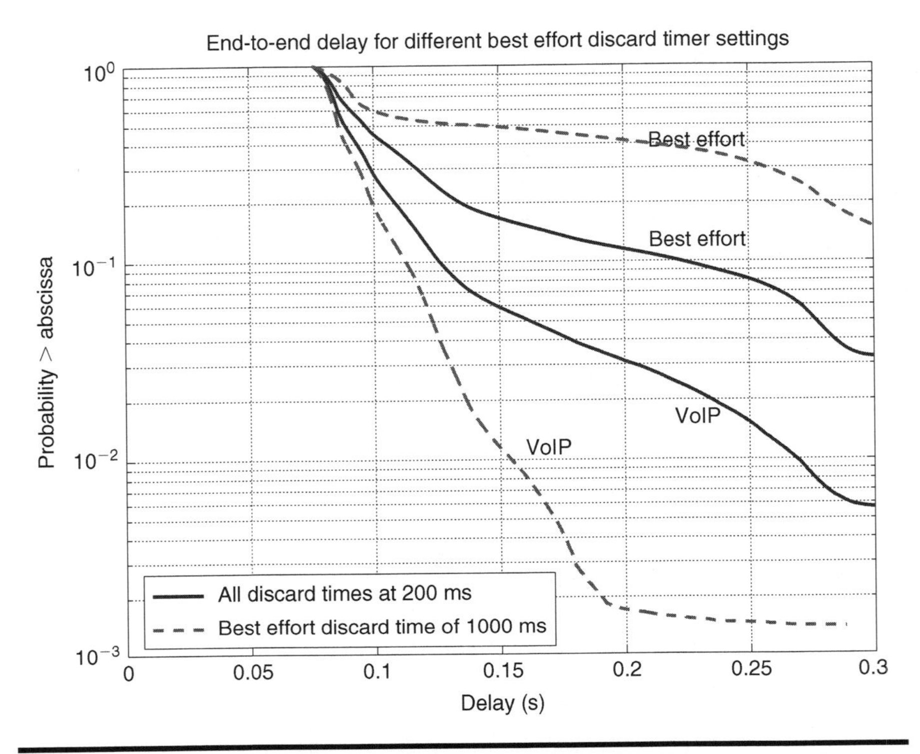

Figure 2.14 Complementary CDF of the delay experienced for two different settings of the best effort discard timer.

The number of lost packets during a handover can probably be reduced by modifying the flow control algorithm. This will ensure that less packets are forwarded to the Node-B in some cases.

2.6.3 Comparison of Scheduling Algorithms

All schedulers described in this section are run without any priority distinctions. After discussing the RR and the C/I schedulers, we will only discuss the PF scheduler with $\alpha = 0.100$ and the TTL scheduler with the 20 percent target of the discard timer of 200 ms. In Figure 2.15, we have collected the delay statistics for four scheduling algorithms. Considering the VoIP results, the results of the C/I and the TTL scheduler generate an end-to-end user delay that is not acceptable for most packets. Packets are delayed too much, and the percentage of packets that do not even arrive at all is also too high. The RR and PF results are close and have a high user satisfaction: those that are received at the UE have a delay of a maximum of 200 ms, and

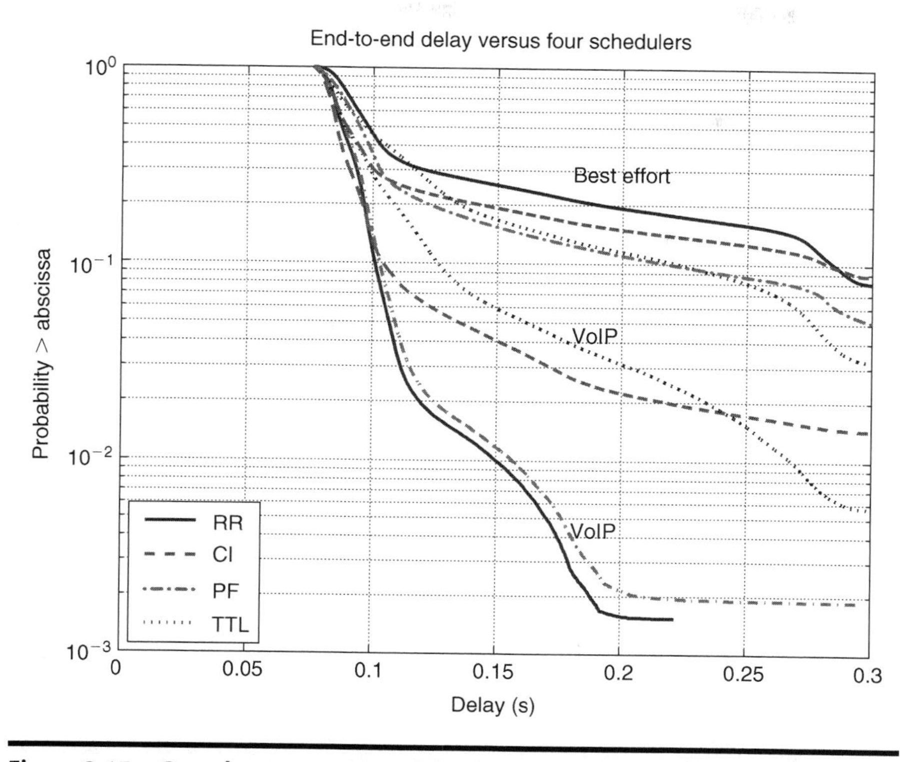

Figure 2.15 Complementary CDF of the delay experienced for four different schedulers.

the percentage of packets that do not arrive is around 0.1 percent for both schedulers.

Besides the probability of scheduling traffic at a TTI, it is interesting to consider system occupancy in relation to the flow control periodicity. In the default scenario the Iub band-width is very large (622 Mbps) in relation to the offered traffic (average 1.57 Mbps). As a result the Iub delivers the traffic within a short interval at the Node-B. Figure 2.16 shows how the occupancy per TTI relates to the timing of the flow control. It shows the number of HS-SCCHs that are simultaneously in use in the complete system versus the TTI number after the arrival of the packets at the Node-B. TTI "1" refers to the first scheduling moment after arrival of packets over the Iub in a flow control period, and the "5" refers to the last scheduling moment of the flow control period. All results are generated with a flow control period of five TTIs. The theoretic limit of the number of HS-SCCHs simultaneously in use in the simulated system equals 36 (four HS-SCCHs, three Node-Bs, and three cells). It is clear from Figure 2.16 that the amount of users scheduled is highest directly after the flow control. In particular, for the C/I

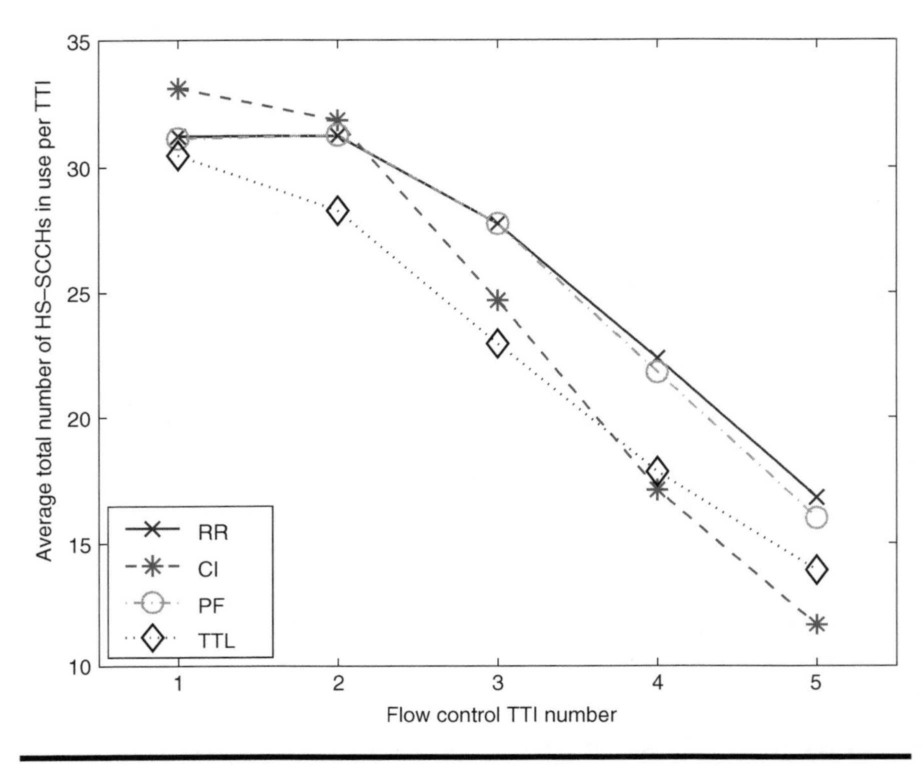

Figure 2.16 Number of HS-SCCHs in use per TTI for four different schedulers.

scheduler, the users that experience the best channel are scheduled first, and therefore the number of users is highest there.

Finally, it is clear from Figure 2.16 that during the first TTI of the flow control period the system is used very efficiently. The C/I scheduler will schedule the best users with the minimum amount of power, and therefore has the highest value at the initial TTI. Later on, during the flow control period, the Node-B runs out of packets or can only schedule to users with a low channel quality. Therefore, the number of HS-SCCHs in use per TTI decreases during a flow control cycle. It is also clear that the TTL scheduler selects users that have on average more packets waiting, and therefore on average transmits more MAC-d PDUs per MAC-hs PDU than the other schedulers. As a result, the number of HS-SCCHs in use per TTI is lower than the other schedulers.

2.6.4 Consumption of System Resources

Considering the VoIP results, the results of the C/I and the TTL schedulers generate an end-to-end user delay that is not acceptable for most packets;

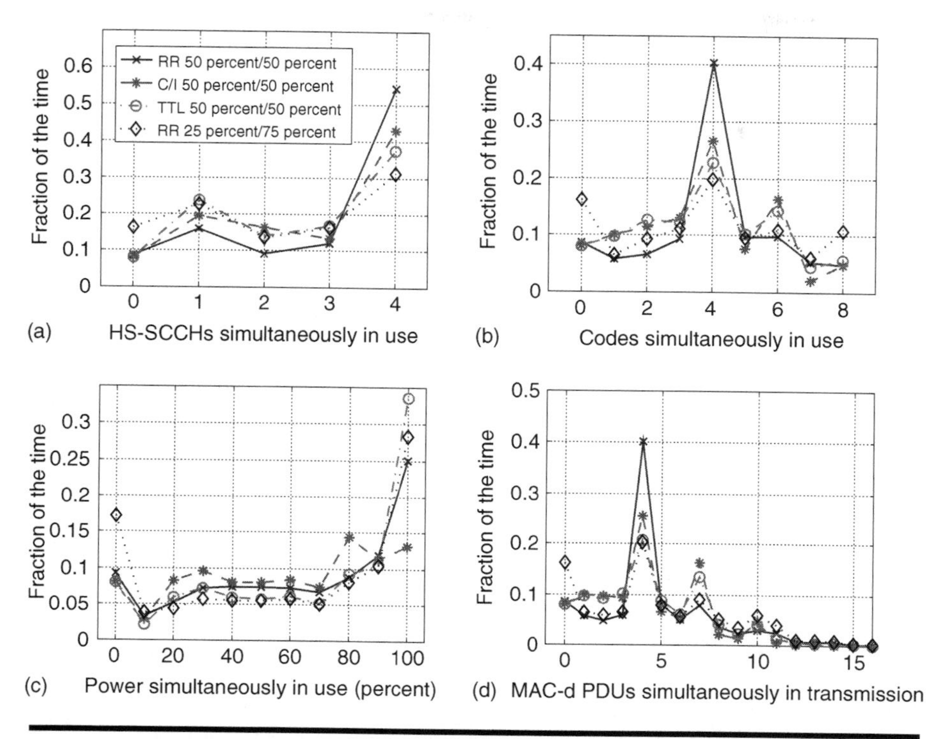

Figure 2.17 Resource consumption (per TTI, per cell) for all three schedulers, as well as another traffic mix (the 50 percent/50 percent mix is changed to a 25 percent/75 percent for the VoIP/best effort traffic). (a) Number of HS-SCCHs, (b) number of HS-PDSCH codes, (c) percentage of HS-DSCH power, and (d) number of MAC-d PDUs simultaneously transmitted.

see Figure 2.17. Packets are delayed too much, and the percentage of packets that do not even arrive at all is also too high. The RR results have a high user satisfaction: those that are received at the UE have a delay of a maximum of 200 ms and the percentage of packets that do not arrive is around 0.1 percent. It is clear that the time the system is idle is about the same for all three schedulers (9 percent). The RR results display that over 50 percent of the time the maximum number of HS-SCCHs is in use (see Figure 2.17a); this corresponds to the fair way this scheduler assigns the users to be scheduled. The other two schedulers have a lower value of around 40 percent. Parallel to this, for the same reason the RR scheduler displays a very high peak for the time moments that four codes are in use (see Figure 2.17b), as well as for the number of MAC-d PDUs (see Figure 2.17d). The HS-DSCH power is collected in Figure 2.17c and shows that the TTL scheduler has consumed the full power for 34 percent of the time, whereas this is only 25 and 13 percent for the RR and C/I schedulers, respectively. For all schedulers, Figure 2.17d displays peaks at 4 MAC-d PDUs due to the parallel scheduling of four VoIP packets. Two other peaks are visible for 7 and 10 MAC-d PDUs. These have to do with the fact that Iub flow control provides the best effort packets in bursts of four PDUs to the Node-B.

Finally, the sensitivity with respect to the current traffic mix is also investigated for the RR scheduler. The total traffic load imposed on the system is kept the same, but the VoIP/best effort contribution is changed from 50 percent/50 percent to 25 percent/75 percent. The idle time is even higher (16 percent) because the best effort traffic can be scheduled more efficiently. Because more of the packets are best effort, the peak at 4 HS-SCCHs has been halved. The same conclusion also holds for the peak at four codes and at 4 MAC-d PDUs. Moreover, the limited number of HS-SCCHSs in the resource allocation is now the bottleneck for only 31 percent of the scheduling moments (compared to 55 percent in the RR results with equal traffic load).

2.7 Conclusion

In this chapter, we have analyzed the performance of VoIP and best effort traffic over HSDPA. As VoIP is a real-time service, the analysis considers delay of packets as well as the percentage of lost packets as the key performance indicators. For best effort traffic, some delay can be tolerated; however, there is a stronger demand that the data transmitted should not be lost.

One of the main improvements of HSDPA, as an extension to the WCDMA standard, is the shorter transmission time interval, which enables the so-called fast scheduling. The control information received from the UE may be used by the Node-B in the scheduling process to allow for an efficient allocation of system resources and an optimization of the user satisfaction. The user velocity assumed throughout this chapter is 120 km/h, and as a result, the control information (on the channel quality) that is generated, is outdated. In particular, the schedulers whose underlying idea is to schedule a user at a good channel condition, turn out to have a high error rate because at the actual time of scheduling the signal quality has been decreased again. This mainly applies to the C/I and TTL schedulers. An analysis at different user velocities falls beyond the scope of this chapter, but it should be made very clear that a velocity of 3–50 km/h (pedestrians or vehicles moving at moderate speeds) or 300 km/h (as used for high-speed train scenarios) will generate different error statitics. In particular for a real-time service like VoIP, H-ARQ retransmissions, resulting in an inevitable extra delay, should be avoided as much as possible.

As VoIP generates a regular stream of small packets, a RR-like scheduling algorithm is very profitable for VoIP, but harmful for traffic creating larger bursts (e.g., HTTP). In the first half of the chapter, all investigations have been done with respect to the RR scheduler, and mainly involved the optimization of the handover process. Because the choice of the scheduler, and also assumptions on, for example, the traffic load and priority assignments might have a strong impact on the simulation results, these issues have also been addressed.

High load conditions lead to additional delay. In case the limiting resource is the number of HS-SCCHs or HS-DPSCH codes, the impact of the overload is moderate. Lack of HS-SCCH resources can be overcome by collecting more MAC-d PDUs into a single MAC-hs PDU. In case power is the most limiting resource, overload is handled in a less straightforward manner. For the VoIP-only scenario, until the resource usage of one of the resources reaches circa 90 percent, hardly any impact of load is found on the performance of the VoIP service. In case of high load conditions, a 10 percent higher load results in approximately 20 ms additional end-to-end delay.

It turns out that VoIP can be carried by the HS-DSCH effectively. An H-ARQ adds a considerable delay (24 ms), as only after the third transmission the residual BLER is below the required packet loss ratio. As VoIP produces a regular stream of small packets that should be transmitted at low delay, the main resource bottleneck in the system is the number of HS-SCCHs. Power is the main bottleneck when the best effort data is scheduled. The prescribed AMR of 0.7 percent for VoIP is achievable as long as the handover delay is chosen carefully. Handovers can contribute significantly to the delay. The main contribution is from the handover delay, that is, the time between the moment the RNC stops transmitting PDUs to the old cell, until the time the first PDUs can be transmitted into the new cell. This delay should be below 100 ms.

Using only C/I as the scheduling mechanism is not a good principle for any service, as users in relatively bad conditions can get very long delays and high loss ratios. Based on the results described in this chapter, the two variants of the C/I scheduler do not satisfy the required end-to-end performance. The target threshold of 0.7 percent is not reached for the VoIP users. A PF scheduler can balance the channel condition and the load. The PF scheduler assumes semistatic radio conditions, that is, the condition of the channel changes slowly compared to the granularity level of TTIs at which the scheduling takes place. However, this assumption is not realistic in many cases, like the user velocity of 120 km/h assumed in this chapter.

The TTL scheduling mechanism can balance the scheduling of packets from VoIP and high bit-rate services. The deadline, introduced by means of a target TTL, is a natural way to create a system where the scheduled traffic mirrors the offered traffic. It also enables the possibility to distinguish between priorities through setting different target delays. To some extent, a TTL scheduler considers the radio conditions. A drawback in the TTL scheduler is that it tends to let packets wait until a set dead-line. Thus, the goal of the scheduler is twofold: schedule users that experience good channel conditions or packets that are about to be droppped. However, unless the load is very small, the scheduler is mainly dealing with the second goal, creating a situation that is dominated by scheduling the packets that should be immediately saved from being discarded. The TTL scheduler is considered the most promising alternative as it enables balancing between different traffic streams. When only VoIP performance is important, the RR scheduler is better. Various variants of the TTL scheduler are discussed. The results are not very sensitive with respect to the exact value of the TTL target. Also the effect of different values of the discard timer, in particular for the best effort traffic have been considered. Some alternative implementations of the TTL scheduler may be interesting to study.

As expected, when including two different priority classes, the VoIP users are far better off at the cost of the longer delay that is experienced by the best effort users. When comparing the priority results for the three schedulers, it becomes clear that the impact of the specific scheduler becomes less important. The bottleneck in the resource consumption now depends on the traffic load, and more particularly, the type of traffic. When there is a relative high contribution of VoIP users, the number of HS-SCCHs is the limiting factor, because only four users can be served in parallel. When the contribution of VoIP is not too high, the power consumption becomes the main bottleneck.

2.8 Open Issues

Although UMTS and HSDPA are already being rolled out, network simulations provide an add-on value because these allow for the execution of controlled experiments, addressing bottlenecks that are not feasible yet in real networks or testbeds. The results described in this chapter are the result of network simulations of multimedia traffic over HSDPA with a fully loaded system. Compared to many other studies, the multicell setting has provided the ability to study handover effects that otherwise would not have been possible to model. Although several processes have been modeled in a realistic way, there are always shortcomings that should be investigated before drawing too decisive conclusions. Most of these apply to network simulations in general.

To avoid too many effects influencing the system at the same time, we have only assumed the UM radio link control. This also allowed for a better comparison between all results for different parameter settings. As a next step, it would be interesting to consider the behavior of TCP, in particular during a handover: Can the TCP settings be tuned such that a handover can be taken smoothly?

Also assuming nonperfect control information would be a clear add-on to the current work. It would be interesting to see the effect of errors in the (negative) acknowledgment of data packets, the CQI, and the downlink control information indicating which user has been scheduled.

Finally, it should be stressed that recently the uplink variant, High-Speed Uplink Packet Access (HSUPA), has also gained interest. It would be interesting to investigate how the combination of HSDPA and HSUPA performs under circumstances of a fully loaded UMTS system.

Acknowledgments

The foundation of the EURANE simulator has been laid in the IST project IST-2001-34900 Simulation of Enhanced UMTS Access and Core Networks (SEACORN). The project aimed at creating simulation tools for the study of enhancements to the UMTS standards. The simulator is used worldwide throughout academia and industry and has regularly been extended with many HSDPA features. Extending the EURANE simulator toward a multicell network performance analysis tool was realized in the international ITEA Easy wireless IP 03008 project. The detailed analysis of the performance of multimedia over HSDPA has been performed in close collaboration between the Twente Institute for Wireless and Mobile Communications and Huawei Technologies Co., Ltd. The authors would like to thank Xiaoxia Wang, Fangfu Guo, Jingrong Zhang, Chaowei Xie, and Simon Oosthoek for their valuable feedback during the course of this work.

References

1. Ericson, M., L. Voigt, and S. Wänstedt, Providing Reliable and Efficient VoIP over WCDMA. *Ericsson Review*, 82(2), 2005.
2. Pedersen, K.I., P.E. Mogensen, and T.E. Kolding, Overview of QoS Options for HSDPA. *IEEE Communications Magazine*, 44(7), 100–105, 2006.
3. Holma, H. and A. Toskala, *HSDPA/HSUPA for UMTS*. John Wiley & Sons, New York, 2006.
4. Simulation Results on VoIP Performance over HSDPA, 3GPP TSG-RAN WG2, R2-052832, November 2005.
5. EURANE (Enhanced UMTS Radio Access Networks Extensions for ns-2), http://www.ti-wmc.nl/eurane.
6. de Bruin, I., F. Brouwer, N. Whillans, Y. Fu, and Y. Xiao, Performance Analysis of Hybrid-ARQ Characteristics in HSDPA. *Wireless Personal Communications*, 42, 337–353, 2007.

7. de Bruin, I.C.C., Network-level Simulation Results of Fair Channel-Dependent Scheduling in Enhanced UMTS. *IFIP-TC6 Ninth International Conference on Personal Wireless Communications*. Delft, 2004.
8. 3GPP TS 25.214 V5.5.0, Physical layer procedures (FDD), Release 5.
9. Technical Specification Group Radio Access Network; Physical layer procedures (FDD), 3GPP TS 25.214 V6.6.0, June 2005.
10. Support of RT services over HSDPA, 3GPP TSG-RAN RP-050106, March 2005.
11. Brouwer, F., I. de Bruin, J.C. Silva, N. Souto, F. Cercas, and A. Correia, Usage of Link Level Performance indicators for HSDPA Network-Level Simulations in E-UMTS. *Spread Spectrum Techniques and Applications*, Sydney, 2004.
12. Revised HSDPA CQI Proposal, 3GPP TSG-RAN Working Group 4, R4-020612, 2002.
13. de Bruin, I., F. Brouwer, N. Whillans, Y. Fu, and Y. Xiao, Performance evaluation of VoIP over HSDPA in a multi-cell environment. *Fifth International Conference on Wired/Wireless Internet Communications WWIC, Lecture Notes in Computer Science (LNCS) 4517*, pp. 177–188, Coimbra, Portugal, 2007.
14. de Bruin, I., G. Heijenk, M. El Zarki, and J.L. Zan, Fair Channel-Dependent Scheduling in CDMA Systems. *12th IST Summit on Mobile and Wireless Communications*, pp. 737–741, Aveiro, Portugal, June 2003.
15. Kolding, T.E., Link and System Performance Aspects of Proportional Fair Packet Scheduling in WCDMA/HSDPA. *IEEE Proceedings of Vehicular Technology Conference*, pp. 1717–1722, Orlando, USA, September 2003.
16. Laneri, J.C., Scheduling Algorithms for Super 3G. *Master's Degree Project KTH*, Stockholm, 2006.
17. Improved HSDPA Re-pointing Procedure, 3GPP TSG-RAN WG2 R2-061298, 2006.

Chapter 3

Interactive Mobile TV Technologies: An Overview

Ramakrishna Vedantham and Igor D.D. Curcio

Contents

3.1 Introduction 88
3.2 Characteristics and Requirements of Interactive Mobile TV Systems 90
3.3 Mobile TV Usage Scenarios 92
3.3.1 Broadcast/Multicast Streaming 92
3.3.2 Broadcast/Multicast Download and Play 93
3.3.3 Broadcast/Multicast Streaming with User Interaction 93
3.4 System Design Principles and Challenges 93
3.4.1 Supported Media Types and Bit Rates 93
3.4.2 Tune-In Delay 94
3.4.3 Reliable Media Transport 94
3.4.4 Roaming and Handovers 95
3.5 Mobile TV Standards 95
3.5.1 3GPP MBMS 95
3.5.2 3GPP2 BCMCS 96
3.5.3 IP Datacast over DVB-H 98

3.5.4 Other Standards 100
3.5.5 OMA BCAST 101
3.6 Media Delivery Framework 104
3.6.1 User Service Announcement 104
3.6.2 File Download Protocols 104
3.6.3 Streaming Protocols 105
3.6.4 User Interaction Protocols 106
3.6.5 Reliability of File Download Sessions 107
3.6.6 Reliability of Streaming Sessions 108
3.6.7 Quality of Experience Framework 110
3.6.8 Channel Zapping 110
3.7 Media Codecs and Formats 112
3.7.1 Media Codecs 112
3.7.2 Dynamic and Interactive Multimedia Scenes 113
3.7.3 Container File Formats 113
3.8 Security 114
3.9 Conclusion and Open Issues 116
References 116

Ubiquitous access and easy user interaction are the key differentiators for the mobile TV services when compared to the traditional TV offerings. Mobile TV services are already deployed in some markets and are being piloted in many other markets. The enabling technologies for mobile TV services are currently fragmented based on the underlying wireless network and infrastructure. On the one hand, mobile TV is offered as an add-on service over the maturing third-generation (3G) wireless networks, for example, 3rd Generation Partnership Project (3GPP) Multimedia Broadcast Multicast Service (MBMS) and 3GPP2 Broadcast Multicast Service (BCMCS). On the other hand, mobile TV is geared to create separate service by itself, for example, digital video broadcasting for handheld (DVB-H) Internet Protocol Datacasting (IPDC), and proprietary technologies such as Media Forward Link Only (MediaFLO). We also notice some convergence of these two. For example, there are some devices that can receive mobile TV through DVB-H IPDC and can interact via 3G cellular networks. This chapter discusses the common technology enablers for the myriad of mobile TV standards. They include media codecs, transport protocols, service guide, user interaction protocols, and service/content protection issues. It also discusses the key design considerations for the choice of selected technologies.

3.1 Introduction

TV services are progressively moving from analog to digital. Today, many countries have transition plans to the digital TV world, and analog TV transmissions will even be discontinued at some point of time to leave space to fully digital transmissions and services. First, digital TV services allow a better audiovisual quality. Second, they allow an easier management of content from generation to production to transmission. Finally, digital TV services enable end users to consume content in a more flexible and personalized way. For example, video on demand is only possible with digital TV services. Such flexibility is fully exploited when giving end users the possibility of interacting with the service in a way that is more suitable to the end user's preferences. This is "digital interactive TV."

Mobile devices, such as phones or personal digital assistants (PDAs), have become more powerful because of the ever increasing mobile processors speeds, miniaturization of components that allow integrating on the mobile device of several gigabytes of read only memory/random access memory (ROM/RAM) and external memory cards. Mobile devices with these characteristics have become real mobile multimedia computers with the capabilities of fast data transfer, audiovisual playback and recording, global positioning systems (GPSs) and navigation, frequency modulation (FM) radio, e-mail, browsing, instant messaging, etc.

Mobile networks have evolved from the first generation of analog networks to 2G networks (e.g., Global System for Mobile communications [GSM], fully digital), 2.5G packet networks (e.g., the General Packet Radio System [GPRS]), and 3G networks (e.g., Universal Mobile Telecommunication System [UMTS], which offers higher speed and multimedia services). Current and forthcoming mobile networks, together with the latest mobile devices generations, offer or will be able to offer to the end users a wide range of multimedia services that range from mobile streaming to video telephony, Multimedia Messaging Service (MMS), and other services.

As mobile phones are very much at the center of everyone's life today, and watching TV is also a widespread practice in the world, the natural evolution of modern TV services is to go mobile. For this reason, one of the services that has attracted much attention in the recent years is that of "mobile TV."

Mobile TV offers end users the pleasant experience of watching TV programs while on the move. This means that a new life paradigm for media consumption is now in the process of being established. Irrespective of the user's location, the new paradigm allows watching and consuming media while traveling in the car, bus, train, plane, or simply while walking on the streets or resting in a park. This is a relatively new concept for end users, but trends indicate that there is an increasing interest for mobile TV services.

Such services are of interest for a complete ecosystem, which includes nonexhaustively content creators, service providers and network operators, mobile device manufacturers, and media and advertisement companies. Technologies for mobile TV have been developed in the past years and many initiatives have also been standardized in different organizations, or they are the fruit of proprietary efforts. Examples of these services and technologies are the MBMS, BCMCS, DVB-H, Terrestrial Digital Media Broadcasting (T-DMB), and MediaFLO. Satellite-based systems such as Satellite-Digital Multimedia Broadcast (S-DMB) and Digital Video Broadcast-Satellite (DVB-S) also exist.

These services have been originally developed almost exclusively as one-way broadcast/multicast services, but are rapidly evolving to increase the level of user satisfaction and personalization through means that allow more user interactivity. In other words, we are now in presence of a new technology and a revolutionary user paradigm of the twenty-first century: the "interactive mobile TV."

This chapter discusses the interactive mobile TV technologies. It discusses the common technology enablers for the different mobile TV standards. They include media codecs, transport protocols, user interaction protocols, security and digital rights management (DRM) issues. We also discuss the key mobile TV usage scenarios and system design principles for selected technologies.

3.2 Characteristics and Requirements of Interactive Mobile TV Systems

To understand the key drivers of mobile TV systems, it is important to answer the question: What are the characteristics of a typical mobile TV system? The most important ones are as follows:

1. *Mass distribution*. It should be possible to distribute media content to a large number of end users. Typically, these range from thousands to several millions.
2. *Simultaneous media content consumption*. It should be possible for end users to watch mobile TV programs as in the traditional analog TV. This means that several millions of end users are enabled to watch the same program at the same time.
3. *Nonsimultaneous media content consumption*. It should be possible for end users to consume media content not at the same time as other end users, but in a more flexible way.
4. *Flexible user reachability*. It should be possible to reach all the users in some cases, but in other cases it should be possible to reach a smaller set of the whole network population.
5. *Interactivity*. It should be possible for end users to interact with the mobile TV service provider to get additional value and flexibility from the service.

These system characteristics generate a set of system requirements that are necessary when building a mobile TV system.

The first characteristic imposes the requirement of "scalability." A mobile TV system must be scalable and support the delivery of media streams to a large number of users. If the system is not scalable, it does not support mass distribution, and fails in achieving the most important characteristic of the system design.

The second characteristic leads to the usage of multicast/broadcast mobile networks. Because the media content can be consumed by a large number of users simultaneously, the idea of utilizing point-to-point connections is pointless. For example, in the 3GPP world, it would be extremely expensive to set up a mobile broadcast network utilizing unicast point-to-point bearers. This would lead to the need of increasing the size of the network to support all the users who have planned to be in the system at any point of time. Network resources are very precious and any attempt to save resources is a gain for network operators and indirectly for end users. Therefore, mobile broadcast networks are the optimal choice. In these networks, the radio layer and the network cells are designed to support broadcast transmissions, and the usage of point-to-point bearers is not required. In fact, mobile broadcast networks are point-to-multipoint networks natively. "Point-to-multipoint" means that the transmission goes from a source (the "point") to many destinations (the "multipoint"). In broadcast bearers, the transmission is also typically unidirectional, that is, from the sender source (e.g., the service provider) to the receivers (e.g., the end users), and there is no backward communication channel to ensure a simpler system design.

The third system characteristic introduces more flexibility compared to the typical analog TV systems. Users may not want to watch programs or news at the time when they are broadcasted, but later at a more suitable time. This is realized by means of a broadcast download function that is made available in most of the mobile broadcast networks. With this function, for example, the latest news is distributed in a broadcast fashion to a large number of end users, but each user watches the news at a potentially different time. The broadcast download function is introduced in addition to the broadcast streaming function, where the former is typically used for delivering relatively small media file (e.g., news and weather forecasts), which are up to a few megabytes large, to the end users. Instead, the latter is typically used for delivering media streams of larger size (e.g., TV series and movies), which last from tens of minutes up to several hours.

The fourth system characteristic means that there should certainly be not only a way for reaching all the users in the network but also a way for addressing a smaller set of users. Two architectural concepts are usually defined in the design of broadcast mobile networks: broadcast and multicast. When the transmission must reach all the users in the system, it is a "broadcast" transmission. This is the same concept as in the analog TV. In contrast, when the transmission must reach only a smaller group of users than the entire system population, it is a "multicast" transmission. Multicast transmissions enable to address specific groups of users that have something in common, and may have different interests than the rest of the system population. For example, a multicast group called "Stocks Nasdaq" may only receive news about the Nasdaq Stock Exchange, whereas the rest of the system population does not receive such multicast transmission. Multicast groups often require a user subscription. The concept of multicasting enables to partition the system population into groups according to specific criteria. This is yet another step toward the personalization of mobile TV services.

The fifth and the last characteristic calls explicitly for interactivity. This introduces a new requirement on a mobile TV system because, as already stated earlier, the system is usually unidirectional with no available backward channel. This architectural restriction greatly simplifies the design and implementation of both network and mobile devices. However, the requirement of user interactivity generates new challenges on the whole mobile TV system because the system must be carefully designed and engineered to guarantee scalability, especially in presence of user interactivity. From the user's perspective, interactivity means, for example, enabling voting, selecting choices, giving feedback, and any other operation where the user's opinion counts to increase the value of the service. Interactivity, from the system perspective, means adding a set of additional features to increase the flexibility of the system and its reliability. In Section 3.6.4, we will explain how this requirement can be satisfied.

3.3 Mobile TV Usage Scenarios

After having analyzed the main characteristics and requirements of mobile TV systems and networks, and after looking at their main system design principles, it is now possible to draw examples of mobile TV usage scenarios. The examples presented in this section are not exhaustive, but represent the most typical usage scenarios in mobile TV [1]. Each of the following three usage scenarios can be applied both in broadcast and multicast scenarios. We will not distinguish here between the two for simplicity reasons.

3.3.1 Broadcast/Multicast Streaming

This is the usage scenario that is closer to the current analog TV service. In this scenario, a user browses an electronic service guide (ESG) containing a list of programs and their broadcast time, and tunes in at a specific time to receive the TV program while it is being broadcasted. An example of broadcast streaming is the transmission of a soap opera episode at a specific time of the day.

3.3.2 Broadcast/Multicast Download and Play

In broadcast download and play, a typically short-time media file is pushed to the receiver terminals at a specified time of the day (e.g., according to an ESG). The file is stored on the mobile TV terminal of the user, and it can be consumed at a later point of time, according to the user's preference. This usage scenario is currently not feasible in analog TV, although it is already possible to record a TV program and watch it at a different time on a video cassette recorder (VCR) or digital video disk (DVD) player. But the difference here is that with analog TV the recording time is as long as the program duration, whereas in broadcast download and play, the recording time is typically shorter. In fact, for example, it is possible that a flash news clip of the duration of five minutes is delivered to the broadcast users in just a few seconds (depending on the available network bandwidth). Therefore, there is a lot of resource saving here because the users' terminals do not need to be tuned and keep listening to the broadcast network for the real duration (in the playback time dimension) of the content. Other data broadcasting scenarios that include text notification services (e.g., stocks text alerts), software download, and carousel services also come under this usage scenario [1].

3.3.3 Broadcast/Multicast Streaming with User Interaction

This usage scenario is an extension of the broadcast streaming usage scenario, and it is not possible in analog TV. In this scenario, a mobile TV program is streamed to the users, but each individual user has the possibility of interacting with the TV program (and the service provider too) to send feedback, votes, etc. For example, the user interaction may include giving a score for the finals of the Miss Universe beauty contest.

3.4 System Design Principles and Challenges

In addition to the general system characteristics and requirements described earlier, other issues are also taken into account when designing an interactive mobile TV network and system. The remainder of this section describes the most relevant ones.

3.4.1 Supported Media Types and Bit Rates

A mobile TV system should be capable of supporting a multitude of media types. Multiple media types enable flexibility in the system and more opportunities for the success of the service. The most typical media types range from speech to audio/music, video, and also synthetic audio, still images, bitmap graphics, vector graphics, and text and timed text. Typical media bit rates supported by a mobile TV system must range from a few kilobits per second (kbps) to several megabits per second (Mbps). Media types and codecs are further described in Section 3.7.

3.4.2 Tune-In Delay

In traditional analog TV, people turn on the TV on a specific channel or switch channel by means of a remote control. The tuning-in to a channel or the switch from the current channel to another is something that users want to be virtually instantaneous. The delay to tune the receiver in to a (new) frequency is called "tune-in delay." In analog TV the tune-in delay is a fraction of a second, whereas in mobile TV systems it is one of the most difficult challenges because error protection mechanisms may not allow such a fast tune-in time and, at the same time, perfect media quality under all radio conditions. This is an important design aspect and, in the future, mobile TV networks and systems that will offer the lowest tune-in delays will certainly be a success. Section 3.6.8 analyzes the factors affecting the tune-in a.k.a. channel zapping and discusses some techniques to reduce this delay.

3.4.3 Reliable Media Transport

One of the main characteristics of digital TV, compared to analog TV, is that the offered media quality is superior. Digital TV should offer nearly error-free media transmission and an incomparable experience to the end users. Mobile TV, being digital by nature, carries the same type of requirement and end-user expectation. However, the additional constraints placed by mobile networks, compared to fixed digital TV networks, make mobile TV media quality more subject to errors and potential user dissatisfaction. The reasons for potential dissatisfaction are all derived by the inherent lossy characteristics of radio networks because of interferences, fading, weather conditions, distance of the mobile terminal from the transmitter base station, and in general any condition that has an impact on the radio signal received at the mobile terminal during a mobile TV session. For these reasons, special attention must be paid when designing a mobile TV system. This must be able to deliver error-free media to the end users. In practice, such a design principle imposes strict transport and coding rules for data delivery. For example, forward error correction (FEC) or other file repair mechanisms can be used to provide (nearly) error-free transmission to the end users. These aspects are analyzed in detail in Sections 3.6.5 and 3.6.6.

3.4.4 Roaming and Handovers

A mobile TV system is designed over a cellular network, and given the users move freely within a country or across countries, it is important that the system offers continuity of service in case of handovers and roaming. "Handovers" happen whenever a user moves and the mobile terminal changes the cell within the same operator's cellular network. "Roaming" happens when not only the cell is changed, but also the operator is changed during the handover phase. Handovers between different systems are also possible (within the same operator's domain or not). These are also known as "vertical" or "intersystem handovers." For example, a mobile TV session may start over an MBMS system and, at some point of time, continue over a DVB-H system because in some area, where the user is moving, the MBMS coverage is not available, but DVB-H coverage is. Handovers and roaming cause a short transmission interruption. However, the service must be designed in such way that the mobility effect is totally hidden from the end user. This is possible by carefully designing the radio and transport/application layers of the mobile TV protocol stacks in a way that the system is resilient to short-time interruptions caused by user mobility. The reliability mechanisms such as FEC and file repair described in Section 3.6.5 ensure a smooth handover at the application layer to a certain extent. Section 3.6.4 describes how roaming MBMS users receive the content.

3.5 Mobile TV Standards

This section briefly discusses the various standards for the current and upcoming mobile TV deployments, which include 3GPP MBMS, 3GPP2 BCMCS, and DVB-H IPDC. Other standards and proprietary solutions for mobile TV are not addressed here due to space constraints. Finally, we give an overview of Open Mobile Alliance Broadcast (OMA BCAST) architecture that attempts to encompass various fragmented mobile TV standards that operate over Internet Protocol (IP).

3.5.1 3GPP MBMS

The MBMS standardized in 3GPP [2] is the European solution for mobile TV services. Figure 3.1 depicts the main architectural components.

MBMS is a point-to-multipoint service that works in both broadcast and multicast mode, and enables an efficient usage of radio network resources. The multicast mode, different from the broadcast mode, requires subscribing, joining, and leaving a particular service. The block components denote the network elements and the user equipment (UE), whereas the lines that connect the different blocks denote the interfaces between the elements. The content source may be located inside or outside an MBMS network (content provider/multicast broadcast source). MBMS services are offered

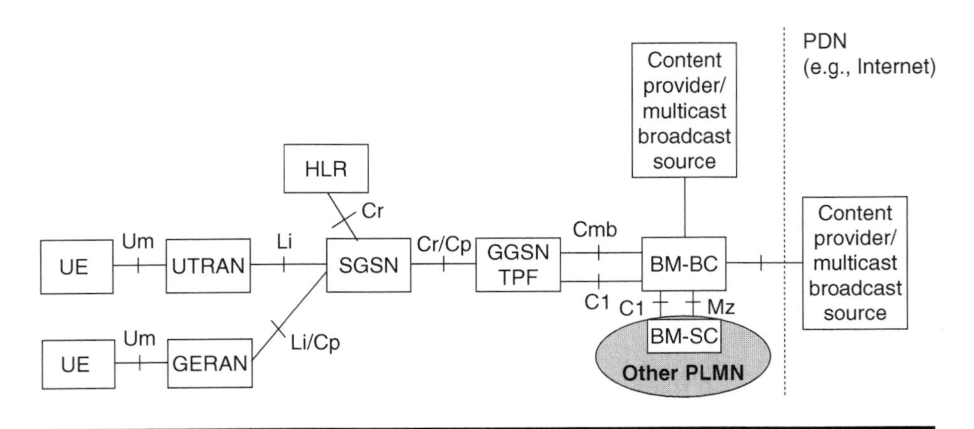

Figure 3.1 Reference architecture of a 3GPP MBMS system.

over different radio access networks (RANs). For instance, the services can be deployed over 2.5G (GSM EDGE radio access network; GERAN) or 3G (UMTS radio access network; UTRAN) bearers.

Of particular relevance in the architecture depicted in Figure 3.1 is the Broadcast–Multicast Service Centre (BM-SC), which provides the different functions to offer the MBMS user services [2]. Among the functions of the BM-SC are the service announcement and the session and transmission functions. The service announcement function enables to send the UEs the media description information about the MBMS sessions (e.g., what audio and video codecs and what bit rates are used). The session and transmission function is used to schedule the multicast/broadcast transmissions to the receivers.

The quality of service (QoS) offered by MBMS services is of the background or streaming traffic classes [2,3]. The background traffic class is essentially a best-effort class with no guaranteed delays and bit rates. The streaming traffic class offers guaranteed delays and bit rates. Transmissions are typically not error-free: IP packet error rates range between 1 and 10 percent [2]. Residual error rates will have to be removed by other means (e.g., application layer-FEC [AL-FEC] or point-to-point repair; see Sections 3.6.5 and 3.6.6). The maximum bit rate supported by MBMS services is 384 kbps [1], depending on the RAN and the network configuration. Minimum delays for streaming of mobile TV sessions are in the order of 300 ms [3]. Figure 3.2 shows the reference protocol stack for 3GPP MBMS. The technical specifications of MBMS can be downloaded from the link provided in Ref. 4.

3.5.2 3GPP2 BCMCS

Similar to MBMS in 3GPP standard, BCMCS is introduced in 3GPP2 standards for providing broadcast multicast services over 3G CDMA2000

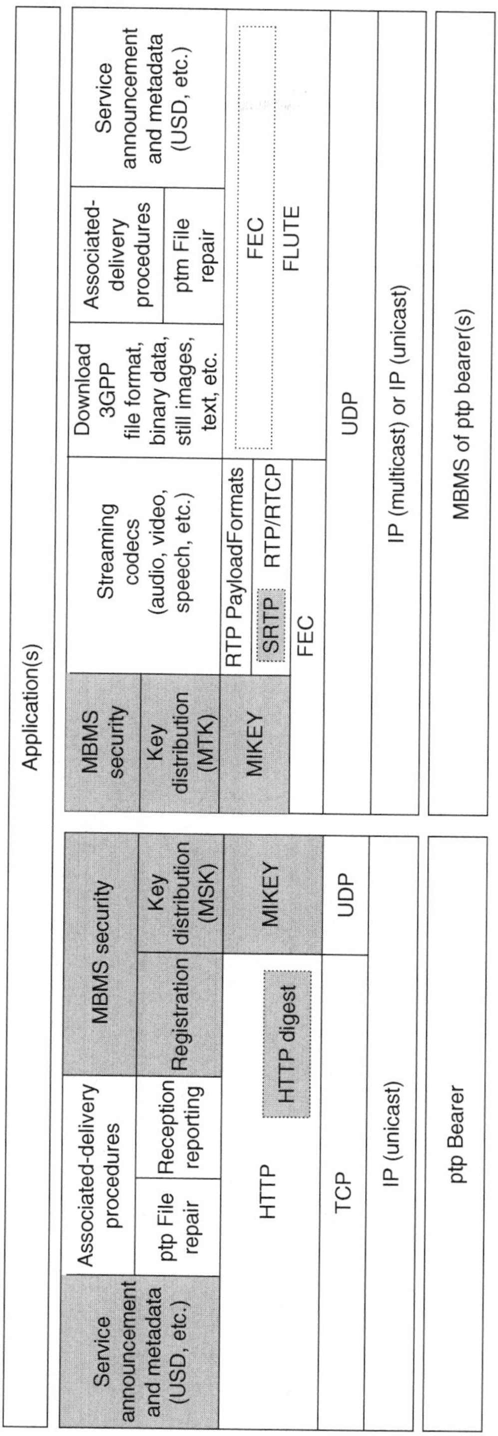

Application(s)

Service announcement and metadata (USD, etc.)
Associated-delivery procedures
ptp File repair
Reception reporting
MBMS security
Registration
Key distribution (MSK)
HTTP
HTTP digest
MIKEY
TCP
UDP
IP (unicast)
ptp Bearer

MBMS security
Key distribution (MTK)
MIKEY
Streaming codecs (audio, video, speech, etc.)
RTP PayloadFormats
SRTP
RTP/RTCP
FEC
Download 3GPP file format, binary data, still images, text, etc.
Associated-delivery procedures
ptm File repair
Service announcement and metadata (USD, etc.)
FEC
FLUTE
UDP
IP (multicast) or IP (unicast)
MBMS of ptp bearer(s)

Figure 3.2 3GPP MBMS protocol stack.

1x evolution data optimized (EVDO)-based cellular systems. BCMCS introduces two new features when compared to MBMS. They are optional link-layer encryption and low-delay outer FEC code at the link layer. Link-layer encryption is optional in BCMCS. If encryption is applied, encryption may be performed either at the link layer or a higher layer (e.g., Secure Real-Time Transport Protocol [SRTP]), but not both. In addition to the FEC at physical layer of CDMA2000 networks, an outer FEC code at the link layer is introduced in BCMCS. This link-layer FEC is based on Reed–Solomon (RS) erasure codes over Galois field (GF)(256) and can be chosen from any of the following (N, K) combinations: for $N = 32$, $K = 32$, 28, 26, and 24 and for $N = 16$, $K = 16$, 14, 13, and 12. The N value is kept low to reduce the buffering delay. An overview of the 3GPP2 BCMCS network architecture and air interface design can be found in Refs 5 and 6. The technical specifications [7–9] contain the details of various layers of the BCMCS protocol stack. They can be downloaded from the link provided in Ref. 10.

3.5.3 IP Datacast over DVB-H

DVB-H makes use of the existing infrastructure and radio frequencies of the terrestrial digital video broadcasting network (DVB-T). DVB-H introduces some techniques to the DVB-T transmission system to tailor it for DVB-H terminals, that is, to extend the receiver's battery life, improve the radio frequency (RF) performance for mobile single-antenna reception, mitigate the effects of high levels of noise in a hostile environment, efficient handover, and fast service discovery. Figure 3.3 shows the coexistence of DVB-T and DVB-H. A comprehensive treatment of DVB-H transmission system can be found in Refs 11 and 12.

IPDC is a service where digital content formats, software applications, programming interfaces, and multimedia services are combined through IP with digital broadcasting. It makes use of the DVB-T networks to broadcast any IP-based data. In contrast to DVB-T, where usually audio-video elementary streams were directly packetized to MPEG-2 transport stream (TS) packets, DVB-H is primarily designed for the carriage of IP datagrams. To maintain compatibility with DVB-T, IP datagrams are packetized to Multi-Protocol Encapsulation (MPE) sections, which are then carried over MPEG-2 TS packets.

MPE-FEC and time slicing [11] are the most important enhancements introduced by DVB-H. The objective of the MPE-FEC is to improve the C/N and Doppler performance in the mobile channel, and to improve the tolerance to impulse interference. The addition of an optional, multiplexer-level, FEC scheme means that DVB-H transmissions can be even more robust. This is advantageous when considering the hostile environments and poor but fashionable antenna designs typical of handheld receivers. MPE-FEC data protection is done by computing the RS codes for the IP datagrams. Figure 3.4 shows the structure of an MPE-FEC frame. Computation of the

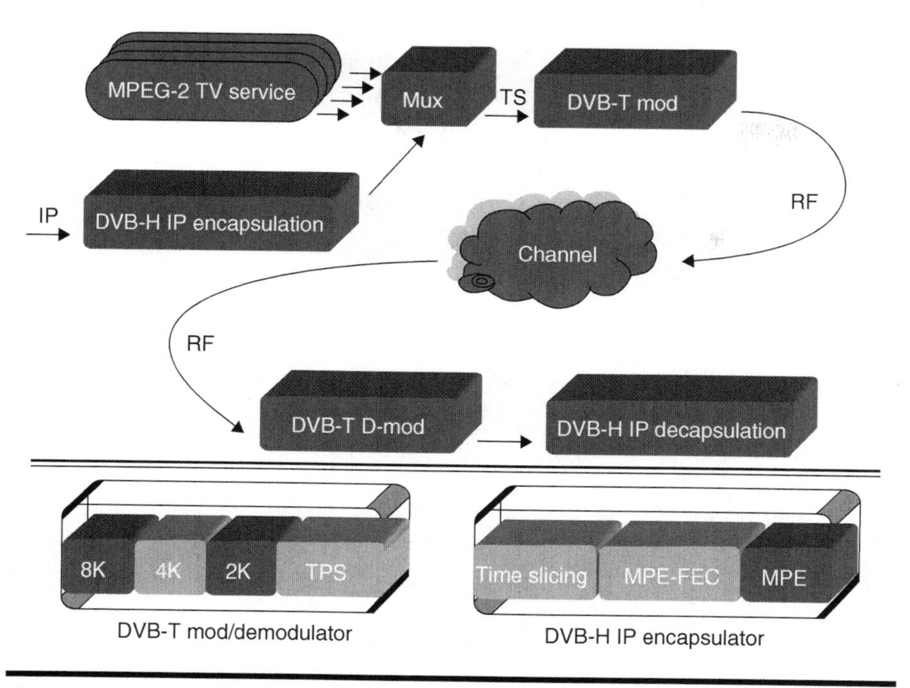

Figure 3.3 Coexistence between DVB-T and DVB-H (TPS – Transmission parameter signaling).

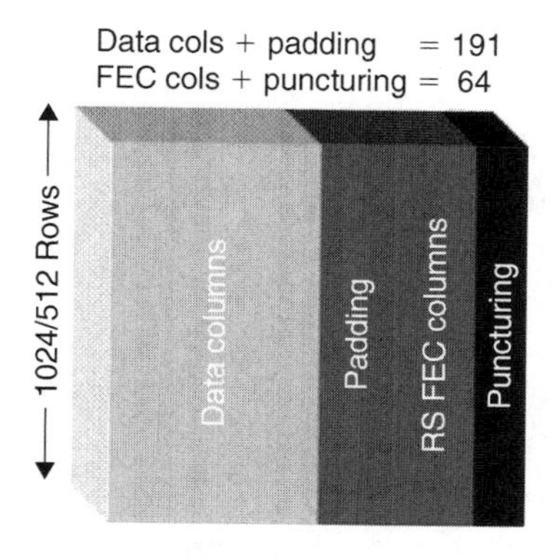

Figure 3.4 MPE-FEC frame with data, padding, FEC, and puncturing.

RS code is done using the same time-slicing buffer. The RS data is packed into a special FEC section, so that an MPE-FEC-ignorant receiver can simply ignore the MPE-FEC sections. This is to enablenon-MPE-FEC receivers to receive and decode MPE-FEC datagrams. Time slicing is applied to enable power saving, so that one MPE-FEC frame is transmitted in one time-slice burst. The time-slice bit rate during the burst is significantly higher than the service bit rate, and the receiver can turn off its radio parts between the bursts to save power. The frame size, transmission bit rate, and

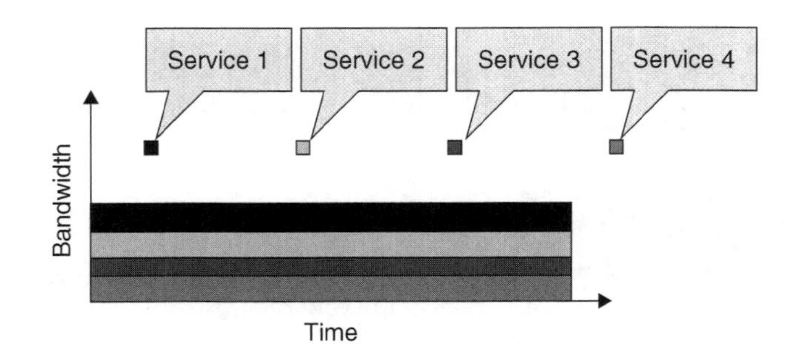

Figure 3.5 Services in DVB-T channel with no time slicing.

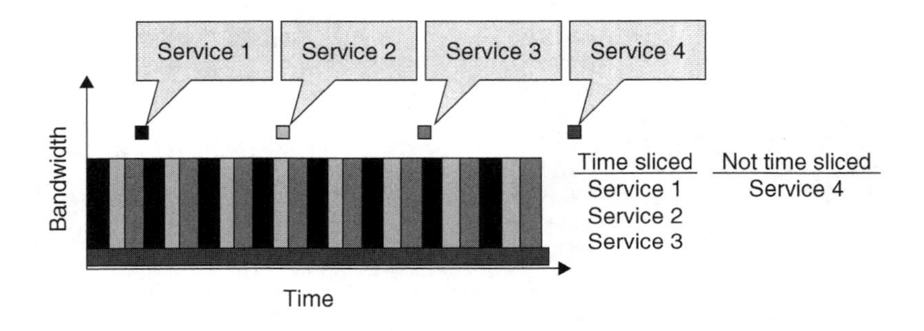

Figure 3.6 Services in DVB-H and its compatibility with DVB-T with time slicing.

off-time between bursts are parameters that affect the service bit rate, service switching time, and power saving. The data is received in bursts so that the receiver, utilizing control signals, remains inactive when no bursts are to be received. Research done to evaluate the savings obtained by this method show almost 90 percent savings in most cases. Figures 3.5 and 3.6 show how time slicing is used in DVB-H to multiplex channels.

Figure 3.7 shows the reference protocol stack for DVB-H IPDC. Note that DVB-H IPDC specifies a self-contained end-to-end system including all protocol layers. At the physical and link layers, it specifies the DVB-H transmission system described earlier. At the application layer, it specifies file delivery protocols and streaming protocols [13], service guide [14], and security issues [13]. The simplified architecture shown in Figure 3.8 illustrates the content distribution in IPDC over DVB-H. The latest DVB-H IPDC specifications can be downloaded from Ref. 15.

3.5.4 Other Standards

Other open standards such as Integrated Services Digital Broadcasting-Terrestrial (ISDB-T) [16] and Terrestrial-Digital Multimedia Broadcast (T-DMB) [17] are popular in some parts of Asia and are expanding to other

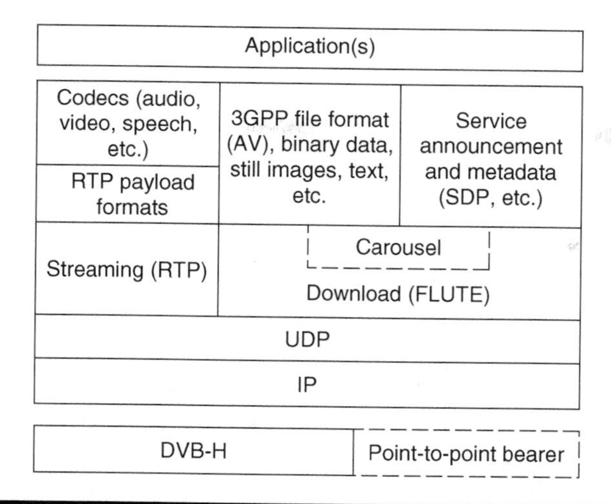

Figure 3.7 DVB-H IPDC protocol stack.

regions also. MediaFLO [18] is a proprietary technology that is currently deployed in the U.S. market.

3.5.5 OMA BCAST

MBMS and BCMCS introduce efficient support for broadcast/multicast transport in mobile networks, whereas OMA BCAST specifies broadcast/multicast-related service-layer functionalities. It is agnostic of the underlying broadcast distribution scheme (BDS), which could be MBMS, BCMCS, or DVB-H.

The OMA BCAST 1.0 supports the following top-level functions:

- Service guide [19]
- File and stream distribution
- Notifications
- Service and content protection [20]
- Service interactivity [21]
- Service provisioning
- Terminal provisioning
- Roaming and mobility support
- Specification on back-end interfaces for each function
- Broadcast distribution system adaptations for 3GPP MBMS, 3GPP2 BCMCS, and "IP datacast over DVB-H"

The functions specified by OMA BCAST and the assumed protocol architecture [22] are shown in Figure 3.9. For more details, refer to the draft specifications [19–23] that can be downloaded from Ref. 24.

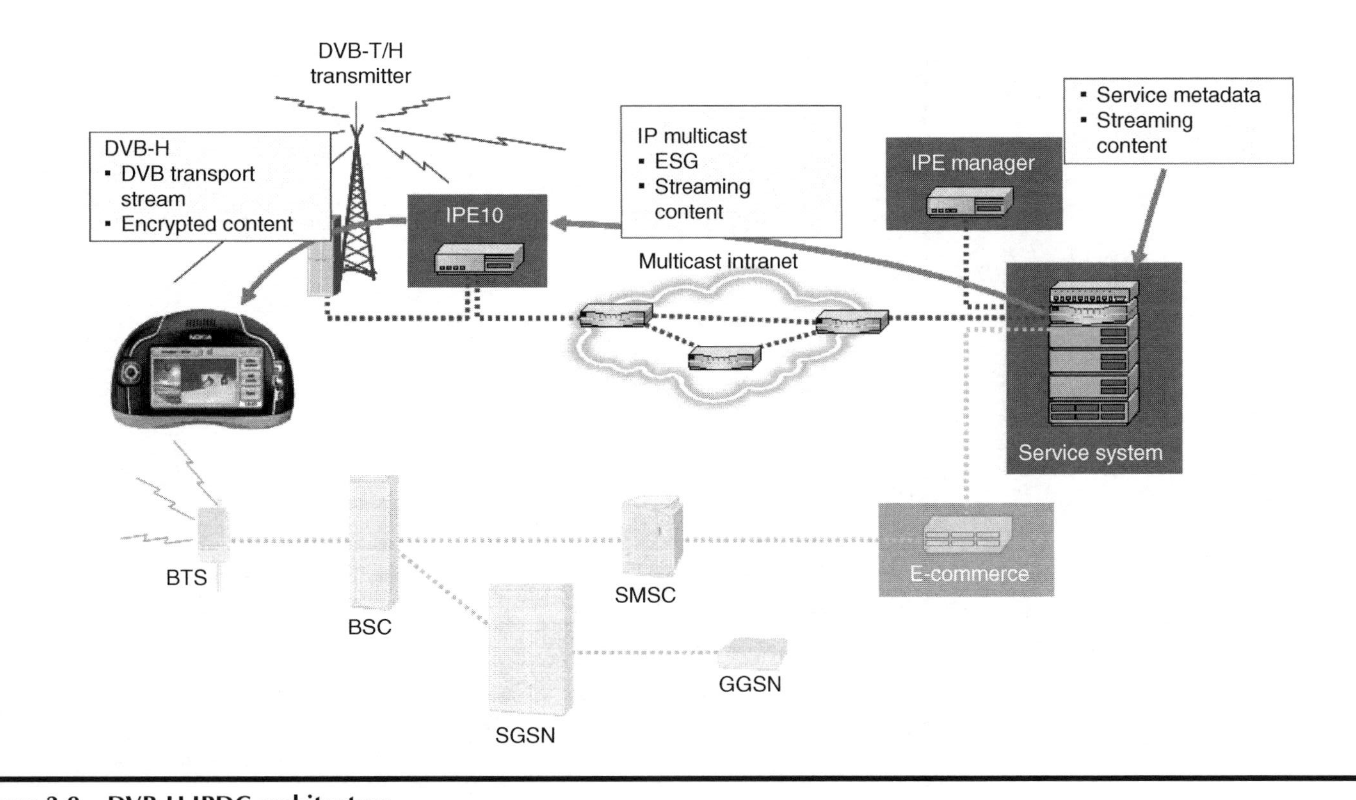

Figure 3.8 DVB-H IPDC architecture.

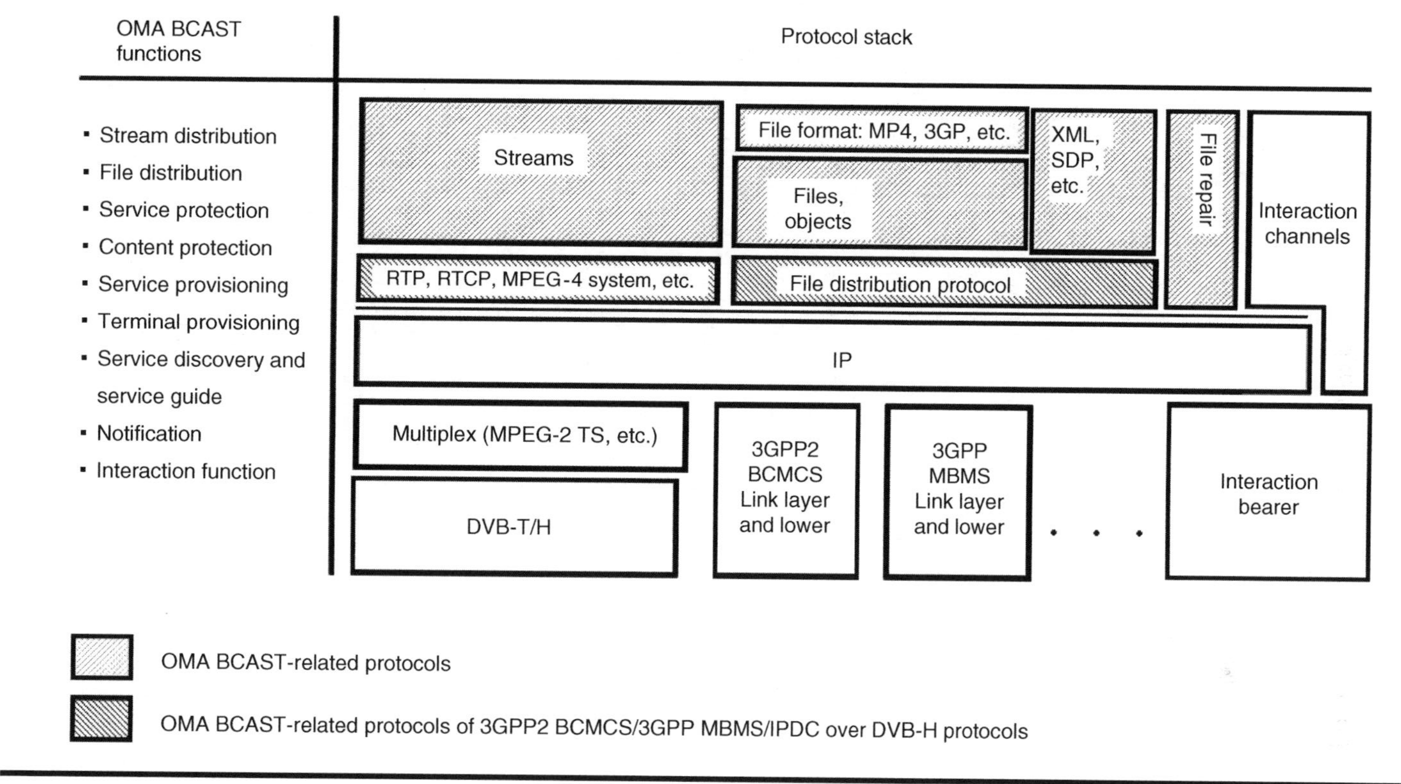

Figure 3.9 OMA BCAST protocol stack.

3.6 Media Delivery Framework

This section discusses the various transport protocols for mobile TV. A dominant design principle among most of the mobile TV standards is that they are IP-based, thus enabling the use of open Internet Engineering Task Force (IETF) protocols. This has allowed in aligning the main architectural principles of the three main standards for mobile TV described so far, MBMS, BCMCS, and DVB-H IPDC. These standards are essentially based on the same protocol families described in this section. However, this section takes a close look at the MBMS standard for mobile TV. In particular, file download protocols such as File Delivery over Unidirectional Transport (FLUTE) [25], the Real-Time Transport Protocol (RTP) framework for streaming, user interaction protocols, reliability methods such as FEC and point-to-point repair, quality of experience (QoE) metric framework, and channel zapping are discussed.

3.6.1 User Service Announcement

Every mobile TV service must be announced to all the receivers before these can receive the content. The idea is that the user is notified in advance about when a particular program is scheduled, irrespective of whether this is delivered via a streaming session or via a file download. The notification is done by means of "service announcements." With this information, the users can choose what programs they wish to receive and subscribe to their reception, if desired.

In MBMS, service announcements are delivered to the users via different means [26]:

1. *MBMS bearer.* The announcement is delivered as a file using the FLUTE protocol (see Section 3.6.2).
2. *Interactive pull bearer.* The announcement is delivered on request via the Hypertext Transfer Protocol (HTTP).
3. *Point-to-point push bearer.* The announcement is delivered via Short Messaging Service (SMS) or HTTP.

3.6.2 File Download Protocols

Traditional file download protocols over Transmission Control Protocol (TCP)/IP such as HTTP cannot be used for mobile broadcast/multicast file download due to scalability issues. A separate transport-layer protocol is required for broadcast/multicast file download. Because the content is carried over User Datagram Protocol (UDP)/IP, which is inherently unreliable, the transport protocol should provide mechanisms for "reliability." In addition, the clients may join the download session at any time (asynchronous download). The protocol has to be "scalable" to support any number of clients. Real-time and in-order delivery of data is not required for these contents. Asynchronous Layered Coding (ALC) [27] is a content delivery protocol framework that satisfies all of these requirements. ALC is a massively scalable and reliable content delivery protocol. ALC combines a layered coding transport (LCT) [28] building block, a congestion control building block, and an FEC [29] building block to provide congestion-controlled reliable asynchronous delivery of content to an unlimited number of concurrent receivers from a single sender. ALC is carried over UDP/IP, and is independent of the IP version and the underlying link layers used. ALC uses the LCT building block to provide in-band session management functionality. ALC uses the FEC building block to provide reliability for the broadcast channel. The FEC building block allows the choice of an appropriate FEC code to be used within ALC, including the use of "no-code FEC" that simply sends the original data using no FEC encoding. In addition, it provides header fields for packet ordering that help in identifying the missing packets. The latest versions of the ALC, LCT, FEC and FLUTE drafts, and subsequent request for comments (RFCs) can be found under IETF reliable multicast transport (RMT) charter [30].

FLUTE [25] defines a specific transport application of ALC, adding the following specifications: definition of a file-delivery session built over ALC including transport details and timing constraints, in-band signaling of the transport parameters of the ALC session, in-band signaling of the properties of delivered files, and details associated with the multiplexing of multiple files within a session. Figure 3.10 shows the building block structure of ALC and FLUTE. 3GPP MBMS and DVB-H IPDC mandate the usage of FLUTE, whereas it is optional in 3GPP2 BCMCS. In the Rel-7 of 3GPP MBMS, only a single FLUTE channel is allowed, whereas multiple FLUTE channels are allowed in IPDC over DVB-H [13]. 3GPP2 BCMCS mandates the usage of ALC for file download, whereas the OMA BCAST ESG [19] is used for managing the file metadata.

3.6.3 Streaming Protocols

RTP is the preferred protocol for the broadcast/multicast streaming of continuous media. All the major standards such as MBMS, BCMCS, DVB-H

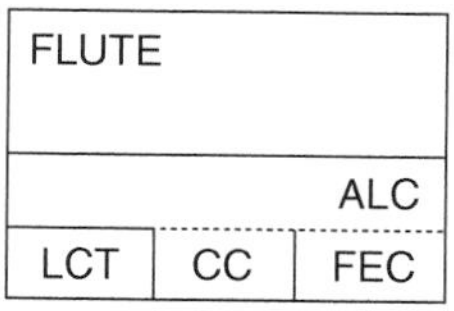

Figure 3.10 Building block structure of ALC and FLUTE.

IPDC, and OMA BCAST specify the usage of RTP for broadcast/multicast streaming. For scalability reasons, the RTCP receiver reports (RR) are disabled. The only modifications to MBMS streaming are (1) the use of FEC as a stream and (2) stream bundling for the FEC purpose. These concepts and the FEC usage for MBMS streaming are elaborated in Sections 3.6.6 and 3.6.8.

3.6.4 User Interaction Protocols

Protocols for user interactivity have been under development since the first release of the MBMS specifications. User interactivity is a very important feature because it can offer added value to the users. However, from the system perspective, this feature must guarantee an efficient scalability.

Given that the majority of the mobile TV networks are designed to support one-way point-to-multipoint transmissions, the only way to support user interactivity is by means of dedicated point-to-point bearers. As mentioned earlier, scalability is the main concern when dealing with user interactivity. Special scalability mechanisms have to be implemented into the system to make sure that the feedback does not cause a collapse of the whole system. Ideally, if the entire broadcast system user population wants to interact with the broadcast system at the same time, the network resources are not sufficient to handle the uplink load and the system will collapse.

A mechanism that is employed successfully to guarantee system scalability is that of spreading the user interactivity requests over a time window reasonably large so that this ensures that the system is not under any risk of collapse [26]. For example, if the users are requested to express a vote for the election of Miss Universe, the system can be configured in such a way that the UE sends the user vote not immediately, but within a time window of 30 s. This mechanism is implemented in the user application, and the user does not need to know when exactly the feedback information is really sent. The dedicated interactivity bearers can be established and kept alive only for the duration of the time they are required to provide interactivity functionality requested by the user. The lifetime of such bearers can be, for instance, of just a few seconds. In other cases, if long interactive sessions are required, these bearers are kept alive for a longer time.

In MBMS, interactivity messages are carried over the HTTP [26]. Currently, there are only two types of interactive functions offered: the file repair function and the reception reporting procedure. The possibility of introducing more user interactivity functions is the subject of the development of future MBMS releases. The file repair function is analyzed in detail in Section 3.6.5.

The reception reporting procedure is used to report the complete reception of files or to report statistics on the reception of a stream delivery. It is possible that only a subset of the whole system population is requested to report complete reception of files. In this case, a "sample percentage" attribute can be specified to determine what percentage of the broadcast users must report the successful reception of files. Reception reports are packed into eXtensible Markup Language (XML) messages and transmitted to the BM-SC via HTTP.

It is expected that MBMS will be initially deployed in selected areas with dense user population, where the usage of broadcast/multicast bearers is more efficient. When MBMS subscribers are roaming in areas where MBMS coverage is not available, MBMS content needs to be delivered to them using unicast bearers. For unicast distribution, MBMS streaming content is delivered using 3GPP PSS [31], which uses Real-Time Streaming Protocol (RTSP) [32] for session control and RTP/UDP/IP for media delivery. Similarly for unicast distribution, MBMS download content is delivered using a combination of Wireless Application Protocol (WAP) Push [33] and HTTPs.

Mobile TV content is distributed using MBMS streaming delivery method, whereas the interactive content is simultaneously distributed using the MBMS download delivery method, that is, via FLUTE. This interactive content may consist of targeted advertisements with links for user interaction, SMS and MMS templates for user response, etc. The UE may download the interactive content and display to the UE at scheduled times during the MBMS streaming session. The user's response to the interactive content may be transported using, for example, HTTP, SMS, or MMS messages. OMA BCAST service [21] describes this process in detail.

3.6.5 Reliability of File Download Sessions

Because transmissions on the mobile TV network are typically not error-free, during a broadcast file transmission, some fragments of the file can be missing. With missing parts the file is unusable. When AL-FEC is used in the broadcast transmission, the sender transmits the original file (source symbols) and some redundant information (FEC symbols) that is derived from the original file by applying the FEC encoding algorithm. If the receiver gets enough number of symbols (source + FEC), then the missing fragments of the file can be recovered by applying the FEC decoding algorithm.

The FEC schemes at this layer are used for recovering from packet losses (erasure recovery), unlike the traditional error correction codes (e.g., convolutional codes, turbo codes, and low-density parity check [LDPC] codes) used at the physical layer of the wireless networks. The FEC decoders of error correction codes need to detect the location of the errors in a code word and then correct the error, whereas the FEC decoders of erasure correction codes have prior knowledge about the location of the errors in the code word. This prior knowledge is provided to the FEC decoder by the FEC header fields defined in the FEC building block of FLUTE or ALC.

According to the current protocol architectures of the wireless networks, only error-free packets are allowed to pass through the intermediate layers of the protocol stack. Erroneous or incomplete packets are discarded at the lower layers if the lower-layer error recovery mechanisms fail. Thus, the packets at the application layer are either error-free or marked as lost (erasures).

The FEC encoder chops the file into one or more source blocks, divides the source blocks further into source symbols, and creates FEC symbols from the source symbols by applying an FEC code. The source and FEC symbols are stuffed into packets, which are assigned unique identifiers that can be used to identify the missing symbols at the receiver. The more the redundancy used for FEC, the more the number of receivers will be able to successfully decode the file, but network resources are tied up for longer duration. The amount of redundancy to be transmitted is decided by the operator based on the estimated channel conditions of a majority of receivers. In MBMS and DVB-H IPDC, raptor codes are used for providing AL-FEC for file download applications.

Even if FEC is used, a minority of the users (who experience severe channel conditions) will not be able to decode the entire file during the broadcast transmission. Therefore, it is essential that the network offers the possibility of interactively repairing a file that has not been completely received during the first broadcast transmission. The target is that files should be 100 percent complete at the end of a file repair session.

The file repair procedure is one of the interactivity functions offered by the system, and it has to be implemented with special care to feedback implosion avoidance, downlink network channel congestion avoidance, and file repair server overload avoidance. The typical sequence of steps for performing a file repair is the following:

1. Identification of the end of the first broadcast transmission.
2. Identification of the missing parts of the file(s).
3. Calculation of a back-off time to guarantee scalability, and selection of a file repair server among a list to guarantee load balancing.
4. Send a file repair request to the selected server at a specified time.
5. The designated server responds with a repair response message that contains the missing data or, if the server is unavailable, a redirection message to another available server.

The file repair request and response messages are carried over the HTTP.

3.6.6 Reliability of Streaming Sessions

Because retransmission is not a scalable solution for providing reliability in the broadcast/multicast streaming, other techniques are required to enable reliable delivery of the media streams. Physical-layer FEC is used in the wireless networks to recover from bit/symbols errors. Examples include convolutional codes, turbo codes, and LDPC codes. DVB-H uses an outer FEC code in the link layer to recover from the residual errors that could not be recovered by the physical-layer FEC. This is known as MPE-FEC and is based on (255,191) RS codes over GF(256). In DVB-H, AL-FEC is useful only if the source block data spans across multiple time-slice bursts. Hence, AL-FEC is used only for downloading large files that span across multiple bursts. For streaming applications and small file downloads, AL-FEC is not used in DVB-H IPDC. BCMCS also employs outer FEC at the link layer. This is based on variants of (32,28) RS codes and (16,12) RS codes in GF(256). The applicability of AL-FEC for BCMCS streaming applications is being evaluated by the standards body.

MBMS, however, does not use any FEC at the link layer. Instead, it uses only AL-FEC to protect the media streams. Figure 3.11 summarizes the MBMS streaming framework that defines the use of AL-FEC for the protection of media streams. It was shown via simulations [34] that using

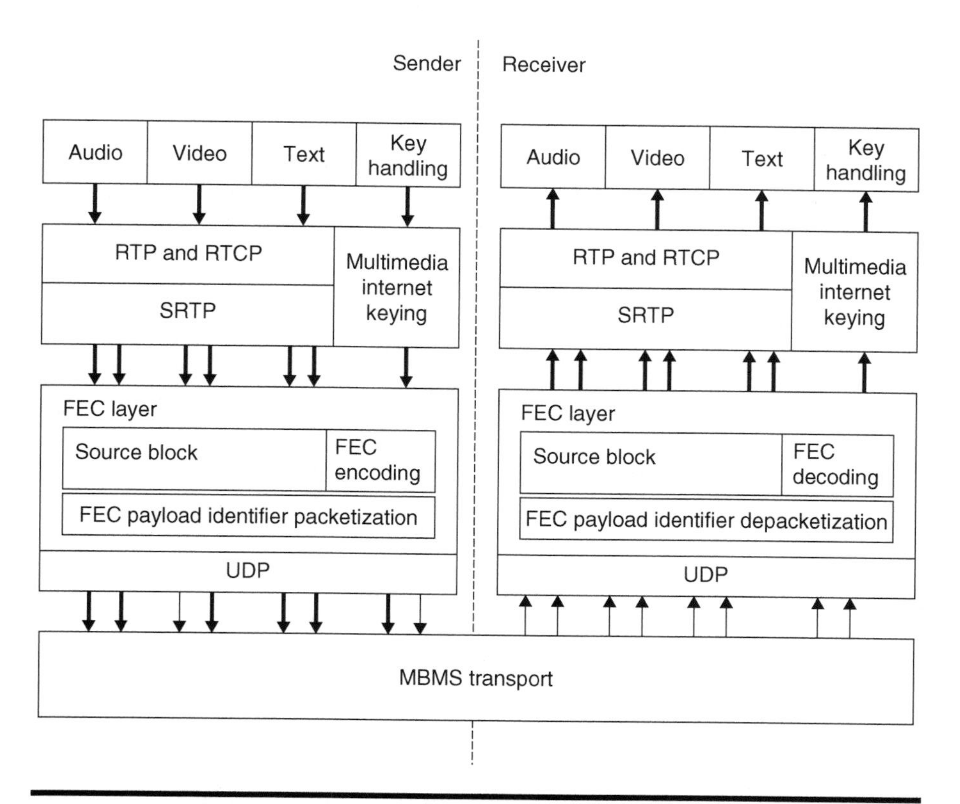

Figure 3.11 MBMS streaming framework.

AL-FEC reduces the required power of MBMS bearers, thus improving the overall capacity of the 3G transmission systems.

MBMS specified the use of raptor codes [26] for both streaming and file download applications. Raptor codes [35] are rateless codes used for erasure recovery. "Rateless property" means that the number of FEC symbols need not be predetermined, as many FEC symbols can be generated as required from the same set of source symbols, thus providing flexibility at the sender side. Raptor codes are of low decoding complexity and their error correction performance is close to that of an ideal FEC.

3.6.7 Quality of Experience Framework

Among the features that are a part of the interactive domain is the one that allows to measure the QoE perceived by the users during a streaming session. This tool can be used by operators in the first phases of the mobile TV service deployment.

The QoE is reported as a part of the reception reporting described in Section 3.6.4 and is measured according to well-defined metrics in MBMS [26]:

- *Corruption duration*. It is used to measure how long a media corruption lasts during a streaming session.
- *Rebuffering duration*. It is defined as the time the continuous playback is stalled due to an involuntary event at the UE side.
- *Initial buffering duration*. It is defined as the time between the first RTP packet received and the time this is played back.
- *Successive loss of RTP packets*. It indicates the number of RTP packets lost in succession (except FEC packets) for a media.
- *Frame rate deviation*. It indicates by how much the frame rate deviates from a predefined value given by the system.
- *Jitter duration*. Jitter happens when the absolute difference between the actual playback time and the expected playback time is greater than a predefined value, which is 100 ms. This metric measures the duration of such jitter.

All the metrics are reported only once at the end of the streaming session. In addition, it is possible to report the metrics only for a specified part of the media stream(s) (e.g., between 100 and 200 s of a video stream).

3.6.8 Channel Zapping

Channel zapping is an important differentiator in the user experience delivered by various mobile TV services. For standard-based mobile TV technologies, the channel zapping involves different layers of the protocol stack. At the physical layer, it may involve switching to a different frequency or time slots. At the application layer, it involves the delays caused by buffering for AL-FEC, media decoders, and security mechanisms. The efficiency of AL-FEC improves with increasing block lengths, which is achieved by buffering the media streams for a longer duration. However, long buffer length for AL-FEC results in longer tune-in delays and hence longer channel-zapping delays. Suppose that the first few packets of a source block are lost. To recover them, the receiver needs to wait for the reception of the entire source block and the corresponding FEC packets before it can start FEC decoding. If the receiver starts receiving the data in the middle of a source block, and the FEC protection is not enough to recover the missing packets, then the receiver has to discard on an average half of the source block. It can start FEC decoding only after receiving the next source block. Thus on an average, AL-FEC results in a delay of 1.5 times the length of a source block.

In addition, the media decoders impose some delays on the tune-in. The video coders typically employ predictive coding to achieve high compression efficiency. The bitstream consists of a series of group of pictures (GOP) that have an independently coded intra (I) frame and a series of predictively coded (P) frames that are compressed with reference to the previous frame(s). The I-frames provide random access points into the video bitstream. The P-frames cannot be decoded as such without having their corresponding reference frames. The size of the I-frames is much larger compared to the P-frames. To achieve faster tune-in to a compressed video bitstream, it should have more frequent random access points, that is, I-frames. But this will result in high bit rate. In addition, the random access points to the media streams must be aligned to the AL-FEC source block boundaries, to eliminate additional delay due to media decoding. In the example mentioned earlier, if a source block does not contain any I-frame, then the entire source block will have to be discarded.

There are various techniques for improving the tune-in delays in MBMS. Instead of applying separate AL-FEC to each individual media in the streaming session, it is advantageous to periodically form a single source block for all media (audio, video, timed – text, etc.) and apply AL-FEC to this source block. In this way, the source block length is increased and hence the efficiency of the AL-FEC is improved. We can take a step further and periodically form a single source block for all media streams of all channels being broadcast in the streaming session. This is called stream bundling. The advantage of this approach is that the user can switch instantaneously between the bundled streams or channels without waiting for AL-FEC buffering. The disadvantage is that the UE will have to receive and process all the bundled streams continuously even if the user is consuming only one stream/channel. The media-decoding step is the only step that can be avoided on the remaining channels that the user is not consuming.

In DVB-H, because no AL-FEC is used for streaming applications, the problems mentioned earlier do not exist. The channel-zapping delay caused by media decoding is reduced by using a separate dynamic zapping stream. This stream consists of a slide show of intracoded frames of all regular streams. Whenever the user wishes to zap a channel, the receiver picks the corresponding I-frame from the dynamic zapping stream and starts decoding the subsequent P-frames from the regular stream. The slight degradation in the decoded picture quality due to mismatch between zapped I-frame and regular P-frame lasts for a few seconds until the next I-frame is received in the regular stream.

3.7 Media Codecs and Formats

3.7.1 Media Codecs

As there is limited opportunity for negotiation between the server and the mobile TV clients, it is desirable to have a few single mandatory codec configuration per media that is supported by the mobile TV clients. In addition, for mobile TV services offered over cellular networks, it is natural to select codecs from the list of codecs already supported by the handset. The selected codecs should be bandwidth efficient to utilize the usually costly broadcast spectrum, should be of low complexity to run on even low/medium-cost terminals, and should have error-resiliency tools to cope with the widely varying channel conditions experienced by the clients. MBMS [26] supports a wide variety of media types that are very similar to the ones supported by the existing 3GPP PSS (unicast streaming service) [31]. The supported media types and the corresponding codecs and their profiles/levels are summarized in Table 3.1. Note that an MBMS client need

Table 3.1 Media Codecs Supported by MBMS Rel-6 Clients

Media Type	*Codec*
Speech	AMR narrowband and AMR wideband
Audio	Enhanced aacPlus and extended AMR wideband
Synthetic audio	Scalable polyphony MIDI
Video	H.264 (AVC) baseline profile level 1 b and H.263 profile 0 level 45 decoder
Still images	ISO/IEC JPEG and JFIF
Bitmap graphics	GIF87a, GIF89a, and PNG
Vector graphics	SVG Tiny 1.2
Text	XHTML Mobile Profile, SMIL 2.0 (text referenced with *text* element and *src* attribute), UTF-8, and UCS-2 encodings
Timed text	3GPP TS 26.245

not support all the media listed in Table 3.1. However, if it supports any of these media types, then it shall support the codec level/profile listed. For video, the state-of-the-art video codec H.264 [36] is strongly recommended for MBMS, whereas H.263 is already supported by 3GPP PSS clients. The H.264 baseline profile means support of 4:2:0 chroma format with 8-bit pixel depth and the use of the following video coding concepts: I- and P-slices, multiple reference frames, in-loop deblocking filter, context adaptive variable length coding (CAVLC) entropy coding, flexible macroblock ordering, arbitrary slice ordering, and redundant slices. Level 1b means decoding capability of up to 128 kbps in baseline profile, which translates to quarter common interchange format (QCIF) (176 × 144) 15 fps video. Note that H.264 video codec support is being upgraded to baseline profile level 1.2 [36,37] in Rel-7.

However, DVB-H IPDC supports only video, audio, and the timed-text formats. Unlike MBMS, which mandates a specific profile/level for a given codec, DVB-H recommends the support of at least one of the video codecs H.264 and video codec-1 (VC-1) [38]. For each codec, five different profiles/levels are selected that correspond to increasing decoder capabilities (A–E). The current DVB-H terminals are likely to fall into capability B. For H.264, capability B means baseline profile level 1.2, that is, decoding capability of up to 384 kbps in baseline profile, which translates to quarter VGA (QVGA) (320 × 240) 20 fps or CIF (352 × 288) 15 fps video. Similarly for VC-1, capability B means simple profile at ML (main level). For audio, DVB-H terminals shall support at least one of the MPEG-4 HE AAC v2 and Extended AMR-WB (AMR-WB+).

3.7.2 Dynamic and Interactive Multimedia Scenes

Rel-7 of 3GPP introduces a new media-type dynamic and interactive multimedia scenes (DIMS) [39] for use in the existing 3GPP services including MBMS, PSS, and MMS. DIMS is a dynamic, interactive, scene-based media system that enables display and interactive control of multimedia data such as audio, video, graphics, images, and text. It ranges from a movie enriched with vector graphics overlays and interactivity (possibly enhanced with closed captions) to complex multistep services with fluid interaction/interactivity and different media types at each step. DIMS is also considered as a superior alternative to Synchronized Multimedia Integration Language (SMIL) [40].

3.7.3 Container File Formats

For the download and play use case of mobile TV, the server may distribute a container file using the FLUTE protocol. The format of this container file enables the storage of the compressed media bitstream together with some metadata about the media tracks, random access information and synchronization details, etc. If a terminal is branded as compliant to a specific profile of these container formats, then it shall be able to decode and render all the media formats supported by that file format.

International Standards Organization (ISO) base media format [41] is a common format that includes all the media types and their codecs. 3GP [42] and 3G2 [43] container formats are defined as extensions of the ISO base media format. They support only a subset of the media types and an overlapping set of codecs that are specific to 3GPP and 3GPP2. They define specific constraints on the media types and codecs supported by the ISO base media format. For example, the "basic profile" of the 3GP file format imposes the following constraints:

- There shall be no references to external media outside the file, that is, the file shall be self-contained.
- The maximum number of tracks shall be one for video, one for audio, and one for text.
- The maximum number of sample entries shall be one per track for video and audio, but unrestricted for text.

Similarly, the "extended presentation profile" of a 3GP file may include an extended presentation that consists of media files in addition to tracks for audio, video, and text. Examples of such media files are static images, for example, JPEG files, which can be stored in a 3GP container file. A 3GP container file that includes an extended presentation must include a scene description that governs the rendering of all the parts of the file.

An MBMS client shall support the basic profile and the extended presentation profile of the 3GPP file format 3GPP TS 26.244 [42]. The DVB-H terminals shall support the basic profile of Rel-6 version of the 3GPP file format TS 26.244. In addition, DVB-H has defined its own version of a file format [44], which is again an extension of the ISO base media file format [41] with some direct constraints. The BCMCS client is expected to support the latest version of the 3GPP2 file format [43].

3.8 Security

The security of MBMC provides different challenges compared to the security of services delivered over point-to-point services. In addition to the normal threat of eavesdropping, there is also the threat that it may not be assumed that valid subscribers have any interest in maintaining the privacy and confidentiality of the communications, and they may therefore conspire to circumvent the security solution (e.g., one subscriber may publish the decryption keys enabling nonsubscribers to view broadcast content). Countering this threat requires the decryption keys to be updated frequently in a manner that may not be predicted by subscribers while making efficient use of the radio network.

The service and content protection functions in OMA BCAST provide a BDS-agnostic way of protecting both content and services delivered within mobile broadcast services. However, 3GPP MBMS, DVB-H IPDC, and 3GPP2 BCMCS have their own protection mechanisms [13,45,46] that are not fully aligned with OMA BCAST. A discussion on the definitions and requirements of content protection and service protection follows. For specific details of how service or content protection is provided, see Refs 20 and 23.

Content protection. This involves the protection of content (files or streams) during the complete lifetime of the content. Content providers require securing the content not only at the present time of broadcasting but also in the future. Some content providers might want to determine postacquisition usage rules or the so-called digital rights. These can be obtained on an individual basis by the end user. Other content providers have content to offer, for which they do not require technical restrictions but limit it to fair use cases and rely on copyright acts.

Service protection. This involves the protection of content (files or streams) during its delivery. Service providers require a secure access mechanism. They are only concerned with managing access to the content at the time of broadcasting. This is independent of the offered content and the presence of digital rights for certain types of content. Only an access/no-access mechanism is required to distinguish between subscribed and not-subscribed users.

Figure 3.12 illustrates the difference between service and content protection. Note that in the case of content protection, the content is controlled by the content provider, whereas in case of service protection, it is controlled by the broadcast operator.

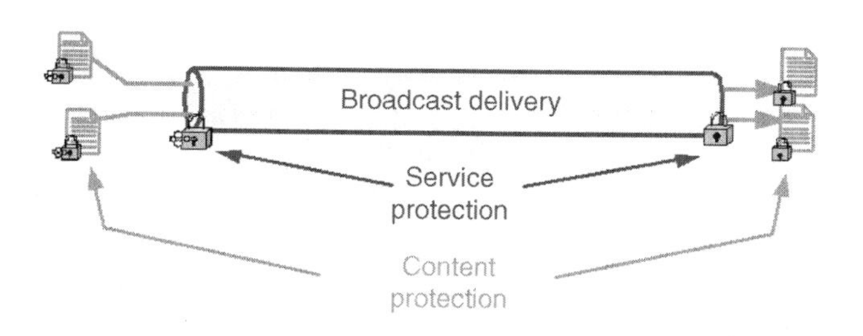

Figure 3.12 Roles of service and content protection.

3.9 Conclusion and Open Issues

Mobile TV services are recently being deployed. Currently, mobile TV services over 3G cellular networks do not use broadcast/multicast transport technologies described in this chapter. They make use of unicast transport that does not scale to serve a large number of simultaneous users. It is expected that the service providers deploy broadcast/multicast services in regions of heavy user density, whereas in other regions unicast transport may be used. Currently, there is limited interaction opportunity for mobile TV services, and the media types are limited to audio, video, and timed-text formats only. A rich mobile TV service should offer new media combinations including interactive graphics, animation, and multimodal interaction. Apart from cellular networks, other wireless broadband networks such as wireless local area network (WiFi) and Worldwide Interoperability for Microwave Access (WiMAX) [47] are being deployed worldwide. Offering mobile TV service over these new bearers and ensuring bearer-agnostic delivery of mobile TV service pose some interesting engineering challenges to the standardization fora. Achieving channel-zapping times comparable to regular television, providing novel interaction opportunities, techniques to offer collaborative mobile TV viewing experience, and seamless transfer of TV viewing experience between fixed and mobile gadgets are some open research issues.

Mobile TV is something personal that is close to the user and always accessible to the user; thus it provides the possibility for a very personalized interaction with the service. This possibility provides the service providers an opportunity for more targeted advertising and the end user an opportunity to view the content of interest anywhere, anytime, and on any device.

References

1. 3GPP, Multimedia Broadcast/Multicast Service (MBMS); User Services Stage 1, 3GPP TS 22.246 V. 7.2.0, Mar. 2007.
2. 3GPP, Multimedia Broadcast/Multicast Service (MBMS); Architecture and Functional Description, 3GPP TS 23.246 V. 7.2.0, Mar. 2007.
3. 3GPP, Quality of Service (QoS) Concept and Architecture, 3GPP TS 23.107 V. 6.4.0, Mar. 2006.
4. 3GPP Specifications, http://www.3gpp.org/specs/numbering.htm.
5. Wang, J., Sinnarajah, R., Chen, T., Wei, Y., and Tiedemann, E., Broadcast and Multicast Services in cdma2000, *Communications Magazine, IEEE*, 42(2), 76–82, 2004.
6. Agashe, P., Rezaiifar, R., and Bender, P., cdma2000 High Rate Broadcast Packet Data Air Interface Design, *Communications Magazine, IEEE*, 42(2), 83–89, 2004.

7. 3GPP2, cdma2000 High Rate Broadcast-Multicast Packet Data Air Interface Specification, 3GPP2 C.S0054-A v1.0, Mar. 2006.
8. 3GPP2, Broadcast and Multicast Service in cdma2000 Wireless IP Network, X.S0022-A v1.0, Apr. 2007.
9. 3GPP2, BCMCS Codecs and Transport Protocols, C.P0070-0 (work in progress).
10. 3GPP2 Specifications, http://www.3gpp2.org/Public_html/specs/index.cfm.
11. Digital Video Broadcasting (DVB), Transmission System for Handheld Terminals (DVB-H), ETSI EN 302 304 V1.1.1, Nov. 2004.
12. Faria, G., Henriksson, J., Stare, E., and Talmola, P., DVB H: Digital broadcast services to handheld devices, *Proceedings of IEEE*, 94(1), 194–209, 2006.
13. Digital Video Broadcasting (DVB), IP Datacast over DVB-H: Content Delivery Protocols, ETSI TS 102 472 V1.2.1, Dec. 2006.
14. Digital Video Broadcasting (DVB), IP Datacast over DVB-H: Electronic Service Guide (ESG), ETSI TS 102 471 V1.2.1, Nov. 2006.
15. DVB-H, http://www.dvb-h.org/technology.htm.
16. ISDB-T, http://en.wikipedia.org/wiki/ISDB#ISDB-T.
17. T-DMB, http://en.wikipedia.org/wiki/T-DMB.
18. MediaFLO, http://www.qualcomm.com/mediaflo/index.shtml.
19. Open Mobile Alliance, Service Guide for Mobile Broadcast Services, OMA-TS-BCAST_ServiceGuide-V1_0-20080103-D, Jan. 2008 (work in progress).
20. Open Mobile Alliance, Service and Content Protection for Mobile Broadcast Services, OMA-TS-BCAST_SvcCntProtection-V1_0-20071218-D, Dec. 2007 (work in progress).
21. Open Mobile Alliance, Mobile Broadcast Services, OMA-TS-BCAST_Services-V1_0-20080103-D, Jan. 2008 (work in progress).
22. Open Mobile Alliance, Mobile Broadcast Services Architecture, OMA-AD-BCAST-V1_0-20071218-D, Dec. 2007 (work in progress).
23. Open Mobile Alliance, OMA DRM v2.0 Extensions for Broadcast Support, Open Mobile Alliance, OMA-TS-DRM-XBS-V1_0-20060425-D, Apr. 2006 (work in progress).
24. Open Mobile Alliance, http://member.openmobilealliance.org/ftp/Public_documents/bcast/Permanent_documents/.
25. Paila, T., Luby, M., Lehtonen, R., Roca, V., and Walsh, R., FLUTE—File Delivery over Unidirectional Transport, draft-ietf-rmt-flute-revised-05, Internet Draft, IETF, Oct. 2007 (work in progress).
26. 3GPP, Multimedia Broadcast/Multicast Service (MBMS); Codecs and Transport Protocols, 3GPP TS 26.346 V. 7.3.0, Mar. 2007.
27. Luby, M., Watson, M., and Vicisano, L., Asynchronous Layered Coding (ALC) Protocol Instantiation, draft-ietf-rmt-pi-alc-revised-05, Internet Draft, IETF, Nov. 2007 (work in progress).
28. Luby, M., Watson, M., and Vicisano, L., Layered Coding Transport (LCT) Building Block, draft-ietf-rmt-bb-lct-revised-06, Internet Draft, IETF, Nov. 2007 (work in progress).
29. Watson, M., Luby, M., and Vicisano, L., Forward Error Correction (FEC) Building Block, RFC 5052, IETF, Aug. 2007.
30. IETF RMT WG, http://www.ietf.org/html.charters/rmt-charter.html.
31. 3GPP, Transparent End-to-End Packet Switched Streaming Service (PSS); Protocols and Codecs, 3GPP TS 26.234 V. 7.2.0, Mar. 2007.

32. Schulzrinne, H., Rao, A., Lanphier, R., Westerlund, M., and Narsimhan, A., Real Time Streaming Protocol (RTSP) 2.0, draft-ietf-mmusic-rfc2326bis-16.txt, Internet Draft, IETF, Nov. 2007 (work in progress).
33. Open Mobile Alliance, WAP Push Architectural Overview: Wireless Application Protocol, WAP-250-PushArchOverview-20010703-a, July 2001 (work in progress).
34. Luby, M., Watson, M., Gasiba, T., and Stockhammer, T., Mobile Data Broadcasting over MBMS—Tradeoffs in Forward Error Correction, *The 5th International Conference on Mobile and Ubiquitous Multimedia*, Stanford University, Stanford, CA, Dec. 2006.
35. Luby, M., Shokrollahi, A., Watson, M., and Stockhammer, T., Raptor Forward Error Correction Scheme for Object Delivery, RFC 5053, IETF, Oct. 2007.
36. ITU-T, Advanced Video Coding for Generic Audiovisual Services, ITU-T Rec. H.264 | ISO/IEC 14496-10, Mar. 2005.
37. H.264, http://en.wikipedia.org/wiki/H.264.
38. VC-1, Society of Motion Picture and Television Engineers, VC-1 Compressed Video Bitstream Format and Decoding Process, SMPTE 421M, 2006.
39. 3GPP, Dynamic and Interactive Multimedia Scenes (DIMS), 3GPP TS 26.142 V. 2.0.0, June 2007.
40. SMIL, http://www.w3.org/TR/2006/WD-SMIL3-20061220/.
41. ISO/IEC, Information Technology—Coding of Audio-Visual Objects—Part 14: MP4 File Format, ISO/IEC 14496-14, 2003.
42. 3GPP, Transparent End-to-End Packet Switched Streaming Service (PSS); 3GPP File Format (3GP), 3GPP TS 26.244 V. 7.1.0, Mar. 2007.
43. 3GPP2, 3GPP2 File Formats for Multimedia Services, 3GPP2 C.P0050-A v1.0, Apr. 2006.
44. Digital Video Broadcasting (DVB), Specification for the Use of Video and Audio Coding in DVB Services Delivered Directly over IP Protocols, ETSI TS 102 005 V1.2.1, Apr. 2006.
45. 3GPP, 3G Security; Security of Multimedia Broadcast/Multicast Service (MBMS) Release 7, 3GPP TS 33.246 V. 7.3.0, Mar. 2007.
46. 3GPP2, Broadcast-Multicast Service Security Framework, S.S0083-A v1.0, Sept. 2004.
47. WiMAX, http://en.wikipedia.org/wiki/WiMAX.

Chapter 4

Multiparty Audioconferencing on Wireless Networks

R. Venkatesha Prasad, Vijay S. Rao, H.N. Shankar, and R. Muralishankar

Contents

4.1 Introduction 120
4.2 Desirable Features of an Interactive Audioconferencing System 123
4.3 Constraints of Interactive Audioconferencing in Wireless Networks 125
4.3.1 Interactivity 126
4.3.2 Customized Mixing 127
4.3.3 Signaling 127
4.3.4 Scalability: Large-Scale Distribution of Participants over a Wide Area 127
4.3.5 Availability of Multicasting 128
4.3.6 Traffic Reduction 128
4.3.7 Quality of the Conference Based on Packet Delay, Packet Loss Percentage, and Delay Jitter 129
4.4 Techniques for Handling Multiple Audio Streams 129

4.5 Mixing Architectures: State-of-the-Art 131
4.5.1 Centralized Mixing Architecture 131
4.5.2 Endpoint Mixing Architecture....132
4.5.3 Hierarchical Mixing....134
4.5.4 Distributed Partial Mixing....135
4.5.5 Distributed Mixing System 136
4.5.5.1 Advantages and Disadvantages of Mixing 137
4.6 The Proposed Architecture 138
4.6.1 Design Requirements....138
4.6.2 Description....139
4.6.3 Selection of Streams....141
4.6.4 Loudness Number....143
4.6.4.1 Safety, Liveness, and Fairness....145
4.6.5 Selection Algorithm Using the Loudness Number 145
4.6.6 Self-Organization of the Entities 146
4.6.7 Reducing Bandwidth Consumption....147
4.7 Open Issues 148
4.8 Conclusions....149
References 150

Internet has been a part and parcel of life for sometime now. With the advent of technology, wireless networking is surging ahead. Multimedia applications are also finding their rightful place in this revolution. This chapter discusses an important multimedia application—audioconferencing. Issues related to media and the wireless networks have been addressed and some existing techniques for conferencing are presented. Learning from their limitations, we formulate the requirements for a better quality conference. We set out with refining the existing Session Initiation Protocol (SIP) architecture by introducing conference servers (CSs). We present a recently proposed metric called "Loudness Number (LN)" that helps in dynamically selecting speakers in a conference. This is aimed at facilitating a smooth transition of speakers while taking turns. Specific to the context of wireless networks, we argue that by adopting the refinements proposed here, some limitations in conferencing are mitigated. More efficient use of the existing bandwidth and reduced computational effort are highlighted, and the chapter concludes with some open issues and possible vistas for investigations.

4.1 Introduction

Concomitant with the Internet maturity and advances in supporting firmware, novel applications are becoming commercially viable at a rapid pace. Multimedia conferencing over Internet Protocol (IP) networks is an application of growing popularity worldwide. It enables many participants to exchange different types of media streams for collaborative work. The idea that collaborative technology is an activity in search of a need should be laid to rest [1]. Collaborative technology carves out a niche among users who spend most of their time in group endeavors, who use computing instruments to do their work, and whose potential for collaborations is otherwise impaired by geographic separation. The trend among conferencing systems has been to draw on real-world interaction protocols, but not necessarily on ensuring an honest-to-goodness virtual replica of a face-to-face conference (except in Refs 2–7).

The key desirable features of any collaborative technology are (a) less cognitive overload for the participants, (b) simplicity, (c) real-time media, and (d) good speech quality [1]. Collaborative work demands close interactions among participants, feedback or corrections for speakers, and mixing of voice streams—all in real-time.

Hindmarsh et al. [8] show that the participants fall back on audio to resolve difficulties with other aspects of collaborative applications such as negotiating a shared perspective. In fact, Doerry [9] demonstrates very marginal improvements in collaborativeness when video is included. Hence, audio as voice is the most important component of any real-time collaborative application. This chapter focuses on the audio medium of a collaborative work platform, that is, on "audioconferencing."

Miniature computing and communication devices, such as motes, PDAs, and mobiles, are increasingly enabling seamless and ubiquitous services across platforms. Exponentially growing volumes, time-to-market, and time-in-market are inevitable concerns placing consumer electronics at its bleeding edge incessantly. Bandwidth bottlenecks are becoming less stringent in terms of their impact, thanks also to the emerging trends in wireless technologies. Users are demanding progressively rich media content in real-time even in handheld, mobile, low-power, and low-cost devices—all in one breath. Multiparty audioconferencing on wireless networks is one such popular demand. Even this application domain has several dimensions to it.

In particular, this trend is accelerating even as it is maturing in broadband wireless networks and *ad hoc* networks. We observe that 802.11a/b/g has changed the way we connect to the external world. In fact, to enable higher bandwidth in the last mile, 5 GHz band at 60 GHz [10] has been reserved to enable, *inter alia*, rich multimedia. The Institute of Electrical and Electronics Engineers (IEEE) 802.16 Working Group on Broadband Wireless Access Standards is working on the deployment of broadband Wireless Metropolitan Area Networks (WiMAX) [11] more pervasively. Devices with IEEE 802.16a/e are already penetrating the market and the IEEE 802.20 version of the mobile broadband wireless networking is in the pipeline [12]. Consequently, from the consumers' perspective, enhanced multimedia applications are rendered easier by the day. In the process of

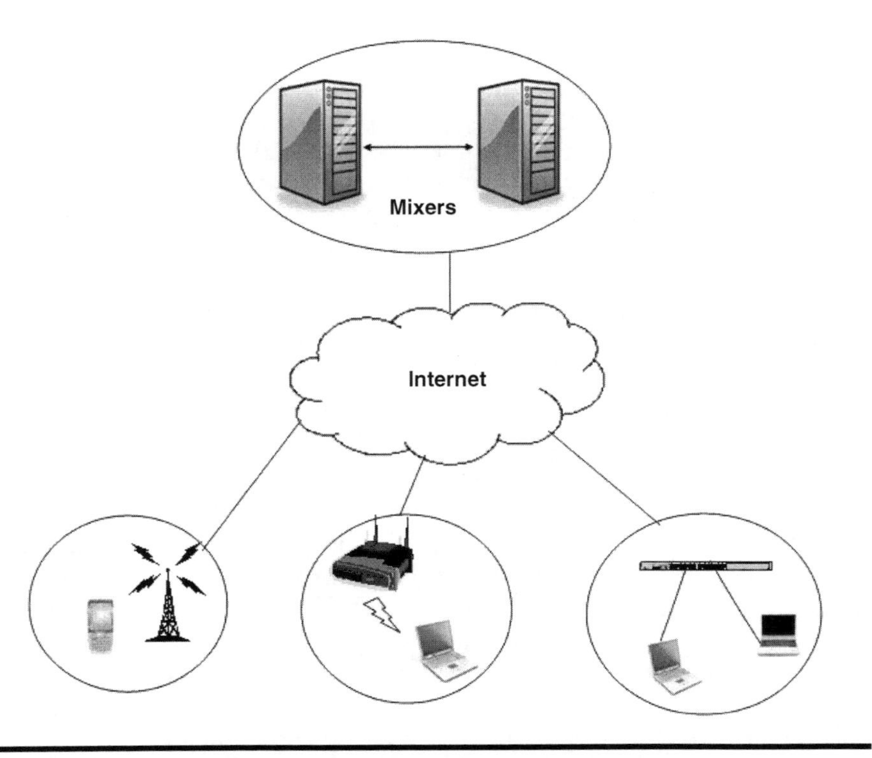

Figure 4.1 Typical scenario of wireless nodes in a session.

launching such applications in the market, however, there are concerns to be addressed. We look at some of them here.

1. Any service provider has to be acutely sensitized to the fact that devices in demand in an aggressively competitive market are functionally heterogeneous (see Figure 4.1). This is a fallout of constraints of computational capabilities, onboard memory, and link capacity having to coexist with mobility. It is strongly desired that such devices have multiple link layers. Then handling multiple linked layers pops up as a problem in its own right. Some devices running interactive multimedia applications, for example, speech conferencing, may not be self-supporting, let alone supporting their peers. This places severe constraints when graduating from devices on wired networks to those on wireless networks.
2. In wired devices characterized by single point of contact, addressing and accessing a device are simple tasks. Such is not the case in the wireless world (except in the case of broadcasting or nonspecific multicasting), wherein mobility across different radios is a key functionality.

3. Wireless devices pose problems as the connectivity that is required to be continuous is affected by many factors. The fact that the dynamics of wireless networks is inherently more random *vis-à-vis* the dynamics of wired networks, not only causes the preceding problems but also goes on to add a further dimension to these problems.
4. Ensuring information security and service reliability in wireless services poses problems that are much more challenging than wired services.
5. Finally, robustness is a tall order in view of the preceding four issues raised.

Design of audio for real-time collaborative activity is constrained by a large number of participants having to share limited bandwidth. Audioconferencing is sensitive to packet loss, latency and delay jitter, and to gaps in the audio caused by the diminished real-time support on host devices as these affect intelligibility.

Section 4.2 outlines the desirable features of an interactive conferencing system and Section 4.3 describes the constraints in designing the conferencing system on wireless networks. Among these constraints, we delve more on mixing architectures and less on session handling. After providing an insight into the existing architectures of mixing, we present an algorithm using the LN and study its effect on the conferencing system. We conclude the chapter with open issues that need further research.

4.2 Desirable Features of an Interactive Audioconferencing System

An interactive audioconferencing system is desired to be characterized by the following features:

i. *Interactivity*. One of the main features in an interactive conference is to allow multiple speakers at any time without any control on "who can speak," "what can be spoken," and "who can hear what." For instance, gaming and other virtual reality services [13] may allow audio from every participant. Supporting most natural audio communication between participants allows spontaneity in the way people interact using the audio medium. A case in point is a face-to-face conference wherein all sounds from participants are heard at all times. The fact that conferencing is as much about human factors as it is about the underlying technology can never be overemphasized. It is crucial to devise collaborative systems with an understanding of how sociological and psychological factors impact group work, especially because mismatched expectations in group-based systems have resulted in serious groupware failures [14].

ii. *Spatialization.* Mixing of audio streams refers to the weighted addition of several input streams into one output stream. Mixing makes a session interactive and reduces the number of streams. Mixing is crucial to devices that are starved for resources. To customize mixing is to choose the individual weights suitably at different terminals. Spatialization refers to the experience of the listener of the customized mixed stream. Hendrix and Barfield [15] consider the presence and absence of spatialized sound and addition of spatialized and non-spatialized sounds to a stereoscopic display. It is reported that the addition of spatialized sound significantly increases one's sense of presence in the virtual environment.

iii. *Signaling.* Signaling involves the exchange of control packets to establish, maintain, alter, and terminate an audioconference. Signaling must contribute to a reliable, robust, and scalable conferencing system. Signaling is also required to keep the terminals updated of the network status. This in turn helps in choosing across codecs, setting parameters such as sampling time of audio packets, and initiate packet recovery algorithms and time-scaling algorithms.

iv. *Scalability.* Collaborative work demands are rapidly growing across the globe. A scalable conferencing system should adapt to different types of network connections and terminals besides supporting a large number of participants. Expectedly, scalability is much more crucial than ever before.

v. *Availability of multicasting.* In multicasting, a number of processes join to form a group; all processes in the group receive all the data dispatched to that multicast group. This reduces the end-to-end delivery time and network traffic. Multicasting in wireless networks is difficult compared to its wired counterpart. Even majority of the routers in the Internet are not multicast-enabled. A common recipe operating on a dynamically evolving infrastructure must context-sensitively leverage multicasting, if and where available.

vi. *Traffic reduction.* To reduce traffic is to reduce the number of audio streams in the network for a given number of participants. Reducing traffic at the application layer contributes to a more efficient use of network resources. Traffic may be reduced at the network layer by using multicast techniques. A few other techniques are listed as follows:

   a. *Compressing audio streams.* Some compression techniques such as ITU-T G.729, G.726, and G.723 and iLBC coding [16–18] result in low bit rates up to 6 kbps per participant.

   b. *Reducing traffic using voice activity detection (VAD).* Speech inherently has several pauses. Dropping the silence segments results in a reduction in the bandwidth requirement [19,20] of up to 50 percent for a single stream.

c. *Reducing the number of streams by mixing necessary streams.* Some of the audio streams may not be relevant in a conference. The conferencing system must be sensitive to the changes in the participants' speech and intelligently select relevant streams to mix.

vii. *Quality of the conference.* The "interactivity" in an audioconference is achieved when the mouth-to-ear delay is below 150 ms (in fact, it can be even up to 600 ms) with permissible packet loss and jitter. The three factors—delay, packet loss, and jitter—influence the quality of the speech and user satisfaction [21]; they should be kept as low as possible.

Watson and Sasse [22] have shown that speech quality depends not only on the percentage loss of the transmitted packets but also on volume settings, microphone settings, and echo. Context sensitive and dynamic customized mixing of audio is one of the solutions to achieve good spatialization through volume and microphone settings. It enables participants to emphasize or deemphasize specific audio streams as desired.

4.3 Constraints of Interactive Audioconferencing in Wireless Networks

Some of the problems of desirable features are listed here and an overview of the existing implementations that aim to mitigate them is presented. While designing a conferencing system in the context of wireless networks, one should keep in mind that

i. A client or terminal* may not have the capability to decode the multiple streams and mix them.

ii. Clients may join or leave the conference because of ephemeral network connectivity, which results in frequent change in the point of contact due to mobility.

iii. The clients experience higher packet loss and, thus, require higher error correction or redundancy.

The implications of these must be handled judiciously.

* We use the terms client, terminal, and participant interchangeably.

The issues in supporting conferencing can be divided into media- and network-related.

A. Media-related issues
   - Quality of the conference is based on interactivity, that is, participants must be able to freely acquire an opportunity to speak to other participants.
   - Mixing and customized mixing of speech for the participants.
   - Signaling.

B. Network-related issues
   - Large-scale distribution of participants over a wide area (scalability)
   - Enabling multicasting when the necessary infrastructure is available
   - Traffic reduction
   - Quality of the conference dependent on packet delay, packet loss, and delay jitter

These issues are explained briefly in the following sections.

4.3.1 Interactivity

"Floor" is a virtual entity that allows a participant to speak. With some form of floor controller [23–26] explicitly telling the participants when to speak and also reserving or waiting for one's turn to speak would evidently result in a "gagging" feeling for participants [2,3,27]. This means that the system's view of "speaking" is not the same as that of the participants. In addition, it even hinders the natural urge to interrupt. This makes a static and explicit floor control strategy [23,24,28] inappropriate in supporting most natural audio communication. Mixing speech from all participants as soon as they occur facilitates communicating the concerns of listeners to the current speaker, which is equivalent to allocate floor to every participant. Without this capability, collaborative platform would be imposing an unnatural curb on the participants. Thus, the requirement of impromptu speech without explicit permission from any controller (deciding who should speak and when) is necessary to mimic a real-life conference. Mixing all streams is a solution. However, it may not be necessary or, for that matter, even desirable, to transmit all speech streams. Floors that are allocated should be relinquished promptly by the participants. Floor control for clients connected through wireless links is difficult to achieve because the point of contact is ephemeral. The floor control as a technique to handle multiple streams is discussed in Section 4.4.

4.3.2 Customized Mixing

Customized mixing is one of the desirable features discussed in Section 4.2. Discussions on mixing *vis-à-vis* Real-Time Transport Protocol (RTP) can be sought in Ref. 29, and some interesting discussion on this with respect to wireless networking is in Ref. 30. Customized mixing at the clients is usually not possible due to the limited capability of wireless devices. Higher bandwidth requirement for transmitting individual streams to enable customized mixing is another constraint here. With the advent of 60 GHz indoor networking, there is a possibility of supporting this in the near future.

4.3.3 Signaling

Wireless and *ad hoc* networks throw many challenges. The centralized entities are to be minimal because the connectivity to a centralized entity is not guaranteed. Further, the dynamic nature of the nodes introduces further constraints. However, for the sake of out-band signaling in wireless broadband networks, we may assume the existence of infrastructure that can support authorization and provide some initial support.

There are two major signaling protocols for IP multimedia conferencing—ITU-T H.323 [31] and Internet Engineering Task Force's (IETF's) SIP [32]. H.323 does not scale for large number of conferences, although the latest version has some improvements. However, the interest in the industry and academia has now been with the SIP and it has established itself as the major signaling standard. Although SIP is very useful and simple, it has not been able to address multiparty conferencing comprehensively. Some variants of SIP, a draft proposal by Mark/Kelley [33], have been useful to a limited extent. Core SIP [32] offers much less for peer-to-peer (P2P) conferencing solutions. Thus, the protocol or any multiparty conferencing solution should workaround these deficiencies. For detailed discussions, see Refs 32–35. We have proposed here just a plausible solution to the signaling problem without going into its in-depth analysis.

4.3.4 Scalability: Large-Scale Distribution of Participants over a Wide Area

Schooler [36] identifies many issues with respect to scalability, *viz.*, (1) session models and its protocols, (2) multicast address management, (3) techniques for bandwidth reduction, and (4) codification of heterogeneity based on voice quality. Handley et al. [37] highlight session scaling based on one or more of the preceding aspects including security and authentication besides network support and reservations.

The key to managing heterogeneity over unmanaged, large-scale networks lies in providing distributed, rather than centralized, solutions.

Centralized solutions are easy to build and maintain [38]. They are woefully inadequate when the participants are geographically far apart because of poor response for time-sensitive operations. They impose heavier traffic at the servers. Distributed solutions that deal with local traffic are suited to large-scale setting of the Internet but are difficult to implement and manage.

4.3.5 Availability of Multicasting

One of the main advantages of packet networks over circuit-switched networks is the ability of the former to support multicasting/broadcasting. With wireless systems, broadcasting in a smaller domain is relatively easy because all the single hop nodes (devices) can listen to the radio frequency (RF) signal. A single multimedia stream can be distributed to a large number of subscribers. The impact of multicasting when the enabling infrastructure—such as wireless broadband connectivity between the multiple participants on networks—is available can never be underestimated. There are no fixed routers on exclusively *ad hoc* wireless networks. Thus, as the end-to-end routes are nonstatic, implementing multicasting is nontrivial.

Deering's [39] monumental work on multicasting offers an efficient multipoint delivery mechanism. A single packet can be sent to an arbitrary number of receivers by replicating packets only at the forks of the network. Transmissions from one to many are accomplished without packet duplication using a common group address. Multicast is based on the formation of groups whereby a number of processes may join a multicast group to receive all the data dispatched to that group. Wireless network must evolve to support IP multicast. This depends on the device capability and implementation. In fact, availability of multicasting in the Internet itself cannot be taken for granted. Thus, if and wherever available, the conferencing architecture must make use of multicasting. However, even if not available, the architecture should support conferences. The solution for operating on a dynamically evolving infrastructure is to context-sensitively leverage multicasting, even as it must not be contingent on its availability. In a wireless domain, many routing protocols do not automatically support multicasting. Multicasting support needs to be explicitly implemented so that one can utilize it. For more details on multicasting in wireless network, see Ref. 40.

4.3.6 Traffic Reduction

Reduction in traffic can be achieved as discussed in Section 4.2. Audio streams consume more bandwidth, which can be reduced by mixing. Depending on the nature of the processor and its speed, wireless devices may implement some or all of the techniques described earlier. One of the challenges is to implement high compression codecs on these wireless and handheld devices with limited capabilities. We revisit this aspect in our discussions on architecture as well as on mixers.

4.3.7 Quality of the Conference Based on Packet Delay, Packet Loss Percentage, and Delay Jitter

Delay of audio packets is due to transmission and queuing at the routers or intermediate nodes in the network. Variable queue sizes seen by audio packets at intermediate routers introduce delay jitter. In a packet speech system, the end-to-end delay (after nullifying the jitter) is always a critical parameter in a real-time voice system. It should, hence, be kept well below 600 ms [41] in the absence of echoes, if conversation patterns are not to break down. Packets are dropped at congested routers whenever the queue overflows. Transportation error is higher in wireless networks compared to their wired counterparts. The extent of packet loss is a primary factor determining whether a network audio stream will be intelligible to the user, and therefore, of any use at all. Delay and jitter play a secondary role and should also be kept under check.

The present-day networks, wired or wireless, do not support these service models. Real-time traffic has to compete for bandwidth with other non-real-time traffic on the best effort network such as IP. In such networks, there can be no assurance about the quality of service when physical limitations are present; packet loss can be kept under tolerable limits only through adaptation. Adaptation to delay jitter and loss is reported in Ref. 42.

Today, as the conference involves a large number of participants, there is a need to efficiently handle the traffic as well as enable free access to the shared resources for the participants. The desirable features also need to be addressed while efficiently handling the multiple streams. The different techniques are discussed in detail in Section 4.4, hinting at the possible architectures for conferencing.

4.4 Techniques for Handling Multiple Audio Streams

In a conference, speech streams of all participants are often not concurrently permitted. If allowed indiscriminately, quality of play out usually declines. This is because there is an upper bound on the number of distinct streams that a given network can handle. In the context of multiparty audioconferencing on wireless networks, the severity of this bottleneck is pronounced. Independently, human comprehension with multiple streams becomes poor due to sharply declining resolution. Some attempts to address such issues are reported in the sequel.

a. *Floor*. Floor control, or turn-taking mechanism, provides a means to mediate access to shared work items. Greenberg [43] recommends that systems should "support a broad range of floor control policies" to suit the needs of the participants. Floor control can be important in many situations, such as shared screens allowing only serial interaction, or systems following strict interaction models similar to a teacher monitoring/controlling the work of students. Roseman and Greenberg tackle many of these aspects on GROUPKIT building. For collaborative environments, several types of floor control policies such as explicit release, free floor, round robin, first-in-first-out, preemptive, and central moderator [43] are available. However, these methods are beset with difficulties in supporting impromptu communication. In a setup with floor control, each subsystem must decide the level of support to simultaneity (i.e., number of active participants at any time) and the granularity to enforce access control. In its simplest form, floor control enables floor access to only one participant at any given time and the floor is handed over when a request is incident. In the case of audio, floor control introduces a management framework around the audio session that enforces turn taking, thereby removing any potential simultaneity. Consequently, the ambience of the application becomes suffocating or gagging for the users. This can happen even if there are more floors because the person who wants to speak may not have a floor. Although explicit floor control may be suitable for some applications such as a broadcast by a panel, it is inherently difficult to implement and maintain for a system with many-to-many active participants. When large-scale groups are to be allowed, implementation of these techniques is cumbersome. Making a policy for floor allocation without human intervention is not simple in a large conference where all members can request and be granted the floor.

b. *Push-to-talk.* To talk, the pointer has to be placed in the window and the mouse button clicked (like a walkie-talkie). This mechanism is known as "push-to-talk" [44]. Only after this explicit action, the participant is allowed to speak to others. This is the default approach used by many of the MBone-based tools such as "vat" and remote access Trojan/tools (RAT). These tools enable every participant to hear everybody else in the conference simultaneously. This allows the users to make very careful choices whether to speak or not, avoiding significant amounts of simultaneous speaking in many more restrained contexts. Nonetheless, it is liable to remove subtle nonverbal cues and sounds. Further, conversation and interaction becomes slower and unnatural due to a conscious turn-taking activity that is required to be heard. This reduces the spontaneity and interactivity of the conversations. When many participants speak simultaneously, the network gets flooded suddenly, thereby causing disruptions due to packet loss. Silence suppression is another form of handling simultaneous speakers.

c. *Audio processing unit (APU).* Yu et al. [45] propose an APU for conferencing. It is a hardware implementation of multipoint control unit (MCU) of H.323. MCU is used for multipoint audioconferencing–videoconferencing systems. This is a centralized unit, which takes a fixed number of audio sources from participants and distributes to each of them a specific mix of selected streams excluding their own to avoid echo. It, therefore, performs total mixing on a dynamic set of input streams. The APU selects four simultaneous speakers on the basis of individual sample energy. Once selected, a participant holds the floor for a certain minimum interval to circumvent frequent changes, thus hindering interactivity. It is implemented using a DSP chip. Its scalability is very limited.

Different techniques allow natural audio communication between participants to different degrees. More specifically, these techniques have different ways of determining as to which speakers are heard in the conference. All the techniques described earlier limit the number of simultaneous audio streams transmitted in the network.

4.5 Mixing Architectures: State-of-the-Art

As mentioned earlier, mixing is an inevitable process in audioconferencing and can reduce the total bandwidth. As a major task, it has attracted different architectures to support audioconferencing. A few important ones are discussed in the following sections.

4.5.1 Centralized Mixing Architecture

In the simplest centralized mixing approach [46], a single mixing server handles all the streams from the participants of the conference (Figure 4.2). Each participant sends his media stream to the mixer. The mixer then adds all the streams and sends it back to the participants after subtracting their own streams to avoid far end echo. If there are some participants who are listen-only, they would be getting the complete mix. When end terminals are not capable of handling many streams, introducing a server for mixing reduces the load on the end terminals as also the number of audio streams in the network.

The advantages. Reduced media streams in the network, at least on the client/participant side only one stream is sent and received. The

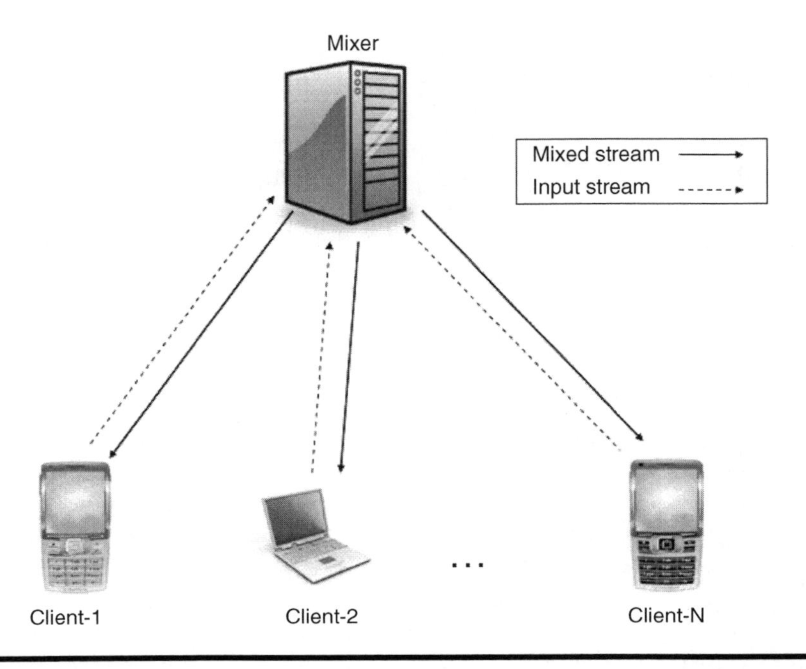

Figure 4.2 Centralized audio mixing.

latency is lesser. Because all the streams meet at a point, synchronization is inherently assured.

The disadvantages. Centralization makes it unsuitable for conferencing in *ad hoc* networks. For *ad hoc* wireless networks it is not good enough because the basic assumption of the availability of a server is not valid. Centralized mixer presents a potential bottleneck in the network. Resources are not used optimally because all the computational and network load is dumped on the server. Network growth depends on the mixer's capabilities. Hence, this architecture is not scalable even with the availability of infrastructure connectivity, although it is implementable in any network.

4.5.2 Endpoint Mixing Architecture

This is a mesh architecture where all the media streams are seen by all the participants. It is P2P and easy to implement [47] (Figure 4.3). Packets are mutually exchanged between all clients. Mixing is done by every client, thus allowing personalized mixing for participants. This can be tailored to the individual's equipment, spatialized according to specific location within a shared virtual space, and under individual control. The number of packets in the network would be enormous when there are too many participants without any control. Even with multicast, it is very demanding on network

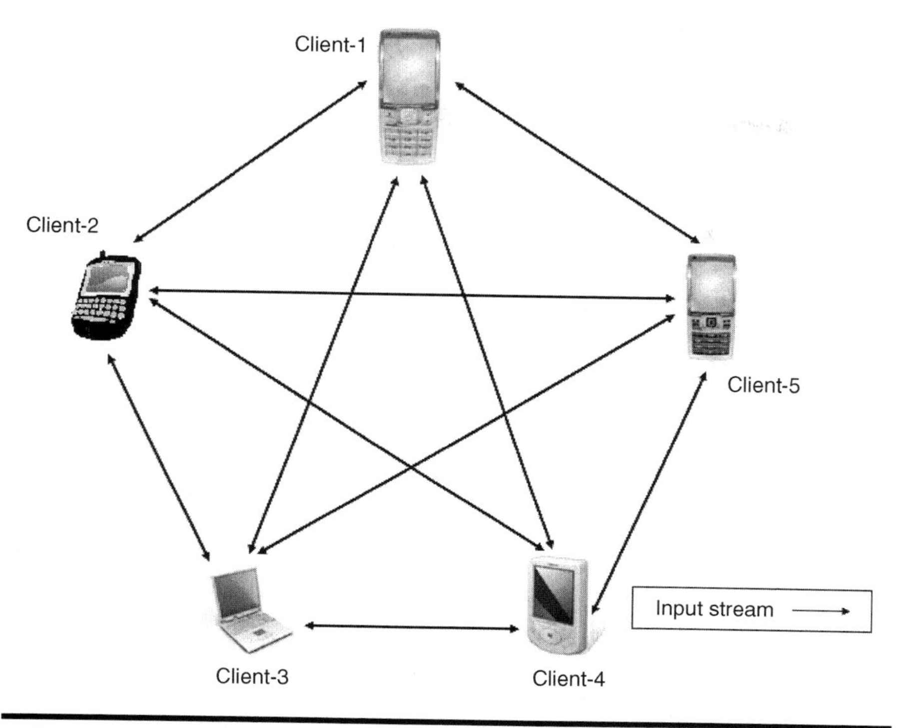

Figure 4.3 Endpoint audio mixing.

resources, particularly bandwidth, apart from computational resources at the end terminals.

The advantages. The architecture is decentralized. Streams are delivered rapidly as there are no intermediate mixers in the end-to-end paths. This also makes synchronization easy. Self-organization is enabled because each participant can mix at its premises and it is faster as there are no intermediate servers.

The disadvantages. Because mixing is done at each client, the duplication of efforts does not lead this architecture to be scalable. Nodes with limited capabilities will not be able to join this sort of conference. Central processing units (CPUs), I/O devices, storage capabilities, codec (dedicated hardware and software), communication protocol, and network interfaces of the participants place limits on the end-system capabilities to process, consume, and generate multimedia data. For example, limited resources in terminals can result in buffer overflow, delay in processing data, and inability to process data. These manifest to the user as unacceptable play out delays, lost audio segments, and poor user interaction. The number of media streams in the network is large, thus overloading the whole network. All the nodes are stressed to the same extent blindly. In fact, the heterogeneous nature of the wireless devices does not make this architecture useful. This architecture may not work for an *ad hoc* network of end terminals. Thus no assumptions regarding the capability of end terminals are admissible except a few minimal requirements.

4.5.3 Hierarchical Mixing

To support applications involving a large number of simultaneously active audio streams, Rangan et al. [48–50] propose a hierarchical mixing architecture. In such applications, all streams are mixed. They compare a distributed hierarchical mixing architecture with centralized mixing. They exploit the comparative durations required for transporting a packet from an end terminal to a mixer and for mixing two streams at a mixer. Depending on the ratio of these two times, they propose different architectures for media mixing. In this mixing hierarchy, participants constitute leaf nodes and the mixers—nonleaf nodes (Figure 4.4). The mixer at the root of the hierarchy forwards the final mixed packet to each of the leaf nodes. This can also be

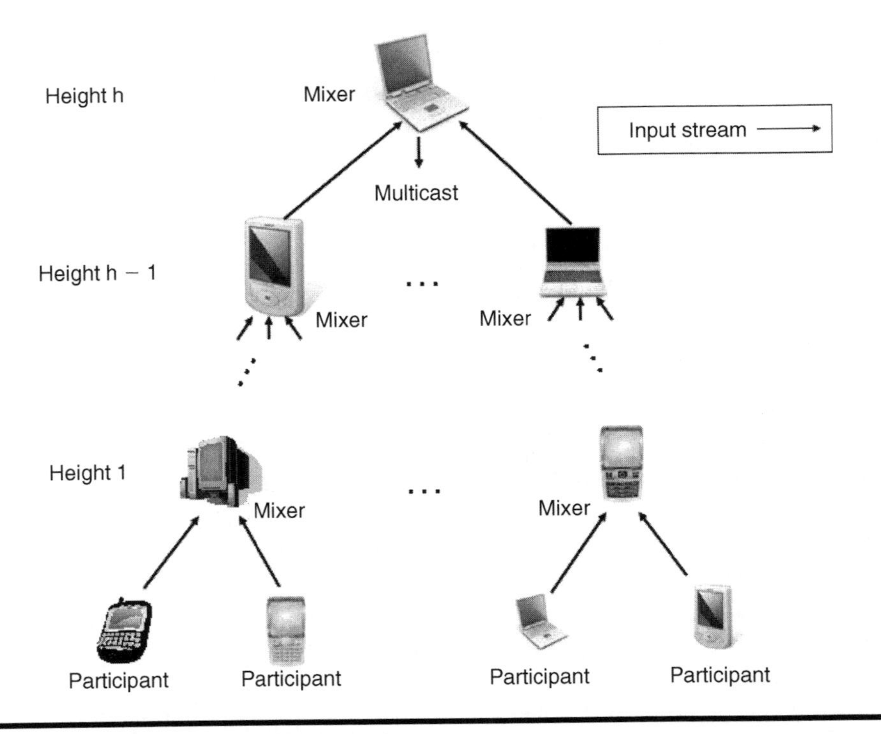

Figure 4.4 Hierarchical audio mixing.

done by multicasting the packet from the root, if supported, to avoid delay during hops. Alternatively, packets can be forwarded through intermediate mixers (application-level multicasting [ALM] in some respect).

The advantages. This tree-based architecture is highly scalable compared to a centralized mixing and completely distributed endpoint mixing. The authors show that it is an order of magnitude more scalable than purely centralized or distributed architectures. The root node mixes lesser number of streams as the intermediate nodes mix streams from their children nodes.

The disadvantages. This approach does not permit the specialization because all streams are mixed at all times as in centralized mixing. The number of hops increases with the height of the tree, albeit slowly. Thus, it is not well suited for interactive and collaborative applications. Mixers are preconfigured and static, and do not adapt to the network conditions. The root node is analogous to a centralized mixer as the final mixing before distribution takes place at the root. If the participants are segregated, then synchronizing their streams, not withstanding the packet loss and jitter, is difficult.

4.5.4 Distributed Partial Mixing

Radenkovic et al. [2,4,27] identify the importance of permitting speech streams from many participants. They consider situations where a large group of people "speak" at the same time. Each participant independently introduces a new audio stream that has to be accommodated by the network and has to be processed by each recipient. Multicasting if used reduces the bandwidth required; however, it puts more pressure on the end terminals in terms of bandwidth and computation. Thus, distributed partial mixing (DPM) reduces traffic on shared links by mixing some streams. An example is shown in Figure 4.5.

Partial mixers (PM) may be dedicated servers or additional roles played by other end terminals. End terminals typically communicate through these PM components, sending their audio to and receiving other participants' audio from a nearby PM. PMs in turn form a completely connected network (e.g., a tree) to achieve end-to-end distribution. PM extends the traditional concept of mixing to render it more dynamic and flexible. Unlike total mixing [2,27], where the entire set of received streams is mixed into a single output stream, partial mixing dynamically chooses only a subset of the available audio streams to mix at a time and forwards it with other unmixed streams. The number of streams mixed in Ref. 29 varies dynamically depending on the number of active participants. Selection of the number of streams for mixing is based on the network conditions. Thus, instead of producing a single output stream in all cases, partial mixing produces

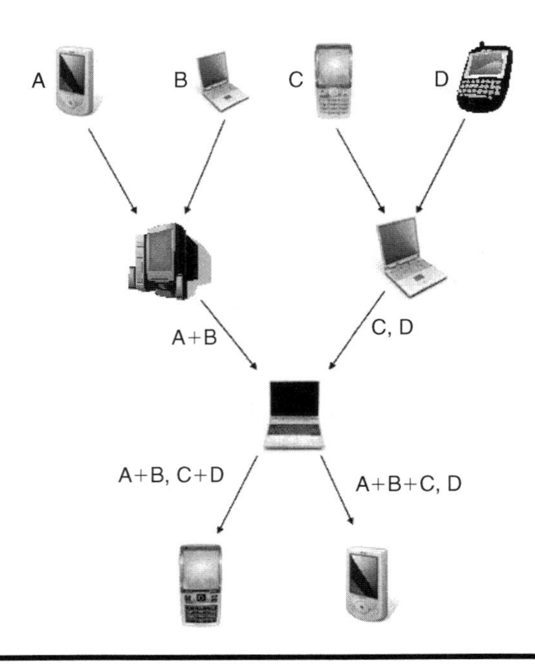

Figure 4.5 Distributed partial mixing.

various number of streams in different situations. Hence, the number of streams in the network would be at least that of hierarchical mixing.

The advantages. It is decentralized and highly scalable. It tries to adapt to the network conditions.

The disadvantages. It introduces many mixers, considerably increasing latency. It leads to fluctuations in the volume of every individual participant causing severe degradation in quality. Customized mixing of streams is not possible when many clients are active. Synchronization is difficult to achieve.

4.5.5 Distributed Mixing System

Khedher et al. [30] propose a modified hierarchical mixing architecture wherein the number of levels is limited to two. Nodes in the first level are mixer nodes (representing cluster heads) and exchange streams among them using multicasting. The other participants in the cluster register to these mixers to avail the facility of mixed streams. Each mixer receives mixed streams from the other mixer nodes in the overlay. It mixes these streams with those received from second-level nodes registered to it. This stream is transmitted to the second-level nodes and to other mixers. In essence, streams from all the nodes are mixed before play out. The idea of electing mixer nodes dynamically based on the availability of resources in the nodes is also proposed. Authors extend the clustering concept of the wireless *ad hoc* networks to form clusters and elect at least one mixer in a cluster.

The advantages. The architecture is a complete-mesh topology with only two levels of mixing; thus the delay is kept under check. Bandwidth reduction is possible because all the streams are mixed. Synchronization can be achieved by adapting the play out so that streams originating at the same time are played out at the same time. The distributed mixing architecture significantly reduces the number of media streams in the network.

The disadvantages. Users lose spatialism. Limitations of mixing many streams apply here. The limitations of mixing streams are discussed in the following section.

4.5.5.1 Advantages and Disadvantages of Mixing

The pros and cons of mixing architectures are summarized.

4.5.5.1.1 Benefits of Any Mixing-Based Approaches

1. Mixing can drastically reduce network traffic for many simultaneous speakers. This makes it efficient with its support for efficient distribution of audio streams in the network. See Table 4.1, where mixing at an intermediate server can reduce the number of streams drastically.
2. Mixing imposes no constraints on individual speakers unlike gagging as in floor control.
3. The system's view of "speaking" is the same as that of the user; naturally, mixing is more effective for supporting the most natural audio communication.

Table 4.1 Order of Traffic in Different Conference Architectures (with and without Multicasting)

Endpoint	*Centralized*	*Hierarchical*	*Cluster (c) Based [30]*	*Domains (d) with N_{max}*
Unicast Communication				
$M(M-1)$	$M-M$	$2(M-1)$	$c(c-1)$	$N_{max}d(d-1)$
Multicast Communication				
M	$M+1$	M	c	$N_{max}d$

Note: For cluster-based and our proposal of domain-based scenario, we have not considered the streams exchanged within the cluster/domain because we think it is completely localized to a cluster.

4.5.5.1.2 Limitations of Mixing

Mixing of audio streams digitally involves summing up all streams at each sampling instant with some weight for each stream. The weights sum up to unity to avoid signal saturation. Mixing has the following problems [2,27]:

1. The mixed signal has a lower signal-to-noise ratio (SNR) and a lower dynamic range than individual streams.
2. Fidelity is poor as it includes noise from all the separate streams in fewer bits.
3. Mixing may not work well with some ultralow bandwidth codecs due to their compression algorithms.
4. Mixing in stages introduces additional delay as audio streams have to be processed by the mixers.
5. Spatialization and other aspects of control of individual listener over each audio stream are lost.
6. While mixing digitally, the volume level of individual streams is reduced to avoid saturation.
7. Mixing is irreversible and the individual streams cannot be recovered at a later stage.
8. Mixing introduces more components (servers) and increases the complexity in the network with the corresponding hardware and real-time control requirements.

These limitations expose the need for a new scalable architecture either with no mixers or with less intermediate mixers so as to keep a check on the bandwidth usage. Of course, user experience is the most important aspect of the architecture.

4.6 The Proposed Architecture

We propose an architecture drawing lessons from the detailed study done hitherto.

4.6.1 Design Requirements

- *Audio mixing*. In an audioconference, streams from all the clients need not be mixed. Actually, mixing many arbitrary streams [29] from clients degrades the quality of the conference due to the reduction in the volume (spatial aspect of speech). There is a threshold on the number of simultaneous speakers above which increasing the number of speakers becomes counterproductive to conference quality. Fixing the maximum number of simultaneous speakers is dealt in Refs 51 and 52 using ethnomethodology, and is conjectured to be three. Thus, it is advisable to honor that constraint. Therefore, the proposed architecture should not mix more than three streams.

- *Less mixers.* There must not be many intermediate mixers (similar to CSs as in Ref. 53) in stages as in Ref. 49 because it brings in inordinate delays by increasing the number of hops and is not scalable with interactivity in focus. In Ref. 2, to allow impromptu speech, mixing is not done when the network can afford high bandwidth requirements for sending/receiving all the streams, but it is unnecessary [52].
- *Floor control.* Floor control for an audioconference (even video-conference) with explicit turn-taking instructions to participants renders the conference essentially a one-speaker-at-a-time affair, not a live and free-to-interrupt one. This way, the conference becomes markedly artificial and its quality degrades. Schulzrinne et al. [29] assume that only one participant speaks at a time. In this case, if applications are implemented with some control [24], the service becomes "gagging" for the users. Generally, floor control and, in particular, explicit floor control needs to be avoided.
- *Scalability.* For large conferences [53,54], a centralized conferencing solution cannot scale up. With multicasting, clients will have to parse many streams, and traffic on a client's network increases unnecessarily. Therefore, distributed architecture is a necessity; however, mixing should be done just before the play out.

4.6.2 Description

Two issues must be taken care of when building a Voice-over-IP (VoIP) conferencing system: (i) the front end, consisting of the application program running on the end-users' computers and (ii) the back end that provides other application programs that facilitate conferencing and the conference. The participating users are grouped into several "domains." These domains are local area networks (LANs), typically, corporate or academia networks. This distributed assumption demands distributed controlling and distributed media handling solutions, qualifying it to support large conferences. More explicitly, in each domain, we can identify several relevant logical components of a conferencing facility (Figure 4.6).

- An arbitrary number of end users (clients) take part in at most one audioconference at a time. Every user is in only one domain at a given instant, but can switch domains (nomadism). In our conferencing environment, these clients are regular SIP user agents (SIP UAs), as defined in Ref. 32 so as to gain in interoperability with other existing SIP-compatible systems. These clients are, thus, not aware of the complex setting of the backbone servers enabling the conference.

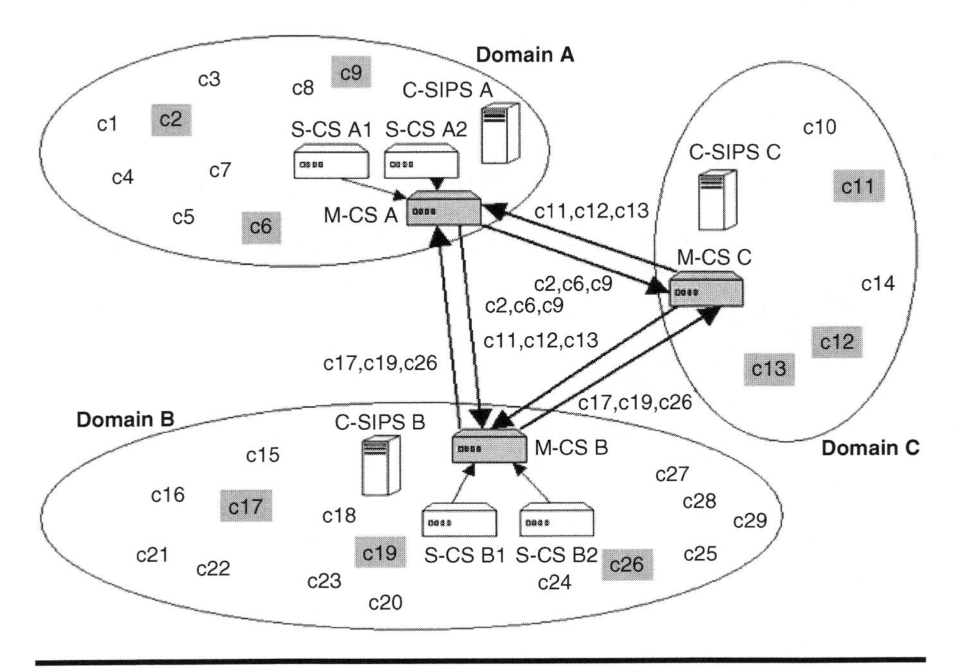

Figure 4.6 Example of a two-level hierarchy of CSs; the shaded clients are the one selected by the M-CS and will be sent to other domains' CSs.

- One SIP server (SIPS) per domain* is set up to take care of the signaling aspects of the conference (clients joining, leaving, etc.) [55]. In particular, it is considered as a physical implementation encompassing different logical roles, namely, a SIP proxy server, a SIP registrar server, a SIP redirect server, and a SIP B2BUA (back-to-back UA) [32]. This physical implementation enables the handling of incoming/outgoing SIP messages by one or another logical entity according to the needs. SIPS is entrusted with maintaining the total service for the following advantages: (a) it works as a centralized entity that can keep track of the activities of the UAs in a conference; (b) it does all the switching for providing private bank exchanges (PBX) features; (c) it locates the UAs and invites them for a conference; and (d) it does the authentication and billing, if required. SIPSs in different domains communicate with one another using standard SIP messages as described in Ref. 55. If the load on a particular SIPS

* Domains in the context of wireless networks may also mean "clusters." It may have a cluster head, which can act as SIPS or CSs. However, formation of clusters is out of the scope of this chapter [60]. We use the term domain to differentiate between the approach in Ref. 30, which is referred to as cluster based (see also Table 4.1), and ours.

increases beyond its capability, it can create another SIPS in the same cluster/domain to share the load.

- One master CS (M-CS) (simply a CS) for each conference is created by the local SIPS when a conference starts. This server will be used for handling media packets for the clients in its domain. The M-CS can create a hierarchy of CSs inside a domain by adding one or more slave CSs (S-CSs) to accommodate all the active clients and prevent its own flooding at the same time. Its mechanism is described in Section 4.6.3.

The entities described here are exhaustive and conform to the SIP philosophy. Thus, the use of SIP makes this architecture more useful and interoperable with any other SIP clients or servers.

4.6.3 Selection of Streams

Similar to SipConf in Ref. 56, a CS [5] has the function of supporting the conference; it is responsible for handling audio streams using RTP. It can also double to convert audio stream formats for a given client if necessary and can work as translators/mixers of RTP specification behind firewalls [29]. The design of CS is similar to that in H.323 Multipoint Processor (MP) [31]. In brief, the CS receives audio streams from the endpoints and processes them and returns them to the endpoints. An MP that processes audio prepares N_{max} audio outputs from M input streams after selection, mixing, or both. Audio mixing requires decoding the input audio to linear signals (pulse code modulation [PCM] or analog), performing a linear combination of the signals and reencoding the result in an appropriate audio format. The MP may eliminate or attenuate some of the input signals to reduce noise and unwanted components.

The limitation of H.323 is that it does not address the scalability of a conference. The architecture proposes a cascaded or daisy chain topology [53], which cannot scale up to a large conference. A CS serves many clients in the same conference. Multiple CSs may coexist in a domain when there are several conferences under way. Signaling-related messages of CSs are dealt in Ref. 6.

The working of a CS is illustrated in Figure 4.7. Without loss of generality we select CS-1. For each mixing interval, CS-1 chooses the "best" N_{max} audio packets out of the M_1 (using a criterion termed "LN," described in Section 4.6.4). It may possibly receive and send packets to these CSs: CS-2 to CS-P. The set of packets sent is denoted by "ToOtherCSs." In the same mixing interval, it also receives the best N_{max} audio packets (out of possibly M_2) from CS-2, similarly the best N_{max} (out of possibly M_p) from CS-P. For simplicity, we ignore propagation delay between CSs, which indeed can be taken into account; it is beyond the scope of this chapter. The set of

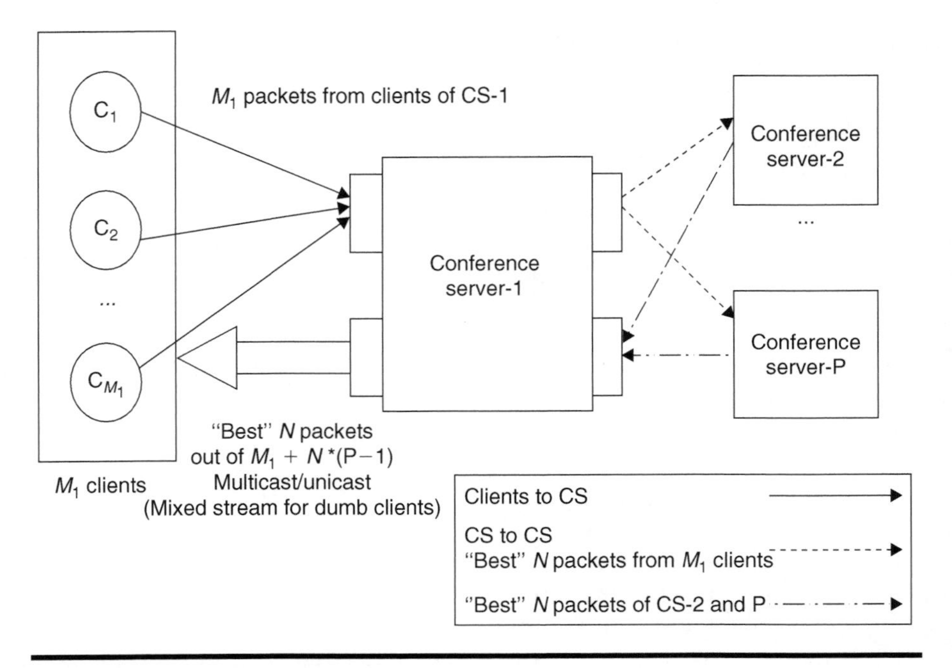

Figure 4.7 Schematic diagram of a CS.

packets received is denoted as "FromOtherCSs." Finally, it selects the best N_{max} packets from the set {ToOtherCSs ∪ FromOtherCSs} and passes these packets to its own group.

It can be seen that the set {ToOtherCSs ∪ FromOtherCSs} is the same at all CSs. This ensures that any client in the conference finally receives the same set of packets for mixing. Hence, all clients obtain a common view of the conference.

Similarly, for each time slot (packet time), a subset, F, of all clients is selected (using the same criterion) from the pool of packets from all other CSs plus the N_{max} clients selected locally. Their packets are mixed and played out at the clients. According to Refs 51 and 52, the cardinality of F, $|F|$, is N_{max} and is fixed at "three." The total number of streams in the network is slightly higher compared to distributed mixer architecture case (see Section 4.5.5). It is only N_{max} times that of a simple cluster-based approach of Ref. 30 (see Table 4.1); however, LN (described later in Section 4.6.4) can be used effectively to reduce the traffic to just N_{max} as explained in Section 4.6.7.

There are cases wherein the processing capacity of an M-CS is exceeded due to an excess of packets—from local and remote domains—to process. In this case, the M-CS will create one or more S-CS (Figure 4.6) and transfer some of its own clients as well as the new clients to it. This implies that a maximum of two intermediate entities exist for each audio packet instead of two in the conventional setup. As the extra hop happens inside the LAN (LANs usually have a high-speed connectivity), it should not prevent us from using this hierarchy of CSs when there is a need to do so. In this configuration, the algorithm outlined earlier will be slightly modified—the audio packets will go from clients to their dedicated S-CS that will select $N_{\max}$ packets to send to the local M-CS, which will then select $N_{\max}$ packets from all its S-CSs in the domain before sending them to the remote domains. The incoming packets from other domains will be received by the M-CS, which selects $N_{\max}$ of them and sends them directly to the domain clients, bypassing the S-CSs.

4.6.4 Loudness Number

A basic question to be answered by the CSs is, how in a mixing interval it should choose $N_{\max}$ packets out of the M it might possibly receive. One solution is to rank the M packets according to their speech energies (this energy should not be confused with the signal strength), and choose the top $N_{\max}$. However, this is usually found to be inadequate because random fluctuations in packet energies can lead to poor audio quality. This indicates the need for a metric different from mere individual energies. The metric should have the following characteristics [7]:

- A speaker (floor occupant) should not be cut off by a spike in the speech energy of another speaker. This implies that a speaker's speech history should be given some weight. This is often referred to as "persistence" or "hangover."
- A participant who wants to interrupt a speaker will have to (i) speak loudly and (ii) keep trying for a little while. In a face-to-face conference, body language often indicates the intent to interrupt. But in a blind conference under discussion, a participant's intention to interrupt can be conveyed effectively through LN.

A floor control mechanism empowered to cut off a speaker forcefully must be ensured to avoid the occupation of floors by one client. Otherwise, in a well-behaved conference, requirements are met by LN [7], which changes smoothly with time so that the selection (addition and deletion) of clients is graceful.

LN (λ) is a function of the amplitude of the current audio stream plus the activity and amplitude over a specific window in the past. It is updated on a packet-by-packet basis. The basic parameter used here is packet amplitude, which is calculated as root mean square (rms) of the energies in audio samples of a packet, and denoted by X_K. Three windows are defined as shown in Figure 4.8.

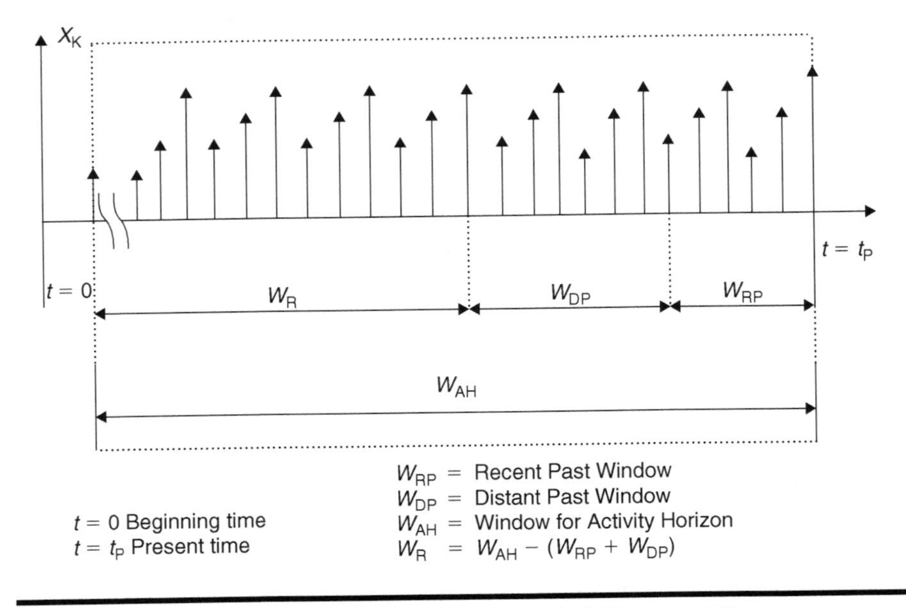

Figure 4.8 Illustration of various windows used in LN computation.

The present amplitude level of the speaker is found by calculating the moving average of packet amplitude (X_K) within a window called Recent Past Window starting from the present instant to some past time. The past activity of the speaker is found by calculating the moving average of the packet amplitude (X_K) within a window called Distant Past Window, which starts at the point where the Recent Past window ends and stretches back in the past for a predefined interval. The activity of the speaker in the past is found with a window called Activity Horizon, which spans the recent past window as well as the distant past window and beyond if necessary. Although the contribution of the activity horizon looks similar to the contribution of the recent past and distant past windows, past activity is computed from activity horizon window differently.

Define the quantities during these three intervals as L_1, L_2, and L_3. L_1 quantifies the Recent Past speech activity, L_2 the Distant Past speech activity, and L_3 gives a number corresponding to the speech activity in the Activity Horizon window quantifying the activity of the speaker in the past few intervals. L_3 yields a quantity that is proportional to the fraction of packets having energies above a predefined threshold (Equation 4.3). The threshold is invariant across clients.

$$L_1 = \frac{1}{W_{\mathrm{RP}}} \sum_{K=t_{\mathrm{P}}}^{t_{\mathrm{P}}-W_{\mathrm{RP}}+1} X_K \tag{4.1}$$

$$L_2 = \frac{1}{W_{DP}} \sum_{K=t_P-W_{RP}}^{t_P-W_{RP}-W_{DP}+1} X_K \tag{4.2}$$

$$L_3 = \frac{1}{W_{AH}} \sum_{K=t_P}^{t_P-W_{AH}+1} \Theta I_{\{X_K \geq \theta\}} \tag{4.3}$$

where

$$I_{\{X_K \geq \theta\}} = \begin{cases} 1 & \text{if } X_K \geq \theta \\ 0 & \text{otherwise} \end{cases}$$

The threshold θ is a constant. θ is set at 10–20 percent of the amplitude of the voice samples of a packet in our implementation here. LN λ for the present time instant (or the present packet) is calculated as

$$\lambda = \alpha_1 L_1 + \alpha_2 L_2 + \alpha_3 L_3 \tag{4.4}$$

Here, α_1, α_2, and α_3 are chosen such that

$$0 < \alpha_1, \quad \alpha_2 < 1, 0 < \alpha_1 + \alpha_2 < 1, \quad \text{and} \quad \alpha_3 = 1 - (\alpha_1 + \alpha_2) \tag{4.5}$$

Here, α_1 is the weight given to the recent past speech, α_2 the weight given to distant past speech, and α_3 the weight given to speech activity in the activity horizon window considered.

4.6.4.1 *Safety, Liveness, and Fairness*

The λ parameter has some memory depending on the spread of the windows. After one conferee becomes silent, another can take the floor. Also, as there is more than one channel, interruption is enabled. A loud conferee is more likely to be heard because of elevated λ. This ensures fairness to all participants. After all, even in a face-to-face conference, a more vocal speaker grabs special attention. All these desirable characteristics are embedded into the LN. A comprehensive discussion on the selection of the various parameters and the dynamics of LN are beyond the scope of this chapter.

4.6.5 Selection Algorithm Using the Loudness Number

Following the developments in Sections 4.6.3 and 4.6.4, we present the simple algorithm that runs at each M-CS (Algorithm 1). This algorithm is based on the discussions in Section 4.6.3. The globally unique set F is found using this procedure. The mechanism proposed here is also depicted in Figure 4.6, where a single conference takes place between three domains.

The shaded clients are the ones selected in their local domains; their audio streams will be sent to other CSs.

ALGORITHM 4.1: The Selection Algorithm at CSs
Input: *Streams from all Clients*
Output: *Selected* N_{max} *streams based on LN*
for each Time *Slot* ***do***

1. *Get the packets from all the Clients;*
2. *Find at most* N_{max} *Clients that have maximum* λ *out of* M *streams out of the received streams;*
3. *Store a copy of packets from those* N_{max} *Clients in database DB1;*
4. *Send these* N_{max} *packets to other M-CSs (on Unicast or Multicast, depending on the configuration and availability);*
5. *Similarly, receive packets from all other M-CSs and store them in database DB2;*
6. *Now compare the packets in DB1 and DB2 on the basis of* λ *and select a maximum of* N_{max} *amongst them (to form set* F*) that should be played out at each Client;*
7. *Send the* N_{max} *packets in set* F *to the Clients in its domain;*
8. *Mix these* N_{max} *audio packets in set* F *after linearizing and send it to dumb Clients who can't mix in the domain;*

end

4.6.6 Self-Organization of the Entities

This architecture is an improved version of the one proposed in Ref. 30. Similar to the clients self-organizing in Ref. 30, SIPS could give a list of CSs to the clients initially when a client joins the conference. Clients may also learn by listening to the transmissions about the CSs in the vicinity. Signaling with the use of standard SIP signaling is simpler compared to the media handling. We present a cursory discussion here on some of the maintenance and self-organizational issues mainly from the point of view of wireless networking.

First, although we showed SIPS and CSs as two different entities, we make the amendments now that SIPS and CS can be combined together to bifurcate the signaling and media handling components of the conferencing. In the following discussions, for the sake of simplicity we assume both the servers as one entity and we simply refer to them as servers handling all the conference-support activity.

Signaling is needed to send conference invitation and maintenance; the client that wants to start the conference should also act as a server in an *ad hoc* wireless network and send invitations or advertisements periodically to contact potential participants. However, in a broadband network one can set up dedicated servers. Particularly, for complete *ad hoc* networks without infrastructure support we refer to Ref. 30 for setting up of the self-organizing conferencing mixers, which can also take care of the signaling. However, this sort of conference maintenance is limited to small and *ad hoc* structures. For a large conference with clients distributed all over we proceed to allocate some fixed servers in the infrastructure. In fact, the clients themselves may become servers if the resources permit or in the initial stages when the invitations are sent. We propose to use a Service Discovery Protocol (SDP) depending on the scenario (a survey as in Ref. 57) at the clients, although servers advertise the conference services offered. In some scenarios, individual invitations may be sent to all the participants. Once the service information is available, clients can join the servers that are near to them from the list of available servers. For each new client that joins the conference subsequently, they would also get the indication of how much load a particular server is handling which are in the vicinity. This is enabled because the servers talk to one another for exchanging media packets. This helps the clients to decide to whom to connect depending on aspects such as number of hops. Because the servers support limited number of streams, we can also have an S-CS that can take some load off the M-CS. An S-CS can be invoked in the clients as and when required.

The observed improvement in the perceived quality of the conference service is due to the following reasons. (1) Limiting the number of concurrent speakers to a low number such as three. Generally, in a conference if more than two participants speak, intelligibility is lost. Conversational analysis demonstrates that there would be a repair mechanism [51] in such a case. (2) Audio stream between any two clients passes through at most two CSs, thus, reducing the end-to-end delay. (3) As the streams are mixed only at the clients, or at the CS just before sending it to the participants (in case of participant's device having less resources), a customized mix of streams can be achieved. Individual tuning of mixing with different weights guarantees preservation of spatialism. Echo produced when a self-stream is selected can be avoided by reducing the weight. Nonetheless, feedback helps in reassuring a speaker that he or she is heard by all.

4.6.7 Reducing Bandwidth Consumption

The major problem identified in many of the mixing architectures is that when more than one stream is mixed, intelligibility of the speech reduces drastically. Thus, we select only N_{max} streams to be mixed at the end.

However, to reduce bandwidth consumption, one can use VAD at the clients [19,20]. Thus, many streams that are unnecessarily sent can be stopped because in a conference, by and large, only one participant would be speaking at a time. Although this architecture reduces the bandwidth on the links, it is slightly higher than distributed mixing (Section 4.5.5) as shown in Table 4.1.

We can use the characteristics of LN to avoid exchanging many streams between the CSs. As mentioned earlier in Section 4.6.3, each CS selects the best N_{max} from its domain and sends it to others. However, at every time slot it finds the best N_{max} considering the packets from all the CSs. Because it knows the best N_{max} streams that are selected in the previous time slot, the streams that are selected from its domain, the next time slot should have the LNs equal to or greater than the ones in the previous time slot. Thus, many of the CSs may not forward streams to other CSs. In a steady state, the number of streams exchanged across the CSs will be approximately N_{max} only. Thus, this architecture with LN saves significant bandwidth. One can note here that time slots are of 20–60 ms and thus, an improper selection due to time delay may last only for one or two time slots. See Ref. 5 for an in-depth analysis.

4.7 Open Issues

There are several open issues with respect to multimedia conferencing on wireless networks. In fact, many of the issues are prevalent in wired networks too. We list here some of the issues needing attention.

1. *Selection of CS (allocation problem).* We assumed that the CSs can be found by a client, or allocated by a SIPS. We also made an assumption that clients would find the nearest CS based on some criterion, for example, the number of hops. However, none of these aspects were addressed in this chapter. Optimal allocation of clients to servers is a generic facility locator problem. It is found to be an NP-hard problem. However, depending on the particular case, some constraints can be relaxed and an approximate solution found. There exist some heuristic solutions for such problems [58]. There are many avenues for approximation and randomized algorithms tuned to this application. Many of the solutions proposed for this class of problems may not work here because the capacity of the devices to serve others is limited and the mobility-induced constraints are not easy to address.
2. *Mobility.* Distributed and dynamic compositions of wireless networks also pose the problem of ephemeral contacts. Thus, while supporting mobility some self-organization schemes need to be addressed here

to reduce breaks in service. The signaling protocol needs to be reliable and robust to maintain and stabilize the network. The protocol should take care of allocating a nearest CS when a node reconnects to the service. Such a protocol needs to be developed.

3. *Session mobility.* If the clients having multiple link layers move around and change their L_2/L_3 point of contact, handover takes place. Thus, from the point of view of application, connectivity and seamless session mobility have to be implemented in L_2/L_3 layers. For a large multimedia conferencing with distributed servers, these are very crucial for the success of multimedia applications. An improvement is to select the best link layer when there are multiple connection possibilities. Implementation on a multitude of devices calls for the selection of an appropriate link layer. With different radio links, gateway and related tasks will have to be looked into. A notable effort in this direction is IEEE 802.21 [59].
4. *Quality improvement.* There is a large body of research with respect to handling the delay, jitter, and packet loss for multimedia applications on wired links. In a wireless domain, these aspects are not the same. Further, adaptive delay algorithms, forward error correction (FEC), and coding for wireless multimedia are a few important needs.
5. *Enhancing LN.* LN has some limitations at present, such as it does not take into account the prosody of speech. There is a higher scope to improve this metric for a dynamic and auto floor control. Moreover, the effective use of LN yields better reduction in the bandwidth, which needs to be studied in greater detail.
6. *Implementation.* Implementation on devices with a variety of capacities and connectivities, and tuning of the devices for enhanced quality application are other challenges for a successful deployment of conferencing service on wireless devices.

4.8 Conclusions

This chapter described issues, existing solutions, and proposed an architecture for multiparty audioconferencing on wireless networks in detail. Network and audio-related issues were introduced. Mixing, an essential ingredient of multiparty conferencing, and its effects on the overall quality were discussed. Because the state-of-the-art solutions are not satisfactory, a distributed architecture was proposed based on a parameter called LN to select an appropriate subset of speakers. This architecture provides better quality of conferencing with less bandwidth consumption. We have indicated some directions for a better collaborative future.

References

1. Schooler, E. M., S. L. Casner, and J. Postel, Multimedia Conferencing: Has It Come of Age? *Proceedings of 24th Hawaii International Conference on System Sciences*, 3, 707–716, Jan 1991.
2. Radenkovic, M. M. and C. Greenhalgh, Multi-Party Distributed Audio Service with TCP Fairness, *Proceedings of the 10th ACM International Conference on Multimedia (MM'02), Juan-les-Pins* (France), pp. 11–20, Dec 2002.
3. Radenkovic, M. and C. Greenhalgh, Supporting Collaborative Audio in the Internet, *WSES Conference*, Greece, pp. 190–196, Sep 2002.
4. Radenkovic, M. and C. Greenhalgh, Supporting Collaborative Audio in the Internet, *Proceedings of WSEAS ICOMIV 2002* (Skiathos, Greece), pp. 3441–3447, Sep 2002.
5. Venkatesha Prasad, R., R. Hurni, H. S. Jamadagni, and H. N. Shankar, Deployment Issues of a VoIP Conferencing System in a Virtual Conferencing Environment, *ACM symposium on Virtual Reality and Software Techniques* (Osaka, Japan), Oct 2003.
6. Venkatesha Prasad, R., J. Kuri, H. S. Jamadagni, H. Dagale, and R. Ravindranath, Control Protocol for VoIP Audio Conferencing Support, in *International Conference on Advanced Communication Technology* (MuJu, South Korea), pp. 419–424, Feb 2001.
7. Venkatesha Prasad, R., J. Kuri, et al., Automatic Addition and Deletion of Clients in VoIP Conferencing, *6th IEEE Symposium on Computers and Communications* (Hammamet, Tunisia), pp. 386–390, July 2001.
8. Hindmarsh, J., M. Fraser, et al., Fragmented Interaction: Establishing Mutual Orientation in Virtual Environments, in *CSCW 98* (Seattle, WA), pp. 217–226, ACM Press, 1998.
9. Doerry, E., An Empirical Comparison of Copresent and Technologically-mediated Interaction based on Communicative Breakdown, PhD thesis, Graduate School of the University of Oregon, Eugene, OR, 1995.
10. Flament, M. and A. Svensson, Virtual Cellular Networks for 60 GHz wireless Infrastructure, *IEEE International Conference on Communications*, 2, 1223–1227, May 2003.
11. Eklund, C., B. Roger, K. L. Marks, and S. W. Stanwood, IEEE standard 802.16: A technical overview of the wirelessMAN air interface for broadband wireless access, *IEEE Communications Magazine*, 40(6), 98–107, 2002.
12. Bolton, W., Y. Xiao, and M. Guizani, IEEE 802.20: Mobile broadband wireless access, *IEEE Wireless Communications*, 14(1), 84–95, 2007.
13. Greenhalgh, C., S. Benford, and M. Craven, Patterns of network and user activity in an inhabited television event, *Proceedings of the ACM symposium on Virtual Reality Software and Technology* (London, UK), pp. 34–41, 1999.
14. Schooler, E. M., Conferencing and collaborative computing, *Multimedia Systems*, 4(5), 210–225, 1996.
15. Hendrix, C. and W. Barfield, Presence in Virtual Environments as a Function of Visual and Auditory Cues, *Virtual Reality Annual International Symposium (VRAIS'95)* (Research Triangle Park, NC), p. 74, March 1995.
16. Coding of Speech at 8 kbit/s Using Conjugate-structure Algebraic-code-excited Linear-prediction (CS-ACELP), ITU-T Rec. G.729, http://www.itu.int/itudoc/itu-t/rec/g/g729.html, 1996.

17. 40,32,24,16 kbit/s Adaptive Differential Pulse Code Modulation (ADPCM), ITU-T Rec. G.726, http://www.itu.int/itudoc/itu-t/rec/g/g726.html, 1996.
18. Andersen, S., A. Duric, et al., Internet Low Bit Rate Codec (iLBC), RFC 3951, 2004.
19. Hardman, V., M. A. Sasse, and I. Kouvelas, Successful multiparty audio communication over the Internet, *Communications of the ACM*, 41, 74–80, 1998.
20. Prasad, R. V., R. Muralishankar, et al., Voice Activity Detection for VoIP: An Information Theoretic Approach, in *Proceedings of IEEE Globecom* (San Francisco), Dec 2006.
21. Park, K. S. and R. V. Kenyon, Effects of Network Characteristics on Human Performance in a Collaborative Virtual Environment, *Proceedings of the IEEE Virtual Reality Conference* (University College London, London, UK), pp. 104–111, 1999.
22. Watson, A. and A. Sasse, The Good, the Bad, and the Muffled: The Impact of Different Degradations on Internet Speech, *Proceedings of MM 2000* (Los Angeles, CA), 2000.
23. Dommel, H. P. and J. Garcia-Luna-Aceves, Networking Foundations for Collaborative Computing at Internet Scope, *Interactive and Collaborative Computing (ICC 2000)* (Wollongong, Australia), Dec 2000.
24. Dommel, H. and J. J. Garcia-Luna-Aceves, Floor control for multimedia conferencing and collaboration, *Multimedia Systems Journal (ACM/Springer)*, 5(1), 23–28, 1997.
25. Dommel, H. P. and J. Garcia-Luna-Aceves, Network Support for Turn-Taking in Multimedia Collaboration, *IS&T/SPIE Symposium on Electronic Imaging: Multimedia Computing and Networking* (San Jose, CA), Feb 1997.
26. Gonzalez, A. J., A Distributed Audio Conferencing System, MS thesis, Department of Computer Science, Old Dominion University, Norfolk, VA, July 1997.
27. Radenkovic, M., C. Greenhalgh, and S. Benford, Deployment issues in multiparty audio for CVEs, in *Proceedings of ACM VRST 2002* (Hong Kong), pp. 179–185, ACM Press, Nov 2002.
28. Gonzalez, A. J., A Semantic-based Middleware for Multimedia Collaborative Applications, PhD thesis, Computer Science, Old Dominion University, Norfolk, VA, May 2000.
29. Schulzrinne, H., S. Casner, R. Frederick, and V. Jacobson, RTP A Transport Protocol for Real-Time Applications, IETF RFC 3550, July 2003.
30. Khedher, D. B., R. H. Glitho, and R. Dssouli, Media handling aspects of Multimedia Conferencing in Broadband Wireless *ad hoc* Networks, *IEEE Network*, 20(2), 42–49, 2006.
31. ITU-T Rec. H.323, Packet Based Multimedia Communications Systems, http://www.itu.int/itudoc/itu-t/rec/h/h323.html, 1998.
32. Rosenberg, J., H. Schulzrinne, G. Camarillo, A. Johnston, J. Peterson, R. Sparks, M. Handley, and E. Schooler, SIP: Session Initiation Protocol, RFC 3261, June 2002.
33. Mark/Kelley, Distributed Multipoint Conferences using SIP, IETF Internet Draft draft-mark-sip-dmcs-00.txt, March 2000.
34. Khlifi, H., A. Agarwal, and J. C. Gregoire, A framework to use SIP in *ad-hoc* networks, *IEEE Canadian Conference on Electrical and Computer Engineering (CCECE)*, 2, 4–7, May 2003.

35. Fu, C., R. Glitho, and R. Dssouli, A Novel Signaling System for Multiparty Sessions in Peer-to-Peer *Ad Hoc* Networks, *IEEE WCNC 2005* (New Orleans, LA), Mar 2005.
36. Schooler, E. M., The impact of scaling on a multimedia connection architecture, *ACM Journal of Multimedia Systems*, 1(1), 2–9, 1993.
37. Handley, M., J. Crowcroft, C. Bormann, and J. Ott, Very large conferences on the Internet: The Internet multimedia conferencing architecture, *Journal of Computer Networks*, 31, 191–204, 1999.
38. Hac, A. and D. chen Lu, Architecture, design, and implementation of a multimedia conference system, *International Journal of Network Management*, 7, 64–83, 1997.
39. Deering, S., Host Extensions for IP Multicasting, IETF RFC 1054, May 1988.
40. Varshney, U., Multicast over Wireless Networks, *Communications of the ACM*, 45(12), 31–37, 2002.
41. Hardman, V. J., M. A. Sasse, A. Watson, and M. Handley, Reliable Audio for Use Over the Internet, in *Proceedings of INET95* (Honolulu, Oahu, Hawaii), Sept 1995.
42. Agnihotri, S., Improving Quality of Speech in VoIP using Time-Sacle Modification, Master's thesis, Indian Institute of Science, Bangalore, India, 2001.
43. Greenberg, S., Personalizable Groupware: Accomodating Individual Roles and Group Differences, in *Proceedings of the European Conference of Computer Supported Cooperative Work (ECSCW '91)* (Amsterdam), pp. 17–32, Kluwer Academic Press, Dordrecht, Sept 1991.
44. Hardman, V. and M. Iken, Enhanced Reality Audio in Interactive Networked Environments, in *Proceedings of the FIVE '96 Framework for Immersive Virtual Environments, the 2nd FIVE International Conference* (London, UK), pp. 55–66, 1996.
45. Yu, K.-Y., J.-H. Park, and J.-H. Lee, Linear PCM Signal Processing for Audio Processing unit in Multipoint Video Conferencing System, in *Proceedings ISCC*, pp. 549–553, Jul 1998.
46. Rangan, P. V., H. M. Vin, and S. Ramanathan, Communication architectures and algorithms for media mixing in multimedia conferences, *IEEE/ACM Transaction on Networks*, 1, 2030, 1993.
47. Yang, S., S. Yu, J. Zhou, and Q. Han, Multipoint communications with speech mixing over IP network, *Computer Communication*, 25, 4655, 2002.
48. Vin, H. A., P. V. Rangan, and S. Ramanathan, Hierarchical Conferencing Architectures for Inter-group Multimedia Collaboration, *ACM Conference proceedings on Organizational Computing Systems* (Atlanta, GA), pp. 43–54, Nov 1991.
49. Ramanathan, S., P. V. Rangan, and H. M. Vin, Designing communication architectures for interorganizational multimedia collaboration, *Journal of Organizational Computing*, 2(3&4), 277–302, 1992.
50. Ramanathan, S., P. V. Rangan, H. M. Vin, and T. Kaeppner, Optimal Communication Architectures for Multimedia Conferencing in Distributed Systems, *ICDCS* (Yokohama, Japan), pp. 46–53, 1992.
51. Venkatesha Prasad, R., H. S. Jamadagni, and H. N. Shankar, Number of Floors for a Voice-Only Conference on Packet Networks—A Conjecture, *IEE Proceedings on Communications, Special Issue on Internet Protocols, Technology and Applications (VoIP)*, 2004.

52. Venkatesha Prasad, R., H. S. Jamadagni, and H. N. Shankar, On the Problem of Specifying Number of Floors in a Voice Only Conference, *IEEE ITRE*, 2003.
53. Koskelainen, P., H. Schulzrinne, and X. Wu, A SIP-based Conference Control Framework, in *Proceedings of the 12th international workshop on Network and operating systems support for digital audio and video (NOSSADAV)* (Miami, FL), pp. 53–61, May 2002.
54. Rosenberg, J. and H. Schulzrinne, Models for Multi Party Conferencing in SIP, Internet Draft, *IETF*, Jul 2002.
55. Venkatesha Prasad, R., R. Hurni, H. S. Jamadagni, A Proposal for Distributed Conferencing on SIP using Conference Servers, *The Proceedings of MMNS 2003* (Belfast, UK), Sep 2003.
56. Thaler, D., M. Handley, and D. Estrin, The Internet Multicast Address Allocation Architecture, RFC 2908, *IETF*, Sep 2000.
57. Zhu, F., M. W. Mukta, L. M. Ni, Service Discovery in Pervasive Computing Environments, *Pervasive Computing, IEEE*, 4(4), 81–90, 2005.
58. Venkatesha Prasad, R., H. S. Jamadagni et al., Heuristic Algorithms for Server Allocation in Distributed VoIP Conferencing, *IEEE Symposium on Computers and Communications* (Cartegena, Spain) 2005.
59. IEEE 802.21, http://www.ieee802.org/21/.
60. Gu, Y., R. Venkatesha Prasad, W. Lu, and I. Niemegeers, *Clustering in Ad Hoc Personal Network Formation, Workshop on Wireless and Mobile Systems (WMS 2007)*, Springer Lecture Notes in Computer Science, vol. 4490, pp. 312–319, International Conference on Computational Science (4), Beijing, 2007.

MULTIMEDIA OVER *AD HOC* AND SENSOR NETWORKS

Chapter 5

Routing for Video Communications over Wireless *Ad Hoc* Networks

Shiwen Mao, Y. Thomas Hou, Hanif D. Sherali, and Scott F. Midkiff

Contents

5.1 Introduction 158
5.2 Problem Formulation 161
  5.2.1 Application-Layer Performance Measure 162
  5.2.2 Video Bit Rates and Success Probabilities 164
  5.2.3 The Optimal Double-Path Routing Problem 166
5.3 A Metaheuristic Approach 168
  5.3.1 Genetic Algorithm-Based Solution Procedure 168
    5.3.1.1 Solution Representation and Initialization 169
    5.3.1.2 Evaluation 170
    5.3.1.3 Selection 170
    5.3.1.4 Crossover 170
    5.3.1.5 Mutation 171

5.3.2 Performance Comparisons 172
5.3.2.1 Comparison with Trajectory Methods....172
5.3.2.2 Comparison with Network-Centric Approaches 174
5.4 A Branch-and-Bound-Based Approach....175
5.4.1 The Branch-and-Bound Framework 177
5.4.2 Problem Reformulation....178
5.4.3 Problem Linearization....181
5.4.4 A Heuristic Algorithm 182
5.4.5 An ε-Optimal Algorithm....183
5.4.5.1 Initialization and Relaxation 183
5.4.5.2 Node Selection....183
5.4.5.3 Local Search....184
5.4.5.4 Partitioning 184
5.4.5.5 Bounding....185
5.4.5.6 Fathoming....185
5.4.6 Numerical Results 185
5.4.6.1 Performance of the Proposed Algorithms 186
5.4.6.2 Comparison with *k*-Shortest Path Routing....187
5.5 Considerations for Distributed Implementation 188
5.6 Conclusion 189
Acknowledgments 190
References....190

Multihop wireless networks (e.g., *ad hoc* networks) are characterized by infrastructure independence and user mobility, making them an excellent match for applications that demand great simplicity and flexibility in deployment and operations. However, the resulting fragile multihop wireless paths and dynamic topology pose significant technical challenges for enabling multimedia services in such networks. This chapter studies the problem of cross-layer multipath routing for multiple description (MD) video in multihop wireless networks. In particular, we provide practical solutions to the following key questions: (i) How to formulate a multimedia-centric routing problem that minimizes video distortion by choosing optimal paths? (ii) What are the performance limits? (iii) How to design an efficient solution procedure? (iv) How to implement the proposed algorithms in a distributed environment? We show that cross-layer design is imperative and effective to meeting the stringent quality of service (QoS) requirements of multimedia applications in highly dynamic multihop wireless networks.

5.1 Introduction

Multihop wireless networking (e.g., *ad hoc* networking) is becoming an increasingly important technology for providing ubiquitous access to mobile users and for quick-and-easy extension of local area networks (LANs) into a wide area. Such networks are characterized by infrastructure independence and user mobility, making them an excellent match for important military and civilian applications, all of which demand great simplicity and flexibility during deployment and operations. In addition, multihop wireless networks are highly robust to disruptions due to their distributed nature and the absence of single point of failure.

Multihop wireless networking, in particular *ad hoc* networking, has been the subject of intensive research over the years. As a result, there has been a flourish of protocols and mechanisms for efficient operations and service provisioning, and more importantly, there have been significant advances in understanding the fundamental properties of such networks, such as capacity, connectivity, coverage, interference, among others. As such developments continue, there is a compelling need for content-rich multimedia applications in such networks. For mission-critical applications, the capability of multimedia communications is vital to thorough situational awareness and shortening of decision cycle. For commercial applications, such capability would enable multihop wireless networks to provide similar capabilities as other alternative access networks, thus catalyzing the deployment of such networks.

Apart from its great advantages such as simplified deployment and operation and robustness to disruption, multihop wireless networks also have brought about some unique challenges, such as fragile wireless routes, and dynamic network topology, all of which have added considerable difficulties in the provisioning of multimedia services (e.g., video) in such networks. Indeed, the end-to-end path quality is one of the most important factors that determine the received video quality. Many efficient protocols have been proposed for QoS routing in multihop wireless networks, for example, see Refs 1–5, among others. These previous efforts mainly focus on the optimization of one or more network-layer metrics, such as throughput, delay, loss, or path correlation, and can be termed as network-centric routing throughout this chapter. Although these protocols are shown to be quite effective for certain applications, they may not produce good video quality due to the fact that video distortion is a highly complex function of multiple metrics at the network layer [6–10]. Optimizing these network-layer metrics does not necessarily guarantee optimal video quality [10–12].

Recent advances in MD coding have made it highly suitable for multimedia communications in multihop wireless networks [7,10–15]. MD coding is a technique that generates multiple equally important descriptions, each giving a low, but acceptable video quality [13,15]. Unlike traditional compression schemes that aim at achieving the lowest bit rate for a target video quality, MD coding provides a trade-off between coding efficiency and error resilience, that is, most MD coding schemes introduce a certain amount of redundancy to make the video stream more robust to transmission errors. Considering the improved quality for reconstructed video in a lossy environment, such redundancy is usually well justified. The decoding independence among the descriptions permits a reconstruction of video from any subset of received descriptions, achieving a quality commensurate with the number of received descriptions. This feature makes MD video an excellent match for multihop wireless networks, where multiple paths between any source–destination pair, albeit unreliable, are readily available.

This chapter studies the important problem of multipath routing for MD video in multihop wireless networks. We present an optimal routing approach that aims to explicitly minimize the distortion of received video (termed multimedia-centric routing throughout this chapter). Specifically, we present practical solutions to address the following questions:

- How to characterize and model the multimedia-centric multipath routing problem via cross-layer design and optimization?
- What are the performance limits for video quality for a given multihop wireless network?
- How to design efficient solution procedures to the cross-layer optimization problem?
- How to implement the proposed multimedia-centric routing algorithms in a distributed environment?

To answer these questions, we first follow a cross-layer design approach in problem formulation by considering the application-layer performance (i.e., average video distortion) as a function of network-layer performance metrics (e.g., bandwidth, loss, and path correlation). For the formulated problem, we show that the objective function is a complex ratio of high-order polynomials. This motivates us to pursue effective approximation algorithms. We develop efficient metaheuristic-based solution procedures for the multimedia-centric routing problem. In particular, we find that Genetic Algorithms (GAs) [16] are quite suitable in addressing such type of complex cross-layer optimization problems, and in outperforming Simulated Annealing (SA) [17] and Tabu Search (TS) [18] in our simulation studies. We also observe that multimedia-centric routing achieves significant performance gains over network-centric approaches.

Furthermore, we develop a branch-and-bound-based approach, predicated on the powerful reformulation linearization technique (RLT), which can provide highly competitive paths with bounded optimality gap (i.e., the gap between the global optimal and an approximate solution). This is a major step forward as compared with the heuristic approaches in the literature. Finally, we show that the multimedia-centric routing can be incorporated into many existing distributed routing protocols, including both the classes of proactive [19] and reactive protocols [20]. In fact, the multimedia-centric routing algorithms could significantly improve the performance of the existing routing protocol with regard to supporting MD video with slightly increased routing overhead.

In the following, we will use the terms multimedia-centric and cross-layer interchangeably, referring to the approach that optimizes application-layer media (e.g., video) quality through multiple layers. Specifically, we consider the joint design of application and network layers in this chapter, assuming that lower layer dynamics, such as Media Access Control (MAC)-layer contention and physical-layer (PHY) interference, could be translated into network-layer performance metrics [21]. It is possible to extend this framework to take more layers (e.g., MAC and PHY) into consideration, which we leave as future work for this research.

This chapter is organized as follows: Section 5.2 presents the cross-layer formulation for double-description (DD) video over multihop wireless networks. An efficient metaheuristic-based solution procedure for this formulated problem is developed in Section 5.3. Section 5.4 introduces a branch-and-bound and RLT-based technique. Section 5.5 presents possible approaches to distributed implementations of the proposed algorithms. Section 5.6 concludes this chapter.

5.2 Problem Formulation

A multihop wireless network, for example, an *ad hoc* network, can be modeled as a stochastic directed graph $\mathcal{G}\{V, E\}$, where V is the set of vertices and E the set of edges. We assume that nodes are reliable during the video session, but links may fail with certain probabilities. Accurate and computationally efficient characterization of an end-to-end path in a multihop wireless network with the consideration of mobility, interference, and the time-varying wireless channels is extremely difficult and remains an open problem. As an initial step, we focus on the network-layer characteristics in this chapter, assuming that the PHY and MAC-layer dynamics of wireless links are translated into network-layer parameters. For example, we could characterize a link $\{i, j\} \in E$ by

- b_{ij}—The available bandwidth of link $\{i, j\}$. We assume that the impact of other traffic sessions are accounted for through the link available bandwidth.
- p_{ij}—The probability that link $\{i, j\}$ is "up."
- l_{ij}—Average burst length for packet losses on link $\{i, j\}$.

In practice, these parameters can be measured by every node, and distributed in the network in routing control messages (e.g., the Topology

Control (TC) message in Optimized Link State Routing Protocol (OLSR) [19]). A similar approach is taken in Ref. 21 that models the *ad hoc* network as a matrix, each element being the packet loss probability of the link between two of the nodes. We focus on the bandwidth and failure probabilities of a path, because these are the key characteristics for data transmission, as well as the most important factors that determine video distortion (see Equation 5.2). Other link characteristics, such as delay, jitter, congestion, and signal strength, could be incorporated into this framework as well (e.g., see Ref. 22). Table 5.1 lists the notations used in this chapter.

5.2.1 Application-Layer Performance Measure

Throughout this chapter, we use DD coding for MD video. We consider DD video because it is most widely used [7,8,23,24,14]. In general, using more descriptions and paths will increase the robustness to packet losses and path failures. However, more descriptions may increase the video bit rate for the same video quality [13]. The study in Ref. 25 demonstrates that the most significant performance gain is achieved when the number of descriptions increases from one to two, with only marginal improvements achieved for further increases in number of descriptions.

For two descriptions generated for a sequence of video frames, let d_h be the achieved distortion when only description h is received, $h = 1, 2$, and d_0 the distortion when both descriptions are received. Let R_h denote the rate in bits/pixel of description h, $h = 1, 2$. The rate-distortion region for a memoryless *i.i.d.* Gaussian source with the square error distortion measure was first introduced in Ref. 26. For computational efficiency, Alasti et al. [6] introduce the following distortion-rate function for studying the impact of congestion on DD coding (also used in this chapter).

$$d_0 = \frac{2^{-2(R_1+R_2)}}{2^{-2R_1} + 2^{-2R_2} - 2^{-2(R_1+R_2)}} \cdot \sigma^2, \quad d_1 = 2^{-2R_1} \cdot \sigma^2, \quad d_2 = 2^{-2R_2} \cdot \sigma^2 \tag{5.1}$$

where σ^2 is the variance of the source (i.e., the distortion when both descriptions are lost). Let P_{00} be the probability of receiving both descriptions, P_{01} the probability of receiving description 1 only, P_{10} the probability of receiving description 2 only, and P_{11} the probability of losing both descriptions. The average distortion of the received video can be approximated as

$$D = P_{00} \cdot d_0 + P_{01} \cdot d_1 + P_{10} \cdot d_2 + P_{11} \cdot \sigma^2 \tag{5.2}$$

Finding the rate-distortion region for MD video is still an open research problem [13,27]. The MD region is well understood only for memoryless

Table 5.1 Notations

Symbols	*Definitions*
$\mathcal{G}\{V, E\}$	Graphical representation of the network
V	Set of vertices in the network
E	Set of edges in the network
s	Source node
t	Destination node
$\mathcal{P}$	A path from s to t
g_i	An intermediate node in a path
$\{i, j\}$	A link from node i to node j
b_{ij}	Bandwidth of link $\{i, j\}$
p_{ij}	Success probability of link $\{i, j\}$
l_{ij}	Average length of loss burst on link $\{i, j\}$
R_h	Rate of description h in bits/sample
d_0	Distortion when both descriptions are received
d_h	Distortion when only description h is received, $h=1, 2$
D	Average distortion
T_{on}	Average "up" period of the joint links
P_{00}	Probability of receiving both descriptions
P_{01}	Probability of receiving description 1 only
P_{10}	Probability of receiving description 2 only
P_{11}	Probability of losing both descriptions
x_{ij}^h	Routing index variables, defined in Equation 5.10
α_{ij}	"Up" to "down" transition probability of link $\{i, j\}$
β_{ij}	"Down" to "up" transition probability of link $\{i, j\}$
p_{jnt}	Average success probability of joint links
p_{dj}^h	Average success probability of disjoint links on $\mathcal{P}_h$
B_{jnt}	Minimum bandwidth of the shared links
θ	GA crossover rate
μ	GA mutation rate
UB	Upper bound in the branch-and-bound procedure
LB	Lower bound in the branch-and-bound procedure
UB_k	Upper bound for subproblem k
LB_k	Lower bound for subproblem k
ε	Optimality gap or tolerance
Ω	The solution space

Gaussian sources with squared-error distortion measure, which bounds the MD region for any continuous-valued memoryless source with the same distortion measure. Although there are several empirical models used in the literature, these models are dependent on the specific video sequence, and the model parameters are determined by using regression techniques [7,8]. Therefore, these models are not entirely suitable for our cross-layer routing algorithm, which is not specific to a particular stored video or MD coding technique. More importantly, we believe that such a model should be robust and effective for live video. Our simulation results show that although the distortion-rate function in Equation 5.1 is an approximation for DD video, significant improvement in received video quality could be achieved over alternative approaches by incorporating Equation 5.2 into the optimal routing problem formulation. It is also worth noting that our formulation does not depend on any specific distortion-rate function. A more accurate distortion-rate function for MD video can be easily incorporated into our formulation, if available in future.

5.2.2 Video Bit Rates and Success Probabilities

As a first step to formulate the problem of optimal multipath routing, we need to know how to compute the average distortion D as a function of link statistics for a given pair of paths, that is, we need to compute the end-to-end bandwidth (or rate) for each stream and joint probabilities of receiving the descriptions (see Equations 5.1 and 5.2).

For a source–destination pair $\{s, t\}$, consider two given paths $[\mathcal{P}_1, \mathcal{P}_2]$ in $\mathcal{G}\{V, E\}$. Because we do not mandate "disjointedness" in routing, $\mathcal{P}_1$ and $\mathcal{P}_2$ may share nodes and links. Similar to the approach in Refs 7 and 8, we classify the links along the two paths into three sets: set one consisting of links shared by both paths, denoted as $\mathcal{J}(\mathcal{P}_1, \mathcal{P}_2)$, and the other two sets consisting of disjoint links on the two paths, denoted as $\overline{\mathcal{J}}(\mathcal{P}_h)$, $h=1,2$, respectively. Then, the minimum bandwidth of $\mathcal{J}(\mathcal{P}_1, \mathcal{P}_2)$, B_{jnt}, is

$$B_{\text{jnt}} = \begin{cases} \min_{\{i,j\}\in\mathcal{J}(\mathcal{P}_1,\mathcal{P}_2)}\{b_{ij}\}, & \text{if } \mathcal{J}(\mathcal{P}_1, \mathcal{P}_2) \neq \emptyset \\ \infty, & \text{otherwise} \end{cases}$$

The rates of the two video streams, R_1 and R_2, can be computed as

$$\begin{cases} R_h = \rho \cdot B(\mathcal{P}_h), & \text{if } \sum_{m=1}^{2} B(\mathcal{P}_m) \leq B_{\text{jnt}}, \quad h = 1, 2 \\ R_1 + R_2 \leq \rho \cdot B_{\text{jnt}}, & \text{otherwise} \end{cases} \tag{5.3}$$

where $B(\mathcal{P}_h) = \min_{\{i,j\}\in\mathcal{P}_h}\{b_{ij}\}$, $h=1,2$, and ρ is a constant determined by the video format and frame rate. For a video with coding rate f frames/s and a resolution of $W \times V$ pixels/frame, we have $\rho = 1/(\kappa \cdot W \cdot V \cdot f)$, where κ is a constant determined by the chroma subsampling scheme. For the quarter common intermediate format (QCIF) (176×144 Y pixels/frame, 88×72 Cb/Cr pixels/frame), we have $\kappa = 1.5$ and $\rho = 1/(1.5 \cdot 176 \cdot 144 \cdot f)$. The first line in Equation 5.3 is for the case when the joint links are not the bottleneck of the paths. The second line of Equation 5.3 is for the case where one of the joint links is the bottleneck of both paths. In the latter case, we assign the bandwidth to the paths by splitting the bandwidth of the shared bottleneck link in proportion to the mean success probabilities

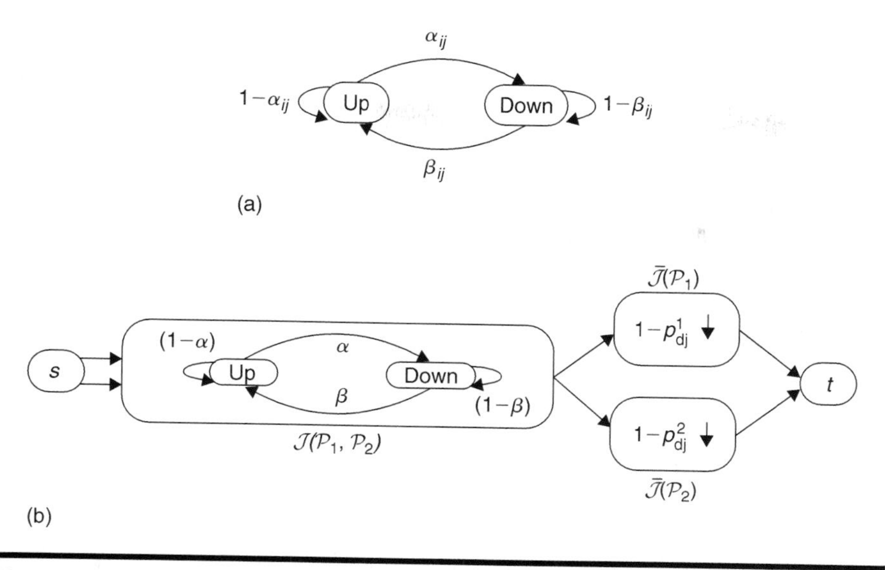

Figure 5.1 Link and path models. (a) The Gilbert two-state link model, (b) a simplified path model for double-description video. (Reprinted from Mao, S., Y.T. Hou, X. Cheng, H.D. Sherali, S.F. Midkiff, and Y.-Q. Zhang. *IEEE Trans. Multimedia*, 8(5): 1063–1074, 2006.)

of the two paths, whereas an alternative approach is to split the bandwidth evenly for balanced descriptions.

We now focus on how to compute the end-to-end success probabilities. For disjoint portion of the paths, it suffices to model the packet loss as Bernoulli events, because losses from the two descriptions are assumed to be independent in the disjoint portions. Therefore, the success probabilities on the disjoint portions of the two paths are

$$p^h_{\mathrm{dj}} = \begin{cases} \prod_{\{i,j\}\in\overline{\mathcal{J}}(\mathcal{P}_h)} p_{ij}, & \text{if } \overline{\mathcal{J}}(\mathcal{P}_h) \neq \emptyset, \quad h = 1, 2 \\ 1, & \text{otherwise } h = 1, 2 \end{cases} \tag{5.4}$$

Losses on the two streams are correlated on the joint portion of the paths. To model such correlation, we model each shared link $\{i, j\}$ as an on–off process modulated by a discrete-time Markov chain, as shown in Figure 5.1a. With this model, there is no packet loss when the link is in the "up" state; all packets are dropped when the link is in the "down" state. Transition probabilities, $\{\alpha_{ij}, \beta_{ij}\}$, can be computed from the link statistics as

$$\beta_{ij} = \frac{1}{l_{ij}}, \quad \alpha_{ij} = \frac{(1 - p_{ij})}{(p_{ij} l_{ij})} \tag{5.5}$$

If there are K shared links, the aggregate failure process of these links is a Markov process with 2^K states. To simplify the computation, we follow the well-known Fritchman model [28] in modeling the aggregate process as an on–off process. Because a packet is successfully delivered on the joint portion if and only if all joint links are in the "up" state, we can lump up all the states with at least one link failure into a single "down" state, while using the remaining state where all the links are in good condition as the "up" state. Let T_{on} be the average length of the "up" period. We have

$$T_{\text{on}} = \frac{1}{1 - \prod_{\{i,j\} \in \mathcal{J}(\mathcal{P}_1, \mathcal{P}_2)} (1 - \alpha_{ij})} \tag{5.6}$$

If the joint link set is not empty, the probability of a successful delivery on the joint links can be written as

$$p_{\text{jnt}} = \begin{cases} \prod_{\{i,j\} \in \mathcal{J}(\mathcal{P}_1, \mathcal{P}_2)} p_{ij} & \text{if } \mathcal{J}(\mathcal{P}_1, \mathcal{P}_2) \neq \emptyset \\ 1, & \text{otherwise} \end{cases} \tag{5.7}$$

Finally, the transition probabilities of the aggregate on–off process are

$$\alpha = \frac{1}{T_{\text{on}}} \quad \text{and} \quad \beta = \frac{p_{\text{jnt}}}{T_{\text{on}}(1 - p_{\text{jnt}})} \tag{5.8}$$

Note that $\alpha = 0$ and $\beta = 0$, if $\mathcal{J}(\mathcal{P}_1, \mathcal{P}_2) = \emptyset$. The consolidated path model is illustrated in Figure 5.1b, where $\mathcal{J}(\mathcal{P}_1, \mathcal{P}_2)$ is modeled as a two-state Markov process with parameters $\{\alpha, \beta\}$, and $\overline{\mathcal{J}}(\mathcal{P}_h)$ is modeled as a Bernoulli process with parameter $(1 - p_{\text{dj}}^h)$, $h = 1, 2$. With the consolidated path model, the joint probabilities of receiving the descriptions are

$$\begin{cases} P_{00} = p_{\text{jnt}} \cdot (1 - \alpha) \cdot p_{\text{dj}}^1 \cdot p_{\text{dj}}^2 \\ P_{01} = p_{\text{jnt}} \cdot p_{\text{dj}}^1 \cdot \left[1 - (1 - \alpha) \cdot p_{\text{dj}}^2\right] \\ P_{10} = p_{\text{jnt}} \cdot \left[1 - (1 - \alpha) p_{\text{dj}}^1\right] \cdot p_{\text{dj}}^2 \\ P_{11} = 1 - p_{\text{jnt}} \cdot \left[p_{\text{dj}}^1 + p_{\text{dj}}^2 - (1 - \alpha) \cdot p_{\text{dj}}^1 \cdot p_{\text{dj}}^2\right] \end{cases} \tag{5.9}$$

5.2.3 The Optimal Double-Path Routing Problem

With the mentioned preliminaries, we now set out to formulate the multipath routing problem for MD video. Consider a DD video session from source node s to destination node t. To characterize any s–t path $\mathcal{P}_h$, we define the following binary variables:

$$x_{ij}^h = \begin{cases} 1 & \text{if } \{i, j\} \in \mathcal{P}_h, \ h = 1, 2 \\ 0 & \text{otherwise} \end{cases} \tag{5.10}$$

With these variables, an arbitrary path $\mathcal{P}_h$ can be represented by a vector x^h of $|E|$ elements, each of which corresponds to a link and has a binary value. We can formulate the problem of multipath routing for MD video (OPT-MM) as follows.
Minimize:

$$D = P_{00} \cdot d_0 + P_{01} \cdot d_1 + P_{10} \cdot d_2 + P_{11} \cdot \sigma^2 \tag{5.11}$$

subject to:

$$\sum_{j:\{i,j\} \in E} x_{ij}^h - \sum_{j:\{j,i\} \in E} x_{ji}^h = \begin{cases} 1, & \text{if } i = s, \quad i \in V, \ h = 1, 2 \\ -1, & \text{if } i = t, \quad i \in V, \ h = 1, 2 \\ 0, & \text{otherwise } i \in V, \ h = 1, 2 \end{cases} \tag{5.12}$$

$$\sum_{j:\{i,j\} \in E} x_{ij}^h \begin{cases} \leq 1, & \text{if } i \neq t, \ i \in V, \ h = 1, 2 \\ = 0, & \text{if } i = t, \ i \in V, \ h = 1, 2 \end{cases} \tag{5.13}$$

$$x_{ij}^1 \cdot R_1 + x_{ij}^2 \cdot R_2 \leq \rho \cdot b_{ij}, \quad \{i, j\} \in E \tag{5.14}$$

$$x_{ij}^h \in \{0, 1\}, \quad \{i, j\} \in E, \ h = 1, 2 \tag{5.15}$$

In problem OPT-MM, $\{x_{ij}^h\}_{h=1,2}$ are binary optimization variables and $\{R_h\}_{h=1,2}$ are continuous optimization variables. Constraints 5.12 and 5.13 guarantee that the paths are loop-free, whereas constraint 5.14 guarantees that the links are stable. For a given pair of paths, the average video distortion D is determined by the end-to-end statistics and the correlation of the paths, as given in Equations 5.1, 5.3, and 5.9.

The objective function 5.11 is a highly complex ratio of high-order exponentials of the x-variables. The objective evaluation of a pair of paths involves identifying the joint and disjoint portions, which is only possible when both paths are completely determined (or can be conditioned on the exceedingly complex products of the binary factors x_{ij}^1 and $(1 - x_{ij}^1)$ with x_{ij}^2 and $(1 - x_{ij}^2)$). Sherali et al. [29] considered a problem that seeks a pair of disjoint paths in a network such that the total travel time over the paths is minimized, where the travel time on a link might be either a constant, or a nondecreasing (or unstructured) function of the time spent on the previous links traversed. Even for a simple special case where all the links except one have a constant travel time (and hence linear objective terms), this problem is shown to be NP-hard. Our problem has much more complex relationships pertaining to the contribution of each individual link to the objective function, which generally depends on the other links that are included in a fashion that has no particular structural property, such as convexity. Hence, it is likely to be NP-hard as well. However, we leave a rigorous proof of this NP-hardness to a separate work.

5.3 A Metaheuristic Approach

5.3.1 Genetic Algorithm-Based Solution Procedure

We suggest that a promising strategy to address the multimedia-centric multipath routing problem is to view the problem as a "black-box" optimization problem and explore an effective metaheuristic approach [30]. Particularly, we find that GAs [16] are eminently suitable for addressing this type of complex combinatorial problems, most of which are multimodal and nonconvex. GAs are population-based metaheuristic inspired by the survival-of-the-fittest principle. It has the intrinsic strength of dealing with a set of solutions (i.e., a population) at each step, rather than working with a single, current solution. At each iteration, a number of genetic operators are applied to the individuals of the current population to generate individuals for the next generation. In particular, GA uses genetic operators known as crossover to recombine two or more individuals to produce new individuals, and mutation to achieve a randomized self-adaptation of individuals. The driving force in GA is the selection of individuals based on their fitness (in the form of an objective function) for the next generation. The survival-of-the-fittest principle ensures that the overall quality of the population improves as the algorithm progresses through generations.

Figure 5.2 displays the flowchart for our GA-based solution procedure to the DD double-path routing problem. In what follows, we use an *ad hoc* network as an example (Figure 5.3a) to illustrate the components in this GA-based approach. The termination condition in Figure 5.2 could be based on the total number of iterations (generations), maximum computing time, a threshold of desired video distortion, or a threshold based on certain lower bounds [10].

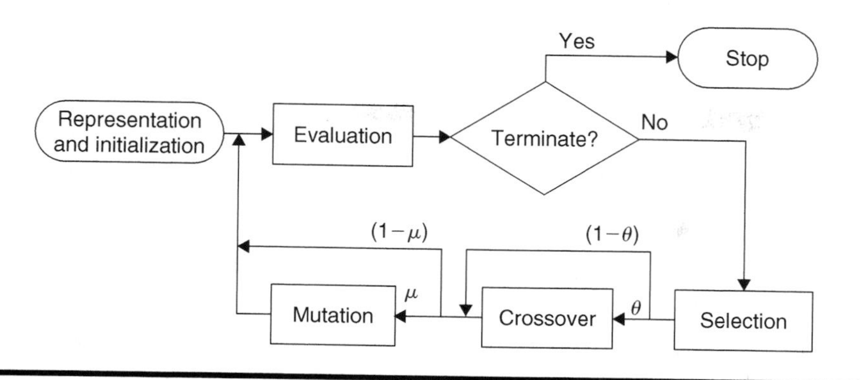

Figure 5.2 Flowchart of the GA-based approach. (Reprinted from Mao, S., Y.T. Hou, X. Cheng, H.D. Sherali, S.F. Midkiff, and Y.-Q. Zhang. *IEEE Trans. Multimedia*, 8(5): 1063–1074, 2006.)

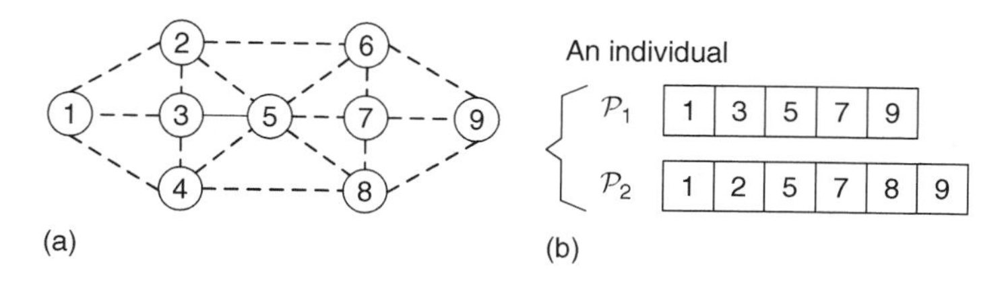

Figure 5.3 Example of network and coding of an individual ($s = 1$ and $t = 9$). (a) Example of an ad hoc network, (b) example of an individual. (Reprinted from Mao, S., Y.T. Hou, X. Cheng, H.D. Sherali, S.F. Midkiff, and Y.-Q. Zhang. *IEEE Trans. Multimedia*, 8(5): 1063–1074, 2006.)

5.3.1.1 *Solution Representation and Initialization*

In GAs, a feasible solution is encoded in the genetic format. For a routing problem, a natural encoding scheme would be to define a node as a gene. Then, an end-to-end path, consisting of an ordered sequence of nodes (connected by the corresponding wireless links), can be represented as a chromosome [31]. For problem OPT-MM, each feasible solution consists of a pair of paths (i.e., a pair of chromosomes), denoted as $[\mathcal{P}_1, \mathcal{P}_2]$. An individual in this case is a pair of vectors containing the nodes on paths $\mathcal{P}_1$ and $\mathcal{P}_2$ (see, e.g., Figure 5.3b).

Before entering the main loop in Figure 5.2, we need to generate an initial population, that is, a set of solutions. A simple approach would be to generate this set of solutions by randomly appending feasible elements (i.e., nodes with connectivity) to a partial solution. Under this approach, each construction process starts with source node s. Then, the process randomly chooses a link incident to the current end-node of the partial path and appends the link with its corresponding head-node to augment the path, until destination node t is reached. It is important to ensure that the intermediate partial path is loop-free during the process. After generating a certain set of paths for s–t independently, a population of individuals can be constructed by pairing paths from this set. Our numerical results show that a properly designed GA is not very sensitive to the quality of the individuals in the initial population.

5.3.1.2 Evaluation

The fitness function $f(\overline{x})$ of an individual, $\overline{x} = [\mathcal{P}_1, \mathcal{P}_2]$, is closely tied to the objective function (i.e., distortion D). Because the objective is to minimize the average distortion function D, we have adopted a fitness function defined as the inverse of the distortion value, that is, $f(\overline{x}) = 1/D(\overline{x})$. This simple fitness definition appears to work very well, although we intend to explore other fitness definitions in our future effort.

5.3.1.3 Selection

During this operation, GA selects individuals who have a better chance or potential to produce "good" offspring in terms of their fitness values. By virtue of the selection operation, "good" genes among the population are more likely to be passed to the future generations. We use the so-called Tournament selection [16] scheme, which randomly chooses m individuals from the population each time, and then selects the best of these m individuals in terms of their fitness values. By repeating either procedure multiple times, a new population can be selected.

5.3.1.4 Crossover

Crossover mimics the genetic mechanism of reproduction in the natural world, in which genes from parents are recombined and passed to offspring. The decision of whether or not to perform a crossover operation is determined by the crossover rate θ.

Figure 5.4 illustrates one possible crossover implementation. Suppose we have two parent individuals $x_1 = [\mathcal{P}_1, \mathcal{P}_2]$ and $x_2 = [\mathcal{P}_3, \mathcal{P}_4]$. We could randomly pick one path in x_1 and one in x_2, say $\mathcal{P}_2$ and $\mathcal{P}_3$. If one or more

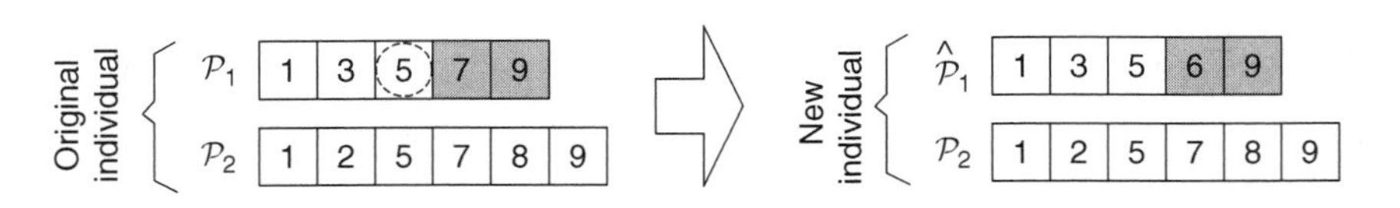

Figure 5.4 An example of the crossover operation. (Reprinted from Mao, S., Y.T. Hou, X. Cheng, H.D. Sherali, S.F. Midkiff, and Y.-Q. Zhang. *IEEE Trans. Multimedia*, 8(5): 1063–1074, 2006.)

common nodes exist in these two chosen paths, we could select the first such common node that exists in $\mathcal{P}_2$, say g_r, where $g_r \notin \{s, t\}$, and we can then concatenate nodes $\{s, \ldots, g_r\}$ from $\mathcal{P}_2$ with nodes $\{g_{r+1}, \ldots, t\}$ in $\mathcal{P}_3$ (where g_{r+1} denotes the next downstream node of g_r in $\mathcal{P}_3$) to produce a new path $\mathcal{P}_{23}$. Likewise, using the first such node $g_{r'}$ in $\mathcal{P}_3$ that repeats in $\mathcal{P}_2$ (which may be different from g_r), we can concatenate the nodes $\{s, \ldots, g_{r'}\}$ from $\mathcal{P}_3$ with the nodes $\{g_{r'+1}, \ldots, t\}$ in $\mathcal{P}_2$ to produce a new path $\mathcal{P}_{32}$. It is important that we check the new paths to be sure that they are loop-free. The two offspring generated in this manner are $[\mathcal{P}_1, \mathcal{P}_{23}]$ and $[\mathcal{P}_{32}, \mathcal{P}_4]$. However, if $\mathcal{P}_2$ and $\mathcal{P}_3$ are disjoint, we could swap $\mathcal{P}_2$ with $\mathcal{P}_3$ to produce two new offspring $[\mathcal{P}_1, \mathcal{P}_3]$ and $[\mathcal{P}_2, \mathcal{P}_4]$.

5.3.1.5 Mutation

The objective of the mutation operation is to diversify the genes of the current population, which helps prevent the solution from being trapped in a local optimum. Just as some malicious mutations could happen in the natural world, mutation in GA may produce individuals that have worse fitness values. In such cases, some "filtering" operation is needed (e.g., the selection operation) to reject such "bad" genes and to drive GA toward optimality.

Mutation is performed on an individual with probability μ (called the mutation rate). For better performance, we propose a schedule to vary the mutation rate within $[\mu_{\min}, \mu_{\max}]$ over iterations (rather than using a fixed μ). The mutation rate is first initialized to $\mu_{\max}$; then as generation number k increases, the mutation rate gradually decreases to $\mu_{\min}$, that is,

$$\mu_k = \mu_{\max} - \frac{k \cdot (\mu_{\max} - \mu_{\min})}{T_{\max}} \tag{5.16}$$

where $T_{\max}$ is the maximum number of generations. Our results show that varying the mutation rates over generations significantly improves the online performance of the GA-based routing scheme. In essence, such schedule of μ is similar to the cooling schedule used in SA, and yields better convergence performance for problem OPT-MM.

Figure 5.5 illustrates a simple example of the mutation operation. In this example, we could implement mutation as follows: First, we choose a path $\mathcal{P}_h$, $h = 1$ or 2, with equal probabilities. Second, we randomly pick an integer value k in the interval $[2, |\mathcal{P}_h| - 1]$, where $|\mathcal{P}_h|$ is the cardinality of $\mathcal{P}_h$, and let the partial path $\{s, \ldots, g_k\}$ be $\mathcal{P}_h^u$, where g_k is the kth node along $\mathcal{P}_h$. Finally, we use any constructive approach to build a partial path from g_k to t, denoted as $\mathcal{P}_h^d$, which does not repeat any node in $\mathcal{P}_h^u$ other than g_k. If no such alternative segment exists between g_k and t, we keep the path intact; otherwise, a new path can now be created by concatenating the two partial paths as $\mathcal{P}_h^u \cup \mathcal{P}_h^d$.

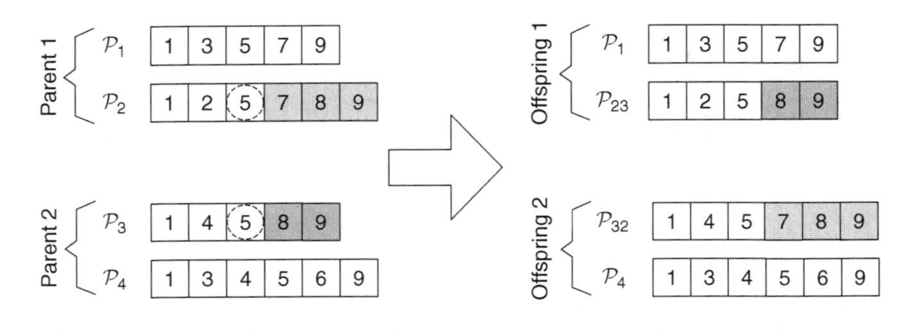

Figure 5.5 An example of the mutation operation. (Reprinted from Mao, S., Y.T. Hou, X. Cheng, H.D. Sherali, S.F. Midkiff, and Y.-Q. Zhang. *IEEE Trans. Multimedia*, 8(5): 1063–1074, 2006.)

5.3.2 Performance Comparisons

The following presents our simulation studies on multimedia-centric routing with the GA-based approach. In each experiment, we generate a wireless *ad hoc* network topology by placing a number of nodes at random locations in a rectangular region, where connectivity is determined by the distance coverage of each node's transmitter. The area is adjusted for networks with different numbers of nodes to achieve an appropriate node density to provide a connected network. The source–destination nodes s and t are uniformly chosen from the nodes. For every link, the failure probability is uniformly chosen from [0.01, 0.3]; the available bandwidth is uniformly chosen from [100, 400] kbps, with 50 kbps increments; the mean burst length is uniformly chosen from [2,6]. A DD video codec is implemented and used in the simulations. A practical distributed implementation architecture of the proposed scheme is presented in Section 5.5.

We set the GA parameters as follows: the population size is 15; $\theta = 0.7$; μ is varied from 0.3 to 0.1 using the schedule described in Section 5.3; σ^2 is set to 1, because it does not affect path selection decisions. The GA is terminated after a predefined number of generations or after a prespecified computation time elapsed. The best individual found by the GA is prescribed as the solution to problem OPT-MM.

5.3.2.1 Comparison with Trajectory Methods

For the purpose of comparison, we implemented SA [17] and TS [18], both of which have been used in solving certain networking problems. We used the geometric cooling schedule for the SA implementation with a decay coefficient $\omega = 0.99$ [17]. For the TS implementation, we chose a tabu list of five for small networks and ten for large networks [18].

In Figure 5.6, we plot the evolution of distortion values achieved by GA, SA, and TS for a 10-node network and a 50-node network, respectively.

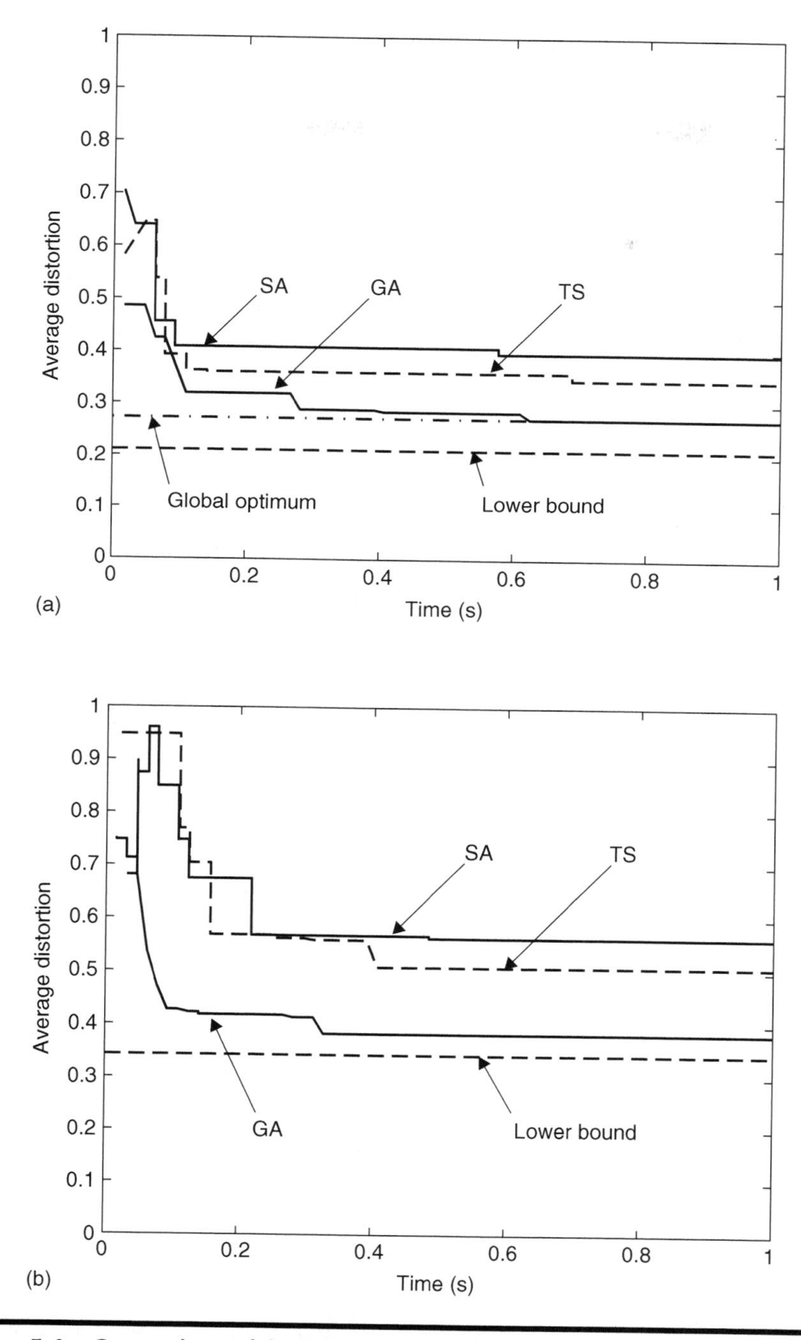

Figure 5.6 Comparison of distortion evolution of three metaheuristic methods. (a) Distortion evolution for a 10-node network, (b) distortion evolution for a 50-node network. (Reprinted from Mao, S., Y.T. Hou, X. Cheng, H.D. Sherali, S.F. Midkiff, and Y.-Q. Zhang. *IEEE Trans. Multimedia*, 8(5): 1063–1074, 2006.)

All the three metaheuristics are terminated after running for 1 s. On termination, GA evolved 210 and 75 generations in Figures 5.6a and 5.6b, respectively; SA ran for 1500 and 700 iterations in Figures 5.6a and 5.6b, respectively; TS ran for 1050 and 550 iterations in Figures 5.6a and 5.6b, respectively. GA had fewer iterations than SA and TS due to its higher computational complexity for each generation (or iteration).

The global optimal value in Figure 5.6a was obtained by an exhaustive search over the entire solution space, whereas the lower bound values in both figures were computed using a lower-bounding heuristic presented in Ref. 10. For both networks, the best distortion values found by GA are evidently much better than those by SA or TS. In Figure 5.6a, GA quickly converges to the global optimal, whereas both SA and TS are trapped at local optima (i.e., no further decrease in distortion value after hundreds of iterations). The same trend can be observed in the 50-node network case shown in Figure 5.6b, although the global optimum is not obtainable here.

An interesting observation from Figure 5.6 is that for GA, the biggest improvement in distortion is achieved in the initial iterations, whereas the improvement gets smaller as GA evolves more generations. The initial population is generated using the random construction method discussed in Section 5.3.1, with no consideration on video performance. The initial solutions usually have high distortion values. The distortion value quickly drops over iterations, indicating that the GA performance is not very sensitive to the quality of the initial population. Also note that the SA and TS curves increase at some time instances (e.g., the TS curve at 0.06 s in Figure 5.6a and the SA curve at 0.08 s in Figure 5.6b), which implies that a nonimproving solution is accepted to escape from local minima. In addition to providing much better solutions, another strength of GA over trajectory methods is that multiple "good" solutions can be found after a single run. Such extra good paths can be used as alternative (or backup) paths if needed.

5.3.2.2 Comparison with Network-Centric Approaches

In the following, we compare the GA-based approach with network-centric routing. Although there are many alternative approaches, we implement two popular network-centric multipath routing algorithms, namely k-shortest path (SP) routing (with $k=2$ or 2-SP) [32] and disjoint-path routing, Disjoint Pathset Selection Protocol (DPSP) [3]. Our 2-SP implementation uses hop count as routing metric such that two SPs are found. In our DPSP implementation, we set the link costs to $\log(1/p_{ij})$, for all $\{i, j\} \in E$, such that two disjoint paths having the highest end-to-end success probabilities are found. We compare the performance of our GA-based multipath routing with these two algorithms over a 50-node network using a video clip.

We choose a time-domain partitioning coding scheme by extending the H.263+ codec [7,8,23,14]. Because our approach is quite general, we conjecture that the same trend in performance would be observed for other video codecs, such as H.264 or MPEG-2 or MPEG-4. The QCIF sequence "Foreman" (400 frames) is encoded at 15 fps for each description. A 10 percent macroblock-level intrarefreshment is used. Each group of blocks (GOB) is carried in a different packet. The received descriptions are decoded and peak-signal-to-noise-ratio (PSNR) values of the reconstructed frames computed. When a GOB is corrupted, the decoder applies a simple error concealment scheme by copying from the corresponding slice in the most recent, correctly received frame.

The PSNR curves of the received video frames are plotted in Figure 5.7. We observe that the PSNR curve obtained by GA is well above those obtained by the aforementioned network-centric routing approaches. Using GA, the improvements in average PSNR value over 2-SP and DPSP are 6.29 and 4.06 dB, respectively. We also experiment with an improved 2-SP algorithm, which also uses link success probabilities as link metric (as in DPSP). In this case, our GA-based routing achieves a 1.27 dB improvement over this enhanced 2-SP version, which is still significant in terms of visual video quality.

5.4 A Branch-and-Bound-Based Approach

This section reconsiders the problem of multimedia-centric multipath routing with a different approach, aiming to develop efficient algorithms and more importantly, providing performance guarantees for multimedia applications. In addition to developing efficient algorithms (as in Section 5.3), we believe it is also important to provide theoretical guarantees on the application performance, that is, bounding the optimality gap (or, the gap between the global optimum and a feasible solution produced by an approximation algorithm) for multimedia applications. Such guarantees are very useful in many cases, such as providing a suitable trade-off between optimality and complexity of computations, as well as providing important insights on performance limits and guidance for designing better algorithms.

We address this problem with a novel RLT and branch-and-bound-based approach [33,34]. We will present a branch-and-bound-based framework, predicated on the RLT [33], which can produce ε-optimal solutions to the optimal multipath routing problem. Here ε is an arbitrarily small number in $(0, 1)$ that reflects the application's tolerance on the optimality of the resulting paths. Based on this rigorous theoretical framework, we develop two algorithms for solving the optimal multipath routing problem:

- A fast heuristic algorithm that can produce highly competitive paths but without guarantees on bounded optimality
- An ε-optimal algorithm that can produce a pair of paths within the ε range of the global optimum

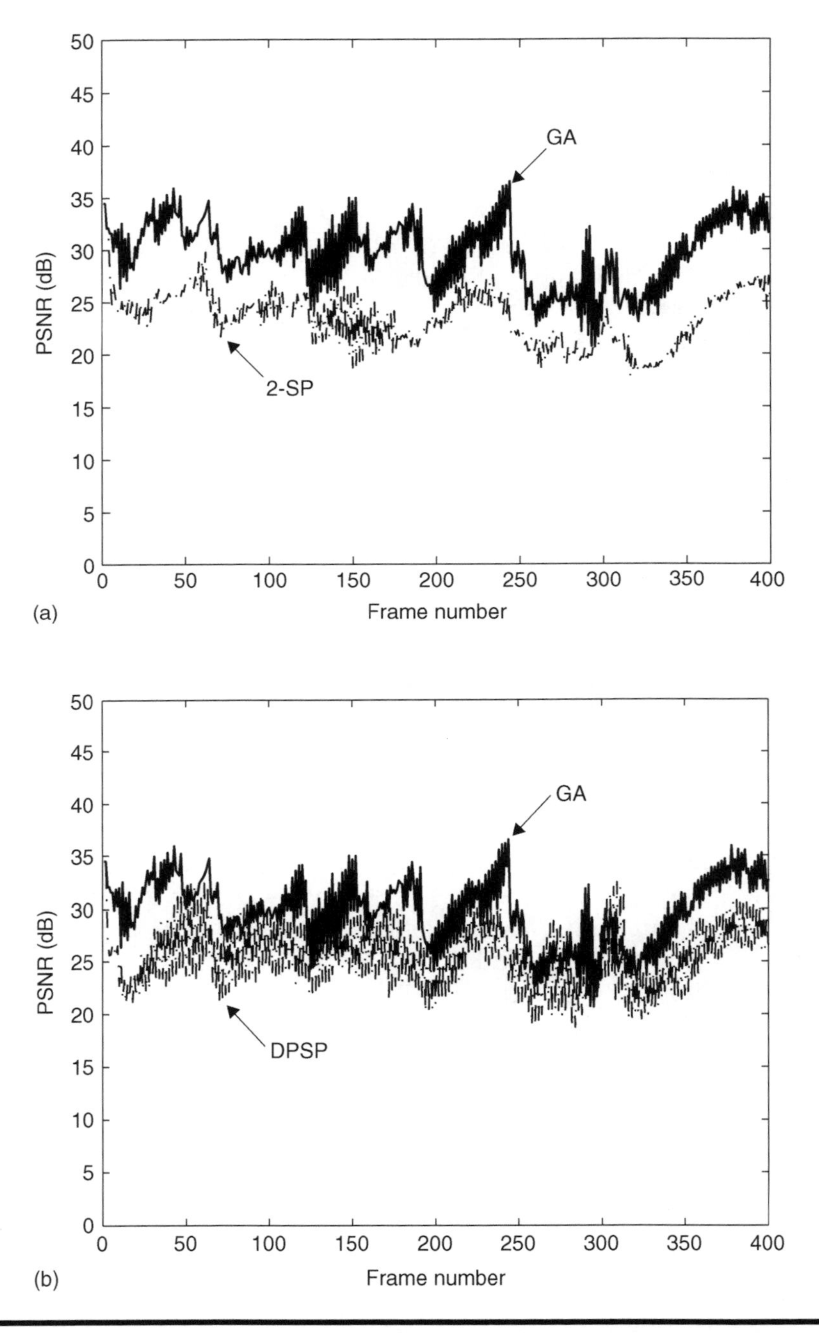

Figure 5.7 PSNR curves of received video sequences. (a) GA-based algorithm versus 2-SP, (b) GA-based algorithm versus DPSP. (Reprinted from Mao, S., Y.T. Hou, X. Cheng, H.D. Sherali, S.F. Midkiff, and Y.-Q. Zhang. *IEEE Trans. Multimedia*, 8(5): 1063–1074, 2006.)

The proposed algorithms are computationally efficient and the ε-optimal algorithm provides an elegant trade-off between optimality and computational complexity.

5.4.1 The Branch-and-Bound Framework

Branch-and-bound is an iterative algorithmic method for solving optimization problems, especially in discrete and combinatorial optimization. It seeks to produce an exact or ε-optimal solution to a nonlinear programming (NLP) problem by partitioning the original solution space into subhyperrectangles [33].

In branch-and-bound, the original problem is first relaxed using a suitable relaxation technique to obtain an easier-to-solve, lower-bounding problem. In our approach, we choose RLT [33,34] to reformulate and linearize OPT-MR into an LP relaxation OPT-MR(ℓ) (discussed in Section 5.4.2). The optimal solution to this LP relaxation provides a lower bound LB for the original problem. Because such an LP relaxation usually yields an infeasible solution to the original problem, a local search algorithm should be employed to obtain a feasible solution to the original problem, using the infeasible lower bounding solution as a starting point. The resulting feasible solution then provides an upper bound UB for the original problem.

The branch-and-bound procedure is based on the idea of divide-and-conquer, that is, the original problem O (or OPT-MR) is partitioned into subproblems, each having a smaller feasible solution space, based on the solution provided by the LP relaxation. New subproblems are organized as a branch-and-bound tree, whereas this partitioning or branching process is carried out recursively to obtain two new subproblems at each node of the tree. The strategy for partitioning the original solution space, that is, the branching rule, will be explained later in this section.

The subproblems are also inserted into a problem list L, which records the active nodes in the branch-and-bound tree structure. More specifically, in the beginning, the problem list L is initialized with the original problem O. At any given iteration, the lower and upper bounds for O are computed as

$$\begin{cases} \text{LB} = \min\{\text{LB}_k \ : \text{Problem } k \in L\} \\ \text{UB} = \min\{\text{UB}_k : \text{all explored nodes } k \text{ thus far}\} \end{cases} \tag{5.17}$$

The method proceeds by choosing the next problem to partition from the problem list. In our approach, the problem $k \in L$ having the smallest LB_k is chosen. This problem k is then partitioned into two subproblems k_1 and k_2, which replace problem k in L. Every time a problem k is added to the list, LB_k and UB_k are computed, and the LB and UB for the original problem O are updated. At any given iteration, if $\text{LB} \geq (1-\varepsilon) \cdot \text{UB}$, the procedure exits with an ε-optimal solution. Also, for any problem k in the problem list, if $\mathrm{LB}_k \geq (1-\varepsilon)\cdot \mathrm{UB}$, no globally optimal solution that improves beyond the ε-tolerance can exist in the subspace of the feasible region represented by this node. Therefore, this node can be removed (or fathomed) from the branch-and-bound tree. In this manner, the branch-and-bound process can fathom certain branches or nodes of the tree, eliminating them from further exploration. The effectiveness of the branch-and-bound procedure depends strongly on that of the employed fathom strategy.

As far as the partitioning process is concerned, the original feasible solution space Ω is decomposed into two corresponding hyperrectangles, based on the so-called branching variable. In RLT, the discrepancy between an RLT substitution variable and the corresponding nonlinear term that this variable represents (e.g., $p^1_{\mathrm{dj}} \cdot p^2_{\mathrm{dj}}$) is called the relaxation error. In our algorithm, the branching variable is chosen to be the one that yields the largest relaxation error. Such a branching rule ensures that all the discrepancies between the RLT substitution variables and the corresponding nonlinear terms will be driven to zero as the algorithm evolves over iterations.

5.4.2 Problem Reformulation

As discussed, the objective function of problem OPT-MR is a highly complex function of exponential terms of the x-variables. Our first goal is to reformulate these terms, which will greatly simplify the objective function as well as the constraints. In the following, we first reformulate problem OPT-MR into a 0-1 mixed-integer polynomial programming problem OPT-MR(p). We then replace all the nonlinear terms as discussed earlier and add the corresponding RLT constraints into the problem formulation to obtain a linear programming relaxation of problem OPT-MR, denoted as OPT-MR(ℓ), in Section 5.4.3.

Our approach is to use the RLT approach in the reformulation and linearization of the original problem. RLT is a relaxation technique that can be used to produce tight polyhedral outer approximations or linear programming relaxations for an underlying nonlinear, nonconvex polynomial programming problem, which, in essence, can provide a tight lower bound on a minimization problem [33,34]. In the RLT procedure, nonlinear implied constraints are generated by taking the products of bounding factors defined in terms of the decision variables up to a suitable order, and also, possibly products of other defining constraints of the problem. The resulting problem is subsequently linearized by variable substitutions, by defining a new variable for each nonlinear term appearing in the problem, including both the objective function and the constraints.

In Equation 5.9, there are four high-order terms that need to be reformulated, namely, $p_{\mathrm{jnt}}, p^1_{\mathrm{dj}}, p^2_{\mathrm{dj}}$, and α. From their definitions in Equations 5.4

and 5.7, we can rewrite the success probabilities as

$$\begin{cases} p_{\text{jnt}} = \prod_{\{i,j\}\in E} p_{ij}^{\{x_{ij}^1 \cdot x_{ij}^2\}} \\ p_{\text{dj}}^1 = \prod_{\{i,j\}\in E} p_{ij}^{\{x_{ij}^1 \cdot (1-x_{ij}^2)\}} \\ p_{\text{dj}}^2 = \prod_{\{i,j\}\in E} p_{ij}^{\{x_{ij}^2 \cdot (1-x_{ij}^1)\}} \end{cases} \tag{5.18}$$

Taking logarithms on both sides, we can convert the high-order terms on the right-hand-side (RHS) of Equation 5.18 into summations of quadratic terms of the x-variables, that is,

$$\begin{cases} \log(p_{\text{jnt}}) = \sum_{\{i,j\}\in E} [x_{ij}^1 \cdot x_{ij}^2 \cdot \log(p_{ij})] \\ \log(p_{\text{dj}}^1) = \sum_{\{i,j\}\in E} [x_{ij}^1 \cdot (1 - x_{ij}^2) \cdot \log(p_{ij})] \\ \log(p_{\text{dj}}^2) = \sum_{\{i,j\}\in E} [x_{ij}^2 \cdot (1 - x_{ij}^1) \cdot \log(p_{ij})] \end{cases} \tag{5.19}$$

Furthermore, we can rewrite α according to Equations 5.6 and 5.8 as

$$\alpha = 1 - \prod_{\{i,j\}\in E} \left[1 - \frac{1 - p_{ij}}{p_{ij} \cdot l_{ij}}\right]^{\{x_{ij}^1 \cdot x_{ij}^2\}} \tag{5.20}$$

Letting $\phi = 1 - \alpha$ and taking logarithms on both sides, we have

$$\log(\phi) = \sum_{\{i,j\}\in E} [x_{ij}^1 \cdot x_{ij}^2 \cdot \log(h_{ij})] \tag{5.21}$$

where $h_{ij} = 1 - (1 - p_{ij})/(p_{ij} \cdot l_{ij})$ is a constant for all $\{i, j\} \in E$.

Having simplified the high-order terms, we now deal with the resulting constraints of the form $y = \log(\lambda)$, as shown in logarithmic terms 5.19 and 5.21. We can linearize this logarithmic function over some tightly bounded interval using a polyhedral outer approximation comprised of a convex envelope in concert with several tangential supports. For instance, if λ is bounded as $0 < \lambda_0 \le \lambda \le 1$, these constraints can be written as follows:

$$\begin{cases} y \ge \dfrac{\log(\lambda_0)}{1 - \lambda_0} \cdot (1 - \lambda) \\ y \le \log(\lambda_k) + \dfrac{\lambda - \lambda_k}{\lambda_k}, \quad k = 1, \ldots, k_{\max} \end{cases} \tag{5.22}$$

where $\lambda_k = \lambda_0 + (1 - \lambda_0) \cdot (k - 1)/(k_{\max} - 1)$, for $k = 1, 2, \ldots, k_{\max}$. A four-point tangential approximation can be obtained by letting $k_{\max} = 4$, as

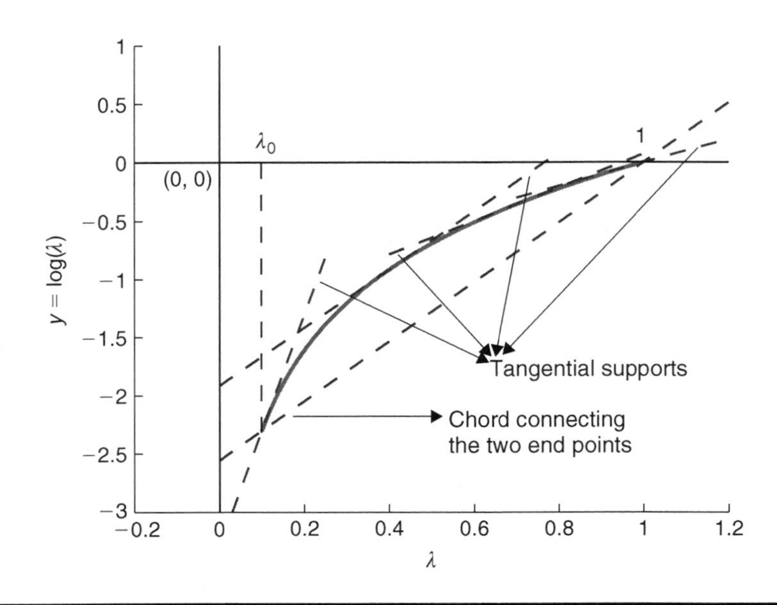

Figure 5.8 Polyhedral outer approximation for $y = \log(\lambda)$ in $0 < \lambda_0 \leq \lambda \leq 1$. (Reprinted from Kompella, S., S. Mao, Y.T. Hou, and H.D. Sherali. *IEEE J. Selected Areas Commun.*, 25(4): 831–840, 2007.)

illustrated in Figure 5.8 [35]. The corresponding convex envelope consists of a chord connecting the two end points, which is used in combination with tangential supports at four points including the two end points. As a result, every logarithmic constraint specified in logarithmic terms 5.19 and 5.21 translates into five linear constraints as in Equation 5.22. Note that such polyhedral outer approximations will be iteratively tightened during the branch-and-bound procedure (see Section 5.4.1).

Substituting ϕ, we can rewrite Equation 5.9 as

$$\begin{cases} P_{00} = p_{\text{jnt}} \cdot p_{\text{dj}}^1 \cdot p_{\text{dj}}^2 \cdot \phi \\ P_{01} + P_{00} = p_{\text{jnt}} \cdot p_{\text{dj}}^1 \\ P_{10} + P_{00} = p_{\text{jnt}} \cdot p_{\text{dj}}^2 \\ P_{00} + P_{01} + P_{10} + P_{11} = 1 \end{cases} \tag{5.23}$$

and the objective function can be rewritten as

$$\begin{aligned} D = P_{00} \cdot d_0 + \left(p_{\text{jnt}} \cdot p_{\text{dj}}^1 - P_{00}\right) \cdot d_1 + \left(p_{\text{jnt}} \cdot p_{\text{dj}}^2 - P_{00}\right) \cdot d_2 \\ + \left(1 + P_{00} - p_{\text{jnt}} \cdot p_{\text{dj}}^1 - p_{\text{jnt}} \cdot p_{\text{dj}}^2\right) \end{aligned} \tag{5.24}$$

This reduces OPT-MR into an approximating 0-1 mixed-integer polynomial programming problem OPT-MR(p). The constraints of OPT-MR(p) include the original constraints 5.12 through 5.15, the reformulated constraint 5.23, and the new constraints derived from reformulating the logarithmic terms 5.19 and 5.21 (in the form of Equation 5.22).

5.4.3 Problem Linearization

Although greatly simplified, problem OPT-MR(p) is still a polynomial programming problem, which is NP-hard in general [33]. This section linearizes problem OPT-MR(p) by defining substitution variables and introducing linear RLT bound-factor constraints, so as to obtain a linear program (LP).

Consider a quadratic product term of the form $(p^1_{\mathrm{dj}} \cdot p^2_{\mathrm{dj}})$. By introducing a new variable $z_0 = p^1_{\mathrm{dj}} \cdot p^2_{\mathrm{dj}}$, we can substitute the $(p^1_{\mathrm{dj}} \cdot p^2_{\mathrm{dj}})$ terms in constraint 5.23 and objective function 5.24 with z_0, thus removing this quadratic term from the objective function and constraints. Assuming p^1_{dj} and p^2_{dj} are each bounded as $(p^1_{\mathrm{dj}})_L \le p^1_{\mathrm{dj}} \le (p^1_{\mathrm{dj}})_U$ and $(p^2_{\mathrm{dj}})_L \le p^2_{\mathrm{dj}} \le (p^2_{\mathrm{dj}})_U$, respectively, we can generate the following relational constraints, which are known as the RLT bound-factor product constraints:

$$\begin{cases} \left\{\left[p^1_{\mathrm{dj}} - \left(p^1_{\mathrm{dj}}\right)_L\right] \cdot \left[p^2_{\mathrm{dj}} - \left(p^2_{\mathrm{dj}}\right)_L\right]\right\}_{\mathrm{LS}} \ge 0 \\ \left\{\left[p^1_{\mathrm{dj}} - \left(p^1_{\mathrm{dj}}\right)_L\right] \cdot \left[\left(p^2_{\mathrm{dj}}\right)_U - p^2_{\mathrm{dj}}\right]\right\}_{\mathrm{LS}} \ge 0 \\ \left\{\left[\left(p^1_{\mathrm{dj}}\right)_U - p^1_{\mathrm{dj}}\right] \cdot \left[p^2_{\mathrm{dj}} - \left(p^2_{\mathrm{dj}}\right)_L\right]\right\}_{\mathrm{LS}} \ge 0 \\ \left\{\left[\left(p^1_{\mathrm{dj}}\right)_U - p^1_{\mathrm{dj}}\right] \cdot \left[\left(p^2_{\mathrm{dj}}\right)_U - p^2_{\mathrm{dj}}\right]\right\}_{\mathrm{LS}} \ge 0 \end{cases}$$

where $\{\cdot\}_{\mathrm{LS}}$ denotes a linearization step under the substitution $z_0 = p^1_{\mathrm{dj}} \cdot p^2_{\mathrm{dj}}$. From the preceding relationships and by substituting $z_0 = p^1_{\mathrm{dj}} \cdot p^2_{\mathrm{dj}}$, we obtain the following RLT constraints for z_0.

$$\begin{cases} \left(p^1_{\mathrm{dj}}\right)_L \cdot p^2_{\mathrm{dj}} + \left(p^2_{\mathrm{dj}}\right)_L \cdot p^1_{\mathrm{dj}} - z_0 \le \left(p^1_{\mathrm{dj}}\right)_L \cdot \left(p^2_{\mathrm{dj}}\right)_L \\ \left(p^1_{\mathrm{dj}}\right)_L \cdot p^2_{\mathrm{dj}} + \left(p^2_{\mathrm{dj}}\right)_U \cdot p^1_{\mathrm{dj}} - z_0 \ge \left(p^1_{\mathrm{dj}}\right)_L \cdot \left(p^2_{\mathrm{dj}}\right)_U \\ \left(p^1_{\mathrm{dj}}\right)_U \cdot p^2_{\mathrm{dj}} + \left(p^2_{\mathrm{dj}}\right)_L \cdot p^1_{\mathrm{dj}} - z_0 \ge \left(p^1_{\mathrm{dj}}\right)_U \cdot \left(p^2_{\mathrm{dj}}\right)_L \\ \left(p^1_{\mathrm{dj}}\right)_U \cdot p^2_{\mathrm{dj}} + \left(p^2_{\mathrm{dj}}\right)_U \cdot p^1_{\mathrm{dj}} - z_0 \le \left(p^1_{\mathrm{dj}}\right)_U \cdot \left(p^2_{\mathrm{dj}}\right)_U \end{cases}$$

We therefore replace the second-order term $p^1_{\mathrm{dj}} \cdot p^2_{\mathrm{dj}}$ with the linear term z_0 in constraint 5.23 and objective function 5.24, and introduce the preceding linear RLT bound-factor constraints for z_0 into the problem formulation.

We also define new variables for all the remaining nonlinear terms that are found in the reformulated problem OPT-MR(p), including $z_1 = p_{\text{jnt}} \cdot p_{\text{dj}}^1$, $z_2 = p_{\text{jnt}} \cdot p_{\text{dj}}^2$, and $z_3 = z_0 \cdot \phi$, and make substitutions in the same manner. We can then rewrite the objective function 5.24 and constraint 5.23 as

$$D = P_{00} \cdot d_0 + (z_1 - P_{00}) \cdot d_1 + (z_2 - P_{00}) \cdot d_2 \\ + (1 + P_{00} - z_1 - z_2) \tag{5.25}$$

and

$$\begin{cases} P_{01} + P_{00} = z_1 \\ P_{10} + P_{00} = z_2 \\ P_{00} + P_{01} + P_{10} + P_{11} = 1 \end{cases} \tag{5.26}$$

The constraints derived from reformulating the logarithmic terms 5.19 and 5.21 (in the form of Equation 5.22) can also be linearized by substituting $z_{ij} = x_{ij}^1 \cdot x_{ij}^2$, and by introducing the corresponding linear RLT bound-factor constraints, for all $\{i, j\} \in E$.

This concludes the reformulation and linearization of problem OPT-MR. As a result, we obtain a linear programming relaxation problem OPT-MR(ℓ), for which many efficient (polynomial-time) solution techniques and tools are available [36].

5.4.4 A Heuristic Algorithm

Before we describe the ε-optimal algorithm, we develop a fast heuristic algorithm in this section (called Heuristic). As shown in Section 5.4.6, Heuristic can compute competitive paths very quickly (e.g., in a few hundred milliseconds for a 100-node network). However, it does not provide theoretical guarantees on the application performance.

Let the solution produced by Heuristic be $\overline{\psi}$. We first solve the relaxed problem OPT-MR(ℓ) to obtain a possibly infeasible solution $\hat{\psi}$. For example, due to the RLT relaxation, the binary x-variables in $\hat{\psi}$ could actually be fractional. If $\hat{\psi}$ is already feasible to problem OPT-MR, we set $\overline{\psi} = \hat{\psi}$ and then obtain a feasible solution in one iteration. Otherwise, we apply a local search algorithm to obtain a rounded feasible solution $\overline{\psi}$ to problem OPT-MR. If necessary, we can also perform a restricted search by making a limited perturbation around the rounded feasible solution $\overline{\psi}$ to obtain an even better solution. The local search algorithm is presented in Section 5.4.5.

Heuristic obtains a feasible solution by solving the root node of the branch-and-bound tree and applying the local search algorithm. Owing to the properly designed RLT relaxations (see Section 5.4.2), the Heuristic solution is highly competitive, and in many cases it achieves ε-optimality, as will be seen in Section 5.4.6.

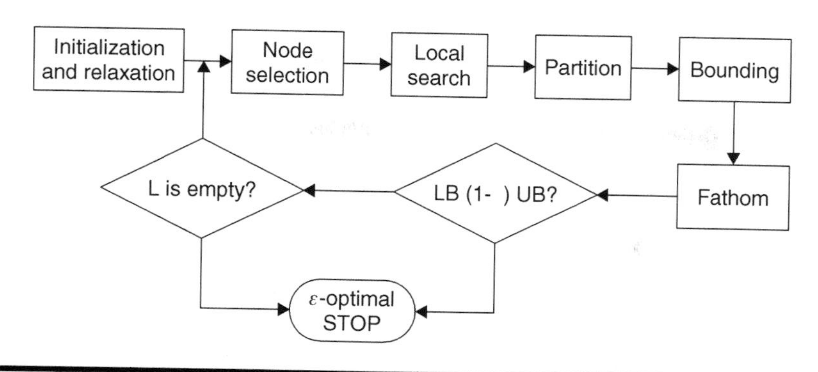

Figure 5.9 Flowchart for the ε-optimal algorithm ALG(ε). (Reprinted from Kompella, S., S. Mao, Y.T. Hou, and H.D. Sherali. *IEEE J. Selected Areas Commun.*, 25(4): 831–840, 2007.)

5.4.5 An ε-Optimal Algorithm

Based on the RLT/branch-and-bound framework, we develop an algorithm, called ALG(ε), that computes ε-optimal paths for DD video streaming. Figure 5.9 shows the flowchart of this algorithm [35]. The iterative branch-and-bound algorithm terminates with an ε-optimal solution when either the lower bound for the original problem O is within ε of the upper bound, that is, $\text{LB} \geq (1-\varepsilon) \cdot \text{UB}$, or the problem list L is empty. The operation of each step in ALG(ε) is described in the following sections.

5.4.5.1 Initialization and Relaxation

We start by initializing the current "best" solution, denoted as ψ^*, with the heuristic solution obtained as described in Section 5.4.4, and the current "best" upper bound UB as the objective value obtained with the heuristic solution. We then initialize the problem list L with the original problem (i.e., problem O). We denote the objective value obtained from the LP relaxation as a lower bound LB_1 for problem O. Also, because this is the only problem in the problem list, we initialize LB_1 as the current "best" lower bound LB for the original problem, that is, set $\text{LB} = \text{LB}_1$.

5.4.5.2 Node Selection

At every iteration, problem k (or the corresponding node in the branch-and-bound tree) that has the minimum LB_k among all the problems $k \in L$ is selected. As discussed earlier, this problem is indicative of the lower bound for the original problem. Subsequent operations of local search, partitioning, and bounding are performed on this problem k.

5.4.5.3 Local Search

As discussed in Section 5.4.1, the solution to the relaxation problem k that is selected in the node selection step is usually infeasible to the original problem O. This is especially true if the original problem involves binary variables (i.e., the x-variables could be fractions). A local search algorithm should be used to find a feasible solution to the original problem starting from the infeasible lower-bounding solution.

Let $\hat{\psi}$ be the infeasible (or fractional) solution obtained by solving the LP relaxation of the original problem. Starting from this fractional solution, we solve for $h=1,2$ in the following SP problem

$$\text{Minimize} \sum_{\{i,j\}\in E} \left(-\hat{x}_{ij}^{h}\right) \cdot x_{ij}^{h} \tag{5.27}$$

subject to the flow constraints. Note that for an optimization variable y, $\hat{y}$ denotes its value in the infeasible solution $\hat{\psi}$. Solving these SP problems provides us with a rounded heuristic solution $\overline{\psi}$ that has a tendency to round up relatively higher valued components of $\hat{\psi}$ and round down relatively lower valued components. The distortion value of the rounded solution $\overline{\psi}$ is an upper bound for this subproblem, that is, UB_k.

5.4.5.4 Partitioning

The objective of the partitioning step is to find the branching variable that will enable us to split the feasible solution space Ω_k of problem k into two subfeasible solution spaces Ω_{k_1} and Ω_{k_2}. In ALG(ε), we need to consider three classes of optimization variables for partitioning, that is, the binary x-variables, the substitution variables (e.g., z_0), and the the logarithm substitution terms (e.g., ϕ in logarithmic term 5.21).

When partitioning based on the x-variables, we need to select a variable that will offer the highest gain in terms of improving the objective value. For this purpose, we should choose the x-variable, which is factional and the closest to 0.5. A strategy that works well is to first find the index variable pair $\{x_{ij}^1, x_{ij}^2\}$, for all $\{i,j\}\in E$ that gives the largest discrepancy between the RLT substitution variable $\hat{z}_{ij}$ and the corresponding nonlinear product $(\hat{x}_{ij}^1 \cdot \hat{x}_{ij}^2)$ (see Section 5.4.3). We then choose x_{ij}^1 or x_{ij}^2 to partition the problem (by fixing it to 0 or 1) depending on which variable is closer to 0.5. We break a tie arbitrarily.

In addition to the x-variables, we also need to examine branching decisions based on the substitution variables such as $z_0 = p_{\text{dj}}^1 \cdot p_{\text{dj}}^2$. For such variables, we first find the maximum relaxation error between the substitution variable and the corresponding product term, say, $|\hat{p}_{\text{dj}}^1 \cdot \hat{p}_{\text{dj}}^2 - \hat{z}_0|$.

We then verify whether the following condition is satisfied.

$$[(p_{\text{dj}}^1)_U - (p_{\text{dj}}^1)_L] \cdot \min\{\hat{p}_{\text{dj}}^1 - (p_{\text{dj}}^1)_L, (p_{\text{dj}}^1)_U - \hat{p}_{\text{dj}}^1\}$$
$$\geq [(p_{\text{dj}}^2)_U - (p_{\text{dj}}^2)_L] \cdot \min\{\hat{p}_{\text{dj}}^2 - (p_{\text{dj}}^2)_L, (p_{\text{dj}}^2)_U - \hat{p}_{\text{dj}}^2\}$$

If this condition holds true, we partition the solution space Ω_k of problem k into two new regions Ω_{k_1} and Ω_{k_2}, by dividing the range $[(p_{\text{dj}}^1)_L, (p_{\text{dj}}^1)_U]$ into two subregions $[(p_{\text{dj}}^1)_L, \hat{p}_{\text{dj}}^1]$ and $[\hat{p}_{\text{dj}}^1, (p_{\text{dj}}^1)_U]$. Otherwise, we partition Ω_k by dividing $[(p_{\text{dj}}^2)_L, (p_{\text{dj}}^2)_U]$ into $[(p_{\text{dj}}^2)_L, \hat{p}_{\text{dj}}^2]$ and $[\hat{p}_{\text{dj}}^2, (p_{\text{dj}}^2)_U]$.

Finally, the branching decisions also include the logarithm substitution terms, for example, ϕ in logarithmic term 5.21. In such cases, we first find the variable that gives the greatest discrepancy between the logarithm value, say, $\log(\hat{\phi})$ and the RHS of the corresponding substitution (e.g., logarithmic term 5.21) among all such terms, and then either bisect the interval of this variable (e.g., $[(\phi)_L, (\phi)_U]$) evenly, or divide this interval at the point $\hat{\phi}$.

5.4.5.5 Bounding

In the bounding step, we solve the RLT relaxations for the two subproblems identified in the partitioning step, and obtain their corresponding lower bounds LB_{k_1} and LB_{k_2}, as well as update the incumbent lower bounding solution. The corresponding upper bounds, that is, UB_{k_1} and UB_{k_2}, are obtained by applying the local search algorithm starting from the relaxation solutions obtained, and the current LB and UB values are updated according to Equation 5.17. If any of the following conditions, $(1-\varepsilon)\cdot \text{UB} > \text{LB}_{k_1}$ and $(1-\varepsilon)\cdot \text{UB} > \text{LB}_{k_2}$, are satisfied, we add the corresponding problem into the problem list L, and remove problem k from the list.

5.4.5.6 Fathoming

For any problem k in the problem list L, if $\text{LB}_k \geq (1-\varepsilon)\cdot \text{UB}$, then the subspace corresponding to this problem does not contain any solution that improves beyond the ε-tolerance of the incumbent solution. Therefore, we can prune this problem from the problem list, such that the computation time can be reduced.

5.4.6 Numerical Results

This section presents our simulation studies of the optimal multipath routing problem. As in Section 5.3.2, we generate a multihop wireless network topology by placing a number of nodes at random locations in a rectangular region, where connectivity is determined by the distance coverage of each node's transmitter. The area is adjusted for networks with different numbers of nodes to achieve an appropriate node density for a connected network. The source–destination nodes s and t are uniformly chosen from the nodes.

Assuming some efficient link statistic measurement and distribution schemes [19,37–39], we focus on the network layer and application layer models [21]. For every link, the failure probability is uniformly chosen from [0.01, 0.3]; the available bandwidth is uniformly chosen from [100, 400] kbps, in 50 kbps steps; the mean burst length is uniformly chosen from [2,6].

An H.263+ like DD video codec is implemented and used in the simulations, which encodes video into balanced descriptions (i.e., $R_1 = R_2 = R$). In all the simulations an ε value of 1 percent is used (i.e., the ALG(ε) distortion is within 1 percent range of the global minimum distortion). The proposed algorithms are implemented in C, and the LINDO API 3.0 is used for solving the LP relaxation problem OPT-MR(ℓ).

5.4.6.1 *Performance of the Proposed Algorithms*

We first examine the performance of the proposed algorithms with different network topologies. Table 5.2 shows the computation time, and the distortion values found by the proposed Heuristic algorithm, and the ε-optimal algorithm ALG(ε). We experimented with six networks with 20, 30, 50 and 100 nodes, respectively. In Table 5.2, the third and fourth columns present the average distortion values found by Heuristic and ALG(ε), and the fifth and sixth columns show the computation time used to run each algorithm, using a Pentium 4 2.4 GHz computer (512 MB memory). Because the optimality tolerance ε is set to 1 percent, the distortion values found by ALG(ε) in the fourth column is within 1 percent range of the achievable minimum distortion for the received video.

We find that the distortion values produced by the two algorithms are very close to one another. For networks I, III, IV, and V, the two distortion values are exactly the same. The maximum difference between the distortion values occurs in network II, which gives a 2.6 percent normalized difference. This indicates that the relaxations made in reformulating and

Table 5.2 Performance of the Proposed Algorithms ($\varepsilon = 0.01$)

Network	*Size*	*Distortion Heuristic*	*Distortion ALG(ε)*	*Time Heuristic*	*Time ALG(ε)*
I	20	0.657	0.657	0.11	0.14
II	30	0.670	0.653	0.25	0.69
III	50	0.656	0.656	0.61	0.83
IV	50	0.615	0.615	0.42	1.87
V	100	0.555	0.555	0.76	2.57
VI	100	0.547	0.537	0.71	1.84

linearizing problem OPT-MR are properly designed (see Section 5.4.2); we do not sacrifice much in optimality by making those relaxations. In addition, this also demonstrates the efficacy of the local search algorithm. The same trend is observed in most of the simulations we performed.

In addition, both algorithms are computationally efficient. Heuristic terminates in a few hundred milliseconds in all the cases, even for the 100-node networks. However, ALG(ε) has a relatively longer execution time than Heuristic (e.g., a couple of seconds for the 100-node networks), but with the strength of providing guarantees on the optimality of the paths found. In practice, Heuristic could be used to compute a pair of highly competitive paths very quickly for a DD video session, whereas for mission-critical applications, ALG(ε) can be used to provide guarantees on optimality. For computationally constrained devices, we can simply choose a larger ε value to reduce the computational complexity.

5.4.6.2 Comparison with k-Shortest Path Routing

This section compares the proposed algorithms with a representative network-centric routing scheme. Specifically, we implement the k-SP routing algorithm (with $k = 2$ or 2-SP) [32]. Our 2-SP implementation uses hop count as routing metric such that two SPs are found.

We transmit DD video along the paths found by the algorithms in a 50-node network, encoded at rates ranging from 64 to 320 kbps. The same time-domain partitioning coding scheme was used as in Section 5.3.2. The QCIF sequence "Foreman" (400 frames) is encoded at 15 fps for each description. A 10 percent macroblock-level intrarefreshment is used. Each GOB is carried in a different packet. When a GOB is corrupted, the decoder applies a simple error concealment scheme by copying the corresponding slice from the most recent, correctly received frame.

The average PSNR values of the reconstructed videos (over all the frames) are plotted in Figure 5.10 for increased description rates. The upper most curve in the figure is for the loss-free case, where the corresponding video only suffers distortion due to the encoder. Hence this curve is an upper bound for the PSNR curve achieved by any routing scheme. We find that the ALG(ε) curve closely follows the loss-free curve, starting from low description rates. For further increased R, the ALG(ε) curve gets flat and the gap between itself and the loss-free curve increases from 0.59 dB when $R = 64$ kbps to 3.66 dB when $R = 320$ kbps. Similar observation can be made for the 2-SP curve, although it is much flatter than the ALG(ε) curve. Again, this is because a large R makes more low bandwidth links ineligible for routing and may therefore force the algorithms to choose high loss, but high bandwidth links. The distortion increase due to higher transmission errors will then dominate the distortion reduction due to the higher rate at the coder, and the net effect is a lower average PSNR value.

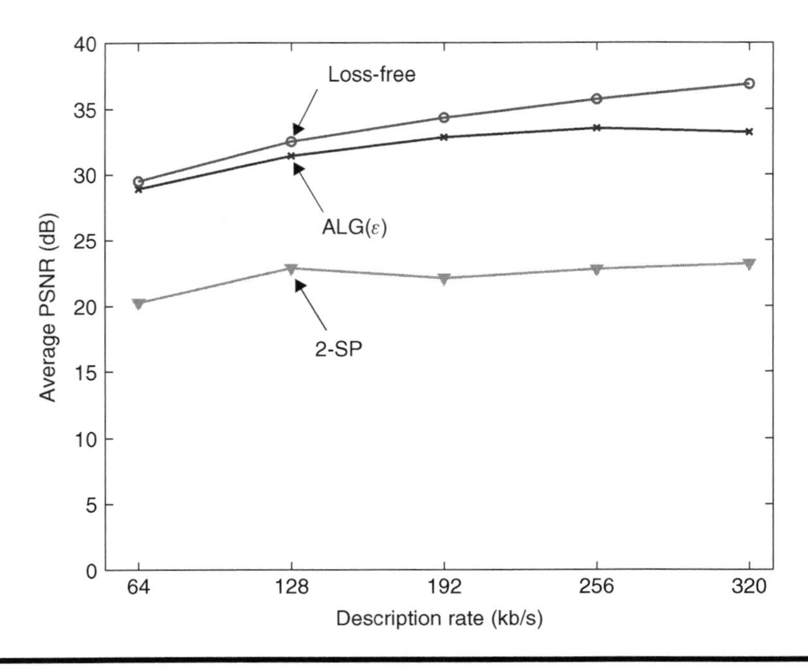

Figure 5.10 Average PSNR values for different description rates.

For all the description rates examined, the ALG(ε) average values are much higher than the corresponding 2-SP values, with differences ranging from 8.57 to 10.75 dB. 2-SP performs poorly when compared with ALG(ε). Although 2-SP finds shortest paths, it does so without much consideration for the application performance and SPs do not necessarily provide good video quality.

5.5 Considerations for Distributed Implementation

This section gives a brief sketch of our approach to distributed implementation of the proposed algorithms. Existing routing protocols can be roughly classified into two categories: (i) source routing and (ii) hop-by-hop routing [40]. In source routing, paths are computed at the source node and carried in the packet header, whereas hop-by-hop routing requires each node set up and maintains a routing table and is usually implemented in a completely distributed manner. Source routing is quite suitable for multimedia applications because it computes routes on the per-session basis, therefore supporting sessions with diverse QoS requirements. In addition, source routing allows the coexistence of several algorithms for different purposes at the source node.

We find that the proposed algorithms are highly suitable for the source routing paradigm. As Dijkstra like algorithms, they require network topology and link metrics. These algorithms are actually complementary to existing network-centric routing protocols for multihop wireless networks, and can be implemented by augmenting these existing protocols, as discussed in the following.

In proactive routing protocols, such as the OLSR [19], an up-to-date network topology is always maintained. However, for QoS routing, each node should measure the link statistics and distribute such statistics by controlled flooding of link state advertisements (e.g., TC messages in OLSR and QOLSR [37]). Once the topology information is available, we can simply replace the routing engine in such protocols with the GA-based algorithm or ALG(ε) to compute highly competitive paths, rather than the SPs in the original protocol.

In reactive routing protocols, such as the Dynamic Source Routing (DSR) protocol [20], a source node discovers network topology by flooding route request (REQ) messages and receiving route replies (REP) from the destination node. When one or more REPs are received, a path selection algorithm will be used to select, say, the shortest path among the discovered paths. To augment this class of protocols, we can piggy-back link statistics in the REP messages, and then substitute the shortest path selection algorithm with either ALG(ε) or Heuristic to select highly competitive paths for DD video. Although the resulting paths may be suboptimal due to incomplete network topology information, such a distributed implementation may have low control overhead and more importantly, can still significantly improve the performance of existing reactive routing protocols.

It is worth noting that both link statistic measurement and distribution are not unique in our implementation; these are actually common components in QoS routing protocols for multihop wireless networks, be it proactive [37] or reactive [4]. For a distributed implementation, we only need to augment these protocols with a more powerful routing engine. Furthermore, such an implementation allows the coexistence of the video-centric routing and the traditional network-centric routing in the nodes. For general data applications, the standard network-centric routing engine will be used, whereas for DD video streaming, the proposed GA or the ε-optimal routing algorithm (or Heuristic) will be used.

5.6 Conclusion

In this chapter, we studied the problem of cross-layer multipath routing for MD video in multihop wireless networks. In particular, we provided practical answers to the following key questions: (i) How to formulate a multimedia-centric routing problem that minimizes video distortion by choosing optimal paths? (ii) What are the performance limits? (iii) How to design efficient solution procedures? (iv) How to implement the proposed algorithms in a distributed environment? We showed that cross-layer design is imperative and effective to meeting the stringent QoS requirements of multimedia applications in highly dynamic multihop wireless networks.

Acknowledgments

This work was supported in part by the National Science Foundation (NSF) under Grants ANI-0312655, CNS-0347390, and DMI-0552676, by the NSF Wireless Internet Center for Advanced Technology (WICAT), and by the Office of Naval Research under Grants N00014-03-1-0521 and N00014-05-1-0179. We thank Mr. Xiaolin Cheng and Dr. Sastry Kompella for obtaining part of the results presented in this chapter.

References

1. Chen, S. and K. Nahrstedt. Distributed quality-of-service routing in ad-hoc networks. *IEEE J. Select. Areas Commun.*, 17(8): 1488–1505, 1999.
2. Lin, C.R. and J.-S. Liu. QoS routing in ad hoc wireless networks. *IEEE J. Select. Areas Commun.*, 17(8): 1426–1438, 1999.
3. Papadimitratos, P., Z.J. Haas, and E.G. Sirer. Path set selection in mobile ad hoc networks. *Proc. ACM Mobihoc*, pp. 1–11, Lausanne, Switzerland, June 2002.
4. Perkins, C.E., E.M. Royer, and S.R. Das. Quality of service in ad hoc on-demand distance vector routing, IETF Internet Draft, Nov. 2001 (submitted).
5. Sivakumar, R., P. Sinha, and V. Bharghavan. CEDAR: a core-extraction distributed ad hoc routing algorithm. *IEEE J. Select. Areas Commun.*, 17(8): 1454–1465, 1999.
6. Alasti, M., K. Sayrafian-Pour, A. Ephremides, and N. Farvardin. Multiple description coding in networks with congestion problem. *IEEE Trans. Inform. Theory*, 47(3): 891–902, 2001.
7. Apostolopoulos, J.G., T. Wong, W. Tan, and S. Wee. On multiple description streaming in content delivery networks. *Proc. IEEE INFOCOM*, pp. 1736–1745, New York, June 2002.
8. Begen, A.C., Y. Altunbasak, and O. Ergun. Multipath selection for multiple description encoded video streaming. *EURASIP Signal Process.: Image Commun.*, 20(1): 39–60, 2005.
9. Link, M., K. Stulmuller, N. Farberand, and B. Girod. Analysis of video transmission over lossy channels. *IEEE J. Select. Areas Commun.*, 18(6): 1012–1032, 2000.
10. Mao, S., Y.T. Hou, X. Cheng, H.D. Sherali, S.F. Midkiff, and Y.-Q. Zhang. On routing for multiple description video over wireless ad hoc networks. *IEEE Trans. Multimedia*, 8(5): 1063–1074, 2006.
11. Mao, S., X. Cheng, Y.T. Hou, H.D. Sherali, and J.H. Reed. On joint routing and server selection for multiple description video in wireless ad hoc networks. *IEEE Trans. Wireless Commun.*, 6(1): 338–347, 2007.

12. Mao, S., X. Cheng, Y.T. Hou, and H.D. Sherali. Multiple description video multicast in wireless ad hoc networks. *ACM/Kluwer Mobile Networks and Appl. J. (MONET)*, 11(1): 63–73, 2006.
13. Goyal, V. Multiple description coding: compression meets the network. *IEEE Signal Process. Mag.*, 18(9): 74–93, 2001.
14. Mao, S., S. Lin, S.S. Panwar, Y. Wang, and E. Celebi. Video transport over ad hoc networks: multistream coding with multipath transport. *IEEE J. Select. Areas Commun.*, 12(10): 1721–1737, 2003.
15. Wang, Y., A.R. Reibman, and S. Lin. Multiple description coding for video delivery. *Proc. IEEE*, 93(1): 57–70, 2005.
16. Back, T., D. Fogel, and Z. Michalewicz, editors. *Handbook of Evolutionary Computation*. Oxford University Press, New York, 1997.
17. Aarts, E. and J. Korst. *Simulated Annealing and Boltzman Machines*. John Wiley & Sons, New York, 1989.
18. Glover, F. and M. Laguna. *Tabu Search*. Kluwer-Academic, Boston, MA, 1997.
19. Clausen, T. and P. Jacquet. Optimized Link State Routing Protocol, IETF RFC 3626. Oct. 2003.
20. Johnson, D.B., D.A. Maltz, and Y.-C. Hu. The dynamic source routing protocol for mobile ad hoc networks (DSR), IETF Internet Draft draft-ietf-manet-dsr-09.txt. Apr. 2003.
21. Ribeiro, A., Z.-Q. Luo, N.D. Sidiropoulos, and G.B. Giannakis. Modelling and optimization of stochastic routing for wireless multi-hop networks. *Proc. IEEE INFOCOM'07*, pp. 1748–1756, Anchorage, AK, May 2007.
22. Mao, S., S. Kompella, Y.T. Hou, H.D. Sherali, and S.F. Midkiff. Routing for concurrent video sessions in ad hoc networks. *IEEE Trans. Veh. Technol.*, 55(1): 317–327, 2006.
23. Chakareski, J., S. Han, and B. Girod. Layered coding vs. multiple descriptions for video streaming over multiple paths. *Proc. ACM Multimedia*, pp. 422–431, Berkeley, CA, Nov. 2003.
24. Gogate, N., D. Chung, S.S. Panwar, and Y. Wang. Supporting image/video applications in a multihop radio environment using route diversity and multiple description coding. *IEEE Trans. Circuits Syst. Video Technol.*, 12(9): 777–792, 2002.
25. Setton, E., Y. Liang, and B. Girod. Adaptive multiple description video streaming over multiple channels with active probing. In *Proc. IEEE ICME*, Baltimore, MD, July 2003.
26. Ozarow, L. On a source coding problem with two channels and three receivers. *Bell Syst. Tech. J.*, 59(10): 84–91, 1980.
27. Venkataramani, R., G. Kramer, and V.K. Goyal. Multiple description coding with many channels. *IEEE Trans. Inform. Theory*, 49(9): 2106–2114, 2003.
28. Fritchman, D.B. A binary channel characterization using partitioned markov chains. *IEEE Trans. Inform. Theory*, 13(2): 221–227, 1967.
29. Sherali, H.D., K. Ozbay, and S. Subramanian. The time-dependent shortest pair of disjoint paths problem: complexity, models, and algorithms. *Networks*, 31(4): 259–272, 1998.
30. Blum, C. and A. Roli. Metaheuristics in combinatorial optimization: overview and conceptual comparison. *ACM Comput. Surveys*, 35(3): 268–308, 2003.
31. Ahn, C.W. and R.S. Ramakrishna. A genetic algorithm for shortest path routing problem and the sizing of populations. *IEEE Trans. Evol. Comput.*, 6(6): 566–579, 2002.

32. Eppstein, D. Finding the *k* shortest paths. *SIAM J. Comput.*, 28(2): 652–673, 1999.
33. Sherali, H.D. and W.P. Adams. *A Reformulation-Linearization Technique for Solving Discrete and Continuous Nonconvex Problems*, Kluwer Academic Publisher, Boston, MA, 1999.
34. Sherali, H.D. and C.H. Tuncbilek. A global optimization algorithm for polynomial programming problems using a reformulation-linearization technique. *J. Global Optim.*, 2(1): 101–112, 1992.
35. Kompella, S., S. Mao, Y.T. Hou, and H.D. Sherali. Cross-layer optimized multipath routing for video communications in wireless networks. *IEEE J. Select. Areas Commun.*, 25(4): 831–840, 2007.
36. Bazaraa, M.S., J.J. Jarvis, and H.D. Sherali. *Linear Programming and Network Flows*. 3rd edition. Wiley-Interscience, New York, 2004.
37. The QoS with the OLSR Protocol homepage. [online]. Available: http://www.ietf.org/html.charters/manet-charter.html.
38. Jain, M. and C. Dovrolis. End-to-end available bandwidth: measurement methodology, dynamics, and relation with TCP throughput. *IEEE/ACM Trans. Networking*, 11(4): 537–549, 2003.
39. Kapoor, R., L.-J. Chen, L. Lao, M. Gerla, and M.Y. Sanadidi. Capprobe: a simple and accurate capacity estimation technique. In *Proc. ACM SIGCOMM*, pp. 67–78, Portland, OR, Oct. 2004.
40. Wang, Z. and J. Crowcroft. Quality-of-service routing for supporting multimedia applications. *IEEE J. Select. Areas Commun.*, 14(7): 1228–1234, 1996.

Chapter 6

Multipath Unicast and Multicast Video Communication over Wireless *Ad Hoc* Networks

Wei Wei and Avideh Zakhor

Contents

6.1 Introduction 194
6.1.1 Unicast 195
6.1.2 Multicast 196
6.2 Related Work 197
6.2.1 Unicast 197
6.2.2 Multicast 198
6.3 Multipath Selection for Unicast Streaming 199
6.3.1 Envisioned Network Model 199
6.3.2 The Optimal Multipath Selection Problem 199
6.3.3 Concurrent Packet Drop Probability of Two Node-Disjoint Paths 200
6.3.4 Computation of PDP over a Link 201

6.3.5 A Heuristic Solution to the Optimum Multipath Selection......203
6.3.6 Simulation Results......204
6.4 Testbed Implementation and Evaluation......206
6.4.1 Software Architecture......206
6.4.2 A Simple Rate Control Scheme......207
6.4.3 Testbed Setup......208
6.4.4 802.11a Wireless *Ad Hoc* Network Result: Static Nodes......208
6.4.5 802.11a Wireless *Ad Hoc* Network Result: Moving Nodes......210
6.5 Multiple-Tree Multicast Video Communication over Wireless *Ad Hoc* Networks......213
6.5.1 Tree Connectivity and Tree Similarity......214
6.5.2 Multiple-Tree Multicast Packet Forwarding......215
6.6 Serial Multiple-Disjoint Trees Multicast Routing Protocol......216
6.7 Parallel Multiple Nearly Disjoint Trees Multicast Routing Protocol......218
6.7.1 Overview of Parallel MNTMR......218
6.7.2 Conditions and Rules......219
6.7.3 Detailed Double Nearly Disjoint Tree Construction......220
6.7.4 Discussion......220
6.7.5 Simulation Results......221
6.8 Conclusion......228
6.9 Appendix......229
6.9.1 Proof of Claim 1......229
6.9.2 Proof of Claim 2......230
References......230

6.1 Introduction

With an increase in the bandwidth of wireless channels and computational power of mobile devices, video applications are expected to become more prevalent in wireless *ad hoc* networks in the near future. Examples of video communication applications over wireless *ad hoc* networks include spontaneous video conferencing at a location without wireless infrastructure, transmitting video in the battlefield, and search and rescue operations after a disaster.

Video communication is fundamentally different from data communication, because they are delay and loss sensitive. Unlike data packets, late arriving video packets are useless to the video decoder. Thus, the retransmission techniques are not generally applicable to video communication applications with low delay requirements, especially in multicast situations.

There are additional challenges for supporting video communication over wireless *ad hoc* networks. Owing to the mobility of wireless nodes, the topology of *ad hoc* networks may change frequently. Thus the established connection routes between senders and receivers are likely to be broken during video transmission, causing interruptions, freezes, or jerkiness in the received video signal. An end-to-end connection route in wireless *ad hoc* networks generally consists of multiple wireless links resulting in higher random packet loss than single-hop wireless connections in wireless networks with infrastructure, such as base stations. Other challenges include lower wireless network capacity compared to wired networks and limited battery life of mobile devices. These constraints and challenges, in combination with the delay- and loss-sensitive nature of interactive video applications, make video communication over wireless *ad hoc* networks a challenging proposition [1].

Multiple description coding (MDC) generates multiple compressed descriptions of the media in such a way that a reasonable reconstruction is achieved if any of the multiple descriptions is available for decoding, and the reconstruction quality is improved as more descriptions become available [2,3]. The main advantage of MDC over layered coding is that no specific description is needed to render the remaining descriptions useful. As such, unless none of the descriptions make it to the receiver, video quality degrades gracefully with packet loss. However, there is a penalty in coding efficiency and bit rate in using MDC as compared to single description coding (SDC) [2,3]. Specifically, for a given visual quality, the bit rate needed for MDC exceeds that of SDC depending on the number of descriptions. In this chapter, we use MDC video in multipath unicast and multicast scenarios.

This chapter introduces new path diversity schemes to provide robustness for both unicast and multicast video communication applications over wireless *ad hoc* networks.

6.1.1 Unicast

In case of unicast, we propose a class of techniques to find two node-disjoint paths, which achieve minimum concurrent packet drop probability (PDP) of all path pairs. For MDC streaming, different descriptions are transmitted on different paths to fully utilize path diversity, and the worst-case scenario occurs when all descriptions are missing. Streaming over the path pair with minimum concurrent PDP (PP_PDP) minimizes the probability of concurrent loss of all the descriptions, thus optimizing the worst-case video quality of all times. Although most of our simulation results refer to MDC, our basic results and conclusions can be easily extended to forward error corrected (FEC) video as well. For FEC streaming, concurrent packet drop over the selected PP_PDP can be shown to be less likely than that of simple node-disjoint paths, resulting in lower unrecoverable probability.

This chapter uses a "conflict graph" [4–6] to model effects of interference between different wireless links. The conflict graph indicates which groups of links interfere with one another, and hence cannot be active simultaneously. We propose a model to estimate the concurrent PDP of two node-disjoint paths, given an estimate of cross traffic flows' rates and bit rate of the video flow. We then propose a heuristic PDP aware multipath routing protocol based on our model, whose performance is shown to be close to that of the "optimal routing," and significantly better than that of the node-disjoint multipath (NDM) routing and shortest widest routing [7].

6.1.2 Multicast

Multicast is an essential technology for many applications, such as group video conferencing and video distribution, and results in bandwidth savings as compared to multiple unicast sessions. Owing to the inherent broadcast nature of wireless networks, multicast over wireless *ad hoc* networks can be potentially more efficient than over wired networks [8].

This chapter introduces an architecture for multiple-tree video multicast communication over wireless *ad hoc* networks. The basic idea is to split the video into multiple parts and send each part over a different tree, which are ideally disjoint with one another so as to increase robustness to loss and other transmission degradations. We then propose a simple Serial Multiple Disjoint Tree Multicast Routing (MDTMR) protocol, which constructs two disjoint multicast trees sequentially in a distributed way, to facilitate multiple-tree video multicast. This scheme results in reasonable tree connectivity while maintaining disjointness of two trees.

However, Serial MDTMR has a larger routing overhead and construction delay than conventional single-tree multicast routing protocols as it constructs the trees in a sequential manner. To alleviate these drawbacks, we further propose Parallel Multiple Nearly Disjoint Trees Multicast Routing (MNTMR) in which nearly disjoint trees are constructed in a parallel and distributed way. By using Parallel MNTMR, each receiver is always able to connect to two trees, regardless of the node density. Simulations show that multiple-tree video multicast with both Serial MDTMR and Parallel MNTMR improve video quality significantly compared to single-tree video multicast; at the same time, routing overhead and construction delay of Parallel MNTMR is approximately the same as that of a single-tree multicast protocol.

The remainder of this chapter is organized as follows. Section 6.2 discusses related work. Section 6.3 proposes a method to estimate the concurrent PDP of two node-disjoint paths, formulate the optimal multipath selection problem for video streaming over wireless *ad hoc* networks, and develop a heuristic PDP aware multipath routing protocol. Section 6.4 presents the testbed implementation and experimental results. Section 6.5 proposes multiple-tree multicast framework. Sections 6.6 and 6.7 present Serial MDTMR and Parallel MNTMR, respectively. Section 6.8 concludes this chapter.

6.2 Related Work

6.2.1 *Unicast*

Dispersity routing by Maxemchuk [9] is one of the earliest works that apply multiple paths in networks, where Gogate and Panwar [10] first use multiple paths and MDC in *ad hoc* networks. Several researchers have also proposed to distribute MDC video flow over multiple paths for multimedia transport [1,11–14]. These efforts have successfully demonstrated that the combination of path diversity and MDC provides robustness in video communication applications. However, most of these either assume that the set of paths is given or simply select two node-/link-disjoint paths.

Only a few recent approaches address the problem of selecting the best path pair for MDC video streaming [15–17]. The path selection model used in Refs 15 and 16 is more suitable for the Internet overlay networks. In Ref. 17, the authors select two paths with minimal correlation for MDC streaming over Internet overlay networks. In contrast, we consider path selection over wireless *ad hoc* networks when interference plays an important role.

Multipath routing for wireless *ad hoc* networks has been an active research area recently [18–23]. Most existing approaches focus on how to obtain multiple node-/link-disjoint paths. In Ref. 23, the authors propose a heuristic algorithm to select multiple paths to achieve the best reliability, assuming that failure probability of different links are independent and given. In contrast, we focus on how to estimate PDP of each link in wireless *ad hoc* networks considering interference.

The problem of finding rate constraints on a set of flows in a wireless *ad hoc* network is studied in Refs 4 and 5. Both papers model the interference between links in *ad hoc* networks using conflict graphs and find capacity constraints by finding the global independent sets of the conflict graph. In Ref. 6, the authors develop a different set of rate constraints using the cliques, that is, complete subgraphs, of the conflict graph.

Our approach differs from previous works in two significant ways. First, our proposed multipath selection model estimates the concurrent congestion probability of two paths by taking into account the interference between different links, which reflects actual constraints of a wireless *ad hoc* network. Second, our proposed heuristic Interference aWare

Multipath (IWM) protocol provides reasonable approximation to the solution of the optimal multipath selection problem for video streaming over wireless *ad hoc* networks.

6.2.2 Multicast

Multicasting MDC video was first introduced in CoopNet [24] in the context of peer-to-peer networks to prevent Web servers from being overwhelmed by a large number of requests. CoopNet uses a centralized tree management scheme, and each tree link is only a logical link, which consists of several physical links and as such is inefficient in wireless *ad hoc* networks. In Ref. 25, the authors propose a genetic algorithm–based solution for multiple-tree multicast streaming, assuming that (a) they obtain each link's characteristics and (b) consecutive links' packet loss rates are independent.

There has also been a great deal of prior work in the area of multicast routing in wireless *ad hoc* networks [26–32]. The On-Demand Multicast Routing Protocol (ODMRP) [26] builds multicast mesh by periodically flooding the network with control packets to create and maintain the forwarding state of each node, when the source has packets to send. It takes advantage of the broadcast nature of the wireless network by forwarding group flooding, which provides a certain amount of diversity. A mesh structure is equivalent to a tree structure with "tree flood" enabled [29]. In the remainder of this chapter, we refer to ODMRP as a single-tree multicast protocol. The Adaptive Demand-Driven Multicast Routing (ADMR) [29] attempts to reduce non-on-demand components within the protocol as much as possible. ADMR does not use periodic networkwide floods of control packets, periodic neighbor sensing, or periodic routing table exchanges. In ADMR, forwarding state is specific to each sender rather than being shared by the entire multicast group. This approach reduces unnecessary forwarding data redundancy. There is also a local subtree repair scheme to detect a broken link by downstream node in ADMR. The Adaptive Core Multicast Routing Protocol (ACMRP) [32] is an on-demand core-based multicast routing protocol that is based on a multicast mesh. A multicast mesh is created and maintained by the periodic flooding of the adaptive core. A core emerges on demand and changes adaptively according to the current network topology. This scheme outperforms ODMRP in multisource scenarios. The Independent-Tree *Ad Hoc* Multicast Routing (ITAMAR) [33] creates multiple multicast trees based on different metrics in a centralized way. ITAMAR constructs multiple edge disjoint or nearly disjoint trees. The main objective of this protocol is to increase the average time between multicast tree failures. The ITAMAR algorithms are basically based on Dijkstra's shortest path first (SPF) algorithm [34], which is a centralized approach, and requires knowledge of network topology. There are two obvious advantages of our proposed techniques as compared to ITAMAR—first, our protocols are distributed, rather than centralized, and hence do not require the knowledge of network topology in advance; second, our protocols' overhead is $O(n)$, rather than $O(n^2)$ of ITAMAR, where n is the number of total nodes.

6.3 Multipath Selection for Unicast Streaming

Our approach is to minimize concurrent PDP of two node-disjoint paths in a wireless *ad hoc* network. As stated earlier, this is equivalent to optimizing the worst-case video quality at clients. The node-disjoint constraint is useful for mobile wireless *ad hoc* networks, because it reduces the correlation of packet drop in different paths significantly.

6.3.1 Envisioned Network Model

A wireless *ad hoc* network can be modeled as a directed graph $G(V,E)$, whose vertices V correspond to wireless stations and the edges E correspond to wireless links. Let $n_i \in V$, $1 \leq i \leq N$ denote the nodes, and d_{ij} denote the distance between nodes n_i and n_j. Each node is equipped with a radio with communication range r and a potentially larger interference range ω. There is a link l_{ij} from vertex n_i to vertex n_j if and only if $d_{ij} < r$. If the transmission over link l_{ij} makes the transmission over link l_{kl} unsuccessful, link l_{ij} interferes with link l_{kl}. We use a model similar to the protocol interference model introduced in Ref. 5 to determine whether two links interfere with one another.

6.3.2 The Optimal Multipath Selection Problem

Let $P^1_{\mathrm{S,D}}$ and $P^2_{\mathrm{S,D}}$ be any two paths connecting nodes N_{S} and N_{D}, $L^1_{\mathrm{S,D}}$ and $L^2_{\mathrm{S,D}}$ denote the set of links on each path, respectively, and $N^1_{\mathrm{S,D}}$ and $N^2_{\mathrm{S,D}}$ denote the set of the nodes on each path, respectively. We define two indication vectors $\mathbf{x} = (\ldots, x_{ij}, \ldots)^T$ and $\mathbf{y} = (\ldots, y_{ij}, \ldots)^T$ to represent $P^1_{\mathrm{S,D}}$ and $P^2_{\mathrm{S,D}}$ respectively, where x_{ij} is set to be 1 if link $l_{ij} \in L^1_{\mathrm{S,D}}$ and is set to be 0 otherwise, and y_{ij} is defined similarly for path 2.

For the optimal multipath selection, we select two node-disjoint paths with minimum concurrent PDP. This corresponds to the following optimization problem:

Minimize

$$P_{\mathrm{drop}}\left(P^1_{\mathrm{S,D}}; P^2_{\mathrm{S,D}}\right)$$

with respect to

$$x_{ij}, y_{mn} \in \{0, 1\}, \quad \forall (i,j), (m,n) \in E$$

subject to

$$\sum_{j:(i,j)\in E} x_{ij} - \sum_{j:(j,i)\in E} x_{ji} = \begin{cases} 1 & i = N_S \\ -1 & i = N_D \\ 0 & \text{otherwise} \end{cases} \tag{6.1}$$

$$\sum_{i:(i,j)\in E} x_{ij} \leq 1 \tag{6.2}$$

$$\sum_{n:(m,n)\in E} y_{mn} - \sum_{n:(n,m)\in E} y_{nm} = \begin{cases} 1 & m = N_S \\ -1 & m = N_D \\ 0 & \text{otherwise} \end{cases} \tag{6.3}$$

$$\sum_{m:(m,n)\in E} y_{mn} \leq 1 \tag{6.4}$$

$$N^1_{S,D} \cap N^2_{S,D} = \{N_S, N_D\} \tag{6.5}$$

Equations 6.1 and 6.2 are flow constraints to guarantee the first path to connect the source N_S and destination N_D. They represent (a) for each node in the first path, except the source and destination, both the number of incoming and outgoing links are 1; (b) for the source node, the number of outgoing links is 1; (c) for the destination node, the number of incoming links is 1. Similarly, Equations 6.3 and 6.4 are flow constraints for the second path. Equation 6.5 is the node-disjoint constraint to ensure that the two selected paths do not share nodes.

We can show the following claim for the optimal multipath selection problem:

> *Claim 1*: The optimal multipath selection over wireless *ad hoc* networks as defined in Equations 6.1 through 6.5 is nondeterministic polynomial (NP)-hard.

The proof is shown in Section 6.9.1.

6.3.3 Concurrent Packet Drop Probability of Two Node-Disjoint Paths

This section shows how to compute the concurrent PDP of any given two node-disjoint paths connecting the same source and destination nodes to solve the optimal multipath selection problem.

We now prove that PDP of two node-disjoint links have low correlation. In a wireless *ad hoc* network, congestion, contention, time-varying wireless channel, and mobility of nodes are four main factors contributing for packet loss. We prove that PDP due to each of these factors is minimally correlated, thus PDP of two node-disjoint links is minimally correlated. First, packet drop due to mobility of two node-disjoint links is independent of one another, assuming that the nodes' movement is independent of one another. Second, PDP due to contention or wireless channel error is generally small because of the 802.11 Media Access Control (MAC) layer retransmission scheme. Thus, we do not need to consider their contributions here. Third, as for congestion, although two node-disjoint links may interfere with one another, causing their PDP to be correlated, we expect that the random backoff scheme in the 802.11 MAC layer protocol reduces the correlation significantly. We have applied network simulator (NS) simulations to verify our conjecture [35]. Specifically, our results show that packet drop over two node-disjoint interfering links have low correlation as long as PDP of each link is small [35].

Because two node-disjoint paths share only the source and destination nodes, packet drop over two node-disjoint paths also have low correlation. Thus we can approximate the concurrent PDP over two node-disjoint paths $P^1_{\mathrm{S,D}}$ and $P^2_{\mathrm{S,D}}$ as

$$\begin{aligned} P_{\mathrm{drop}}\left(P^1_{\mathrm{S,D}}; P^2_{\mathrm{S,D}}\right) &\approx P_{\mathrm{drop}}\left(P^1_{\mathrm{S,D}}\right) \cdot P_{\mathrm{drop}}\left(P^2_{\mathrm{S,D}}\right) \\ &= \left[1 - \prod_{l_{ij} \in L^1_{\mathrm{S,D}}} (1 - P_{\mathrm{drop}}(l_{ij}))\right] \\ &\quad \times \left[1 - \prod_{l_{mn} \in L^2_{\mathrm{S,D}}} (1 - P_{\mathrm{drop}}(l_{mn}))\right] \end{aligned} \tag{6.6}$$

6.3.4 Computation of PDP over a Link

To complete the computation of the concurrent PDP of two node-disjoint paths, we now show how to estimate PDP over one link, assuming that we have already estimated the flow rates F_i over each link l_i. As stated earlier, in a wireless *ad hoc* network, congestion, contention, time-varying wireless channel errors, and mobility of nodes are four main reasons for packet loss. Thus PDP over link l_{ij} can be represented as

$$P_{\text{drop}}(l_{ij}) = 1 - [1 - P_{\text{drop-cong}}(l_{ij})][1 - P_{\text{drop-cont}}(l_{ij})] \times [1 - P_{\text{drop-chan}}(l_{ij})][1 - P_{\text{drop-mob}}(l_{ij})] \quad (6.7)$$

where $P_{\text{drop-cong}}(l_{ij})$, $P_{\text{drop-cont}}(l_{ij})$, $P_{\text{drop-chan}}(l_{ij})$, and $P_{\text{drop-mob}}(l_{ij})$ are packet drop over link l_{ij} due to congestion, contention, wireless channel error, and mobility, respectively. It is possible to apply the broadcast packet technique described by De Couto et al. [36] to estimate PDP due to contention and wireless channel error, and apply results on link availability [37] to estimate the PDP over a link due to mobility. In this chapter, we only focus on PDP due to congestion, because we assume that static scenarios and packet loss caused by channel error and contention are mostly recovered by 802.11 MAC layer retransmissions.

In the remainder of this section, we describe how to compute PDP over link l_{ij} due to congestion $P_{\text{drop-cong}}(l_{ij})$. An "interfering link set" of link l_{ij} is defined to be a set consisting of all links that interfere with it. We partition the interfering link set $I(l_{ij})$ into several disjoint subsets such that each subset is an independent set. An "independent set" denoted by IS is defined to be a set of links, which can successfully transmit simultaneously without interfering with one another. The set of independent sets resulting from partitioning $I(l_{ij})$ is denoted by $PT(l_{ij})$. We define equivalent rate of flows over all links in the kth independent set IS_k as follows:

$$CF_k = \max_{l_m \in IS_k} F_m \quad (6.8)$$

where F_m is the aggregate incoming flow rate over the mth link l_m in the kth independent set IS_k. Because links of the same independent set can transmit simultaneously, the equivalent rate of an independent set denotes link l_{ij}'s channel resource needed by all the links in this independent set per unit time.

Given a partition of the set $I(l_{ij})$, we can estimate the PDP due to congestion of link l_{ij} as follows:

$$P_{\text{drop-cong}}(l_{ij}|PT(l_{ij})) \approx \max\left(1 - \frac{C}{\sum_{IS_k \in PT(l_{ij})} CF_k}, 0\right) \approx P_{\text{drop}}(l_{ij}) \quad (6.9)$$

where C is wireless channel capacity. The last equality reflects our assumption that congestion is the main reason of packet drop.

We name the partition $PT(l_{ij})^*$ that minimizes $P_{\text{drop-cong}}(l_{ij}|PT(l_{ij}))$ as the most efficient partition. Because computing the actual PDP due to congestion is prohibitively compute-intensive, we choose to use its lower bound instead, that is, the PDP of the most efficient partition, as a metric in comparing PDP of two links and subsequently two paths. We note that using the most efficient partition results in underestimating the PDP due to congestion and the total PDP. However, simulations show that it is sufficient to use the lower bound of PDP due to congestion to compare and select paths. Also, with the development of more efficient MAC layer protocols in the future, our underestimation is likely to approach the actual results.

We propose a greedy algorithm to approximately find the most efficient partition. The basic idea behind the greedy partitioning algorithm is to combine as many links as possible with large flow rates to reduce the sum of equivalent flow rates of independent sets, thus minimizing $P_{\text{drop-cong}}(l_{ij}|PT(l_{ij}))$ [35].

Combining Equations 6.6, 6.7, and 6.9, we obtain an estimate of PDP of two node-disjoint paths.

An approach to solve the optimal multipath selection problem described in Section 6.3.2 is to enumerate all possible pairs of node-disjoint paths from a source N_S to a destination N_D, estimate the concurrent PDP for each path pair using the scheme proposed in Section 6.3.4, and choose the best one. We refer to this solution as the Optimal Multipath Routing (OMR). Unfortunately, the computation complexity of the OMR grows exponentially with the size of the network; thus it cannot be run in real time. However, as will be seen later, OMR can be used in non-real-time simulations to provide an upper bound on the performance of other lower complexity heuristic schemes.

6.3.5 A Heuristic Solution to the Optimum Multipath Selection

Because the optimal multipath selection problem is NP-hard, we propose a heuristic solution called IWM Routing, which can be implemented in real time. By assuming that the PDP of each link is small, we can approximate $P_{\text{drop}}(P^1_{\text{S,D}}; P^2_{\text{S,D}})$ in Equation 6.6 as follows:

$$P_{\text{drop}}(P^1_{\text{S,D}}; P^2_{\text{S,D}}) = \sum_{l_{ij} \in L^1_{\text{S,D}}} P_{\text{drop}}(l_{ij}) \cdot \sum_{l_{mn} \in L^2_{\text{S,D}}} P_{\text{drop}}(l_{mn}) \tag{6.10}$$

Our approach is to first determine the first path so as to minimize PDP, and then to choose to minimize the second path's PDP among node-disjoint paths from the first one. Note that this approach is similar to the one proposed in Ref. 17. The main difference is that our metric is PDP and theirs is correlation. Specifically, we apply the techniques described in Section 6.3.4 to compute PDP for each link in wireless *ad hoc* networks.

The optimization problem of finding the first path can be formulated as follows:

$$\underset{\mathbf{x}}{\text{Minimize}} \sum_{l_{ij} \in E} x_{ij} P_{\text{drop}}(l_{ij})$$

such that the flow constraint in Equation 6.1 is satisfied. $P_{\text{drop}}(l_{ij})$ denotes the cost assigned to link l_{ij}, and is estimated using Equations 6.7 and 6.9. To obtain the first path, we solve this optimization problem using the Dijkstra's algorithm, whose complexity is polynomial [34].

After obtaining the first path, we update flow rate over each link by taking into account the allocated video flow into corresponding links. Given the first path, we compute the second path by defining a link cost for each link as follows:

$$C_{mn} = P_{\text{drop}}(l_{mn}) + \text{nd\_cost}_{mn} \tag{6.11}$$

where

$$\text{nd\_cost}_{mn} = \begin{cases} b_1 \gg 1 & \text{destination node of link } l_{mn} \in P^1_{\text{S,D}} \\ 0 & \text{otherwise} \end{cases}$$

is a penalty factor to maintain the node disjointness between the two paths. The optimization problem of finding the second path, which minimizes PDP and is node-disjoint from the first path, can be formulated as follows:

$$\underset{\mathbf{y}}{\text{Minimize}} \sum_{l_{mn} \in E} y_{mn} C_{mn}$$

subject to the constraint, the indicator vector for the second path satisfies Equation 6.3. We also solve the second optimization problem with the Dijkstra's algorithm.

6.3.6 Simulation Results

In this section, we compare the OMR (described in Sections 6.3.2 and 6.3.4), IWM (described in Section 6.3.5), NDM routing [19], and the shortest widest path (SWP) routing [7]. We use a simulation model based on NS-2 [38], and focus on the case of static wireless *ad hoc* networks. Each node's radio range is 250 m and its interference range is 550 m. We consider a grid network consisting of 49 nodes, placed in a 7×7 grid with the distance between neighboring nodes being 200 m.

We randomly choose one video sender and one video receiver. For MDC, we encode one frame into two packets, and the group of pictures (GOP) size is chosen to be 15. Standard Motion Picture Experts Group

(MPEG) Quarter Common Intermediate Format (QCIF) sequence Foreman is coded with a matching pursuit multiple description codec [2] at 121.7 kbps. We insert 20 one-hop cross traffic flows whose bit rates are uniformly distributed in the range [0,200.0] kbps. The bit rates of cross flows are changed every 30 s. We run 30 simulations for different network topologies and select different sender and receiver pair in each scenario. Each simulation lasts 900 s.

We evaluate the performance using the following metrics:

a. *The ratio of bad frames.* The ratio of bad frames is the ratio of the number of nondecodable frames to the total number of video frames that should have been decoded in the receiver. A description of an I-frame is nondecodable if the packet corresponding to the description is not received on time. A description of a P-frame is nondecodable if at the playback deadline, either the packet corresponding to the description is not received or the same description of the previous frame is nondecodable. A frame of an MDC stream is nondecodable if both its descriptions are nondecodable. This metric takes into account the dependency between consecutive frames in a predictive coding scheme, and also reflects the fact that MDC can, to some extent, conceal the undesirable effects caused by missing packets.
b. *The number of bad periods.* A bad period consists of contiguous bad frames. This metric reflects the number of times the received video is interrupted by the bad frames.

Figures 6.1a and 6.1b show the ratio of bad frames and number of bad periods of the four schemes for 30 runs with NS-2 simulations. As seen, the average performance of IWM is very close to that of OMR and is significantly

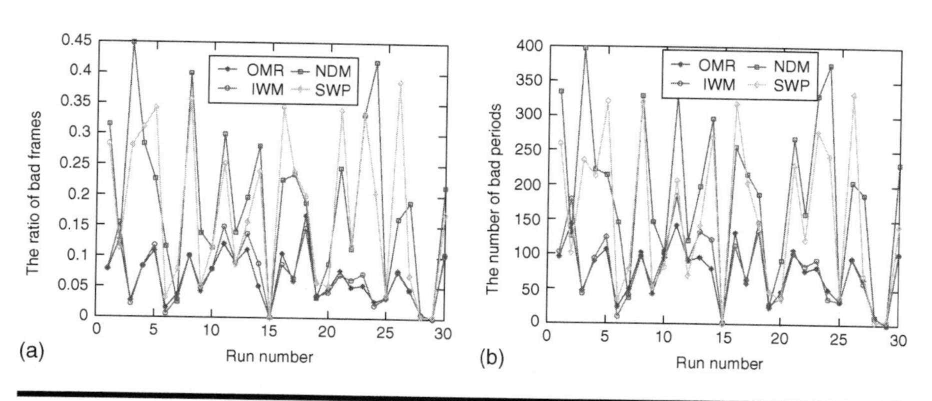

Figure 6.1 Simulation results for the 7 x 7 grid network. (a) The ratio of bad frames, (b) the number of bad periods.

Table 6.1 Summary—The Ratio of Bad Frames

	OMR	*IWM*	*NDM*	*SWP*
Average	0.0655	0.0685	0.1864	0.1755
Number of "best"	29	26	7	8

better than that of NDM and SWP, although its computational complexity is similar to NDM and SWP. We count one protocol as the "best" under a scenario if its ratio of bad frames is not more than 5 percent higher than the lowest among all the other protocols. Specifically, shown in Table 6.1, IWM performs close to the best among all protocols in 26 out of 30 runs. The results show that the relaxation of the optimal multipath selection problem used by IWM is very efficient.

6.4 Testbed Implementation and Evaluation

To demonstrate the feasibility of the proposed multipath selection framework and the IWM, we build a small wireless *ad hoc* network testbed consisting of 11 desktops and laptops. This section summarizes the key components of the testbed and reports the results obtained from the performance study conducted on it.

6.4.1 *Software Architecture*

We implement the proposed IWM protocol in the Mesh Connectivity Layer (MCL), which is an *ad hoc* routing framework provided by Microsoft Research [39]. MCL implements a virtual network adapter, that is, an interposition layer between layer 2 (the link layer) and layer 3 (the network layer). The original MCL maintains a link cache in each node to store loss rate and bandwidth information of each link. Also, the original MCL implements a routing protocol named Link Quality Source Routing (LQSR) to route packets. LQSR is similar to SWP described in Section 6.3.6. The LQSR supports different link-quality metrics, for example, weighted cumulative expected transmission time (WCETT) and expected transmission count (ETX) [39]. In our experiments, LQSR uses WCETT as the link-quality metric.

It may be argued that applying LQSR with WCETT twice can result in two node-disjoint paths with similar performance to that of IWM. However, LQSR attempts to obtain the path with the largest bandwidth, rather than the one with the largest available bandwidth. Unlike IWM, LQSR does not take into account the impact of interference from cross traffic flows and the video flow itself on the path selection. The two paths resulting from LQSR are likely to be close to one another, because the metrics for different paths are computed with the same network parameters. Rather, IWM considers both cross traffic flows and video flow to compute PDP. As such, the two paths obtained by IWM adapt to available bandwidth resources. When there is sufficient bandwidth, the two paths obtained by IWM are likely to be close to one another, otherwise the two paths are distributed within different regions of network to minimize PDP.

We make two major modifications to MCL. First, we implement IWM inside the MCL framework such that it coexists with LQSR in MCL. When forwarding a packet, the MCL uses one bit of information transmitted from the upper layer to decide which routing protocol to use. If the packet is a high priority video packet, MCL uses IWM to route it; otherwise, it uses LQSR. This way, we can run IWM and LQSR simultaneously in the network and compare them under same network conditions. In our experiments, IWM and LQSR are used to route MDC and SDC packets,* respectively. The second modification we have made is to enable the estimation of flow rate of each link to compute the PDP by using the scheme described in Section 6.3.2.

We also implement both MDC and SDC streaming protocol in the application layer. In the streaming protocol, we implement time-stamping, sequence numbering, feedback functions, and the rate control scheme described in Section 6.4.2. User Datagram Protocol (UDP) sockets are used at the transport layer. The deadline of each frame is 2 s after the transmission time. If a packet is received after its deadline, it is discarded.

6.4.2 A Simple Rate Control Scheme

In our multipath selection framework, we assume that there exists a simple rate control scheme to determine the video application's sending rate. This way, the sending rate can be adjusted according to the amount of congestion in the network.

The basic idea behind our rate control scheme is to employ an Additive Increase Multiplicative Decrease (AIMD) algorithm, which is the default congestion control mechanism used in Transmission Control Protocol (TCP) today. The receiver transmits a feedback packet to the sender periodically to inform the sender whether the network is congested or not. PDP due to contention and wireless channel error is generally small, because of 802.11 MAC layer retransmissions. Thus the scheme uses lost packets as a signal to detect congestion. The receiver detects lost packets using sequence numbers carried by each packet. To alleviate the out-of-order problem caused by multipath transmission, the receiver counts packets received from each path separately.

* Recall that SDC rate is about 30 percent lower than that of MDC video due to compression inefficiency of MDC.

At the sender side, after receiving the feedback packet, if the network is not congested, the sender increases the video transmission rate by 1 fps in each time period. If the network is congested, the sender decreases the video frame rate immediately by half. If the sender has not received one feedback packet in a time interval twice that of the feedback period, it triggers a timeout event, and the sender reduces the video transmission rate to the minimum transmission rate.

For simplicity, we change the transmission bit rate through changing the number of transmitted video frames per unit time without dropping a frame. This has the effect of changing the playback duration of a given chip at the receiver. Our motivation for doing so is purely ease of implementation. This way, we do not have to implement fine grain or temporal scalability to compute our metrics, such as ratio of bad frames or bad periods. For a fixed GOP, this method results in the same metrics as modifying the encoding and decoding rate on the fly, that is, applying temporal scalability. For example, assuming a GOP of 15, if frame number 4 is nondecodable, the number of bad frames for both methods is 12.

6.4.3 Testbed Setup

We deploy an 11-node wireless *ad hoc* network testbed on the third floor of Cory Hall, the office building of Electrical Engineering and Computer Sciences (EECS), University of California at Berkeley. The nodes are placed in the offices and aisles, which are separated from one another with floor-to-ceiling walls and solid wood doors.

Each node in the testbed is either a standard desktop or laptop running Windows XP. Each desktop is equipped with either a Linksys 802.11 a/b/g Peripheral Component Interconnect (PCI) card or a Netgear 802.11 a/b/g PCI card. Similarly, each laptop is equipped with either a Linksys 802.11 a/b/g Personal Computer Memory Card International Association (PCMCIA) card or a Netgear 802.11 a/b/g PCMCIA card. All cards operate in the *ad hoc* mode.

All of our experiments are conducted over Internet Protocol (IPv4) using statically assigned addresses. Except for configuring *ad hoc* mode and fixing the frequency band and channel number, we use the default configuration for the radios. All the cards perform autorate selection.

6.4.4 802.11a Wireless Ad Hoc Network Result: Static Nodes

We performed a series of tests in 802.11a wireless *ad hoc* networks. We carried out ten 360 s long experiments with varying cross traffic level. The maximum throughput of each link is 54 Mbps. The senders and receivers

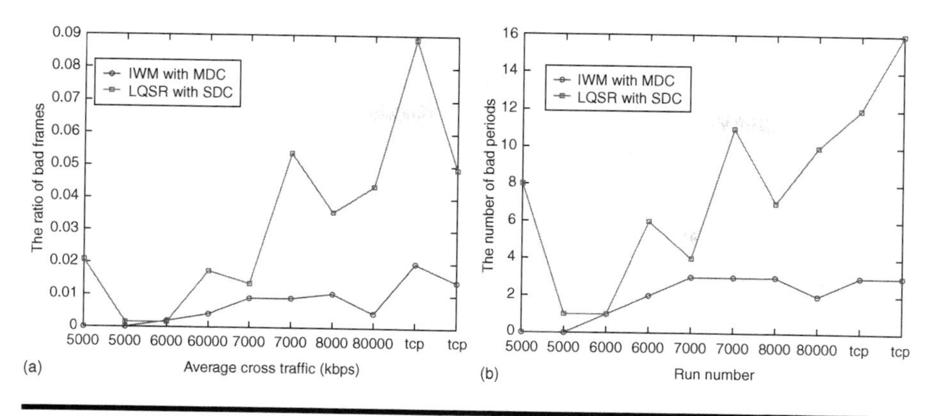

Figure 6.2 Performance evaluation of IWM/MDC over 802.11a. (a) The ratio of bad frames, (b) the number of bad periods.

are the same as those of the previous experiments. In runs 1 through 8, there are two one-hop UDP cross traffic flows whose bit rate is changed every 30 s based on uniform distribution. In runs 9 and 10, the cross traffic is one two-hop TCP connection.

Figures 6.2a and 6.2b show the result of the ratio of bad frames and the number of bad periods of all ten runs, respectively. The horizontal axis shows the average bit rate of combined cross traffic. As seen, IWM/MDC significantly outperforms LQSR/SDC in nine out of ten runs, and the performance gap with IWM/MDC and LQSR/SDC increases as cross traffic increases. Once again, this shows the advantage of path diversity with MDC over single-path transmission of SDC video.

Figure 6.3 compares peak signal-to-noise ratio (PSNR) of two schemes for all ten runs. On average, IWM/MDC outperforms LQSR/SDC by 1.1 dB, and outperforms it in eight out of ten runs. The reason for a slightly worse performance in runs 2 and 3 is the low packet loss rate for both schemes in these runs. As a result, the PSNR of received video in these runs are close to the PSNRs of original MDC and SDC videos, respectively. The PSNR of encoded MDC is slightly lower than that of encoded SDC, because in practice it is very hard to make two video flows achieve the same PSNRs. In general, we would expect performance gain of IWM/MDC over LQSR/SDC to become wider as PDP increases, which is also in agreement with the results in Figure 6.2.

We plot PSNR, loss traces, and frame rate traces of run 7, that is, the first run with cross traffic 8000 kbps, using IWM/MDC and LQSR/SDC in Figures 6.4 and 6.5, respectively. IWM/MDC outperforms LQSR/SDC by 1.1 dB in run 7. As seen in Figure 6.4a, for IWM/MDC, PSNR drops gracefully when there is packet loss in only one substream. As seen in Figure 6.4b, most of the time, packet losses of two substreams do not overlap, thus

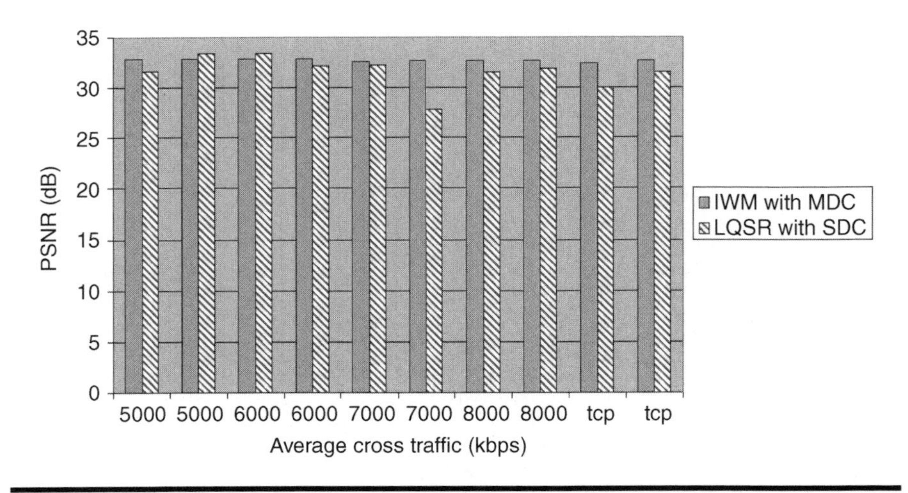

Figure 6.3 PSNR performance evaluation of IWM/MDC over 802.11a.

reducing both the number and amount of PSNR drops. The PSNR curve of LQSR/SDC shown in Figure 6.5a has more frequent and severe drops than that of IWM/MDC, because PSNR drops for every packet drop in SDC video, and would drop severely when there is a burst of packet loss. As seen in Figure 6.4e, our simple rate/frame control scheme adjusts the video rate promptly whenever there is packet drop in any path, and keeps the maximum sending rate whenever there is no packet drop.

6.4.5 *802.11a Wireless* Ad Hoc *Network Result: Moving Nodes*

We also carried out experiments with one moving node in 802.11a wireless *ad hoc* networks. In these experiments, we do not take into account PDP due to mobility although the nodes are slowly moving. During the experiment, we randomly select one laptop, move it to a random position, and repeat the process. The senders and receivers are the same as those of previous experiments. At any time, there is always one laptop moving. Figures 6.6 and 6.7 show the results of three 600 s experimental run.

As seen in Figure 6.6, the ratio of bad frames and the number of bad periods are both greatly reduced for IWM/MDC in all three runs. With the continuous movement of one node, one path is broken from time to time. If the path selected by LQSR is broken during the video transmission, the SDC receiver suffers from packet loss and interruption of video playback. In contrast, even if one path selected by IWM is broken, the received video quality is still acceptable. Figure 6.7 compares PSNR of the two schemes. Averaged over three runs, IWM outperforms LQSR by 2.1 dB.

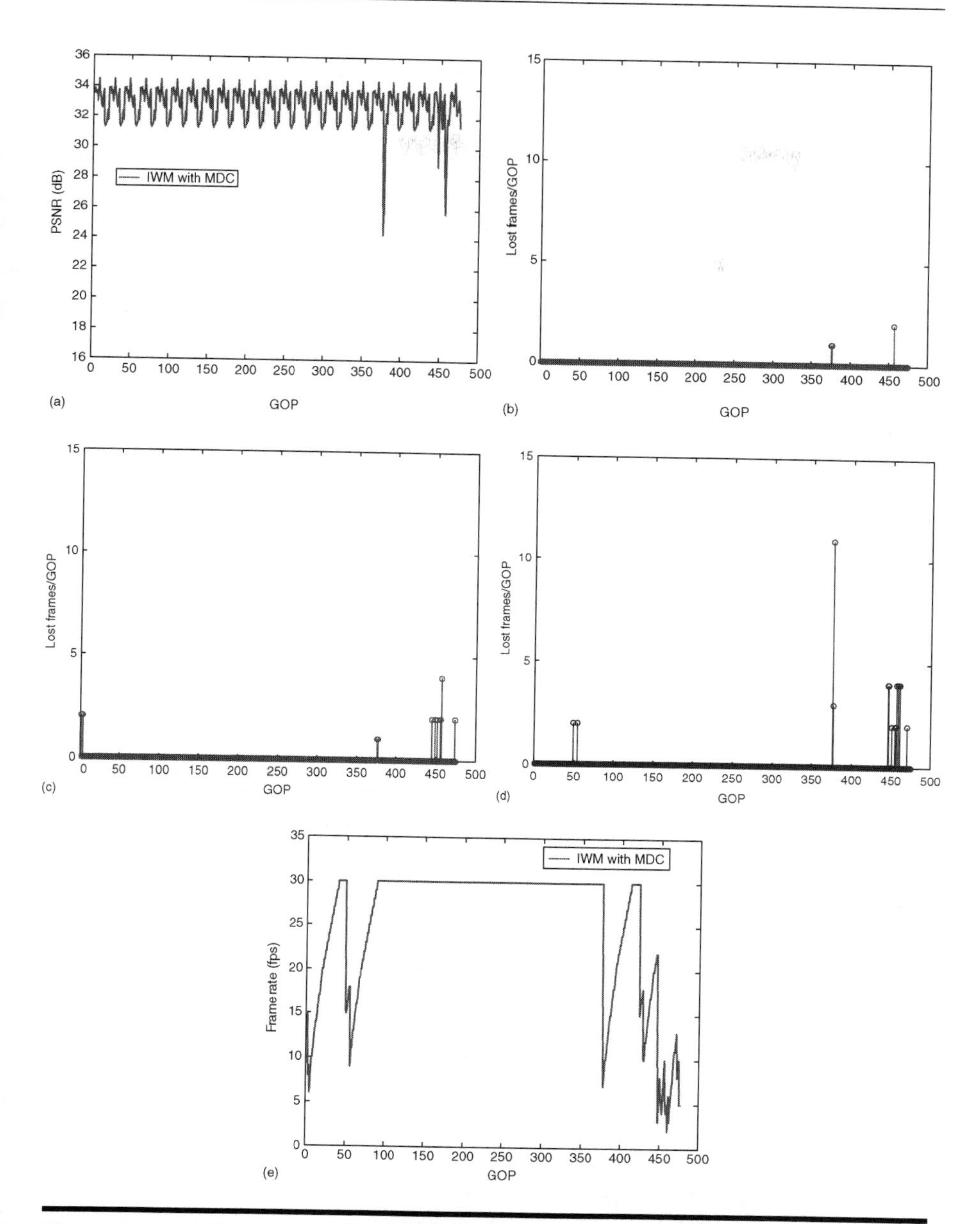

Figure 6.4 Performance evaluation of IWM/MDC over 802.11a. (a) PSNR of the received frames, (b) number of frames lost in both descriptions, (c) lost frames per GOP for substream 0, (d) lost frames per GOP for substream 1, (e) sending frame rate.

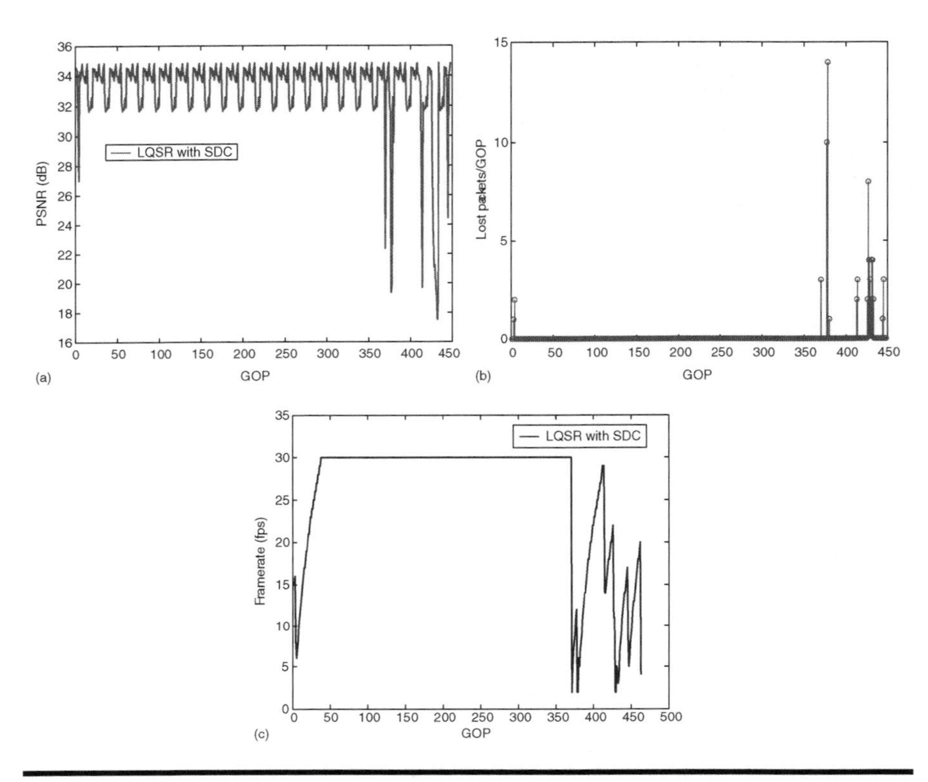

Figure 6.5 Performance evaluation of LQSR/SDC over 802.11a. (a) PSNR of the received frames, (b) lost frames per GOP for the stream, (c) sending frame rate.

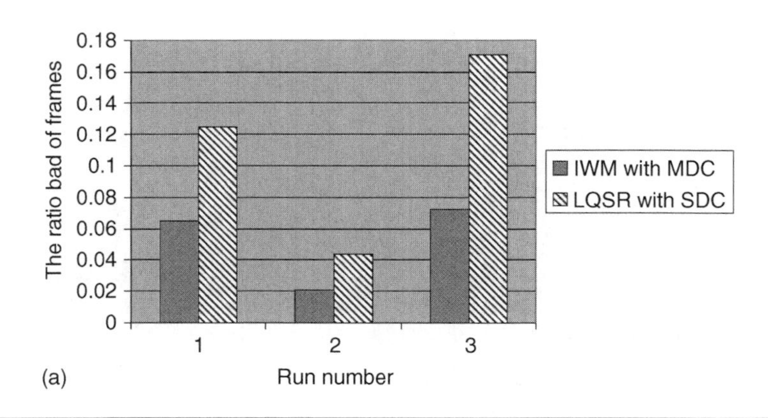

Figure 6.6 Performance evaluation of 802.11a with moving nodes. (a) The ratio of bad frames, (b) the number of bad periods.

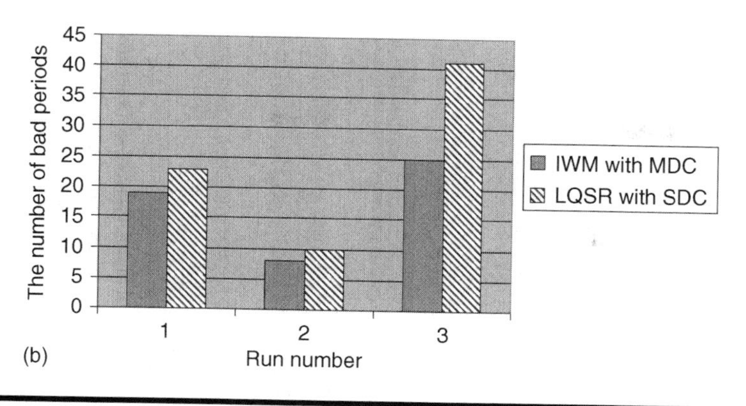

Figure 6.6 **(*Continued*)**

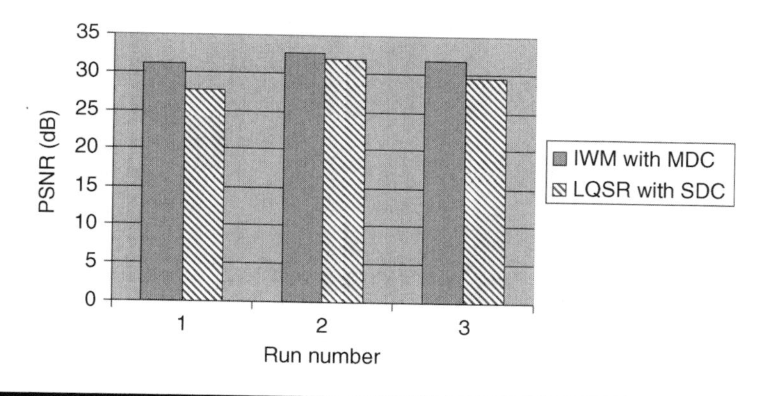

Figure 6.7 PSNR performance evaluation of IWM/MDC for 802.11a with moving nodes.

6.5 Multiple-Tree Multicast Video Communication over Wireless *Ad Hoc* Networks

Our proposed multiple-tree multicast video communication system consists of two parts—a multicast routing protocol to construct multiple trees and a simple scheme to distribute video packets into different trees. For the latter part, we employ MDC video to form multiple video streams and transmit different video streams through different trees. In this section, we assume that the network is lightly loaded, that is, mobility and poor channel condition rather than congestion are the major reasons for packet drop. In this case, multiple-tree multicast with MDC effectively alleviates undesirable effects caused by packet drop due to mobility and poor channels.

In this section, we begin by showing the feasibility of multiple-tree multicast, and then move on to describe ways to forward packets through multiple trees. We describe the proposed multiple-tree protocols in detail in Sections 6.6 and 6.7. Without loss of generality, we limit our discussion to the case of two trees, with each tree carrying one description.

6.5.1 Tree Connectivity and Tree Similarity

To measure the tree construction capability of multicast routing protocols, we define tree connectivity level P as follows [40]:

$$P \triangleq \frac{E[N]}{M} \tag{6.12}$$

where M is the product of the total number of receivers m and the number of trees n_t, $N = \sum_{i=1}^{m} n_i$, with $n_i \leq n_t$ denoting the number of trees that receiver i connects to, and m the number of receivers. Because $N \leq m \times n_t = M$, it can be shown that $0 \leq P \leq 1$. Given a random topology with n nodes, one random sender, and m random receivers, N is the sum of all receivers connected to each multicast tree, and $E[N]$ is the expected value of N over all topologies. Tree connectivity is a measure of the tree construction ability of a multicast routing protocol. Obviously, it is desirable to design a tree construction scheme with as high tree connectivity level, P, as possible.

To measure the level of disjointness of two trees, we define tree similarity, S, between two trees as the ratio of the number of shared nodes to the number of middle nodes of the tree with a smaller number of middle nodes. A node, which is neither the root nor a leaf, is a middle node in a tree. Tree similarity between two disjoint trees is zero and between two identical trees is one. The lower the tree similarity between two trees, the lower the correlated packet drop across two trees, and hence, the more effective multiple-tree video multicasting is in achieving high video quality.

Ideally, we desire a multicast routing protocol to achieve both a high tree connectivity level and a low tree similarity level in video applications space. Intuitively, if the node density is low, it is difficult to construct disjoint trees that connect to all m nodes, and hence either tree connectivity has to be low or tree similarity has to be high. However, for a sufficiently high node density or large radio range, we would expect a routing protocol to be able to achieve both a high tree connectivity level and a low tree similarity level.

Thus, an important issue in multiple-tree video multicasting is whether or not the required node density to obtain disjoint trees with high connectivity is too high to make it feasible in practice. We have developed the following theorem to address this issue for tree connectivity of two disjoint trees. Before stating the theorem, we need to introduce the term "critical density" λ_c. Dousse et al. [41] stated that there exists one critical density λ_c for a wireless *ad hoc* network such that if the density $\lambda < \lambda_c$, all the connected clusters are almost surely bounded; otherwise, almost surely there exists one unique unbounded superconnected cluster.

THEOREM 6.1: *Consider an infinite wireless network with nodes assumed to be distributed according to two-dimensional poisson process. Let D_1 denote the required node density to achieve a given tree connectivity level, P, in a single-tree case. If $D_1 > \lambda_c$, there exists at least one double-disjoint tree whose required node density D_2 satisfies to achieve P.*

$$D_2 - \frac{\ln(\pi D_2 r^2 + 1)}{\pi r^2} \leq D_1 \leq D_2 \tag{6.13}$$

where r is the radio link range.

The detailed proof is included in Ref. 40. We can see from Theorem 1 that the difference between D_1 and D_2 is only a logarithm factor of D_2, which is small compared to the value of D_1 and D_2. The difference is negligible as $D_1, D_2 \to \infty$, which are requirements for keeping the network connected as the number of total nodes $n \to \infty$ [42,43]. Thus we conclude that the required density for double-disjoint tree schemes is not significantly larger than that of single-tree schemes, and that tree diversity is a feasible technique to improve the robustness of multicast video transmission over wireless *ad hoc* networks.

6.5.2 Multiple-Tree Multicast Packet Forwarding

Our approach is to transmit different descriptions of MDC video flow through different trees simultaneously. If packet drop over two trees are not correlated, when some packets in one tree do not arrive at the destination on time, the receiver continues to decode and display packets corresponding to the other description on the other tree, resulting in acceptable video quality without interruption [44].

Our proposed multiple-tree multicast packet forwarding works as follows. The application-layer protocol sets a tree flag in each packet's header to determine the tree to which the packet should be forwarded. The multiple-tree multicast protocol forwards the packet in different trees according to the tree flag as follows: when a node receives a data packet, it checks the node's "forwarding table" for the forwarding status and "message cache" to avoid forwarding duplicate data packet. The node forwards a nonduplicate packet forwarded in tree-y if it is a forwarder for tree-y. Each packet flows along the corresponding tree from the sender to the receivers, but is not constrained to follow preset branches in the tree, as in the tree flood approach [29] or the "forwarding group flooding" approach [26]. Thus our packet forwarding scheme utilizes the broadcast nature of wireless *ad hoc* networks to obtain extra diversity gain without using extra network resources. Our packet forwarding scheme does not support packet forwarding across the trees, because nodes in one tree are unaware of the status of nodes in the other.

6.6 Serial Multiple-Disjoint Trees Multicast Routing Protocol

Because of the nature of MDC, the less the correlated packet drop between two trees, the more robust the video multicast. We assume that the network is lightly loaded, that is, mobility and poor channel conditions rather than congestion are the major causes of packet drop. In this case, if two trees do not share any middle nodes, packet drop over two trees are independent. Thus our main objective in the design of Serial MDTMR is to construct two node-disjoint multicast trees.

The proposed Serial MDTMR constructs two node-disjoint trees in a distributed way. First, we build a shortest path multicast tree. Then after requiring all the middle nodes in the first tree not to be middle nodes of the second tree, we construct another shortest path tree. Because these two trees do not share middle nodes at all, they are node disjoint. As Serial MDTMR is a way of constructing two disjoint multicast trees, it can be easily applied on top of any suitable single-tree multicast routing protocol. Without loss of generality, we design the detailed Serial MDTMR based on ODMRP [26], because ODMRP has been demonstrated to perform well and is well known [27]. By comparing Serial MDTMR and ODMRP, it is easy to quantify the performance gain obtained by the multiple-tree multicast routing. We can also design detailed Serial MDTMR based on other multicast routing protocols [17–23], taking advantage of their individual strengths. For example, we can apply a "local repair" scheme similar to Ref. 29 to maintain the tree structure with less control overhead. When a middle node or receiver detects that it is disconnected from the corresponding multicast forwarding tree, tree-x, where x is 0 or 1, it initiates a local repair process for tree-x, which searches the neighborhood of the middle node or receiver to find a new upstream node to reconnect the middle node or receiver to tree-x. To keep the disjointness between two trees, the middle node or receiver only selects a node, which is not a forwarding node for tree-$(1-x)$, as its new upstream node.

Similar to ODMRP, group membership and multicast trees in Serial MDTMR are established and updated by the source on demand. When a multicast source has packets to send, it periodically triggers a two-step multicast tree construction/refresh process. In the first step, the multicast source broadcasts to the entire network a JOIN REQUEST message, which includes the tree ID. When a node receives a nonduplicate JOIN REQUEST message for the first tree, it stores the upstream node ID and rebroadcasts the packet. When the JOIN REQUEST message reaches a multicast receiver, the receiver unicasts a JOIN ACK message to the multicast source via the reverse shortest path. When a middle node in the reverse path receives a nonduplicate JOIN ACK message, it updates its corresponding forwarding state in the forwarding table and forwards the message to its upstream node. Each middle node of the tree only forwards the JOIN ACK message once in one-tree construction cycle.

After receiving the first JOIN ACK message, the multicast source waits for a short time period before broadcasting another round of JOIN REQUEST message for the second tree to ensure the disjointness of two trees. When a node receives a nonduplicate JOIN REQUEST message, it forwards the packet only if it is not a middle node of the first tree in this round. When the JOIN REQUEST message reaches a receiver, the receiver unicasts back a JOIN ACK message to the multicast source to set up the second tree.

We compare tree connectivity of a single shortest path tree and Serial MDTMR as a function of node density through simulations, as shown in Figure 6.8. Note that for Serial MDTMR, tree connectivity is averaged

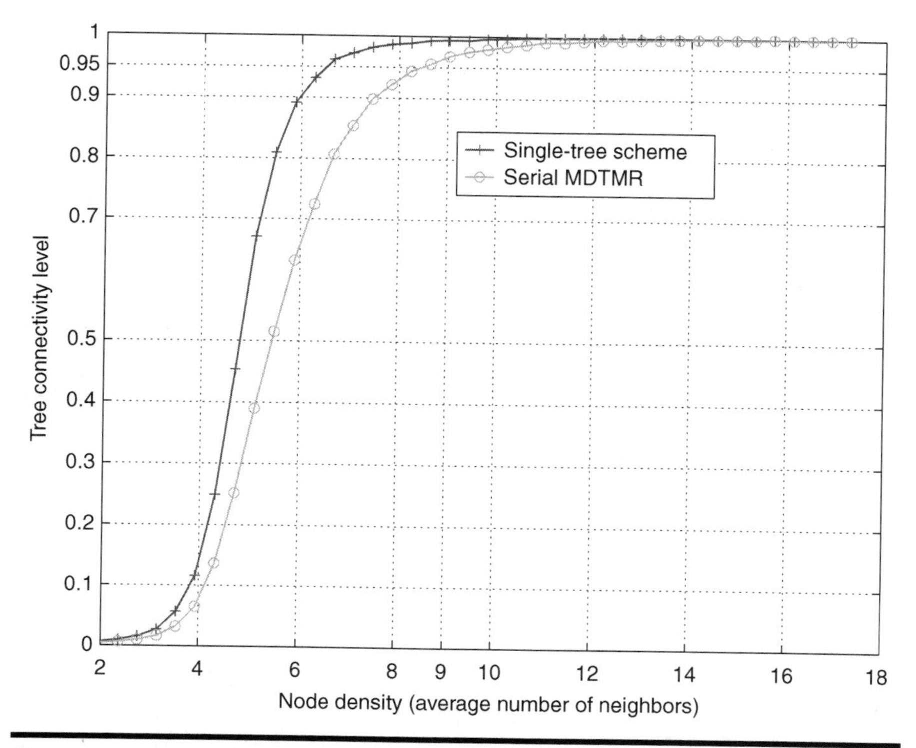

Figure 6.8 Tree connectivity of Serial MDTMR.

across two trees. The total number of nodes is 1000, with 50 receivers. The nodes are randomly distributed according to a two-dimensional poisson process. The results are averaged over 5000 runs. As seen, there is only a small performance gap between the two schemes when node density is larger than 7 nodes/neighborhood. For example, when node density is 8.2 nodes/neighborhood, tree connectivity of a single-tree scheme and Serial MDTMR is approximately 0.99 and 0.95, respectively.

6.7 Parallel Multiple Nearly Disjoint Trees Multicast Routing Protocol

Serial MDTMR achieves reasonable tree connectivity, while maintaining the disjointness of two trees. However, its routing overhead and construction delay are potentially twice as much as that of a parallel scheme that would build two trees simultaneously. In this section, we propose a novel parallel double-tree scheme, named Parallel MNTMR, to overcome the preceding disadvantages of Serial MDTMR. We have three main design goals for the Parallel MNTMR:

- *Low routing overhead and construction delay*. The routing overhead and construction delay of Parallel MNTMR should be similar to that of a typical single-tree multicast protocol.
- *High tree connectivity*. If a receiver is connected to the sender, it should be able to connect to both trees.
- *Near disjointness*. The ratio of the number of shared nodes of two trees to the number of nodes of the smaller tree should be minimized.

During multicast operation, the application-layer protocol sets a tree flag in each packet's header to determine to which tree the packet should be forwarded. The multiple-tree multicast protocol forwards the packet in different trees according to the tree flag. We also apply the tree flood approach [29] to achieve extra diversity gain without consuming extra network resources.

6.7.1 Overview of Parallel MNTMR

In a general single-tree multicast protocol, for example, ODMRP [26], when a multicast source has packets to send, it triggers a multicast tree construction process by flooding a join-query (JQ) message to the network. On receiving the JQ message, each receiver unicasts back a join-reply (JR) message to the sender to construct the multicast tree. In Parallel MNTMR, we apply similar JQ and JR processes to construct two nearly disjoint trees simultaneously.

The basic idea behind parallel tree construction is to first classify all the nodes randomly into one of the two categories—group 0 or 1. We define a "pure JQ message" as a JQ message whose route only consists of nodes in the same group, and a "mixed JQ message" as a JQ message whose route consists of nodes in both groups. The protocol uses a technique based on delay timer to select and forward pure JQ messages with a priority over mixed JQ messages. As will be seen in Section 6.7.4, this improves the disjointness of the constructed trees in the JR process.

We also propose an "upstream node selection rule" so that nodes close to one another tend to select the same upstream node for the same tree, thereby avoiding nodes of the other tree. This rule improves the disjointness of two trees, and forwarding efficiency of the multicast protocol.

6.7.2 Conditions and Rules

To construct two trees with both high tree connectivity and low tree similarity, Parallel MNTMR applies the following conditions and rules at each node to control the flow of JQ and JR messages. Without loss of generality, we assume the current node a is in group x, where x is 0 or 1. For brevity, we call a JQ message with a group-x node as the last hop, a group-x JQ message.

- *JQ message storing condition*. To obtain two loop-free trees in the JR process, each node only stores JQ messages satisfying the "storing condition" into its "JQ message cache." A JQ message satisfies the storing condition if it is the first received JQ message or the following two conditions are satisfied: (a) the number of hops it travelled is no larger than that of the first received JQ message at node a plus one and (b) the JQ message has not been forwarded by node a.
- *JQ message forwarding condition*. A JQ message satisfies the "forwarding condition" if the following two conditions hold true: (a) node a has not forwarded a JQ message in the JQ round and (b) the message's last hop is the sender or a group-x node. The forwarding condition results in "pure group-x JQ messages" to be selected and forwarded with a priority over mixed JQ messages, thus helping the protocol to construct trees that are as disjoint as possible.
- *Upstream node selection rule*. The objective of the upstream node selection rule is to maximize the disjointness of two trees. Let JQM_a denote the set of all the messages in the JQ message cache of node a. If there exist both group-0 and group-1 JQ messages in JQM_a, node a selects last hops of the earliest received group-0 and group-1 JQ messages as upstream nodes for tree-0 and tree-1, respectively. Otherwise, we assume that all the JQ messages in JQM_a are group-y

JQ messages. In this case, if $|JQM_a| > 1$, node a selects last hops of the earliest and the second earliest received JQ messages as upstream nodes for tree-y and tree-$(1-y)$, respectively; otherwise if there is only one message in JQM_a, the last hop of the only JQ message is selected as upstream nodes for both tree-0 and tree-1.

6.7.3 Detailed Double Nearly Disjoint Tree Construction

When a multicast source has packets to send, it triggers a multicast tree construction process by broadcasting a JQ message to its neighbors. When a node receives a group-y JQ message, if the message satisfies the storing condition, the node stores it into the JQ message cache for later use in the JR process, otherwise the message is simply discarded. If the message also satisfies the forwarding condition, the current node forwards the JQ message to its neighbors immediately; otherwise if the JQ message is the earliest received JQ message in the current JQ round, the node sets a JQ-delay timer. When the JQ-delay timer expires, if the node has not forwarded a JQ message in this JQ round, it forwards the earliest received JQ message. The JQ-delay scheme encourages pure JQ messages to be selected and forwarded with a priority over mixed JQ messages in the distributed tree construction process.

When a receiver receives a group-y JQ message, if the message is a pure JQ message, and the node has not initiated a JR message in this JQ round for tree-y, it selects the last hop of this JQ message as its upstream node for tree-y, and unicasts a JR message to the sender via the selected upstream node. All nodes, receiving and forwarding the JR message for tree-y, become middle nodes of tree-y. The receiver also sets a timer on receiving the earliest JQ message. When the timer expires, for each tree for which it has not yet initiated a JR message, the receiver selects an upstream node according to the upstream node selection rule and unicasts a JR message to the sender via the selected upstream node to construct this tree. In the end, we obtain one tree mainly consisting of group-0 nodes and another mainly consisting of group-1 nodes.

6.7.4 Discussion

In this section, we argue that Parallel MNTMR achieves our three design goals. Firstly, the Parallel MNTMR builds two trees simultaneously, and each node forwards the JQ message at once in one JQ round. Therefore, the routing overhead and the construction delay are similar to that of a typical single-tree multicast routing protocol. Secondly, as long as a receiver is connected to the sender, the protocol requires it to send JR messages for both trees; therefore the tree connectivity is the same as that of a single-tree protocol. Thirdly, regarding the disjointness of the two trees constructed by MNTMR, we have the following claim:

> *Claim 2:* Given any two nodes N_a and N_b, which are middle nodes for tree-0 and tree-1, respectively, let JQ_a and JQ_b denote node sets of last hops of JQ messages stored in the JQ message caches of nodes N_a and N_b, respectively. We sort nodes in JQ_a and JQ_b according to the arrival time of corresponding JQ messages. Let nodes N_c and N_d denote upstream nodes obtained by the Parallel MNTMR of nodes N_a and N_b, respectively. We have $N_c \neq N_d$ if the first two nodes of JQ_a and JQ_b are the same.

The proof is shown in Section 6.9.2.

Intuitively, *claim 2* shows that if two nodes in different trees share the same first two JQ messages in their JQ message caches, they will not select the same node as their upstream nodes. Thus for many scenarios, the Parallel MNTMR is likely to maintain disjointness between two trees.

6.7.5 Simulation Results

We use a simulation model based on NS-2 [38]. We only consider the continuous mobility case with zero pause time, and vary the maximum speed from 2.5 to 15 m/s. In each run, we simulate a 50 node wireless *ad hoc* network within a $1500 \times 300\,\text{m}^2$ area. Each simulation is 900 s long, and results are averaged over 30 runs.

We randomly choose one sender and eight receivers. For MDC, we encode one frame into two packets, whereas for SDC, we encode one frame into one packet. We set the frame rate as 8 fps and GOP size as 15. For fairness, we set the PSNR of MDC and SDC to be approximately the same, that is, 33 dB. To achieve similar quality, standard MPEG QCIF sequence Foreman is coded with a matching pursuit multiple description video Codec (MP-MDVC) [2] at 64.9 kbps for MDC, and with matching pursuit codec [45] at 41.2 kbps for SDC sequence. The playback deadline of each packet is set to 150 ms after it is generated.

We evaluate the performance using the following metrics:

a. *The ratio of bad frames.* It is the ratio of the total number of nondecodable frames to the total number of frames that should have been decoded in all the receivers in multicast scenario.
b. *The number of bad periods.* Defined in Section 6.3.6.
c. *Normalized packet overhead.* It is the total number of control packets transmitted by any node in the network divided by the total number of video frames received by all the receivers.

d. *Forwarding efficiency*. It is the total number of data packets transmitted by any node in the network divided by the total number of packets received by all the receivers.

We compare the following four schemes:

- Multiple-tree multicast with Parallel MNTMR and MDC
- Multiple-tree multicast with Serial MDTMR [44] and MDC
- Single-tree multicast with ODMRP [26] and MDC
- Single-tree multicast with ODMRP [26] and SDC

For fair comparison, all the three multicast routing protocols use 3 s for the JOIN REQUEST flooding interval and 4.5 s as a forwarding state lifetime.

Figures 6.9a and 6.9b show the result of the ratio of bad frames and the number of bad periods of the four schemes, respectively. As expected, both the ratio of bad frames and the number of bad periods increase with maximum speed. As seen, performance of multiple-tree multicast with Parallel MNTMR is similar to Serial MDTMR, and they both perform much better than the other two schemes with ODMRP. Shown in Figure 6.10, two trees obtained by Parallel MNTMR only share approximately 8 percent of nodes, which means they are nearly disjoint. This explains the reason for the two multiple-tree protocols performing similarly. The combination of our proposed multiple-tree multicast protocols, for example, Parallel MNTMR or Serial MDTMR and MDC reduces contiguous packet loss caused by broken links of multicast tree, because links of two nearly disjoint trees fail nearly when independent, resulting in much better received video performance than that with ODMRP and MDC. By comparing ODMRP with MDC and ODMRP with SDC, respectively, we conclude that MDC by itself can also reduce scattered packet loss caused by wireless channel error, or packets collision, thus reducing both the ratio of bad frames and the number of bad periods.

Figure 6.11a shows the normalized control packets for the four schemes. Simulation results show that the number of normalized control packets of Parallel MNTMR is similar to that of ODMRP and is about 50 percent lower than that of Serial MDTMR. Figure 6.11b shows that the number of the normalized forwarded data packets is almost the same for all four schemes with Parallel MNTMR being slightly worse. This indicates that the performance gain of Parallel MNTMR and Serial MDTMR is not at the expense of forwarding a packet more than ODMRP, rather by the combined effect of independent trees and MDC.

We plot PSNR and loss traces of a randomly selected receiver using ODMRP with SDC and Parallel MNTMR with MDC in Figures 6.12 and 6.13, respectively. Every node moves randomly with a maximum speed 5 m/s.

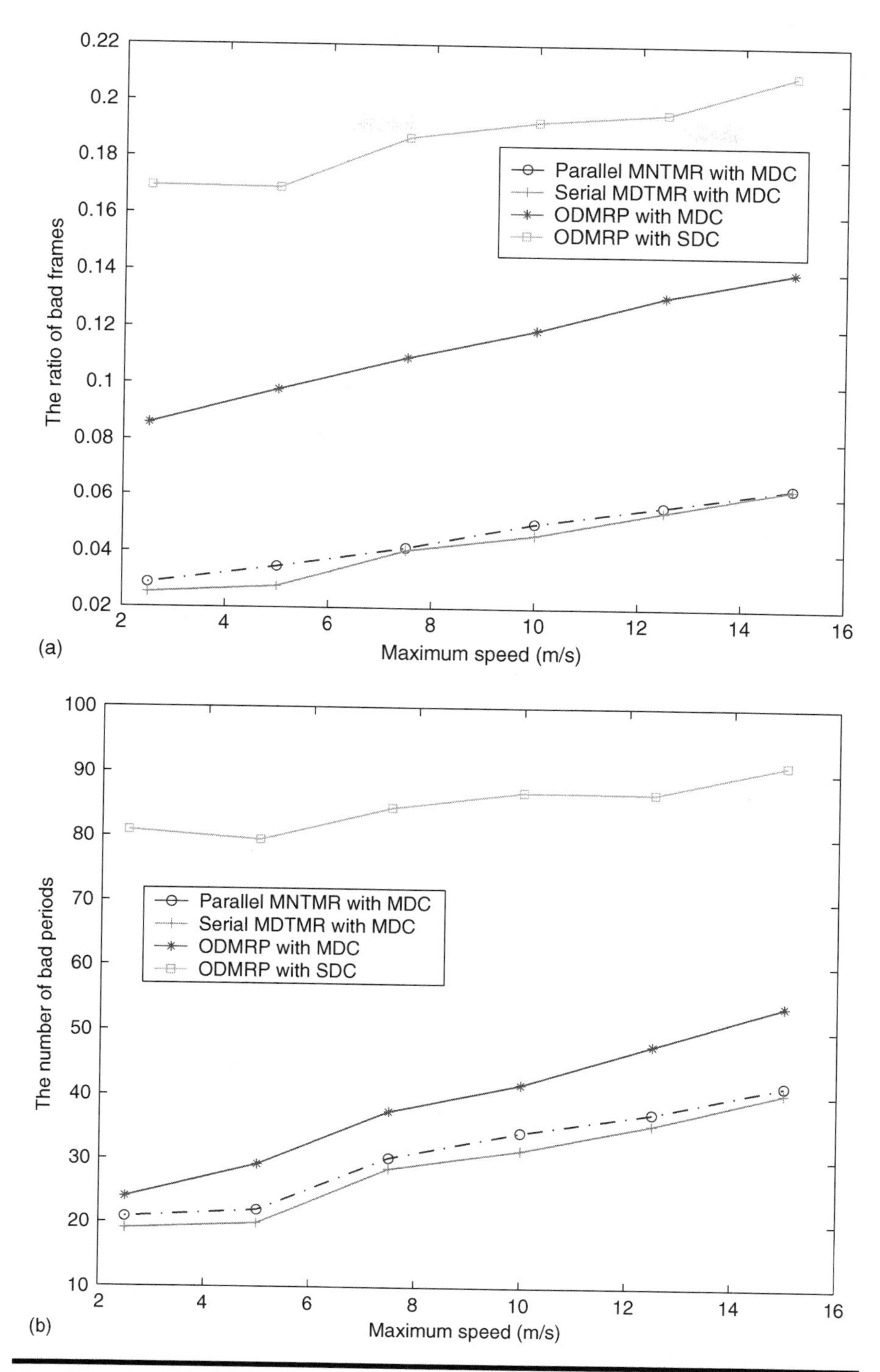

Figure 6.9 Performance evaluation for multiple-tree video multicast. (a) The ratio of bad frames, (b) the number of bad periods.

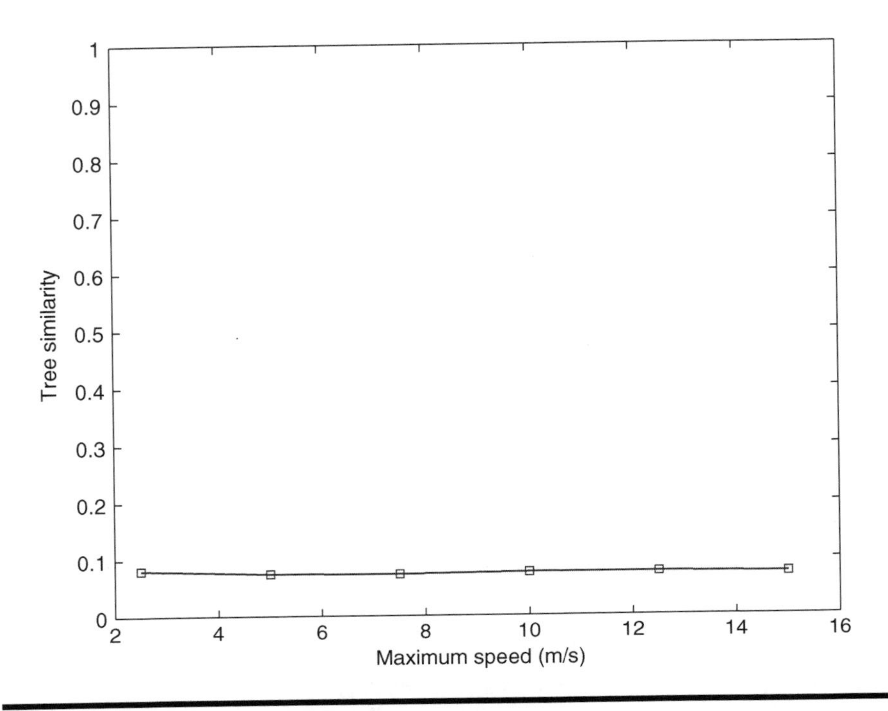

Figure 6.10 Tree similarity of Parallel MNTMR.

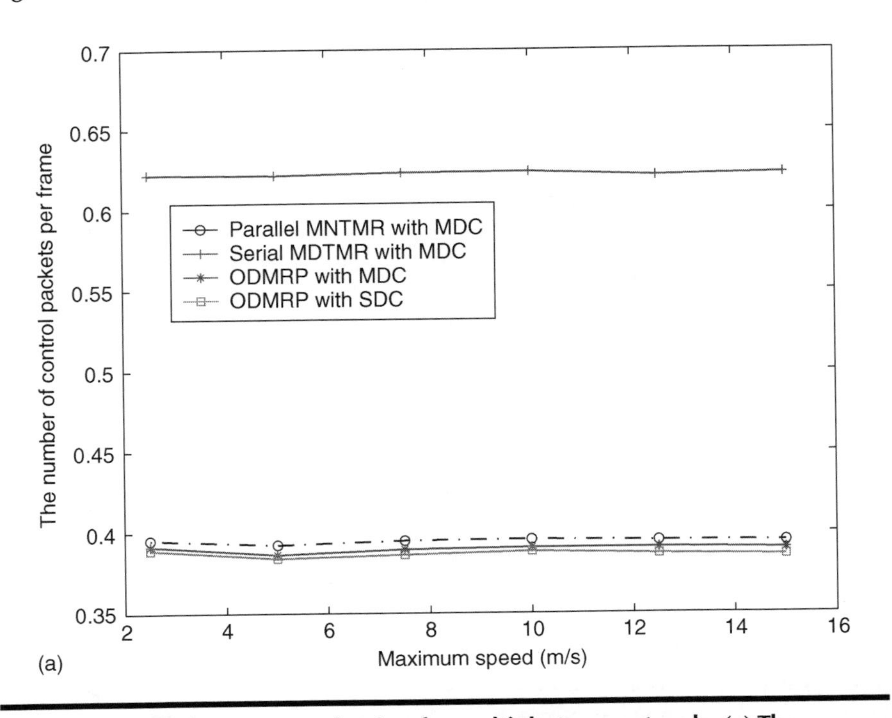

Figure 6.11 Performance evaluation for multiple-tree protocols. (a) The normalized control packets, (b) the normalized forwarded data packets.

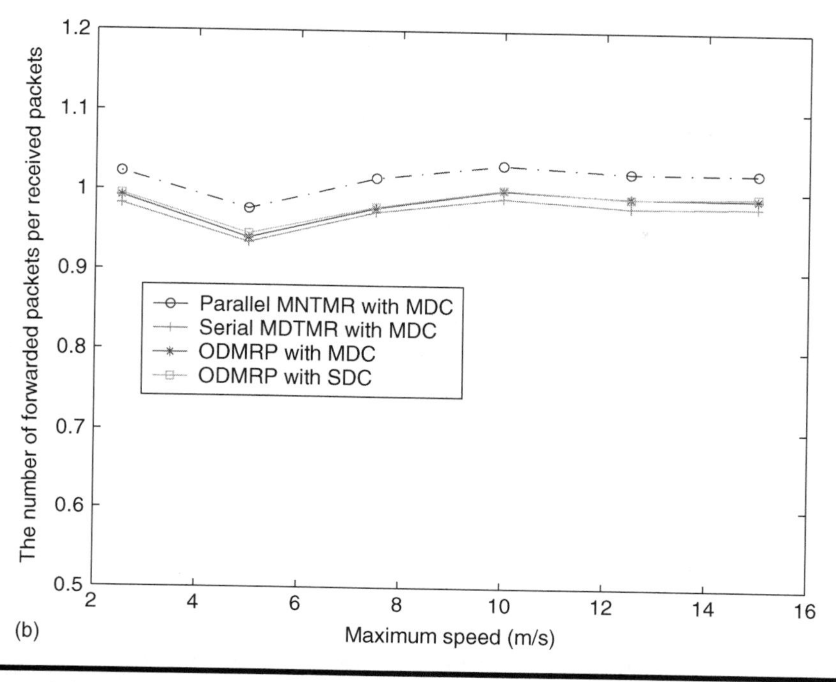

Figure 6.11 (*Continued*)

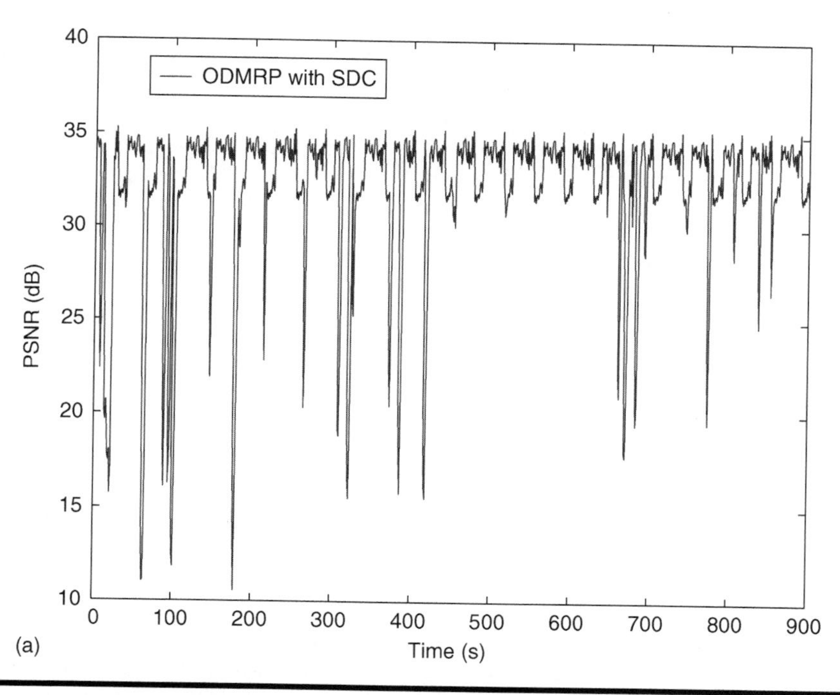

Figure 6.12 Performance evaluation of ODMRP and SDC. (a) PSNR of the received frames, (b) lost packets per second for the stream.

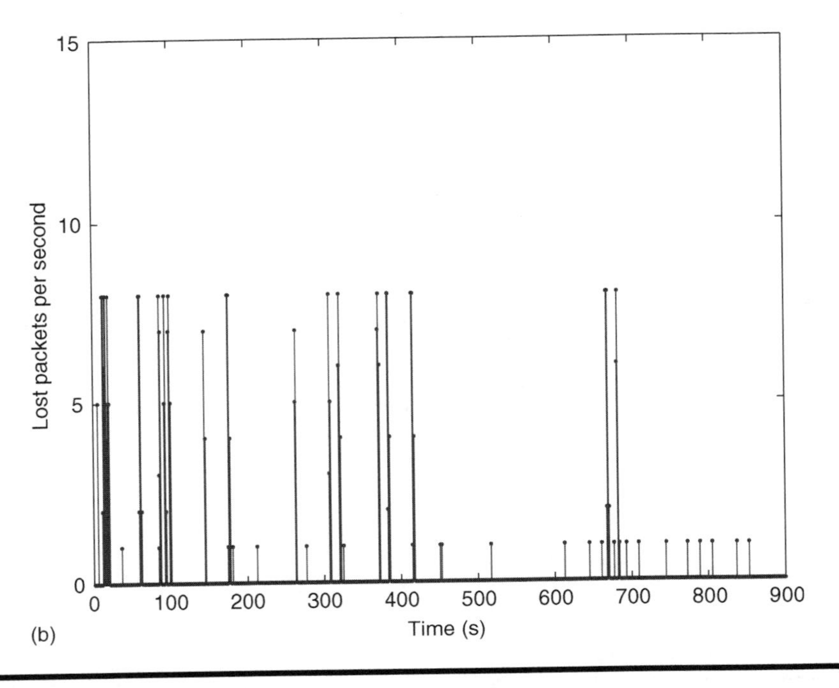

Figure 6.12 (*Continued*)

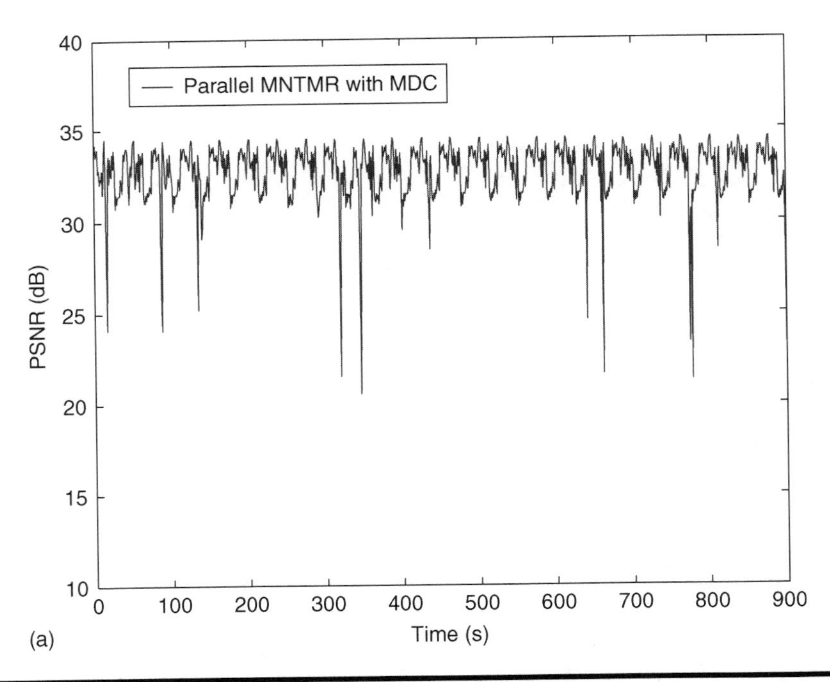

Figure 6.13 Performance evaluation of Parallel MNTMR and MDC. (a) PSNR of the received frames, (b) number of frames that both descriptions are lost, (c) lost packets per second for substream 0, (d) lost packets per second for substream 1.

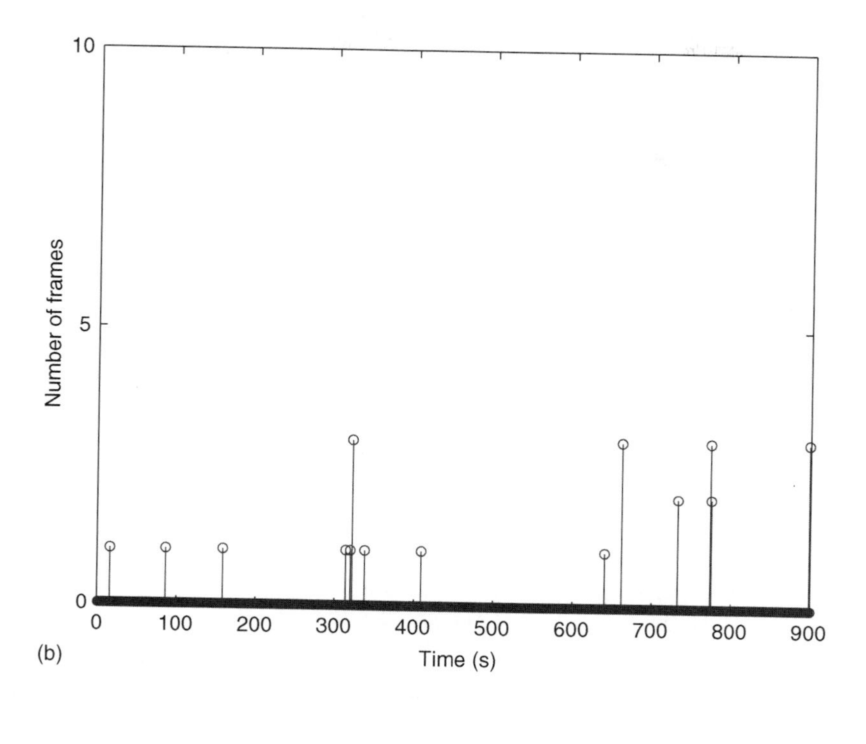

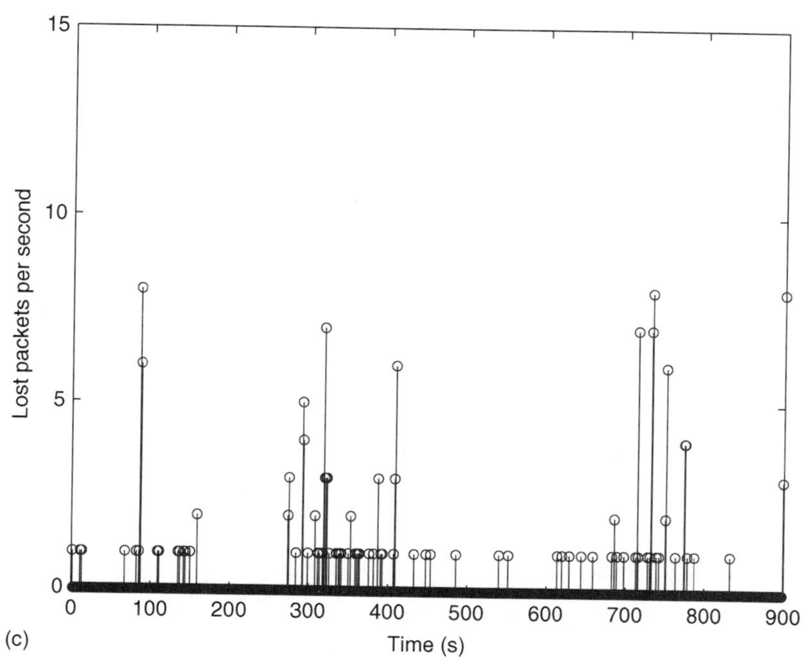

Figure 6.13 ***(Continued)***

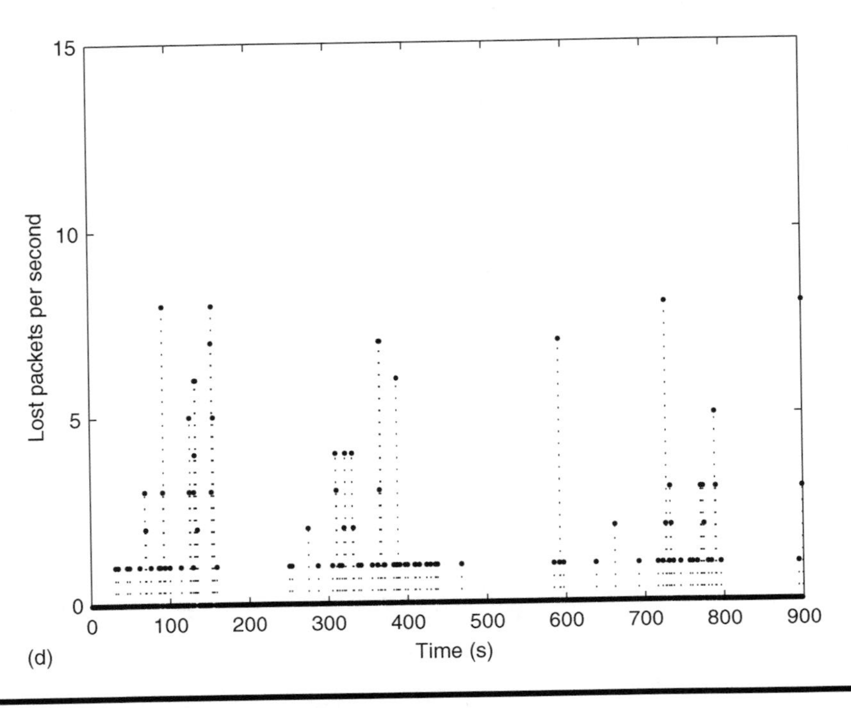

Figure 6.13 ***(Continued)***

For MDC, it can be seen in Figure 6.13a that PSNR drops gracefully when there is packet loss only in one substream. As seen in Figure 6.13, in Parallel MNTMR, most of the time, packet losses of two substreams do not overlap, thus reducing both the number and amount of PSNR drops. The PSNR curve of ODMRP with SDC shown in Figure 6.12a has more frequent and severe drops than that of Parallel MNTMR with MDC, because PSNR drops for every packet drop in SDC video and would drop severely when there is a burst of packet loss. We also visually examine the reconstructed video sequences under different schemes. For the video sequence obtained via Parallel MNTMR with MDC, we experience six short periods of distorted video in 900 s, whereas for the video sequence obtained via ODMRP with SDC, we experience 16 longer periods of more severely distorted video in the same time period.

6.8 Conclusion

In this chapter, we propose multipath unicast and multicast streaming in wireless *ad hoc* networks. In the unicast case, we propose a model to estimate the concurrent PDP of two paths by taking into account the interference between different links, and formulate an optimization problem to select two paths with minimum concurrent PDP. The solution to the optimal problem OMR is shown to be NP-hard. Then we propose a heuristic routing protocol based on our path selection model, whose performance is shown to be close to that of the "optimal routing" and significantly better than that of existing schemes, through both NS simulations and actual experiments in a testbed.

In the multicast case, we propose multiple-tree video multicast with MDC to provide robustness for video multicast applications. Specifically, we first propose a simple distributed protocol, Serial MDTMR, which builds two disjoint trees in a serial fashion. This scheme results in good tree connectivity while maintaining disjointness of two trees. To reduce the routing overhead and construction delay of Serial MDTMR, we further propose Parallel MNTMR, which constructs two nearly disjoint trees simultaneously in a distributed way. Simulation shows that video quality of multiple-tree multicast video communication is significantly higher than that of single-tree multicast video communication, with similar routing overhead and forwarding efficiency.

6.9 Appendix

6.9.1 *Proof of Claim 1*

Let A_i denote the event that $P^i_{\mathrm{S,D}}$ does not drop packets, for $i = 1, 2$ and L_{ij} denote the event that link (i, j) does not drop packets.

$$P_{\mathrm{drop}}\left(P^1_{\mathrm{S,D}}; P^2_{\mathrm{S,D}}\right) = P\,\overline{(A_1 \cup A_2)}$$

$$= P\left(\overline{\left(\bigcap_{(i,j)\in L^1_{\mathrm{S,D}}} L_{ij}\right) \cup \left(\bigcap_{(m,n)\in L^2_{\mathrm{S,D}}} L_{mn}\right)}\right)$$

$$= P\left(\bigcup_{(i,j)\in L^1_{\mathrm{S,D}},(m,n)\in L^2_{\mathrm{S,D}}} \left(\bar{L}_{ij} \cap \bar{L}_{mn}\right)\right) \tag{6.14}$$

By assuming that the PDP of each link is small, we can approximate $P_{\mathrm{drop}}(P^1_{\mathrm{S,D}}; P^2_{\mathrm{S,D}})$ as follows:

$$P_{\mathrm{drop}}\left(P^1_{\mathrm{S,D}}; P^2_{\mathrm{S,D}}\right) \approx \sum_{(i,j)\in L^1_{\mathrm{S,D}}} \sum_{(m,n)\in L^2_{\mathrm{S,D}}} P(\bar{L}_{ij}, \bar{L}_{mn}) \tag{6.15}$$

The simplified optimization problem, which uses Equation 6.15 as its metrics and keeps all the constraints of the original optimization problem in

Table 6.2 All Scenarios of Claim 2

Scenario	*Types of Nodes in JQ_a*	*Types of Nodes in JQ_b*
1	All group 0 nodes	All group 0 nodes
2	All group 0 nodes	Both group 0 and 1 nodes
3	All group 1 nodes	All group 1 nodes
4	All group 1 nodes	Both group 0 and 1 nodes
5	Both group 0 and 1 nodes	All group 0 nodes
6	Both group 0 and 1 nodes	All group 1 nodes
7	Both group 0 and 1 nodes	Both group 0 and 1 nodes

Equations 6.1 through 6.5, has shown to be an integer quadratic programming problem [17,46], which is a known NP-hard problem as proved in Ref. 47. □

6.9.2 *Proof of Claim 2*

We prove claim 2 by enumerating all possible scenarios. We list all the scenarios in Table 6.2.

Let message sets JQM_a and JQM_b denote JQ message caches of nodes N_a and N_b, respectively. In scenarios 2, 6, and 7, according to the upstream node selection rule, N_c is the last hop of the first received group-0 JQ message in JQM_a, and N_d is the last hop of the first received group-1 JQ message in JQM_b. Therefore $N_c \neq N_d$.

In scenario 1, using the upstream node selection rule, N_c is the last hop of the first received group-0 JQ message, which is also the first received JQ message in JQM_a. N_d is the last hop of the second received JQ message in JQM_b. Because the first two JQ messages of JQM_a and JQM_b are the same, $N_c \neq N_d$. Similarly in scenario 3, we arrive at the same conclusion.

In scenario 4, N_c is the last hop of the second received JQ message. Because the first two JQ messages of JQM_a and JQM_b are the same, the first two JQ messages of JQM_b are group-1 JQ messages. Thus, N_d is the last hop of the first group-1 JQ message, which is also the first JQ message. Therefore $N_c \neq N_d$. We could arrive at the same conclusion in scenario 5 in a similar fashion.

Therefore for all seven possible scenarios, $N_c \neq N_d$. □

References

1. Mao, S., Y. Wang, S. Lin, and S. Panwar, Video transport over *ad hoc* networks: multistream coding with multipath transport, *IEEE Journal on Selected Areas in Communications*, 21, 1721–1737, 2003.

2. Tang, X. and A. Zakhor, Matching pursuits multiple description coding for wireless video, *IEEE Transactions on Circuits and Systems for Video Technology*, 12, 566–575, 2002.
3. Wang, Y. and S. Lin, Error-resilient video coding using multiple description motion compensation, *IEEE Transactions on Circuits and Systems for Video Technology*, 12, 438–452, 2002.
4. Luo, H., S. Lu, and V. Bhargavan, A new model for packet scheduling in multihop wireless networks, *ACM Mobicom*, pp. 76–86, Boston, MA, August 6–11, 2000.
5. Jain, K., J. Padhye, V. Padmanabhan, and L. Qiu, Impact of interference on multihop wireless network performance, *ACM MobiCom*, San Diego, CA, September 14–19, 2003.
6. Gupta, R., J. Musacchio, and J. Walrand, Sufficient rate constraints for QoS flows in ad-hoc networks, *Ad-Hoc Networks Journal*, 5(4) 429–443, May 2007.
7. Jia, Z., R. Gupta, J. Walrand, and P. Varaiya, Bandwidth guaranteed routing for *ad hoc* networks with interference consideration, *ISCC*, Cartagena, Spain, June 27–30, 2005.
8. Obraczka, K. and G. Tsuduk, Multicast Routing Issues in *Ad Hoc* Networks, *IEEE International Conference on Universal Personal Communications*, 1998.
9. Maxemchuk, N.F., Dispersity Routing in Store and Forward Networks, Ph.D. thesis, University of Pennsylvania, Philadelphia, PA, May 1975.
10. Gogate, N. and S.S. Panwar, Supporting video/image applications in a mobile multihop radio environment using route diversity, *IEEE ICC '99*, Vancouver, Canada, June 6–10, 1999.
11. Apostolopoulos, J., Reliable video communication over lossy packet networks using multiple state encoding and path diversity, *Visual Communications and Image Processing (VCIP)*, pp. 392–409, San Jose, CA, January 2001.
12. Apostolopoulos, J., T. Wong, W. Tan, and S. Wee, On multiple description streaming in content delivery networks, *IEEE Infocom*, 1736–1745, 2002.
13. Wei, W. and A. Zakhor, Robust Multipath Source Routing Protocol (RMPSR) for Video Communication over Wireless *Ad Hoc* Networks, *International Conference on Multimedia and Expo (ICME)*, Taiwan, China, 2004.
14. Chakareski, J., S. Han, and B. Girod, Layered coding versus multiple descriptions for video streaming over multiple paths, *Multimedia Systems*, (online journal publication) 10(4), 275–285, Springer, Berlin, January 2005.
15. Begen, A., Y. Altunbasak, and O. Ergun, Multipath Selection for Multiple Description Encoded Video Streaming, *IEEE Internationl Conference on Communications (ICC)*, Anchorage, AK, May 11–15, 2003.
16. Mao, S., Y. Hou, X. Cheng, H. Sherali, and S. Midkiff, Multipath routing for multiple description video in wireless *ad hoc* networks, *IEEE Infocom*, Miami, FL, March 13–17, 2005.
17. Ma, Z., H. Shao, and C. Shen, A New Multipath Selection Scheme for Video Streaming on Overlay Networks, *IEEE Internationl Conference on Communications (ICC)*, Anchorage, AK, May 11–15, 2003.
18. Johnson, D.B., D.A. Maltz, and Y. Hu, The dynamic source routing protocol for mobile *ad hoc* network, Internet-Draft, draft-ietf-manet-dsr-09.txt, April 2003.
19. Lee, S.J. and M. Gerla, Split Multipath Routing with Maximally Disjoint Paths in *ad hoc* Networks, *IEEE International Conference on Communications (ICC)*, pp. 3201–3205, Helsinki, Finland, June 2001.

20. Wu, K. and J. Harms, On-demand multipath routing for mobile *ad hoc* networks, *4th European Personal Mobile Communication Conference (EPMCC 01)*, pp. 1–7, Vienna, Austria, February 2001.
21. Pham, P. and S. Perreau, Performance analysis of reactive shortest path and multipath routing mechanism with load balance, *IEEE Infocom*, San Francisco, CA, March 30–April 03, 2003.
22. Nasipuri, A. and S.R. Das, On-demand multipath routing for mobile *ad hoc* networks, *IEEE International Conference on Computer Communication and Networks (ICCCN'99)*, pp. 64–70, Boston, MA, October 1999.
23. Papadimitratos, P., Z.J. Haas, and E.G. Sirer, Path Set Selection in Mobile *Ad Hoc* Networks, *ACM Mobihoc 2002*, Lausanne, Switzerland, June 9–11, 2002.
24. Padmanabhan, V.N., H.J. Wang, P.A. Chou, and K. Sripanidkulchai, Distributing streaming media content using cooperative networking, *ACM NOSSDAV*, Miami Beach, FL, May 2002.
25. Mao, S., X. Cheng, Y. Hou, and H. Sherali, Multiple description video multicast in wireless *ad hoc* networks, *BroadNets*, San Jose, CA, October 25–29, 2004.
26. Lee, S., M. Gerla, and C. Chiang, On-demand multicast routing protocol, *IEEE WCNC'99*, pp. 1298–1302, New Orleans, LA, September 1999.
27. Lee, S., W. Su, J. Hsu, M. Gerla, and R. Bagrodia, A performance comparison study of *ad hoc* wireless multicast protocols, *IEEE INFOCOM 2000*, Tel Aviv, Israel, March 2000.
28. Toh, C., G. Guichala, and S. Bunchua, ABAM: on-demand associativity-based multicast routing for *ad hoc* mobile networks, *IEEE Vehicular Technology Conference, VTC 2000*, pp. 987–993, Boston, September 2000.
29. Jetcheva, J. and D. Johnson, Adaptive demand-driven multicast routing in multihop wireless *ad hoc* networks, *ACM Mobihoc*, Long Beach, CA, October 4–5, 2001.
30. Wu, C. and Y. Tay, AMRIS: a multicast protocol for *ad hoc* wireless networks, *IEEE Military Communications Conference 1999*, pp. 25–29, New York, 1999.
31. Xie, J., R.R. Talpade, A. McAuley, and M. Liu, AMRoute: *ad hoc* multicast routing protocol, *Mobile Networks and Applications*, 7(6), 429–439, 2002.
32. Park, S. and D. Park, Adaptive core multicast routing protocol, *Wireless Networks*, 10(1), 53–60, 2004.
33. Sajama, S. and Z. Haas, Independent-tree *ad hoc* multicast routing (ITAMAR), *ACM Mobile Networks and Applications*, 8(5), 551–566, October 2003.
34. Dijkstra, E.W., A note on two problems in connexion with graphs, *Numerische Mathematik*, 1, 269–271, 1959.
35. Wei, W., Multipath Unicast and Multicast Video Communication over Wireless *Ad Hoc* Networks, PhD thesis, University of California, Berkeley, CA, 2006.
36. De Couto, D., D. Aguayo, J. Bicket, and R. Morris, High-throughput path metric for multihop wireless routing, *ACM Mobicom*, San Diego, CA, September 14–19, 2003.
37. Jiang, S., D. He, and J. Rao, A prediction-based link availability estimation for mobile *ad hoc* networks, *IEEE Infocom*, Anchorage, AK, April 22–26, 2001.
38. NS-2: network simulator. http://www.isi.edu/nsnam/ns/
39. Draves, R., J. Padhye, and B. Zill, Routing in Multiradio, Multihop Wireless Mesh Networks, *ACM MobiCom '04*, Philadelphia, PA, September 2004.
40. Wei, W. and A. Zakhor, Connectivity for Multiple Multicast Trees in *Ad Hoc Networks, International Workshop on Wireless AdHoc Networks (IWWAN)*, Oulu, Finland, June 2004.

41. Dousse, O., P. Thiran, and M. Hasler, Connectivity in *ad hoc* and hybrid networks, *IEEE Infocom*, New York, June 23–27, 2002.
42. Philips, T., S. Panwar, and A. Tantawi, Connectivity properties of a packet radio network model, *IEEE Transactions on Information Theory*, 35(5), 1044–1047, September 1989.
43. Gupta, P. and P. Kumar, Critical power for asymptotic connectivity in wireless networks, *Stochastic Analysis, Control, Optimization and Applications: A Volume in Honor of W.H. Fleming*, (eds., W.M. McEneaney, G.G. Yin and Q. Zhang) Springer, Boston, 1999.
44. Wei, W. and A. Zakhor, Multipath Unicast and Multicast Video Communication over Wireless *Ad Hoc* Networks, *International Conference on Broadband Networks (Broadnets) 2004*, pp. 496–505. San Jose, CA, October 2004 (invited).
45. Neff, R. and A. Zakhor, Very low bit-rate video coding based on matching pursuits, *IEEE Transactions on Circuits and Systems for Video Technology*, 7, 158–171, 1997.
46. Cui, W., I. Stoica, and R. Katz, Backup path allocation based on a correlated link failure probability model in overlay networks, *ICNP*, Paris, France, November 12–15, 2002.
47. Garey, M. and D. Johnson, *Computers and Intractability: A Guide to The Theory of NP-Completeness*, W. H. Freeman Company, San Francisco, CA, 1979.

Chapter 7

Video Communications over Wireless Sensor Networks

Min Chen, Shiwen Mao, Yong Yuan, and
Victor C.M. Leung

Contents

7.1 Introduction 236
7.2 Background..238
7.2.1 Video Transmissions over WLANs 239
7.2.2 QoS Provisioning for Time-Constrained Traffic in WSNs........239
7.2.3 Video Transmissions over WSNs 240
7.3 Multipath-Based Real-Time Video Communications in WSNs..........241
7.3.1 Architecture of Video Sensor Network 241
7.3.2 Multipath-Based Video Transmission Strategy........................243
7.4 Simulation Methodology..247
7.4.1 Simulation Model..247
7.4.2 Performance Metrics 248
7.5 Performance Evaluations 249
7.6 Conclusion 254
References 255

This chapter addresses the problem of real-time video streaming over a bandwidth and energy-constrained wireless sensor network (WSN). Because the compressed video bit stream is extremely sensitive to transmission errors, conventional single-path routing schemes typically based on shortest paths are not very effective to support video transmissions in unreliable and bandwidth-limited WSNs. Considering the constraints in bandwidth, energy in WSNs, and delay in video delivery, we propose to divide a single video stream into multiple substreams, and exploit multiple disjoint paths to transmit these substreams in parallel. These multiple paths will facilitate load balancing, bandwidth aggregation, and fast packet delivery. For efficient multipath routing of these parallel substreams from the source to the sink, we propose a hybrid video stream broadcasting and substreams unicasting scheme based on the construction of an application-specific number of multiple disjoint paths.

7.1 Introduction

With recent advances in WSNs, it is foreseeable that video sensors will be supported in such networks, for applications such as battlefield intelligence, security monitoring, emergency response, and environmental tracking [1].

We investigate H.26L [2,3] real-time video communications in video sensor networks (VSNs), where video streams are transmitted under a number of resource and performance constraints such as bandwidth, energy, and delay. Although a high compression ratio makes H.26L real-time video applications suitable for low bit-rate channels, the received video quality is susceptible to transmission errors. Thus, it remains a challenging problem to deliver H.26L video data with a high quality of service (QoS) in WSNs with bandwidth-limited error-prone wireless channels. Owing to the bandwidth limitation of a VSN, we consider only a small number of video sensor nodes (VNs), which have video capturing capability, taking turn to transmit video to a single sink; that is, only one VN transmits video to the sink at any time.

Because the compressed video bit stream is extremely sensitive to transmission errors due to dependencies between video frames, error control techniques such as forward error correction (FEC) and automatic repeat request (ARQ) are necessary to obtain the high reliability required by video services [4]. Between these two error control mechanisms, FEC is generally preferred for real-time applications due to the strict delay requirements and semireliable nature of media streams [5]. However, links in a WSN may not have adequate bandwidth to satisfy the higher bandwidth requirement of FEC coding. Thus, conventional single-path routing schemes typically based on shortest paths [6,7] are not very effective to support video transmissions in unreliable and bandwidth-limited WSNs, as they will cause either significant degradation in the perceived quality of the video at the sink nodes if FEC coding is not used, or large queuing delays due to insufficient bandwidth if it is used. Furthermore, transmitting a video stream using the shortest path will drain the energy of the nodes along this path and shorten the network lifetime. Thus, considering the constraints in bandwidth, energy in WSNs, and delay in video delivery, we propose to divide a single video stream into multiple substreams, and exploit multiple disjoint paths to transmit these substreams in parallel. The multiple paths will facilitate load balancing, bandwidth aggregation, and fast packet delivery. For efficient multipath routing of these parallel substreams from the source to the sink, we propose a hybrid video stream broadcasting and substreams unicasting scheme based on the construction of an application-specific number of multiple disjoint paths.

In WSNs, multipath routing is used to establish multiple paths between each source–sink pair. Most applications of multipath routing in WSNs aim to increase the reliability for a single flow [8–11]. In contrast, multipath routing is used in the proposed scheme, that is, directional geographical routing (DGR) [12], to support the delivery of multiple flows in a VSN, although the responsibility of reliable data delivery at the routing layer is relieved by the use of FEC coding.

Similar to many previous multipath routing schemes, the proposed DGR mechanism also encounters the route-coupling problem [13], caused by interference between packets transmitted over different paths between the same source–destination pair. If the number of paths is small (e.g., 2 or 3), noninterfering paths may be established. However, if a large number of paths are required by a specific application, noninterfering paths cannot be guaranteed due to the limited spatial size in proximity to the source/sink. In such cases, the best approach is to spatially distribute these paths as evenly as possible.

Given the scenario presented in Section 7.4, Figures 7.1 and 7.2 show the OPNET simulation results of DGR's path construction and illustrate DGR's

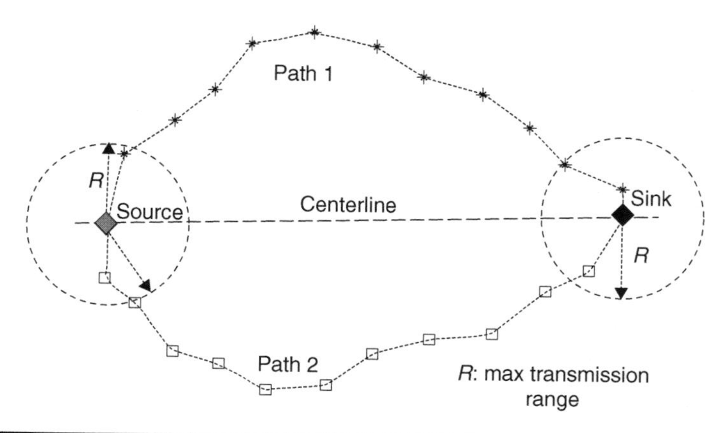

Figure 7.1 Minimum number of paths constructed in DGR.

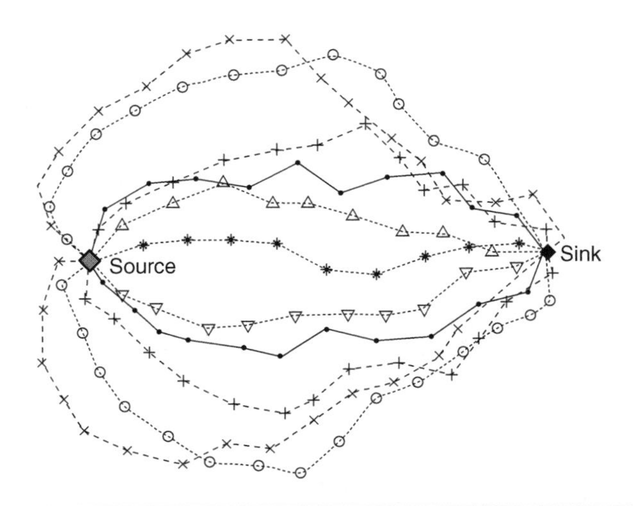

Figure 7.2 Maximum number of paths constructed in DGR.

adaptability to an application-specific path number (PathNum). As an example, with a minimum PathNum of 2, DGR tries to pick two paths that do not interfere with one another (see Figure 7.1). Let N_S be the minimum number of neighbors among the source and sink nodes. In Figure 7.2, $N_S = 11$; therefore, the maximum possible value of PathNum is 11 if it is required that no two paths traverse the same node(s). This maximum PathNum value is only achievable in an ideal WSN, where the node density is sufficiently high. Assuming this is the case, DGR constructs all 11 paths as illustrated by the simulation result in Figure 7.2. In practice, for a large PathNum, DGR spreads the paths in all directions in the proximity of the source and sink nodes, which implies that packets along some paths are likely to be forwarded to a neighbor farther to the sink than the node itself. Thus, DGR differs from traditional geographic routing schemes [14], in which each node forwards packets to a neighbor that is closer to the sink than the node itself until the packets reach the sink.

The remainder of this chapter is organized as follows. Section 7.2 presents the background of video transmissions over WSNs. Section 7.3 describes the proposed multipath-based real-time video transmissions scheme in WSNs. Sections 7.4 and 7.5 present simulation model and experiment results, respectively. Section 7.6 summarizes the chapter.

7.2 Background

Our work is closely related to video transmissions over wireless local area networks (WLANs), QoS provisioning for time-constrained traffic in WSNs, and image/video transmissions over WSNs. We will give a brief review of the existing work in these areas.

7.2.1 Video Transmissions over WLANs

A survey is presented in Ref. 15 on video streaming over WLANs. Bucciol et al. [16] proposed a cross-layer ARQ algorithm for H.264 video streaming in 802.11 WLANs, which gives priority to perceptually more important packets at (re)transmissions. In Ref. 17, a transmission strategy is examined that provides adaptive QoS to layered video for streaming over 802.11 WLANs. In Refs 18 and 19 hybrid transmission techniques that combine ARQ and FEC are proposed for improved real-time video transport over WLANs. However, only single-hop network scenarios are investigated in Ref. 19. However, this chapter considers real-time video transmissions in multihop WSN environments. Mao et al. [20] combined multistream coding with multipath transport, and showed that, in addition to traditional error control techniques, path diversity provides an effective means to combat transmission errors in *ad hoc* networks. In Ref. 20, the Dynamic Source Routing (DSR) protocol is extended to support multipath routing. With their extension, multiple maximally disjoint routes are selected from all the routes returned by a route query. However, only two paths are constructed in their simulation model and only two substreams are considered. In contrast, our scheme is adaptive to an application-specific PathNum that can be greater than two.

7.2.2 QoS Provisioning for Time-Constrained Traffic in WSNs

Many applications of WSNs require QoS provisioning for time-constrained traffic, such as real-time target tracking in battlefield environments and emergent event triggering in monitoring applications. Recent years have witnessed increasing research efforts in this area. SPEED [21] is an adaptive real-time routing protocol that aims to reduce the end-to-end deadline miss ratio in WSNs. MMSPEED [22] extends SPEED to support multiple QoS levels in the timeliness domain by providing multiple packet delivery delay guarantees. Akkaya and Younis [23] proposed an energy-aware QoS routing protocol to support both best effort (BE) and real-time (RT) traffic at the same time, by meeting the end-to-end delay constraint of the RT traffic while maximizing the throughput of the BE traffic. Weighted fair queuing (WFQ)–based packet scheduling is used to achieve the end-to-end delay bound in Ref. 24. Yuan et al. [25] proposed an integrated energy and QoS aware transmission scheme for WSNs, in which the QoS requirements in the application layer, and the modulation and transmission schemes in the data link and physical layers are jointly optimized. Energy-efficient differentiated directed diffusion (EDDD), proposed in Ref. 26, provides service differentiation between BE and RT traffic by deploying BE and RT filters. The BE filter aims to balance the global energy and prolong network lifetime, whereas end-to-end delay is not a primary concern. The RT filter aims to provide better end-to-end delay performance for time-sensitive traffic. In this chapter, we use multiple paths to increase the end-to-end capacity and achieve the QoS requirements in terms of end-to-end latency.

7.2.3 Video Transmissions over WSNs

Recent advances in hardware miniaturization have allowed the fabrication of sensor devices that support the use of specialized add-on modules for imagery applications. As an example, the Cyclops image capturing and inference module [27] is designed for extremely light-weight imaging, and can be interfaced with popular WSN devices such as Crossbow's MICA2 and MICAz [28,29]. The availability of such inexpensive imaging hardware has fostered the development of VSNs that allow collection and dissemination of video streams and still images. From a survey presented in Ref. 1, it is clear that VSN research is a field of growing activity, in which innovations in applied signal processing interact with emerging applications and technology.

He and Wu [30,31] studied the resource utilization behavior of a wireless video sensor and analyzed its performance under resource constraints. Wu and Chen proposed a novel collaborative image coding and transmission scheme for WSNs. Energy reduction is evidenced from the use of this collaborative approach to distributed image compression. A multiagent framework for video sensor–based coordination in surveillance applications was proposed by Patricio et al. Awad et al. addressed the problem of action classification using multiple real-time video signals collected from homogeneous sites. Chow et al. investigated the optimization of energy resources when transmitting visual data on demand to a mobile node via judicious path selection for tracking applications. A distributed protocol requiring only local information was proposed and evaluated through simulations. In addition, mobile agents were found specially useful for image/video transmissions over WSNs. Because the imagery data can occupy large memory spaces, transmitting whole pictures not only consumes a lot of energy, but may not be necessary also if the sink only needs information on a certain region of interest (RoI) within the picture. Figure 7.3 shows an application of mobile agent–based image transmissions over WSNs. Here, a mobile agent carries an image segmentation code and is dispatched to the target region. It visits camera sensors one by one and collects only the RoI by segmenting the image; thus, the amount of imagery data sent by the mobile agent at each target sensor node is substantially reduced. Note that a single kind of image segmentation algorithm may not achieve good performance for all kinds of images to be extracted. Thus, the sink may dispatch multiple mobile agents carrying alternate image-processing codes to the sensors of interest.

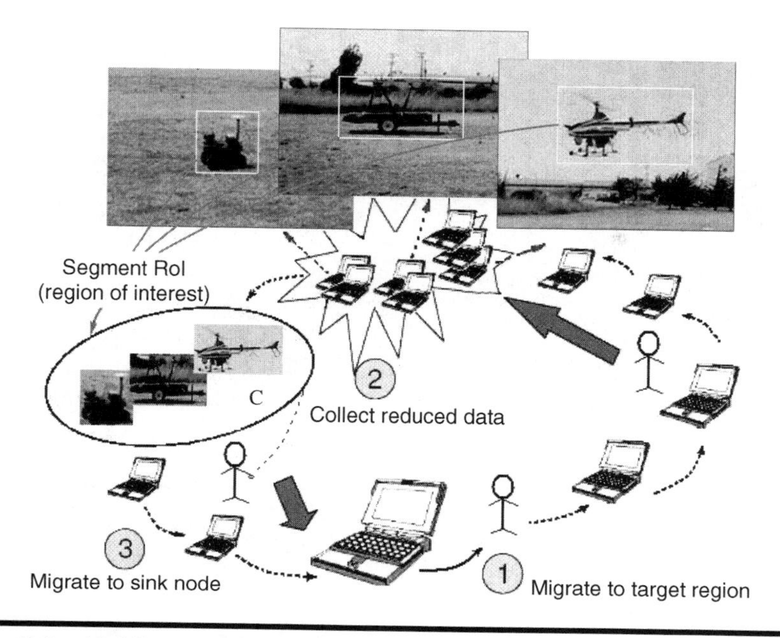

Figure 7.3 Mobile agent–based image sensor querying.

7.3 Multipath-Based Real-Time Video Communications in WSNs

This section describes the architecture of a VSN and the proposed multipath-based video transmission strategy.

7.3.1 Architecture of Video Sensor Network

In a VSN, VNs equipped with video capturing and processing capabilities are tasked to capture digital visual information about target events or situations, and deliver the video streams to a sink node [30]. Generally, a VN should be equipped with a battery of higher energy capacity than an ordinary sensor node, because it is already equipped with a relatively expensive camera that would become useless if the VN ran out of energy.

However, it is economically infeasible and often unnecessary to equip all the sensor nodes with video capturing and processing capabilities, especially for large-scale or dense WSNs. We consider a VSN architecture, as illustrated in Figure 7.4, where a small number of VNs is sparsely deployed among a much larger number of densely deployed low-power sensor nodes. The set of VNs only cover the target regions remotely monitored by the sink. The inexpensive ordinary sensor nodes perform the simple task of forwarding packets that carry sensed video data to the sink. Owing to bandwidth limitation of a typical WSN link, we consider that the VNs

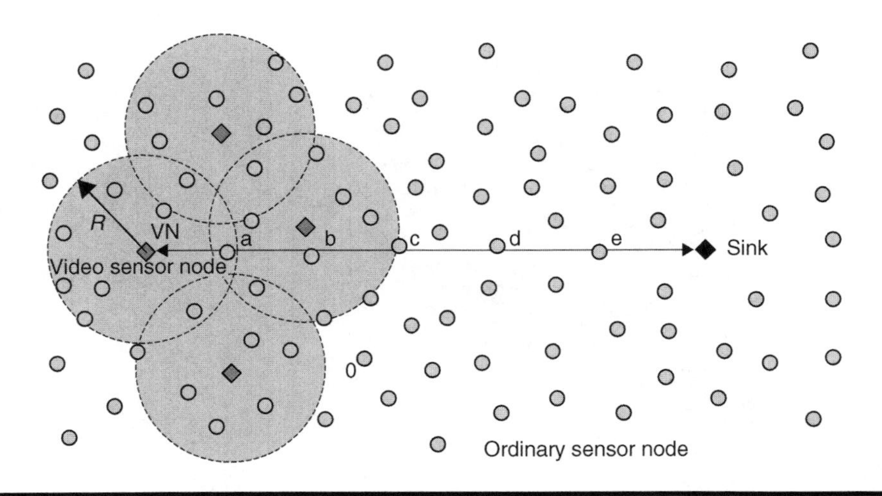

Figure 7.4 System architecture of video sensor network.

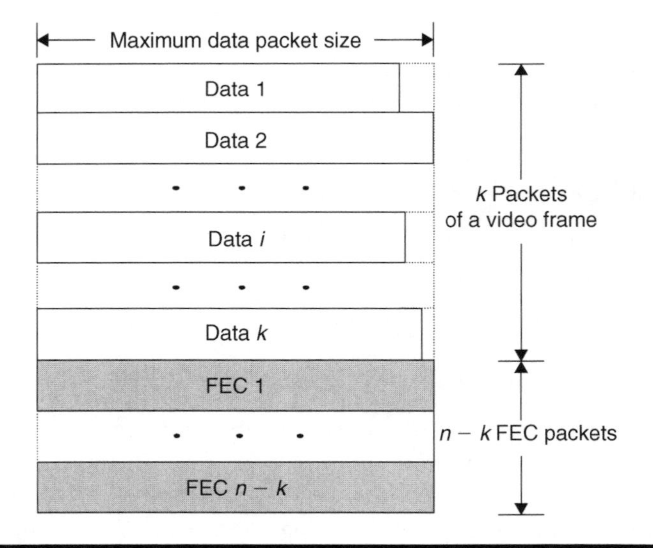

Figure 7.5 FEC-coding scheme.

take turn to send video streams to the sink; that is, at any instance only one of the VNs is actively sending video data to the sink.

To combat unreliable transmissions over the wireless environment and satisfy the strict end-to-end delay requirements, we assume that an FEC coding scheme is employed whereby each VN generates redundant packets to increase error resilience for real-time video transmissions.

We implement the FEC coding scheme proposed in Ref. 32, where $n - k$ redundant packets are generated to protect k data packets of a video frame, as shown in Figure 7.5. The size of each FEC packet is equal to the maximum size of the data packets. If any k of the n packets in the coding block are received by sink, the corresponding video frame can be successfully decoded.

7.3.2 Multipath-Based Video Transmission Strategy

The typical application of multipath routing in traditional WSN designs is to provide path redundancy for failure recovery. In Ref. 8, multiple disjoint paths are set up first, then multiple data copies are delivered using these paths. In Ref. 9, a protocol called ReInForM is proposed to deliver packets at the desired reliability by sending multiple copies of each packet along multiple paths from the source to the sink. The number of data copies (or the number of paths used) is dynamically determined depending on the probability of channel error. Instead of using disjoint paths, GRAB [10] uses a path interleaving technique to achieve high reliability. These multipath routing schemes for WSNs aim at increasing the reliability for a single flow [8–10]. In contrast, this chapter proposes to use multipath routing to support the delivery of multiple flows in a WSN, although the required level of reliability is achieved using FEC. Thus, in applying multipath routing, our goal is to maximize the load-balancing effect by spreading traffic evenly over the network and using all possible paths to maximize the end-to-end capacity.

In DGR [12], using a deviation angle adjustment method, a path can be established successfully using any initial deviation angle specified at the source node. To set up an application-specific number of paths with different initial deviation angles, the source can transmit a series of control packets, each specifying a different deviation angle. For example, in Figure 7.6, the source changes the absolute value of the deviation angle (i.e., α) from 0° to 90° in steps of 22.5°, and sends a different probe (PROB) message with each deviation angle. Thus, in total nine paths are established with α equal to −90°, −77.5°, −45°, −22.5°, 0°, 22.5°, 45°, 77.5°, and 90°, respectively.

To establish a direction-aware path, a PROB message is broadcast initially by the source for route discovery. The next selected hop will continue to broadcast the PROB message to find its next hop, and so forth. A node receiving a PROB will calculate its mapping coordinates based on α and the positions of the node itself, the upstream node, and the sink. Then, DGR will select as the next hop node the neighbor whose mapping coordinates is closest to the strategic mapping location, instead of the neighbor closest to the sink as in traditional geographical routing protocols. Because DGR is an existing geographical routing mechanism that is employed in the video transmission scheme presented in this chapter, we described briefly only the path setup mechanism of DGR. Interested readers can refer to Ref. 12 for a detailed description of DGR.

Source
Sink
|α| = 90
|α| = 67.5
|α| = 45
|α| = 22.5
|α| = 0

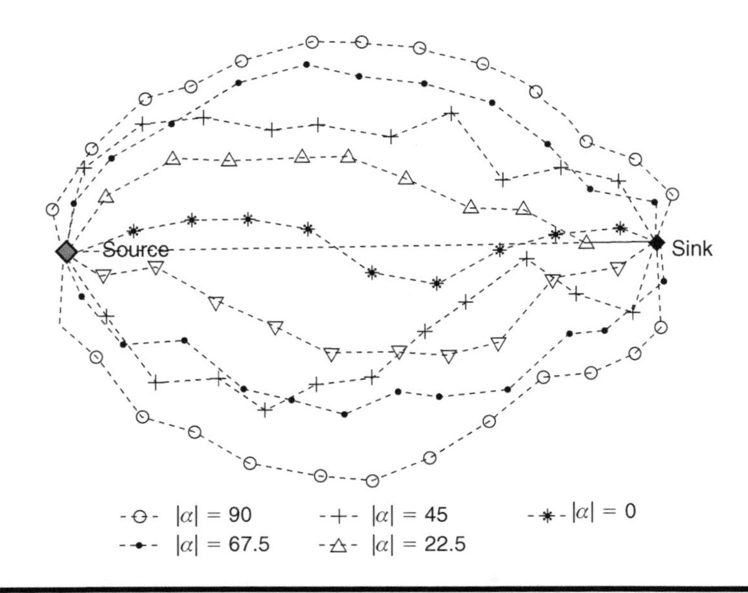

Figure 7.6 Example of nine disjointed paths using DGR.

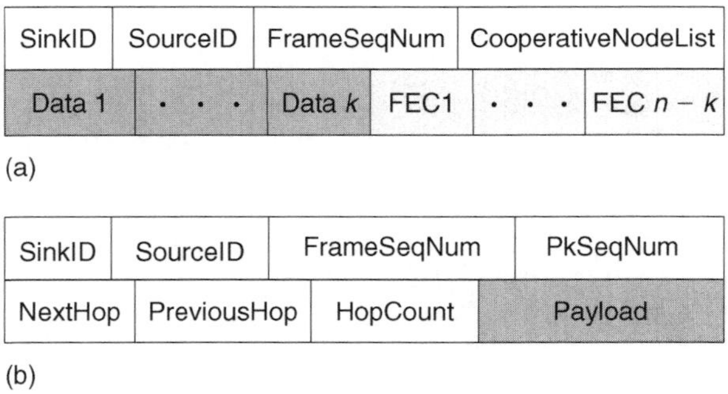

SinkID
SourceID
FrameSeqNum
CooperativeNodeList
Data 1
Data k
FEC1
FEC n − k
(a)
SinkID
SourceID
FrameSeqNum
PkSeqNum
NextHop
PreviousHop
HopCount
Payload
(b)

Figure 7.7 Data packet formats. (a) Concatenated data broadcast by VN, (b) data unicast by ordinary sensor node.

After the construction of multiple disjointed paths between the source VN and the sink as illustrated in Figure 7.6, a hybrid video stream broadcasting and substreams unicasting scheme can be implemented based on these pre-established multipaths. An active VN first broadcasts a request-to-send (RTS) message to its one-hop neighbors where delayed broadcast [33] of clear-to-send (CTS) message is used to solve the hidden terminal problem. Then, the VN will broadcast to its one-hop neighbors a packet concatenating all the data and FEC packets of a video frame, following the structure shown in Figure 7.7a. Those neighboring nodes that are the

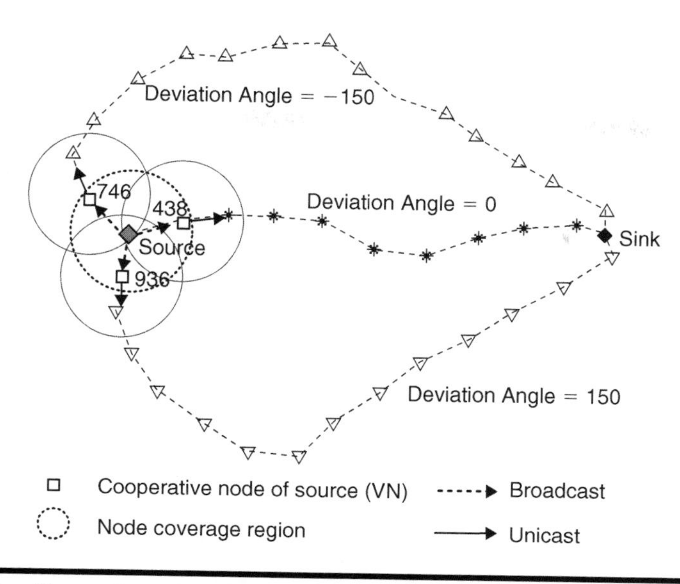

Figure 7.8 Illustration of DGR-based multipath video transmission.

first intermediate nodes of individual paths to the sink are referred as CooperativeNodes.

On receiving the concatenated packet broadcast by the VN, each CooperativeNode selects its own payload according to the Cooperative-NodeList in the concatenated packet. CooperativeNodeList contains the identifiers of the CooperativeNodes, and the sequence numbers of the corresponding packets (denoted by PkSeqNum in Figure 7.7) are assigned to these nodes. Then these CooperativeNodes unicast the assigned packets to the sink via the respective individual paths using the packet structure shown in Figure 7.7b.

A simple example of the proposed transmission architecture is illustrated in Figure 7.8, where the multipath routing layer sets up three paths between the source and the sink. Each path goes through a different CooperativeNode of the VN. To simplify the analysis, we consider that the VN encodes one video frame into two data packets and one FEC packet, and divides the video stream into three-packet substreams: two data flows and one FEC flow. The structure of substream entry is shown in Figure 7.9. The CooperativeNodeList is highlighted in Figure 7.9 and contains the list of NodeIDs and PacketToSend.

In general, the VN can intelligently specify the number of substreams and assign these substreams according to the number of available paths, the path length, and the number of data/FEC packets to be sent for each video frame. If the number of data/FEC packets of a video frame is larger than the number of available paths, some paths will deliver multiple packet

Substream entry of VN

SubStreamSeq	PathSeq	Deviation Angle	NodeID	Packet-To-Send
1	1	−150	746	Seq(Data **1**)
2	2	0	438	Seq(Data **2**)
3	3	150	936	Seq(FEC)

CooperativeNodeList

Figure 7.9 Example of substream entry of the VN (NodeID as in Figure 7.8).

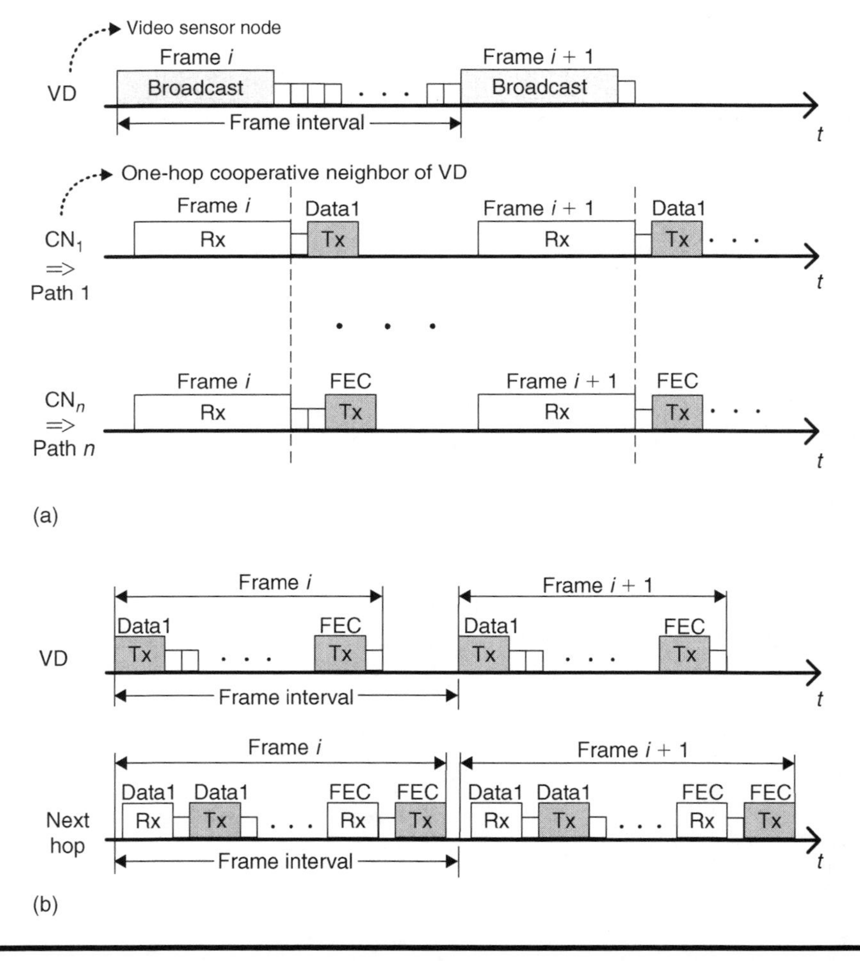

Figure 7.10 Comparison of video transmissions. (a) DGR, (b) traditional scheme.

flows. Otherwise, the VN can select a set of shorter paths to achieve faster delivery. The length of a path can be estimated by the value of the deviation angle. If the number of residual paths is high, the VN can adopt a Round-Robin path scheduling algorithm among the available paths to achieve load balance. To adapt to the fluctuations in channel quality, the VN also can adjust the number of FEC packets for each video frame and the number of paths used according to the feedback information received from the sink.

Assume that the FEC scheme generates $n - k$ redundant packets to protect k data packets of a video frame. If the sink has correctly received the k data packets, it may decode the frame immediately to reduce latency, although the redundant packets are subsequently ignored and dropped. However, if there are errors in the data packets, then the redundant packets are applied in an attempt to correct the errors using the FEC scheme.

In Figure 7.10, the DGR-based transmission scheme is compared with the traditional scheme where the whole video stream is transmitted over the shortest path.

7.4 Simulation Methodology

7.4.1 Simulation Model

To demonstrate the performance of the proposed video transmission scheme employing DGR, we compare it with GPSR [7] via extensive simulation studies. This section presents the simulation settings and performance metrics. The simulation results will be presented in the following section.

We implement the DGR protocol and the video transmission scheme using the OPNET Modeler [34,35]. The network consists of 500 nodes randomly deployed over a 300 m × 500 m field. We assume that all the nodes (VN, sensor nodes, the sink) are stationary. Figure 7.11 illustrates the topology of a set of randomly generated sensor nodes, as well as the VN and the sink node in the network. As in Ref. 6, we use IEEE 802.11 DCF as the underlying Media Access Control (MAC), and the radio transmission range (R) is set to 50 m. As in Ref. 26, the data rate of the wireless channel is 2 Mbps. We employ the energy model used in Refs 12, 26, and 36. To model link failures, we simply block the channel between two nodes with a link failure rate p. Thus, a packet will be lost with a probability p.

The test video sequence is Foreman encoded by the H.26L video coding standard [2,3] in QCIF format (176 × 144 pixels/frame) at a temporal resolution of 20 frames/s. The average bit rate of the video data and the average bit rate after the packet encapsulation is approximately 178 and 200 kbps, respectively. The first frame is intracoded and the remaining frames are intercoded. Each frame is packetized into six data packets. Three FEC packets are transmitted per video frame to protect the video data packets.

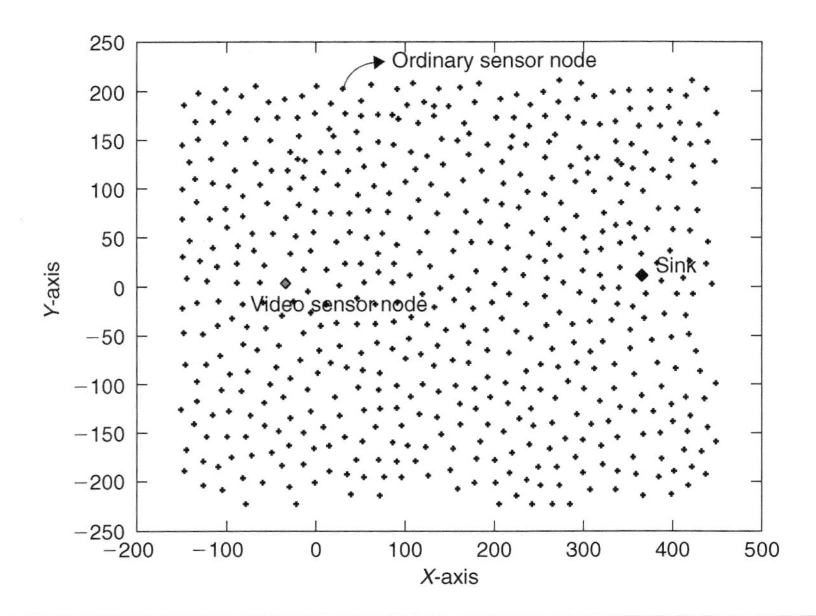

Figure 7.11 Example of network topology in simulation.

7.4.2 Performance Metrics

This section defines five performance metrics as follows:

- *Lifetime*. We believe that the choice of the definition of network lifetime is dependent on the specific application. A lifetime can be measured by the time when the first node exhausts its energy or the network can be declared dead when a certain fraction of nodes die, or even when all nodes die. Sometimes, it is better to measure the lifetime by application-specific parameters, such as the time until the network can no longer relay the video. We define the network lifetime as the time until the first node dies due to energy depletion, for the sake of simplicity.
- *Number of successful frames received by sink before lifetime* (n_{frame}). It is the number of video frames delivered to the sink before network lifetime is reached. Note that due to FEC coding, some packets of a video frame may be lost and the frame can still be correctly received. n_{frame} is an alternate measure of the network lifetime in this chapter.
- *Average end-to-end packet delay*. Let T_{dgr}, T_{gpsr}, and $T_{\mathrm{gpsr}}^{\mathrm{fec}}$ be the average end-to-end packet delay of DGR, GPSR, and GPSR with FEC coding, respectively. They include all possible delays during data dissemination, caused by queuing, retransmission due to collision at the MAC, and transmission time.

- *Energy consumption per successful data delivery* (e). It is given by the ratio of network energy consumption to the number of data packets delivered to the sink during the network's lifetime. The network energy consumption includes all the energy consumed by transmitting and receiving data packets during a simulation. As in Ref. 37, we do not account for energy consumption in the idle state, because this element is approximately the same for all the schemes being compared.
- *Peak signal-to-noise ratio (PSNR)*. It is a measure of the received video quality.

Among the performance metrics defined earlier, we believe that n_{frame} is the most important metric for WSNs, whereas PSNR and T_{ete} are the important indicators of QoS for real-time video transmissions.

7.5 Performance Evaluations

In this section, the simulation results for three video transmission techniques are evaluated; that is, DGR, GPSR (without FEC coding), and GPSR (with FEC coding). In each group of experiments, we change the link failure rate from 0 to 0.3 in step size of 0.05.

In Figure 7.12, T_{dgr} is always lower than T_{gpsr}, although the average path length in DGR is higher than the length of the shortest path. This is because the average bandwidth provided by the shortest path is very close to the

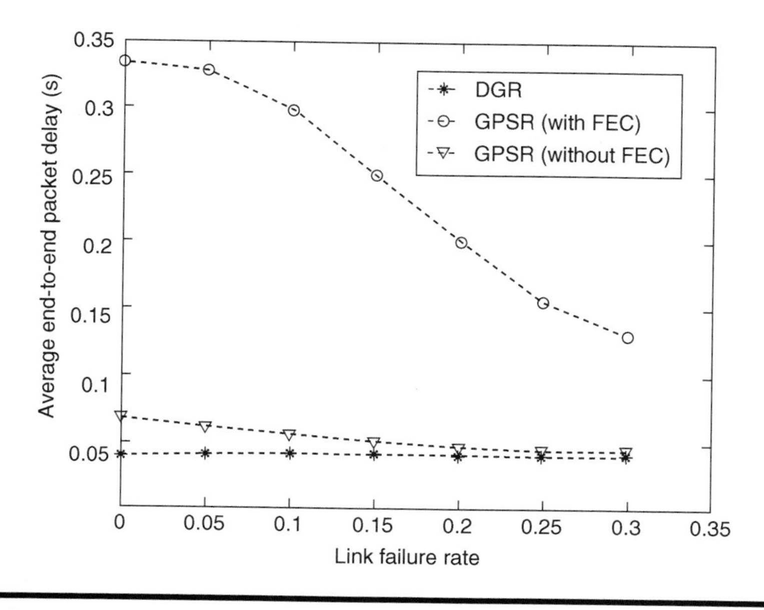

Figure 7.12 Comparisons of T_{ete}.

bandwidth required by a video stream, so that link congestion and video frame corruption due to burst packet losses are inevitable when a single path routing is employed. $T^{\mathrm{fec}}_{\mathrm{gpsr}}$ is much higher than T_{dgr} and T_{gpsr}, especially at low packet loss rates, which shows that the limited link bandwidth cannot accommodate the additional transmission overhead of FEC packets. As the packet loss rate increases, more packets are lost before they reach the sink. The lost packets do not affect the average end-to-end delay (which only accounts for correctly received packets), but they do help to alleviate congestion. Therefore, both T_{gpsr} and $T^{\mathrm{fec}}_{\mathrm{gpsr}}$ show a reduction as the packet loss rate is increased. However, congestion is not a problem for DGR due to the load-balancing effect of multipath routing; therefore, T_{dgr} stays relatively constant as the packet loss rate changes.

Figure 7.13 compares n_{frame} values for DGR, GPSR, and GPSR with FEC coding as the packet loss rate is varied. When the packet loss rate increases, n_{frame} of all the schemes increases because some sensor nodes save the energy of packet transmissions if they fail to receive the packets. DGR has higher n_{frame} values compared with that of GPSR and GPSR with FEC coding, because DGR distributes the traffic load of each video frame evenly over multiple paths. Thus, energy consumption of each path in DGR is much smaller than that of GPSR. In DGR, although nine paths are exploited, only six paths are used to transmit data substreams whereas the remaining three paths are used to transmit FEC substreams. Ideally, DGR should achieve about five times more n_{frame} than that of GPSR (without FEC).

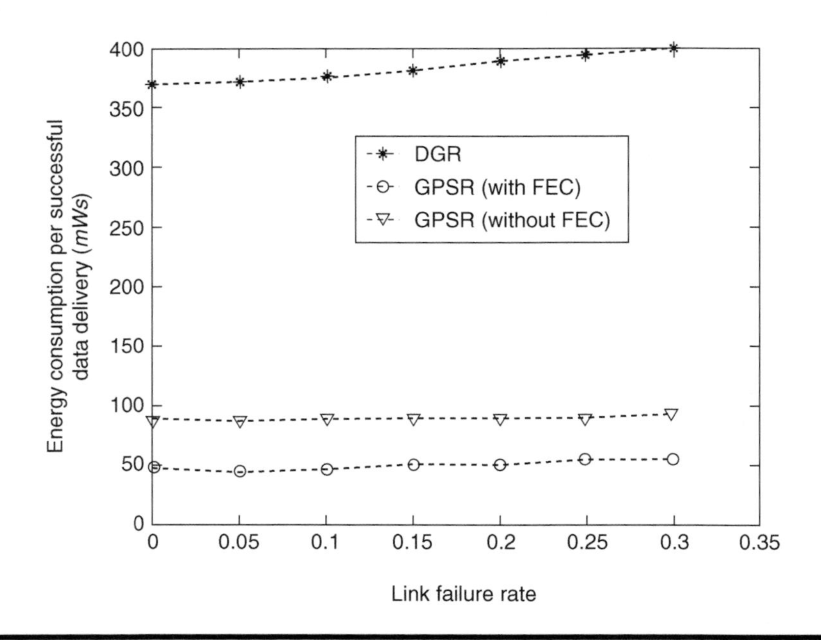

Figure 7.13 Comparisons of n_{frame}.

However, the cooperative neighbors of the VN are the bottlenecks with respect to energy consumption, because they receive the long concatenated packets from the VN, although other sensor nodes receive much shorter packets with only a single-data/FEC payload. The end result is that n_{frame} of DGR is about three times more than that of GPSR. With GPSR, if FEC coding is adopted, more energy is consumed to transmit the FEC packets. Thus, n_{frame} of GPSR with FEC is about 40 lower than that of GPSR.

Figure 7.14 shows the comparison of PSNR for these three schemes. It can be seen that DGR has the highest PSNR, which on average is about 3 dB higher than that of GPSR with FEC and 5 dB higher than that of GPSR. Although the PSNR of GPSR with FEC is higher than GPSR, the delay and lifetime performances of GPSR with FEC are worst, as described earlier.

Figure 7.15 compares the PSNR of each frame resulting from the test sequence Foreman with packet loss rate that equals to 0.05. Because GPSR does not take any measure to prevent error propagations, the PSNR of the reconstructed image decreases rapidly as more frames are received, and the subjective quality of the received video is poor. At a higher packet loss rate (0.2 in Figure 7.16), the received video quality degrades rapidly for all the schemes due to packet losses. Nevertheless, DGR still achieves the highest perceived quality for the video frames received at the sink.

Figure 7.17 shows that e_{dgr} is always higher than that of GPSR and GPSR with FEC under varying packet loss rates. This is because the average path length of DGR, which is equal to total path length (accounting for all the paths used to deliver the video stream) divided by PathNum, is larger

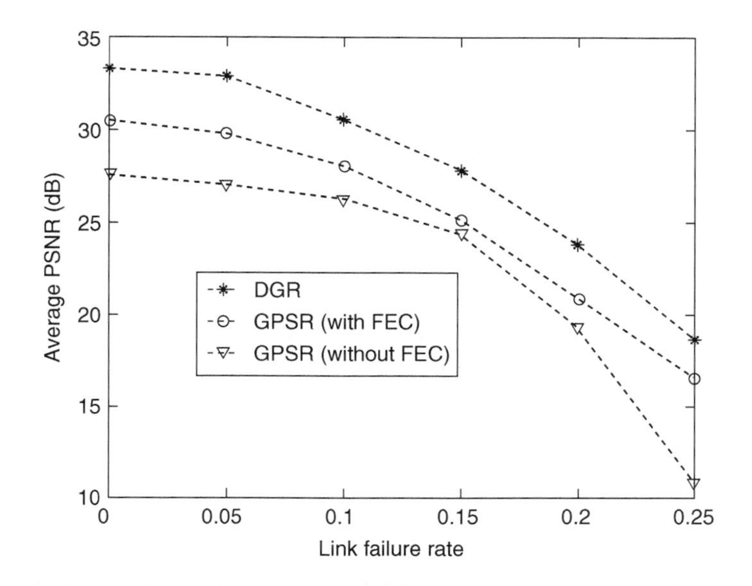

Figure 7.14 Comparisons of PSNR.

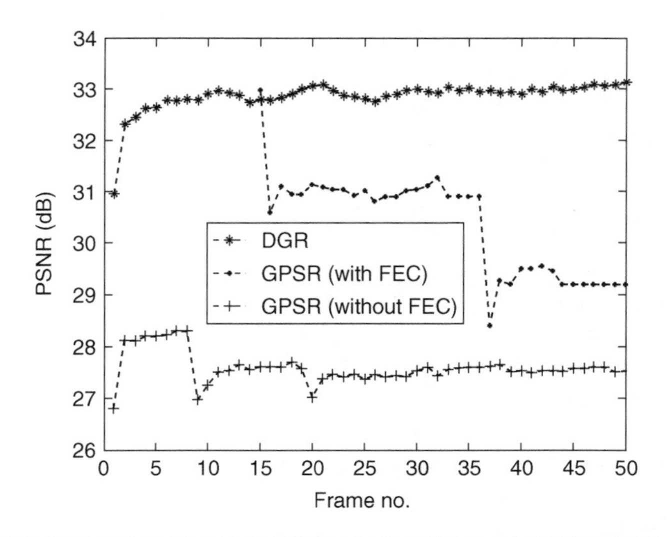

Figure 7.15 Comparisons of PSNR with link failure rate that equals to 0.05.

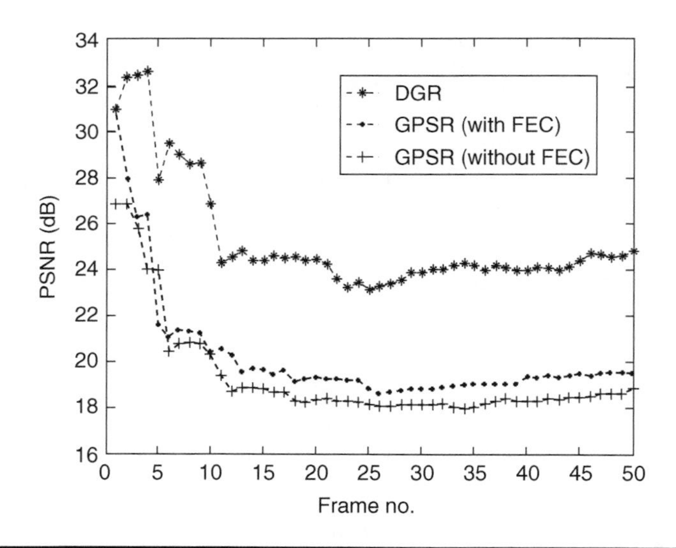

Figure 7.16 Comparisons of PSNR with link failure rate that equals to 0.2.

than that of GPSR. GPSR with FEC has a higher e than GPSR due to the additional energy consumed to transmit the FEC packets. However, for real-time service in WSNs, e is not as important as the other performance metrics.

For comparison purposes, it is important to consider the lifetime, PSNR, and average delay for real-time video applications. Thus, we adopt the

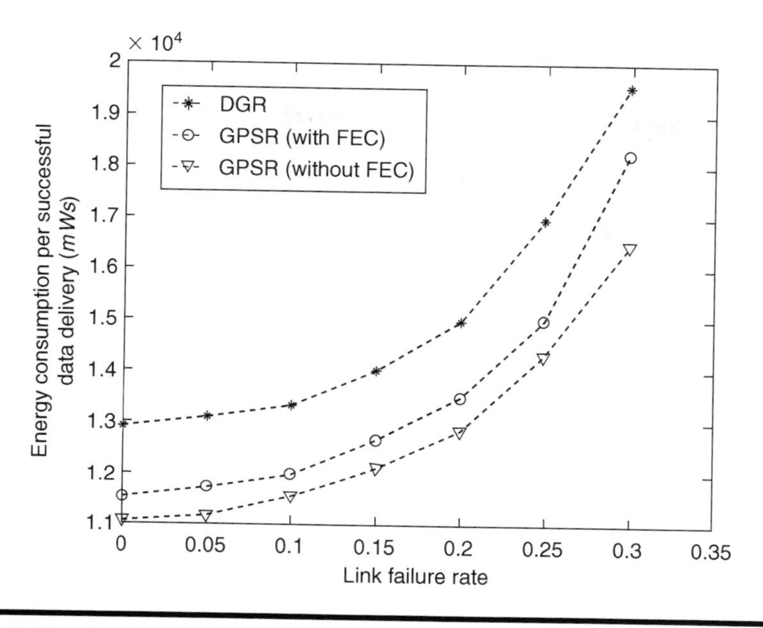

Figure 7.17 Comparisons of e.

following metric to evaluate the integrated performance of n_{frame} as an indicator for lifetime, PSNR, and delay:

$$\eta = \frac{n_{\text{frame}} \cdot \text{PSNR}}{\text{delay}} \tag{7.1}$$

The higher is η, the better is the composite QoS provided by the WSN to support real-time video services. No matter whether FEC coding is adopted or not, DGR achieves a much higher η than GPSR. Because GPSR with FEC obtains small improvements of PSNR by sacrificing the energy efficiency and delay performance, the η of GPSR with FEC is lower than that of GPSR (Figure 7.18).

Note that the mentioned simulation scenarios do not reflect the impact of multiple active video sources. When multiple video sources are active, the complexity of our scheme is higher than that of a single-path routing scheme such as GPSR. We believe that this is a price well paid for improved video quality. Furthermore, DGR may not work efficiently when the number of source–sink pairs is larger, although GPSR is suitable for the general cases where any sensor could be a source. When multiple data sources exist, GPSR is expected to maintain its performance whereas DGR may degrade due to path interference. This poses a severe limitation on the range of applications that can be supported by DGR, such as in a VSN as proposed in this chapter. Owing to the bandwidth limitation of a typical WSN,

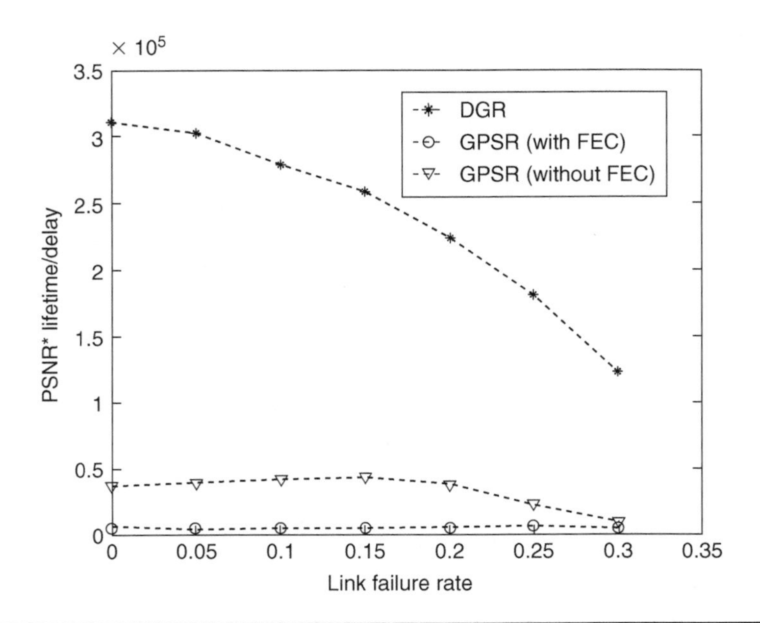

Figure 7.18 The comparison of η.

it is reasonable, as we have assumed in this chapter, that video sources do not transmit data simultaneously to the sink. Instead, they may make requests to the sink when there are video streams to send, and take turn to send video packets when instructed to do so by the sink. The protocol for making/granting these initial requests is a subject for further study.

7.6 Conclusion

This chapter presents a novel architecture for video sensor networks, and investigates the problem of real-time video transmissions over WSNs, in general. Compressed video is susceptible to transmission errors. However, the limited bandwidth in a WSN may not allow VNs to transmit additional FEC packets to protect the video data without subjecting all packets to excessive queuing delay. It is challenging to simultaneously achieve delay guarantees and obtain a high perceived video quality at the sink. To solve this problem, we have proposed a novel video transmission scheme which efficiently combines multipath routing with FEC coding to tackle the natural unreliability of WSNs as well as their bandwidth constraints. After the construction of an application-specific number of multiple disjointed paths, the proposed hybrid video stream broadcasting and substreams unicasting scheme is applied. This study has also provided insights into novel usage of multipath transmissions in WSNs. Instead of the typical application of multipath routing in traditional WSN designs to provide path redundancy for failure recovery, our scheme employs multipath routing to increase aggregate source-to-sink bandwidth and achieve better load balancing. Performance evaluations have shown that in combination with packet-level FEC coding, the proposed multipath-based video transmission scheme simultaneously achieves reliability, energy-efficiency, and timely packet delivery to support real-time video service over WSNs.

References

1. Akyildiz, I.F., T. Melodia, and K. Chowdhury, A survey on wireless multimedia sensor networks. *Computer Networks*, 51(4), 921–960, 2007.
2. ITU-T SG16. Draft, H.26L Video Coding[s], February 1998.
3. Bjontegaard, G., H.26L Test Model Long Term Number 9(TML-9) Draft[s], July 2002.
4. Wang, Y., S. Wenger, J. Wen, and A. Katsaggelos, Error resilient video coding techniques. *IEEE Signal Processing Magazine*, 86, 61–82, 2000.
5. Ma, H. and M. El Zarki, Broadcast/multicast MPEG-2 video over broadband fixed wireless access networks. *IEEE Network Magazine*, 13(6), 80–93, 1998.
6. Intanagonwiwat, C., R. Govindan, and D. Estrin, Directed diffusion: A scalable and robust communication paradigm for sensor networks. *Proceedings of the 6th Annual ACM/IEEE MobiCom*, Boston, MA, August 2000.
7. Karp, B. and H.T. Kung, GPSR: Greedy perimeter stateless routing for wireless networks. *Proceedings of the ACM MobiCom 2000*, pp. 243–254, Boston, MA, August 2000.
8. Ganesan, D., R. Govindan, S. Shenker, and D. Estrin, Highly resilient, energy efficient multipath routing in wireless sensor networks. *Mobile Computing and Communications Review* (*MC2R*), 1(2), 10–24, 2002.
9. Deb, B., S. Bhatnagar, and B. Nath, ReInForM: Reliable information forwarding using multiple paths in sensor networks. *IEEE LCN*, 406–415, October 2003.
10. Ye, F., G. Zhong, S. Lu, and L. Zhang, GRAdient broadcast: A robust data delivery protocol for large scale sensor networks. *ACM WINET* (*Wireless Networks*), 11(3), 285–298, 2005.
11. De, S., C. Qiao, and H. Wu, Meshed multipath routing with selective forwarding: An efficient strategy in wireless sensor networks. *Elsevier Computer Communications Journal*, 26(4), 2003.
12. Chen, M., V.C.M. Leung, S. Mao, and Y. Yuan, Directional geographical routing for real-time video communications in wireless sensor networks. *Computer Communications* (*Elsevier*), 30(17), 3368–3383, 2007.
13. Pearlman, M., Z. Haas, P. Sholander, and S. Tabrizi, On the impact of alternate path routing for load balancing in mobile ad hoc networks. *ACM International Symposium on Mobile Ad Hoc Networking and Computing* (*MobiHOC*), San Diego, CA, 2003.
14. Stojmenovic, I., Position-based routing in ad hoc networks. *IEEE Communications Magazine*, 40(7), 128–134, 2002.
15. Etoh, M. and T. Yoshimura, Advances in wireless video delivery. *Proceedings of the IEEE*, 93, 111–122, January 2005.

16. Bucciol, P., G. Davini, E. Masala, E. Filippi, and J.D. Martin, Cross-layer perceptual ARQ for H.264 video streaming over 802.11 wireless networks. *Proceedings of the IEEE Global Telecommunications Conference (Globecom)*, 5, 3027–3031. Dallas, TX, November/December 2004.
17. Li, Q. and M. Schaar, Providing adaptive QoS to layered video over wireless local area networks through real-time retry limit adaptation. *IEEE Transactions on Multimedia*, 6(2), 278–290, 2004.
18. Majumdar, A., D. Sachs, I. Kozintsev, K. Ramchandran, and M. Yeung, Multicast and unicast real-time video streaming over wireless LANs. *IEEE Transactions on Circuits and Systems for Video Technology*, 12(6), 524–534, 2002.
19. Chen, M. and G. Wei, Multi-stages hybrid ARQ with conditional frame skipping and reference frame selecting scheme for real-time video transport over wireless LAN. *IEEE Transactions on Consumer Electronics*, 50(1), 158–167, 2004.
20. Mao, S., S. Lin, S. Panwar, Y. Wang, and E. Celebi, Video transport over ad hoc networks: Multistream coding with multipath transport. *IEEE Journal on Selected Areas in Communications, Special Issue on Recent Advance in Wireless Multimedia*, 21(10), 1721–1737, 2003.
21. He, T., J. Stankovic, L. Chenyang, and T. Abdelzaher, SPEED: A stateless protocol for real-time communication in sensor networks. *Proceedings of the IEEE ICDCS'03*, May 2003, pp. 46–55.
22. Felemban, E., C. Lee, and E. Ekici, MMSPEED: Multipath multi-SPEED protocol for QoS guarantee of reliability and timeliness in wireless sensor networks. *IEEE Transactions on Mobile Computing*, 5(6), 738–754, 2006.
23. Akkaya, K. and M. Younis, Energy-aware routing of time-constrained traffic in wireless sensor networks, *Journal of Communication Systems, Special Issue on Service Differentiation and QoS in Ad Hoc Networks*, 17, 663–687, 2004.
24. Akkaya, K. and M. Younis, Energy and QoS aware routing in wireless sensor networks. *Journal of Cluster Computing Journal*, 2(3), 179–188, 2005.
25. Yuan, Y., Z.K. Yang, Z.H. He, and J.H. He, An integrated energy aware wireless transmission system for QoS provisioning in wireless sensor network. *Elsevier Journal of Computer Communications, Special Issue on Dependable Wireless Sensor Networks*, 29(2), 162–172, 2006.
26. Chen, M., T. Kwon, and Y. Choi, Energy-efficient differentiated directed diffusion (EDDD) for real-time traffic in wireless sensor networks. *Elsevier Journal of Computer Communications, Special Issue on Dependable Wireless Sensor Networks*, 29(2), 231–245, 2006.
27. Rahimi, M., R. Baer, O. Iroezi, J. Garcia, J. Warrior, D. Estrin, and M. Srivastava, Cyclops: In situ image sensing and interpretation in wireless sensor networks. *Proceeding of the ACM SenSys'05*, Annapolis, MD, November 2005.
28. Crossbow MICA2 Mote Specifications. http://www.xbow.com.
29. Crossbow MICAz Mote Specifications. http://www.xbow.com.
30. He, Z. and D. Wu, Resource allocation and performance limit analysis of wireless video sensors. *IEEE Transactions on Circuits and System for Video Technology*, 16(5), 590–599, May 2006.
31. He, Z. and D. Wu, Performance analysis of wireless video sensors in video surveillance. *Proceedings of the Globecom 2005*, 1, 178–182, November 2005.
32. Rizzo, L., Effective erasure codes for reliable computer communication protocols. *ACM Computer Communication Review*, 27, 24–36, April 1997.

33. Mao, S. and Y. Hou, BeamStar: An edge-based approach to routing in wireless sensor networks. *IEEE Transaction on Mobile Computing*, 6(11), 1284–1296, November 2007.
34. http://www.opnet.com.
35. Chen, M., *OPNET Network Simulation*. Press of Tsinghua University, China, 2004, 352p.
36. Chen, M., T. Kwon, S. Mao, Y. Yuan, and V.C.M. Leung, Reliable and energy-efficient routing protocol in dense wireless sensor networks. *International Journal on Sensor Networks*, 3(2), 2008 (in press).
37. Gao, J. and L. Zhang, Load balancing shortest path routing in wireless networks. *IEEE INFOCOM 2004*, 2, 1098–1107, March 2004.

MULTIMEDIA OVER WIRELESS LOCAL AREA NETWORKS

Chapter 8

Multimedia Quality-of-Service Support in IEEE 802.11 Standards

Zhifeng Tao, Thanasis Korakis, Shivendra Panwar, and Leandros Tassiulas

Contents

8.1 Introduction 262
8.2 Multimedia over Wireless LAN: A Background 264
8.2.1 Requirements and Challenges 264
8.2.2 The IEEE 802.11 Protocol 265
8.2.3 Inadequate QoS Support in Legacy 802.11 267
8.3 Enhanced QoS Capability in IEEE 802.11e 268
8.3.1 HCCA: A Protocol for Parameterized QoS 270
8.3.2 Scheduling Algorithm: A Reference Design 270
8.3.3 EDCA: A Protocol for Prioritized QoS 271
8.3.4 Admission Control 273
8.3.5 EDCA: Performance Evaluation 274
8.4 QoS Features in IEEE 802.11n 283
8.4.1 Overview of IEEE 802.11n 283

8.4.2 MAC Efficiency Improvement.....283
8.4.2.1 Frame Aggregation.....283
8.4.2.2 Block Acknowledgment.....286
8.4.2.3 Reduced Interframe Space.....287
8.4.3 Reverse Direction Protocol.....287
8.5 QoS Mechanisms in IEEE 802.11s Mesh Network.....289
8.5.1 Overview of IEEE 802.11s.....289
8.5.2 Mesh Deterministic Access.....290
8.5.3 Mesh Common Control Channel Protocol.....291
8.5.4 CCC MAC: Performance Evaluation.....292
8.5.4.1 Capacity.....293
8.5.4.2 Delay.....295
8.6 Challenges and Future Perspectives.....299
References.....299

The world today perceives the wireless access enabled by Institute of Electrical and Electronics Engineers (IEEE) 802.11 more as a ubiquitous utility than as the novelty it once was only a few years ago. Untethering them from a desktop, wireless network enables end users to receive multimedia content in a manner approaching that of wireline networks. However, legacy IEEE 802.11 possesses neither sufficient capacity nor adequate QoS capability for multimedia. Thus, several task groups have been established within IEEE 802.11 to address the urgent need of proper QoS support demanded by a vast variety of applications in different deployment scenarios. This chapter intends to provide an exposition and a discussion on key techniques that have been proposed or adopted in 802.11e, 802.11n, and 802.11s.

8.1 Introduction

With the advance of technology in wired networks and an abundance of backbone transport bandwidth at the continuously dropping prices, multimedia applications such as Voice-over-IP (VoIP), peer-to-peer (P2P) video streaming, and mobile TV have turned from a long-time aspiration to a completely embraced reality.

Meanwhile, wireless networks based on the IEEE 802.11 standard started to experience an unprecedented deployment growth. The adoption of IEEE 802.11 has gone far beyond the wireless local area network (WLAN) environment and become truly omnipresent, covering usage scenarios from short-range point-to-point communications, which intend to beam high-definition television (HDTV) signals, to metropolitan mesh networks that serve tens of thousands of users.

However, the burgeoning demand for mobile data, especially multimedia with large bandwidth requirements, and the need for stringent quality of service (QoS) have highlighted some of the inherent constraints of IEEE 802.11, which will ultimately limit its future growth.

Several task groups have been established within IEEE 802.11 to address the urgent need of proper QoS support demanded by a vast variety of applications in many different deployment scenarios. Among these pertinent task groups, IEEE 802.11e [1] was the first to be developed, specifically to improve the QoS capabilities by enhancing the current Distributed Coordination Function (DCF) and Point Coordination Function (PCF) protocols. The ratified amendment has established a new framework for multimedia services, wherein prioritized and parameterized QoS is provided for traffic classes and traffic streams (TSs), respectively.

To fulfill the requirement of delivering a minimum throughput of greater than 100 Mbps above the Media Access Control (MAC) layer and serve as the primary bearer technology for multimedia applications in consumer electronics and home entertainment systems, IEEE 802.11n is another critical amendment, which introduces numerous novel physical (PHY) and MAC innovations. At the MAC layer, in particular, a new feature called Reverse Direction Protocol (RDP), which aims to provide better QoS support for bidirectional multimedia traffic and improved protocol efficiency, is discussed and accepted into the current version of the 802.11n draft [2].

In addition, IEEE 802.11 has recently been adopted as the primary technology for mesh networking, which significantly expands the footprint of a conventional wireless LAN and covers an area as large as a metropolitan area. Multimedia traffic experiences a significant degradation in the QoS as it traverses through multiple hops in a mesh network. Techniques have been proposed in IEEE 802.11s to ameliorate serious QoS deterioration and to improve overall performance in the multihop environment. For instance, mesh deterministic access (MDA) is a novel scheme that allows mesh points (MPs) to access the channel with lower contention probability in selected times, thereby significantly reducing end-to-end delay for multimedia traffic. The idea of using multiple channels has also been carefully examined in the IEEE 802.11s task group, and several MAC protocols have been proposed.

The remainder of the chapter is organized as follows. Section 8.2 provides the necessary background by first summarizing the requirements and challenges imposed by multimedia traffic. In addition, the legacy IEEE 802.11 protocol is briefly described and its QoS problem is analyzed. Problems specific to each related task group (i.e., 802.11e, 11n, and 11s) and the solution proposed thereof are then elaborated in Sections 8.3 through 8.5, respectively. The chapter concludes with a discussion on the remaining issues and future research perspectives in Section 8.6.

8.2 Multimedia over Wireless LAN: A Background

8.2.1 Requirements and Challenges

As demonstrated in Table 8.1, different applications have distinct QoS requirements. For instance, VoIP requires relatively modest bandwidth when compared to other multimedia applications, and can tolerate a certain degree of packet loss. However, the delay and jitter have to be kept under tight control (i.e., normally <100 ms) to achieve an acceptable voice quality. Streaming video, however, can gobble up a significant amount of bandwidth, but is somehow more tolerant of delay and jitter, given a proper buffer provisioning. Others, such as interactive graphics or interactive computing applications, are both delay- and loss-sensitive.

The comparison carried out in Table 8.1 readily reveals that different measures have to be installed to achieve optimal performance for a variety of applications, which is by no means a trivial undertaking for a wired network. The problem is further complicated in a wireless environment in general, and in current 802.11 networks in particular, given the higher packet loss rate, more bursty error patterns, orders of magnitude less bandwidth than wireline network, and lack of inherent QoS features.

In fact, it turns out that the current 802.11 network fails to deliver a satisfactory performance even for simple applications such as VoIP, let alone other multimedia applications with more sophisticated QoS requirements. As reported in Ref. 3, for instance, although an 802.11b network delivers approximately 6 Mbps at the MAC layer, it can only support at most six concurrent VoIP calls due to the delay constraints. This is particularly frustrating, given that a single G711 VoIP stream constitutes only 64 kbps in one direction.

Table 8.1 Applications and Corresponding QoS Requirements

Application	*Reliability*	*Delay*	*Jitter*	*Bandwidth*
E-Mail	High	Low	Low	Low
File Transfer Protocol (FTP)	High	Low	Low	Medium
Web access	High	Medium	Low	Medium
Remote log-in	High	Medium	Medium	Low
Streaming audio	Low	Low	High	Medium
Streaming video	Low	Low	High	High
VoIP	Medium	High	High	Low
Videoconferencing	Low	High	High	High
Mobile TV	High	High	High	High

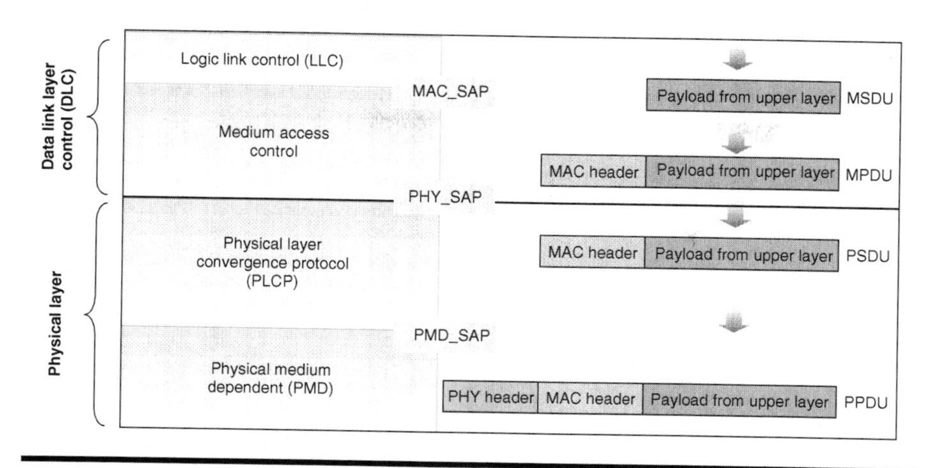

Figure 8.1 IEEE 802.11: protocol stack and concept of data units.

8.2.2 *The IEEE 802.11 Protocol*

Since the ratification of the IEEE 802.11 protocol in 1999 [4], numerous amendments [5–7] have been standardized to enhance the performance and enable new services. As shown in Figure 8.1, the initial baseline and all the subsequent amendments limit their scope to MAC and PHY layers only. This chapter concentrates on MAC-layer QoS mechanisms, and details on PHY layer will not be covered hereafter.

Associated with each protocol layer, a corresponding service data unit (SDU) and protocol data unit (PDU) are defined, as illustrated in Figure 8.1. Generally, an SDU refers to a set of data specified in a protocol of a given layer and consists of protocol control information of that layer, and possibly user data of that layer [8]. Meanwhile, a PDU of a given protocol layer includes the SDU coming from a higher layer and the protocol control information of that layer. At MAC and PHY layer, the SDU and PDU are usually called MAC SDU (MSDU), PHY SDU (PSDU), MAC PDU (MPDU), and PHY PDU (PPDU), respectively. As demonstrated in Figure 8.1, MPDU and PSDU essentially refer to the same data unit, but from the perspective of different protocol layers. Note that the concept of PDU and SDU will be particularly important when we discuss the 802.11n protocol in Section 8.4.

IEEE 802.11 has defined two operational modes, namely, the *ad hoc* mode and the "infrastructure" mode. In the *ad hoc* mode, all the stations within the transmission range of one another form a network spontaneously and exchange packets directly. Meanwhile, stations operating in an infrastructure mode must associate themselves with a centralized station called the access point (AP), and communicate through the AP only, regardless

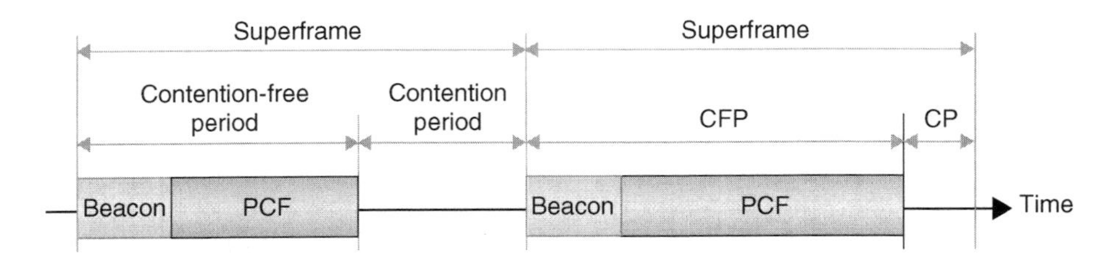

Figure 8.2 IEEE 802.11 superframe structure.

of whether the traffic destination is within the range of direct transmission of the source station or not.

To coordinate the wireless channel access, two distinct functions have been defined within the framework of the 802.11 MAC protocol. One function is a contention-based channel access mechanism called the DCF, which essentially is a refined version of Carrier Sense Multiple Access with Collision Avoidance (CSMA/CA). The other one, namely, the PCF, designates authority for channel control to the AP, which assigns channel access in a contention-free manner by polling each station associated with it.

Owing to its distributed nature, DCF is the only valid access mechanism for an *ad hoc* 802.11 network. For an infrastructure 802.11 network, the channel access is governed by a "superframe" structure, as illustrated in Figure 8.2. Each superframe starts with a beacon frame broadcast by the AP, which is followed by a contention-free period (CFP) and a contention period (CP). The beginning of a new superframe is signaled by the subsequent beacon. Note that the beacon frames are generated at regular intervals, so that every station in the network knows when the next beacon frame will arrive. The expected arrival time of the next beacon is called the target beacon transition time (TBTT), which is announced in the current beacon frame.

- *DCF*. Each station maintains a set of parameters, including the minimum contention window size (CW_{min}), maximum contention window size (CW_{max}), and the maximum number of retries allowed (*retryLimit*). Once a data packet reaches the head of line (HOL) in the buffer, it should wait for the channel to be idle for a DCF interframe space (DIFS) period before performing a random backoff. The station starts the backoff by initializing the backoff counter (denoted as CW) with a random number drawn from the interval $[0, CW_{min} - 1]$, based on the uniform distribution. At the boundary of each time slot, the channel is sensed to determine its status. The backoff counter is decremented for every slot the channel is perceived idle, whereas the backoff is suspended once a transmission by another station is detected. The random backoff resumes when the wireless channel is found idle again for a DIFS period. The packet can only be transmitted when CW becomes zero, and the station still senses the channel idle. After a packet transmission, the existence or lack of an acknowledgment from the intended recipient indicates whether the transmitted packet has been successfully delivered or not. A retransmission will be attempted if no proper acknowledgment has been received and the number of retransmissions (i.e., retryCount) has not exceeded the retryLimit. A similar random backoff rule applies for retransmissions, except that the corresponding backoff counter value is drawn from the interval $[0, CW_{\min} \times 2^{\min[\text{retryCount},\text{retryLimit}]} - 1]$, that is, the retransmission window keeps doubling with each unsuccessful transmission until an upper limit is reached.

To further mitigate the well-known "hidden terminal problem" [9], an optional handshake of "request-to-send (RTS)" and "ready-to-send" (CTS) messages have been introduced into the DCF protocol. The source and destination stations can transmit the actual data packets only if the exchange of the RTS and CTS messages has been successfully completed.

- *PCF.* During the CFP, where PCF is applied, no station associated with the AP can transmit unless it is polled by the AP. To encourage innovation and enable product differentiation, only the polling procedure has been explicitly specified in the standard, whereas the corresponding polling discipline and scheduling algorithm have been left as implementation-dependent. The PCF mechanism was initially designed for those applications that have a stringent requirement for QoS. Indeed, with a properly selected scheduling algorithm, PCF has been shown to provide better support for QoS than DCF [10]. Because IEEE 802.11 WLAN was primarily used for data-centric applications during its infancy, little incentive was found by vendors to implement PCF in the product. Therefore, there are few, if any, PCF products available in the current market.

8.2.3 Inadequate QoS Support in Legacy 802.11

As a wireless replacement to Ethernet, the IEEE 802.11 was initially designed without any special attention paid toward the QoS support. It is apparent that the DCF function does not possess any effective service differentiation capability as it treats all the upper-layer traffic in the same fashion [11]. The PCF mode, unfortunately, also lacks the level of QoS support required by many real-time traffic due to problems such as unpredictable beacon delays and unknown transmission durations of the polled stations [12,13].

- *Unpredictable beacon delay.* An AP schedules to transmit a beacon frame at TBTT if the medium is determined to be idle for at least PCF interframe space (PIFS) time at that moment. As the baseline 802.11 standard [4] does not prohibit a station from transmitting a packet even if it cannot complete before the upcoming TBTT, the scheduled beacon frame may experience an unpredictable delay, depending on whether the wireless channel is idle or busy around the TBTT. The delay of the beacon, which could be in the order of milliseconds in an 802.11a system [12], in turn may severely affect the transmission of time-bounded packets during the CFP interval.

- *Unknown transmission duration.* After a station is granted channel access by the AP, the station can transmit a frame of arbitrary size ranging from 0 to 2304 bytes, at one of the possible rates supported by the corresponding PHY layer. Therefore, the AP has no control over how long the polled station will be transmitting, which eventually defeats any attempt to provide delay or jitter bound to other stations that will be served in the rest of CFP.

Evidently, the legacy 802.11 protocol fails to provide a satisfactory level of service for various QoS-sensitive applications. Therefore, as multimedia applications become increasingly popular in residential and enterprise WLAN deployments, a significant enhancement to the QoS capability of IEEE 802.11 [4] becomes an urgent need.

8.3 Enhanced QoS Capability in IEEE 802.11e

Officially established in March 2000 and finally ratified in 2005, IEEE 802.11e [1] was the first standardization attempt within the 802.11 working group to address the need for enhanced QoS support and service differentiation. As illustrated in Figure 8.3, IEEE 802.11e introduces a new channel access function called hybrid coordination function (HCF), which combines the contention-based enhanced distributed channel access (EDCA)

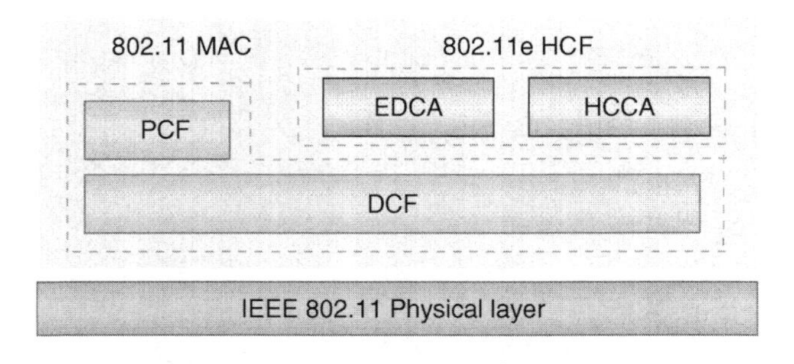

Figure 8.3 IEEE 802.11e MAC protocol stack.

and contention-free HCF controlled channel access (HCCA) methods to provide "prioritized" and "parameterized" QoS, respectively.

As an extension of the basic DCF mechanism, EDCA assigns relative priority to packets, and gives preferential treatment to those with higher priority over those with lower priority, thereby fulfilling the need for a prioritized QoS. In contrast, HCCA is a contention-free channel access scheme developed based on the PCF polling. It handles packets in conformance with more quantitative QoS requirements (e.g., data rate, delay, and jitter bound) negotiated *a priori*, and thus, delivers a parameterized QoS.

Another salient feature of 802.11e protocol is the introduction of transmission opportunity (TXOP), the concept of which is illustrated in Figure 8.4. A legacy 802.11 station always releases the channel once it completes the transmission of one packet. In comparison, however, multiple packets can be transmitted by an 802.11e station within a single-channel access with an SIFS gap between two consecutive packet exchange sequences. The maximum value of a TXOP, regardless of whether it is obtained through contention (e.g., EDCA-TXOP) or being polled (e.g., HCCA-TXOP), is bounded by a value called $TXOP_{limit}$.

Note that although the notion of TXOP is intended for efficiency improvement, $TXOP_{limit}$ can also provide a bound for any transmission time assigned to a station, thereby addressing the QoS issue described in Section 8.2.3.

Sections 8.3.1 and 8.3.2 explain the HCCA and the associated scheduling algorithm, respectively, Section 8.3.3 introduces EDCA, and Section 8.3.4 discusses the admission control in an 802.11e network. Finally, some quantitative results will be presented to shed further light on the QoS performance of EDCA.

Before delving into the details, however, several key concepts frequently used in 802.11e should be defined first. A traffic category (TC) is a label at MAC level for packets that have a distinct user priority (UP), as viewed by higher-layer entities, relative to other packets provided for delivery over the same link. Meanwhile, a TS refers to a set of data packets to be delivered subject to the QoS parameter values provided to the MAC in a particular traffic specification (TSPEC).

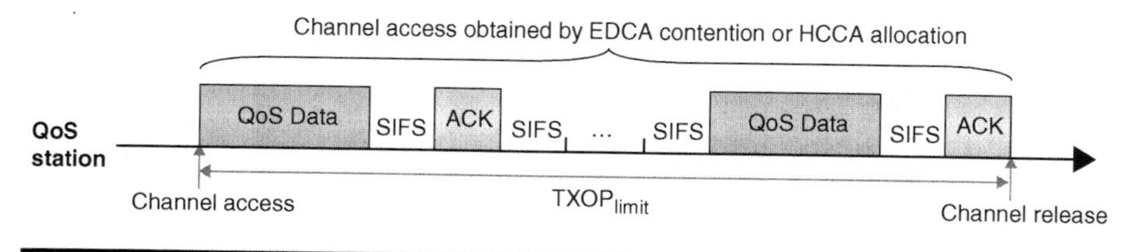

Figure 8.4 EDCA TXOP: an illustration.

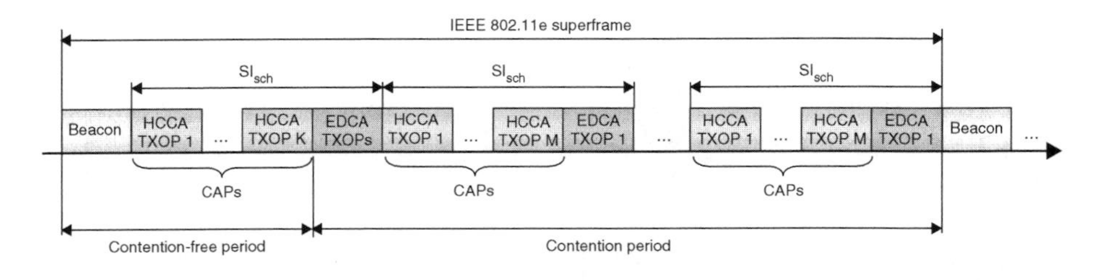

Figure 8.5 IEEE 802.11e superframe structure.

For the sake of brevity, QoS AP (QAP) and QoS station (QSTA) will be used in the ensuing discussion to denote such APs and stations that support the QoS facility specified in the 802.11e standard, respectively.

8.3.1 HCCA: A Protocol for Parameterized QoS

As an extension to PCF, HCCA protocol also relies on a centralized controller to coordinate medium access and allocate limited-duration controlled access period (CAP) for contention-free data transmission. Meanwhile, HCCA has been equipped with several new features so that it can deliver the parameterized QoS.

First and foremost, the HCCA modifies the superframe structure, as illustrated in Figure 8.5, to provide more flexible and timely response to QoS needs. Similar to legacy 802.11, 802.11e starts its superframe with a beacon, which is followed by a CFP and then a CP. Unlike legacy 802.11, an 802.11e QAP can generate multiple CAPs during the CP, when the channel is determined to be idle for a PIFS time. Based on the network wide knowledge of the amount of pending traffic and subject to certain QoS policies, such CAPs are created in a CP on demand to meet the QoS requirement of packets belonging to a particular TS or traffic class.

In addition, HCCA operates based on the notion of TS, on which a sophisticated scheduling algorithm can be applied to ensure proper parameterized QoS treatment. Although the scheduling algorithm to be implemented is intentionally left outside the scope of the standard to encourage further innovation [14–17], a sample reference design is suggested by IEEE 802.11e, which is briefly discussed in Section 8.3.2.

8.3.2 Scheduling Algorithm: A Reference Design

When a TS is initially established, QSTA should negotiate the traffic specification, including mean data rate, nominal MSDU size, and maximum service interval or delay bound with the QAP. The QSTA should also inform the QAP of the maximum required service interval (RSI), which refers to the maximum duration between the start of successive TXOPs that can be tolerated by a requesting application. Given the fact that maximum RSI and delay bound are related, the scheduler would only use the RSI for the calculation of the schedule. The schedule for an admitted stream is then calculated in two steps as follows.

- *Calculation of the scheduled service interval (SI).* The scheduler calculates the "minimum" of all maximum RSIs for all admitted streams, which is represented by RSI_{min}. The scheduler chooses a number lower than RSI_{min}, which is a submultiple of the beacon interval. This value is the scheduled SI (SI_{sch}) for all QSTAs with admitted streams.
- *Calculation of the TXOP duration.* Use $\overline{R}$ and L_i to denote the mean data rate and nominal packet size, respectively. The number of packets (N_i) from a TS i that arrive at the mean data rate during an SI_{sch} is

$$N_i = \left\lceil \frac{\text{SI}_{\text{sch}} \times \overline{R}}{L_i} \right\rceil \tag{8.1}$$

The TXOP duration for the particular TS i can be further computed as

$$\text{TXOP}_i = \max\left[\frac{N_i \times L_i}{R_i} + O, \quad \frac{M}{R_i} + O\right] \tag{8.2}$$

where M and O stand for the maximum allowable packet size and overhead in time units, respectively. The overhead in time includes all the IFS, ACK frames, and poll messages. In addition, R_i represents the PHY-layer transmission rate. $((N_i \times L_i)/R_i) + O$ in Equation 8.2 represents the time needed to transmit N_i packets at rate R_i plus overhead, whereas $(M/R_i) + O$ is the time consumed to transmit one maximum size packet at rate R_i plus overhead.

Note that whenever the network admits a new TS that has a maximum service interval smaller than the current SI_{sch}, the preceding two steps will be repeated to recalculate the SI_{sch} and TXOP.

8.3.3 EDCA: A Protocol for Prioritized QoS

Each IEEE 802.11e EDCA QSTA can have multiple queues that buffer packets of different priorities as depicted in Figure 8.6. Each frame from the upper layer bears a priority value, which is passed down to the MAC layer. Up to eight priorities are supported in an 802.11e QSTA and they are mapped into four different access categories (ACs) at the MAC layer [1]. The mapping between UP and AC is inherited from the specification of IEEE 802.1D [18], and is shown in Table 8.2.

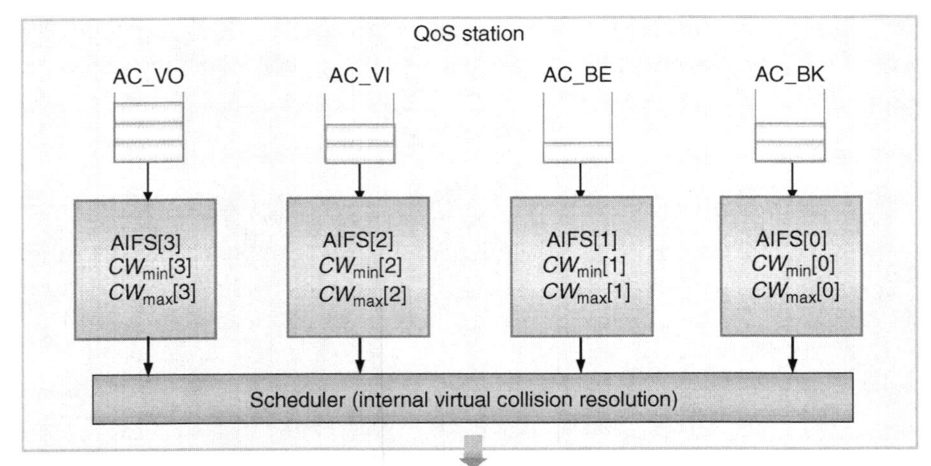

Figure 8.6 Multiple-queue architecture within an IEEE 802.11e QSTA.

Table 8.2 Mapping between UP and AC

UP	*AC*	*IEEE 802.11e Designation (Informative)*	*IEEE 802.1D Designation*
1	AC_BK (0)	Background	BK
2	AC_BK (0)	Background	—
0	AC_BE (1)	Best Effort	BE
3	AC_BE (1)	Best Effort	EE
4	AC_VI (2)	Video	CL
5	AC_VI (2)	Video	VI
6	AC_VO (3)	Voice	VO
7	AC_VO (3)	Voice	NC

A set of EDCA parameters, namely, the arbitration interframe spacing (AIFS[AC]), minimum contention window size (CW_{min}[AC]), maximum contention window size (CW_{max}[AC]), TXOP[AC], and maximum retry limit (retryLimit[AC]) is associated with each AC to differentiate the channel access. The usage of each parameter is demonstrated in Figures 8.6 and 8.7, and summarized as follows.

AIFS[AC] is the number of time slots a packet of a given AC has to wait after the end of a time interval equal to a short interframe spacing (SIFS) duration before it can start the backoff process or transmit. After i ($i \geq 0$) collisions, the "backoff counter" in 802.11e is selected uniformly from [0, $2^i \times CW_{min}[AC] - 1$], until it reaches the "backoff stage" i, such that $2^i \times CW_{min}[AC] = CW_{max}[AC]$. At this point, if a collision occurs, the packet

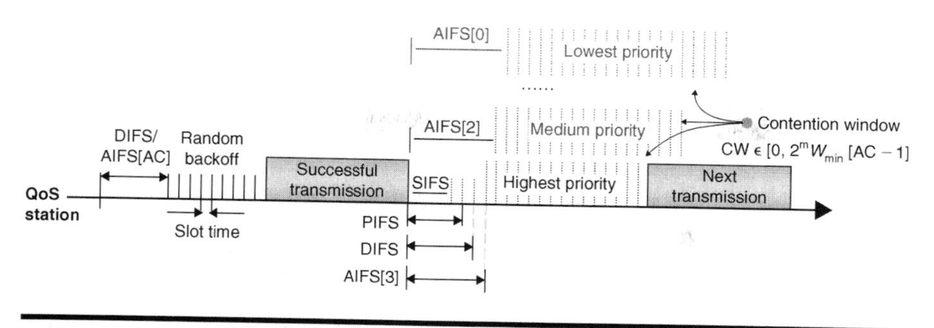

Figure 8.7 EDCA: an illustration.

will still be retransmitted, and the backoff counter is always chosen from the range [0, $CW_{\max}$[AC] – 1], given that the total number of retransmissions is less than or equal to the maximum number of allowable retransmissions (i.e., retryLimit[AC]).

Because multiple priorities coexist within a single station, it is likely for them to collide with one another when their backoff counters decrement to zero simultaneously. This phenomenon is called an "internal collision" in 802.11e and is resolved by letting the highest priority involved in the collision win the contention. Note that the internal collision resolution occurs only among queues residing in the same station and has no effect on the possible collision among different stations.

8.3.4 Admission Control

To prevent the network from being overloaded and to ensure proper QoS to the traffic already admitted, admission control has been introduced into 802.11e for EDCA and HCCA. Generally, an admission control algorithm may take into consideration factors such as available channel capacity, link conditions, retransmission limits, and the scheduling requirements of a given stream. It is worthwhile to note that EDCA can also provide parameterized QoS, when it is used in conjunction with appropriate QoS characteristics of the treated data flow (a.k.a. "traffic specification") for admission control.

- *EDCA*. In EDCA contention-based period, a QSTA shall maintain two variables, namely, "admitted_time" and "used_time," both of which will be set to zero upon association or reassociation. To request for traffic admission, each AC in a QSTA should send a TSPEC to the QAP in a request message, indicating mean data rate, minimum PHY-layer rate, etc. On the receipt of this request message, the QAP determines whether to accept this request or not, and calculates the amount of channel time (known as "medium_time") to allocate. Note that if the traffic is not admitted, the medium_time would be set to zero. The actual algorithm used to decide the admissibility and the channel resource allocation is again out of the scope of the 802.11e standard, and has been discussed in numerous recent papers [19–21]. The QAP sends a response message back to the requesting QSTA, notifying it of whether the traffic is admitted, and the medium_time if the request is accepted. The QSTA then sets its local admitted_time to the value of medium_time contained in the received response message. Local timer used_time records the actual usage of the channel at a QSTA, and is updated on the completion of a transmission by the QSTA. Note that once the used_time exceedes the allocated medium_time, no transmission attempt will be made until the used_time is reset.

- *HCCA*. The IEEE 802.11e standard does not specify any specific algorithm for HCCA admission control. Instead, it only provides a reference design, which is conformant to the normative behavior described in the standard. When a new TS requests for admission, the suggested reference HCCA admission control algorithm takes the TSPEC information as input to compute the $\mathrm{SI}_{\mathrm{sch}}$. It then calculates the number of packets that arrive at the mean data rate during the $\mathrm{SI}_{\mathrm{sch}}$ and the TXOP available for allocation by using the method outlined in Section 8.3.2. Finally, given that there are K streams already in the system, the $(K+1)$th requesting stream can only be admitted if the following inequality is satisfied.

$$\frac{\mathrm{TXOP}_{K+1}}{\mathrm{SI}_{\mathrm{sch}}} + \sum_{i=1}^{K} \frac{\mathrm{TXOP}_i}{\mathrm{SI}_{\mathrm{sch}}} \le \frac{T - T_{\mathrm{CP}}}{T} \tag{8.3}$$

where T and T_{CP} stand for the beacon interval and the time used for EDCA traffic, respectively.

8.3.5 EDCA: Performance Evaluation

With the ratification of IEEE 802.11e in 2005, and the availability of EDCA products certified by the Wi-Fi Alliance [22], a considerable amount of effort has been directed to understanding the behavior of the EDCA protocol and for further optimizing and improving its performance. Recent work either relies on simulation [12,19,23,24] or seeks to analyze the system by decoupling the intrinsic interactions among multiple queues, and thus, simplifying the problem [13,25–27]. It is worthwhile to note that although the modifications introduced in EDCA are incremental and seemingly straightforward, they render the analytical modeling unexpectedly challenging due to the complicated interactions among queues of different priorities residing in the same as well as different stations.

The results reported in Refs 28 and 29 will be used here to further illustrate the impact of each individual EDCA parameter. For the following evaluation, saturation traffic load is assumed.

- *Impact of CW_{min}[AC] and CW_{max}[AC].* Varying contention window size CW_{min}[AC] and CW_{max}[AC] is one of the primary means introduced for traffic prioritization. For a system with parameters listed in Table 8.3, Figure 8.8 demonstrates that appreciable throughput differentiation between high- and low-priority traffic can be achieved once the mechansim of varying CW_{min}[AC] and CW_{max}[AC] is enabled. As illustrated in Figures 8.8a and 8.8b, the enhanced prioritization can also reduce the collision rate in the network, and thus lead to a considerable increase in the network capacity.

  The impact of the contention window size is examined from a different perspective in Figure 8.9, where CW_{min} and CW_{max} for

Table 8.3 EDCA Parameters for Figure 8.8

Scenario	CW_{min}*[1]/[0]*	CW_{max}*[1]/[0]*	*AIFS[1]/[0]*	*RetryLimit[1]/[0]*
Case 1	16/16	32/32	2/2	1/1
Case 2	16/32	32/64	2/2	1/1

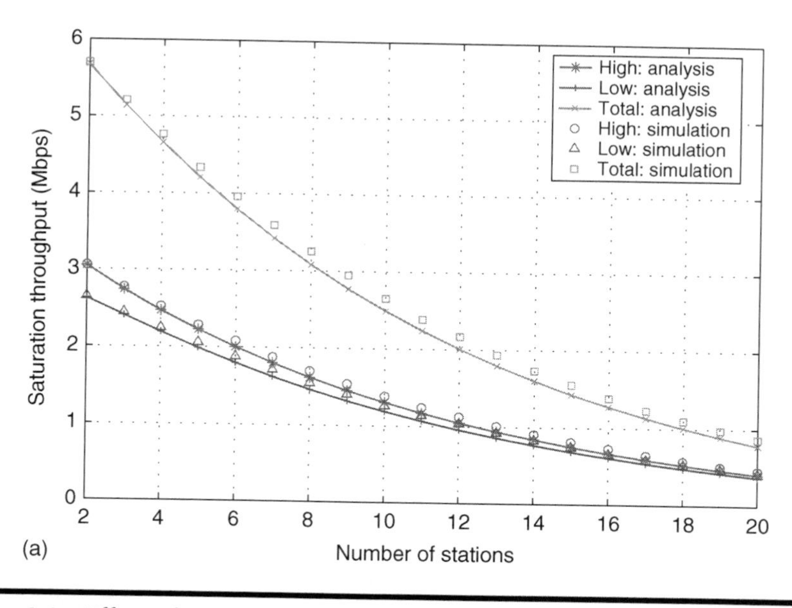

Figure 8.8 Effect of contention window mechanism. (a) Case 1: only internal collision resolution is enabled, (b) case 2: differentiation of contention window is enabled.

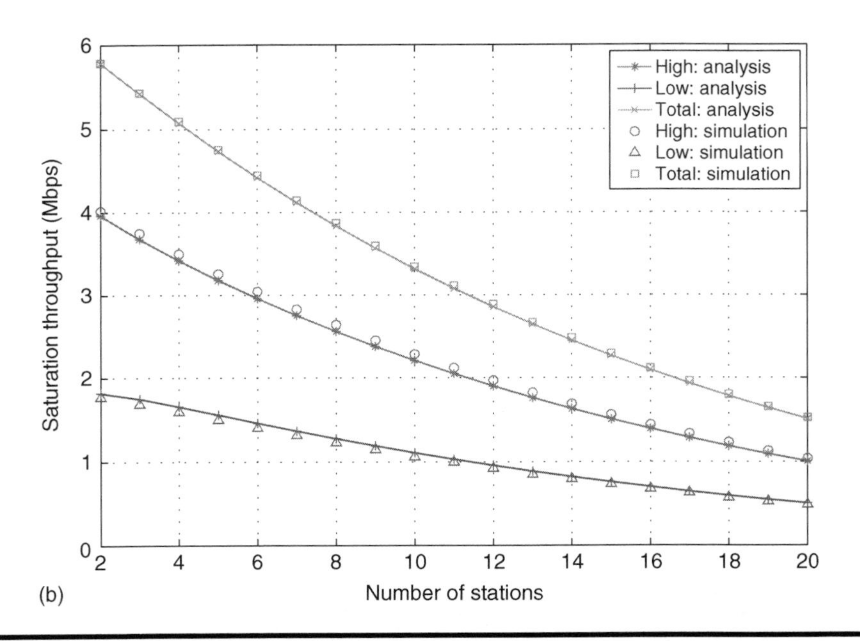

Figure 8.8 ***(Continued)***

high-priority traffic (e.g., $CW_{\min}[1]$ and $CW_{\max}[1]$) are held fixed at 8 and 16, whereas that for low-priority traffic (i.e., $CW_{\min}[0]$ and $CW_{\max}[0]$) are varied. As expected, when the $CW_{\min}[0]$ grows larger with respect to $CW_{\min}[1]$, the low-priority traffic obtains less bandwidth. The parameters of the system studied in Figure 8.9 are listed in Table 8.4.

The ratio of high-priority throughput to the low-priority throughput (i.e., S_1/S_0) can be computed based on Figure 8.9, and is further related to the ratio of minimum contention window sizes, $CW_{\min}[1]/CW_{\min}[0]$ in Figure 8.10. Interestingly, Figure 8.10a indicates that for a network with two nodes, S_1/S_0 increases almost linearly with respect to $CW_{\min}[1]/CW_{\min}[0]$. Figure 8.10b attempts to generalize this linear relation, and plots the corresponding slope for a wider range of network sizes. As suggested in Figure 8.10b, the throughput ratio, S_1/S_0, is approximately in proportion to the ratio $CW_{\min}[1]/CW_{\min}[0]$, with a factor approaching 1.8. This key observation can serve as the basis for a heuristic, which determines the "optimal" window size for a given bandwidth partition among different traffic classes.

- *Impact of AIFS[AC].* AIFS[AC] is another QoS parameter that plays a key role in determining the traffic prioritization. Figure 8.11 compares the throughput for the system when only internal collision

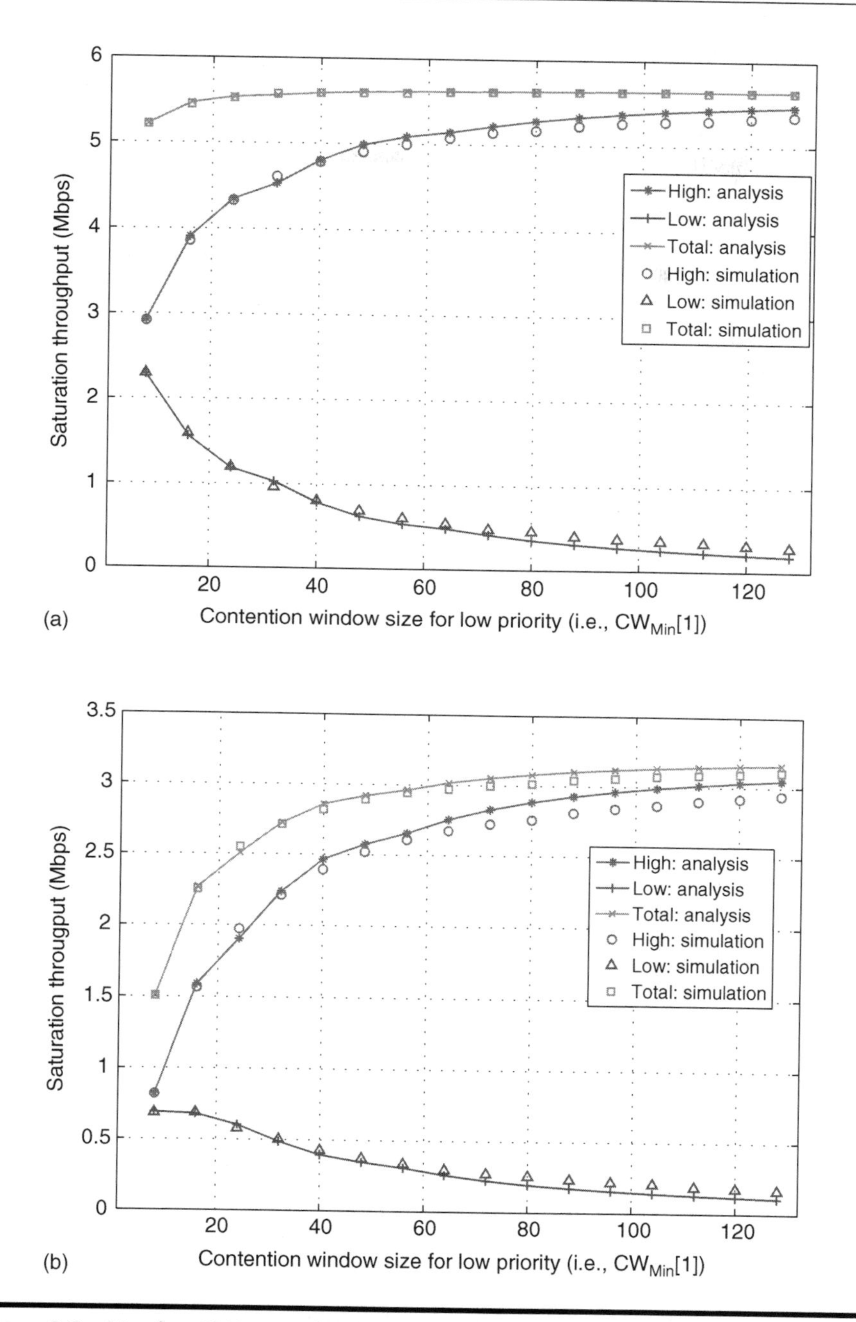

Figure 8.9 Varying CW_{min} and CW_{max} of high-priority traffic. (a) Case 1: a network with 2 stations, (b) case 2: a network with 8 stations.

Table 8.4 EDCA Parameters for Figure 8.9

Scenario	CW_{min}*[1]/[0]*	CW_{max}*[1]/[0]*	*AIFS[1]/[0]*	*RetryLimit[1]/[0]*
Case 1	8/variable	16/variable × 2	2/2	1/1
Case 2	8/variable	16/variable × 2	2/2	1/1

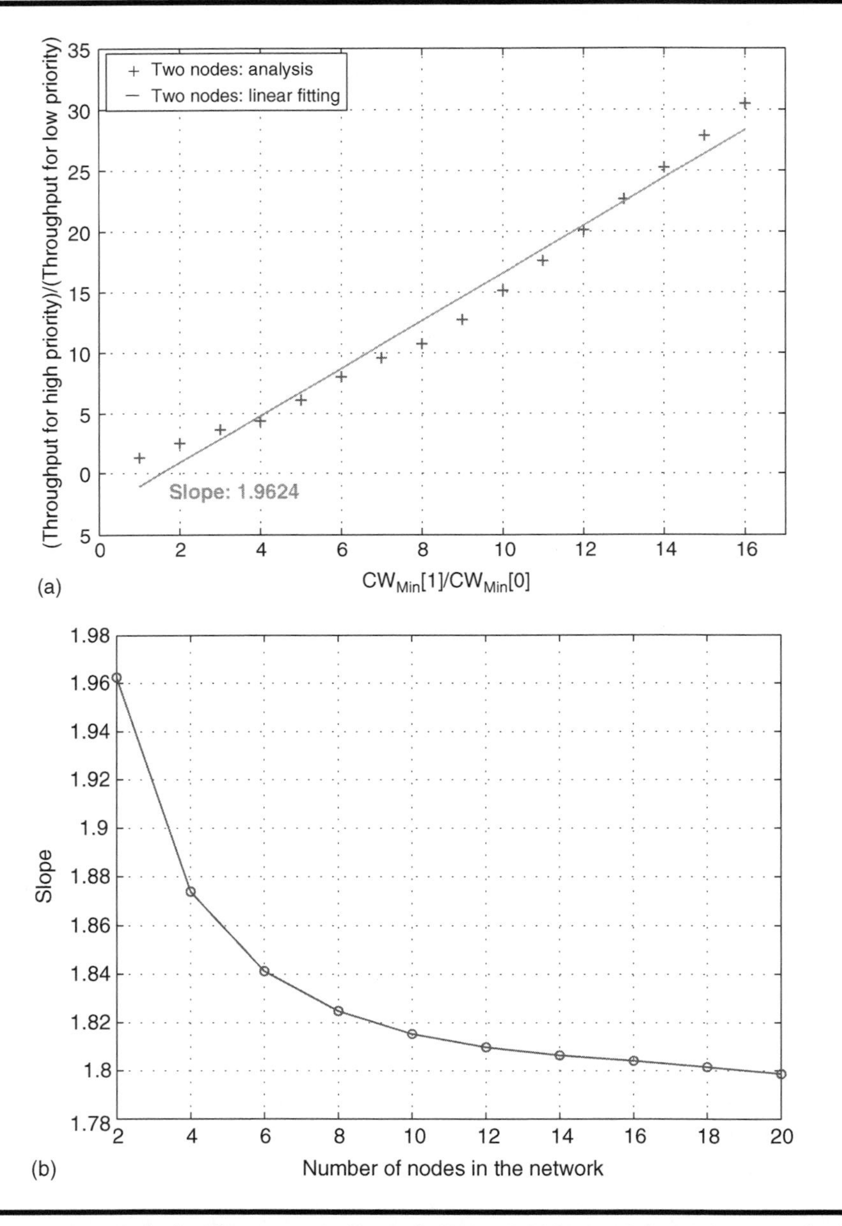

Figure 8.10 Relation between S_1/S_0 and $CW_{\min}[1]/CW_{\min}[0]$. (a) A network of two stations, (b) general study of the slope.

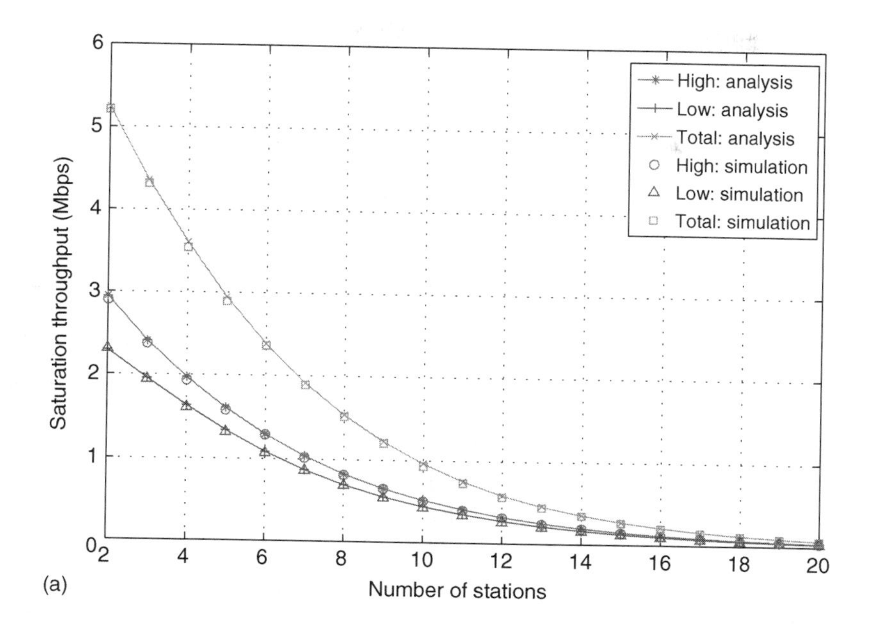

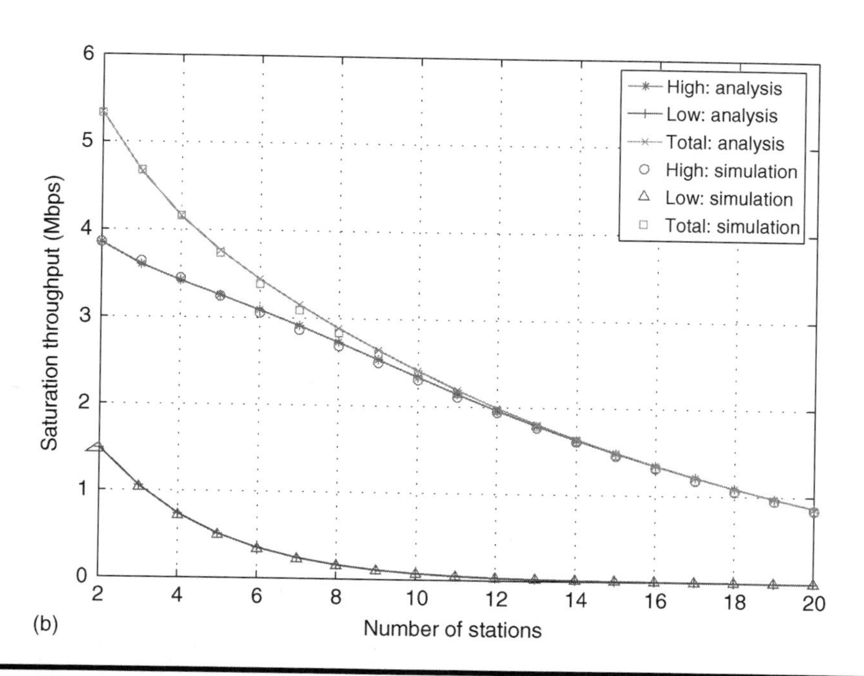

Figure 8.11 Varying AIFS[AC]. (a) Only internal collision resolution is enabled, (b) AIFS differentiation is enabled.

Table 8.5 EDCA Parameters for Figure 8.12a

Scenario	CW_{min}*[1]/[0]*	CW_{max}*[1]/[0]*	*AIFS[1]/[0]*	*RetryLimit[1]/[0]*
Case 1	8/8	16/16	2/2	1/1
Case 2	8/8	16/16	2/3	1/1

Table 8.6 EDCA Parameters for Figure 8.11b

Scenario	CW_{min}*[1]/[0]*	CW_{max}*[1]/[0]*	*AIFS[1]/[0]*	*RetryLimit[1]/[0]*
Case 1	8/8	16/16	2/variable	1/1

resolution is enabled versus when AIFS[AC] is also turned on, and evidently shows that a slightly shorter AIFS value for high-priority traffic can give it an immense advantage over low-priority traffic in terms of bandwidth acquisition. The parameters of the system studied in Figures 8.11a and 8.11b are listed in Tables 8.5 and 8.6, respectively.

A comparison between Figures 8.8b and 8.11b seems to suggest that arbitration interframe spacing exerts more influence on service differentiation than contention window size. Similar qualitative observations have been reported and discussed further in Ref. 30.

In a closer examination of its impact, AIFS[1] is kept fixed for high-priority traffic, whereas for low-priority queues *AIFS*[0] is varied across a wide range of values. The throughput ratio S_1/S_0 versus the ratio AIFS[1]/AIFS[0] is plotted in Figure 8.12, where the vertical axis has a logarithmic scale. It shows that S_1/S_0 grows very large even for relatively small AIFS[1]/AIFS[0].

To conclude, therefore, arbitration interframe spacing is able to provide a much more pronounced differentiation in throughput than contention window size, without compromising the aggregate network throughput.

- *Other traffic scenarios.* Note that all the analytical and simulation results presented hitherto give the "worst-case" maximum throughput for low-priority traffic class because the network operates under the saturation condition. When light or medium traffic load for high priority is applied, we expect the throughput of low-priority traffic to be above these values.

To confirm the preceding conjecture, extensive simulations have been performed for a much wider region of load condition. More specifically, each station now supports three traffic classes, and the traffic generated by each class increases from zero to saturation. Other related parameters used in the simulation are listed in Table 8.7. Figure 8.13 portrays the dynamics in the system throughput, as the load for each traffic class grows simultaneously.

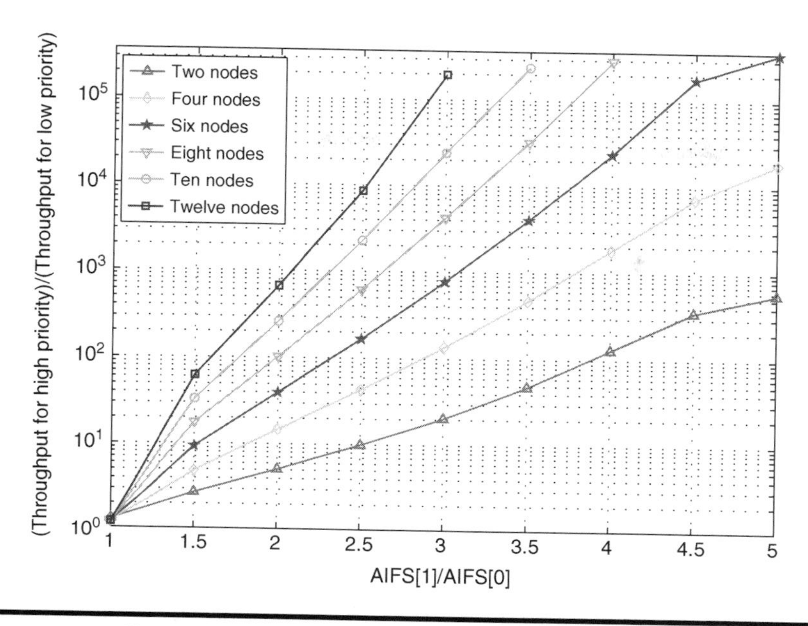

Figure 8.12 Effect on throughput of changing AIFS.

Table 8.7 EDCA Parameters for Figure 8.13

AC/Priority	CW_{min}	CW_{max}	*RetryLimit*	*AIFS*
AC_VO/high	8	16	3	2
AC_VI/medium	16	32	3	2
AC_BE/low	32	1024	6	3

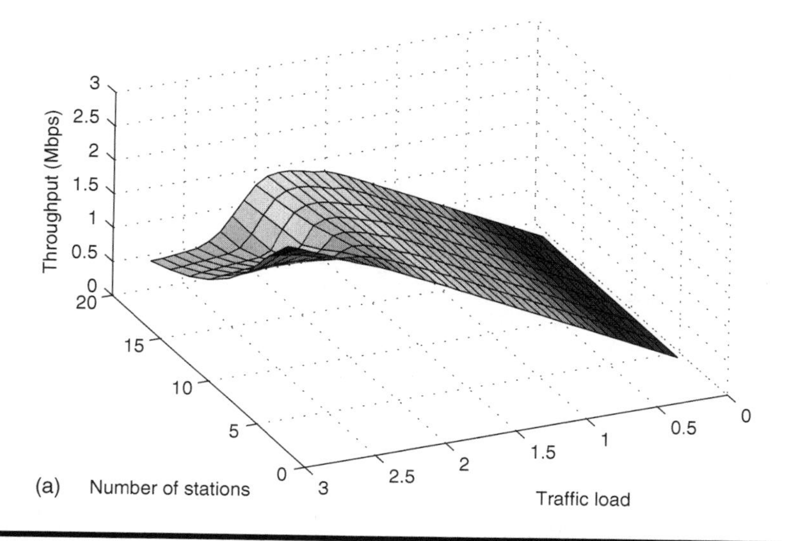

Figure 8.13 A network of three traffic classes. (a) High priority, (b) medium priority, (c) low priority.

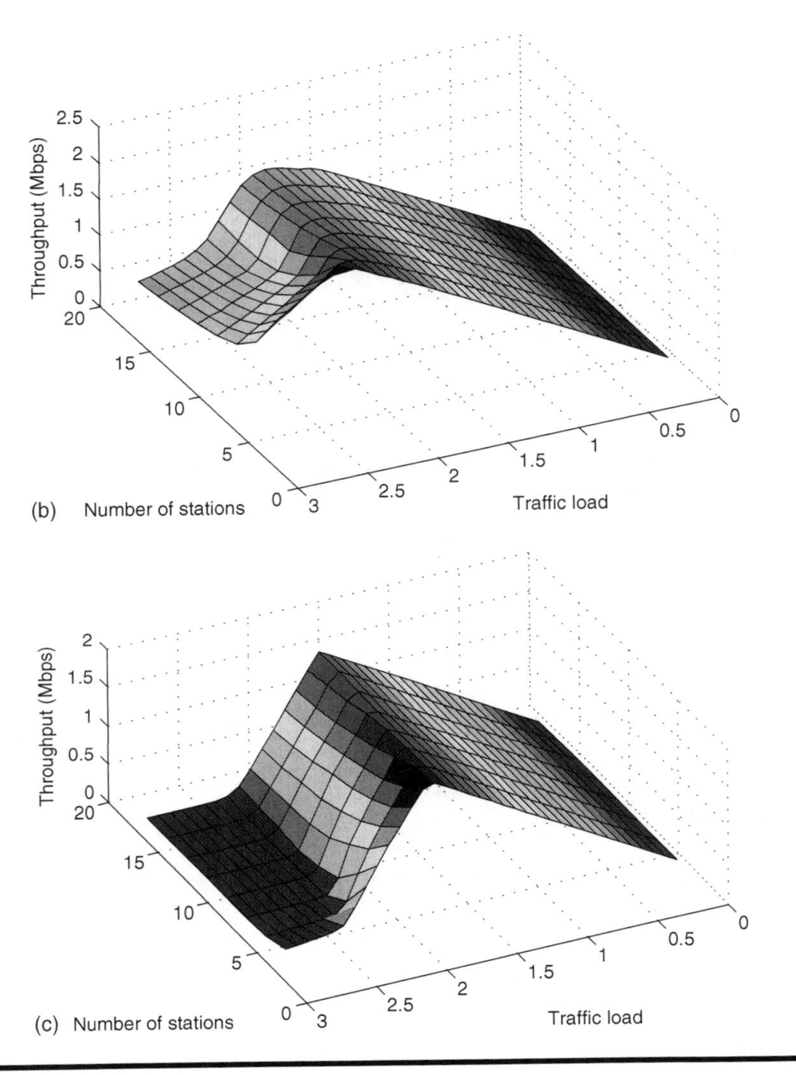

Figure 8.13 ***(Continued)***

Indeed, when the network is not saturated, traffic from all three classes can share the available bandwidth proportionally. However, once the load exceeds the saturation point, the balance can no longer be maintained, as the high-priority traffic tends to occupy more channel resources than the medium- and low-priority ones. Unlike low-priority traffic, which becomes completely suppressed when the load is sufficiently heavy, the medium-priority traffic is still able to seize a nonzero portion of the bandwidth. How the bandwidth would be shared will eventually be determined by the value of the QoS parameters associated with these three traffic classes.

8.4 QoS Features in IEEE 802.11n

8.4.1 Overview of IEEE 802.11n

Great expectations have been generated for the much-hyped IEEE 802.11n [2], which is likely to emerge as a stable standard in 2008 and gradually replace 802.11a/g as the dominant wireless transport technology. To support such a bandwidth-hungry application as the wireless HDTV, 802.11n is designed to deliver up to 600 Mbps at the PHY layer, a rate virtually unthinkable only a few years ago. Nevertheless, when it comes to wireless home entertainment and consumer electronics application, high raw bandwidth is necessary, but far from sufficient. Indeed, both guaranteed bandwidth and QoS are essential for 802.11n to provide wirelike performance regardless of time-varying wireless channel conditions.

IEEE 802.11n has focused on developing enabling technologies at both PHY and MAC layers, including multiple-input multiple-output (MIMO), channel bonding, high-efficiency MAC protocols, and more robust QoS support [2,31,32]. Given the scope of this chapter, the ensuing discussion focuses on the MAC-layer innovations only.

8.4.2 MAC Efficiency Improvement

To achieve a minimum throughput of 100 Mbps at MAC layer, which is its primary standardization requirment, 802.11n cannot solely rely on boosting transmission rate. In fact, it has been proven in Ref. 33 that MAC-layer throughput has a bound, as long as the legacy 802.11 or 802.11e MAC protocol is used, even when PHY-layer rate grows to infinity. It has been revealed further therein that the main culprit of such a throughput limit is the strikingly low protocol efficiency of the current 802.11 MAC. As a remedy, techniques such as frame aggregation, block acknowledgment, and interframe spacing reduction have been proposed or incorporated in 802.11n to enhance efficiency.

8.4.2.1 Frame Aggregation

Aggregation in principle can be performed at MSDU level (i.e., A-MSDU) and MPDU level (i.e., A-MPDU), the format of which is depicted in Figures 8.14 and 8.15, respectively. The average overhead associated with transmitting payload bits can be reduced by aggregating multiple MSDUs into a longer MPDU and multiple MPDUs into a longer PPDU, essentially amortizing the same protocol overhead over more data bits.

Figure 8.14 portrays the MPDU format specified in 802.11n [2], which is extended from that of 802.11e [1]. The QoS control field is a 16-bit field first introduced in 802.11e to identify the traffic class or TS to which the MPDU

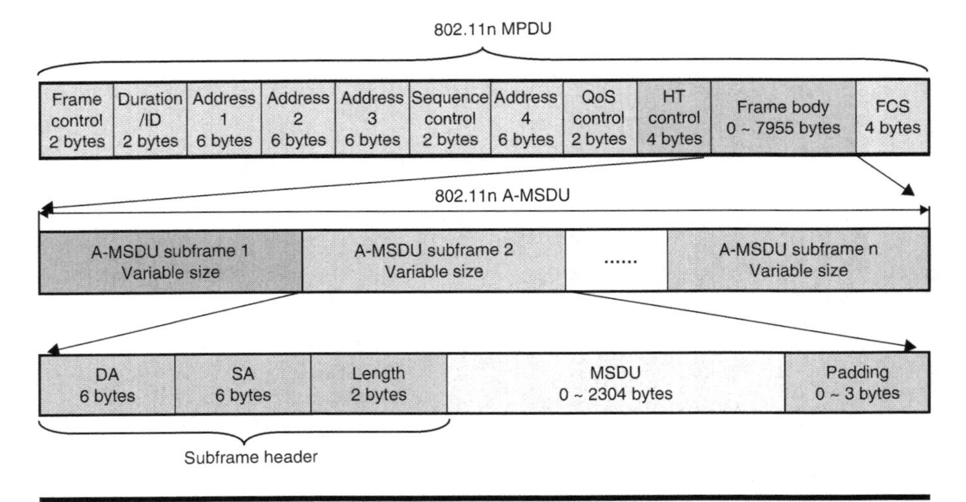

Figure 8.14 Aggregation at MSDU level (A-MSDU).

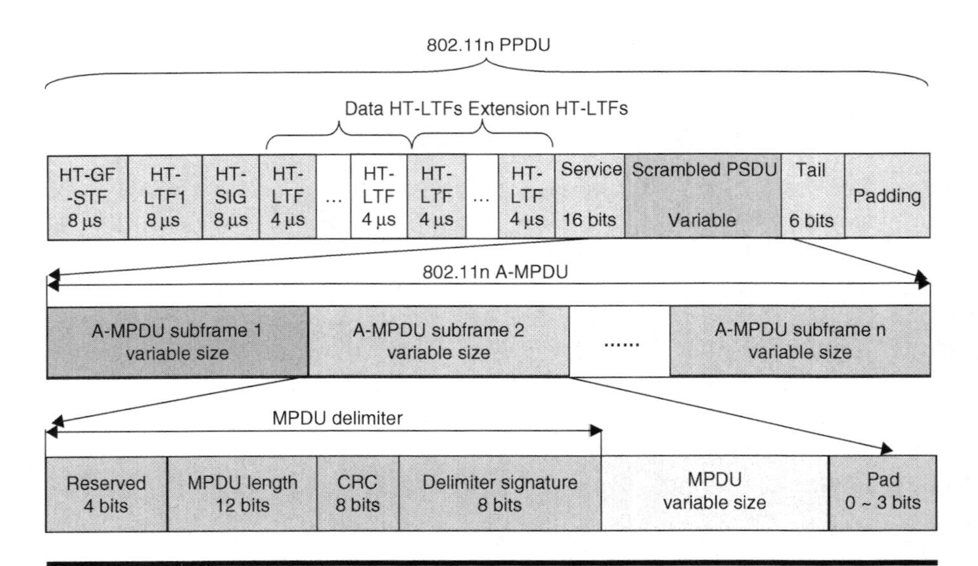

Figure 8.15 Aggregation at MPDU level (A-MPDU).

belongs, and various other QoS-related information about the MPDU. The current 802.11n draft further defines a 4-byte-long high-throughput (HT) control field, which conveys signaling information related to operations particular to 802.11n.

The frame body part of the MPDU can be composed of an MSDU or a sequence of A-MSDU subframes; the latter is referred to as an MSDU-level aggregation. Figure 8.14 also depicts the detailed format of such an A-MSDU subframe, which consists of a subframe header, an MSDU, and 0–3 bytes of padding. Padding bits are attached to each subframe, except the last one to make its length a multiple of 4 bytes. The subframe header further comprises three fields, namely, destination address (DA), source address (SA), and length. The DA and SA fields shall be interpreted in the same way as defined in legacy 802.11 [4], whereas the "length" field contains the length in bytes of the MSDU.

To avoid any interference with QoS management, it is stipulated that only MSDUs from the same traffic class or TS can be aggregated into an A-MSDU, thereby assuring that the channel access policy for an MPDU carrying an A-MSDU is consistent with that of an MPDU carrying only one MSDU of the same TC or TS. In addition, an A-MSDU shall only contain MSDUs that are intended to be received by a single receiver, and all are certainly transmitted by the same transmitter.

As illustrated in Figure 8.15, aggregation can also be implemented at the MPDU level, wherein a sequence of A-MPDU subframes are concatenated to form an A-MPDU. Each A-MPDU subframe consists of an MPDU delimiter, an MPDU, and possibly some padding bytes. The MPDU delimiter has 4 bytes, and its detailed structure is shown in Figure 8.15. The MPDU delimiter is primarily used to delimit MPDUs within the A-MPDU so that the structure of the A-MPDU can be recovered when one or more MPDU delimiters are received with errors. The delimiter signature is an 8-bit pattern that can be used to detect an MPDU delimiter when scanning for an A-MPDU subframe. The A-MPDU, after scrambling and forward error correction (FEC) coding, becomes the payload (i.e., PSDU) of an 802.11n PPDU. A greenfield short-training field (HT-GF-STF), one or more long training field (HT-LTF), a signaling field (HT-SIG), a service field, tail bits, and padding bits are appended in front of or behind the scrambled PSDU to form an 802.11n PPDU. The details of these fields are not discussed here as they are not directly related to the MAC-layer aggregation protocols. Also, note that the PPDU format shown in Figure 8.15 is only used in the "green field" operation, which is a new mode introduced in 802.11n to achieve optimal system performance at the expense of backward compatibility with legacy 802.11b/g/a station. A similar PPDU format has also been specified for mixed mode operations, wherein both 802.11n and legacy stations can coexist.

A performance comparison for A-MSDU and A-MPDU is provided in Figure 8.16, which confirms that both aggregation schemes can enable significant protocol efficiency improvement. In particular, the effect of A-MPDU is more pronounced, as it enables the sharing of considerable amount of PHY-layer overhead (e.g., training sequence, etc.) among multiple MPDUs.

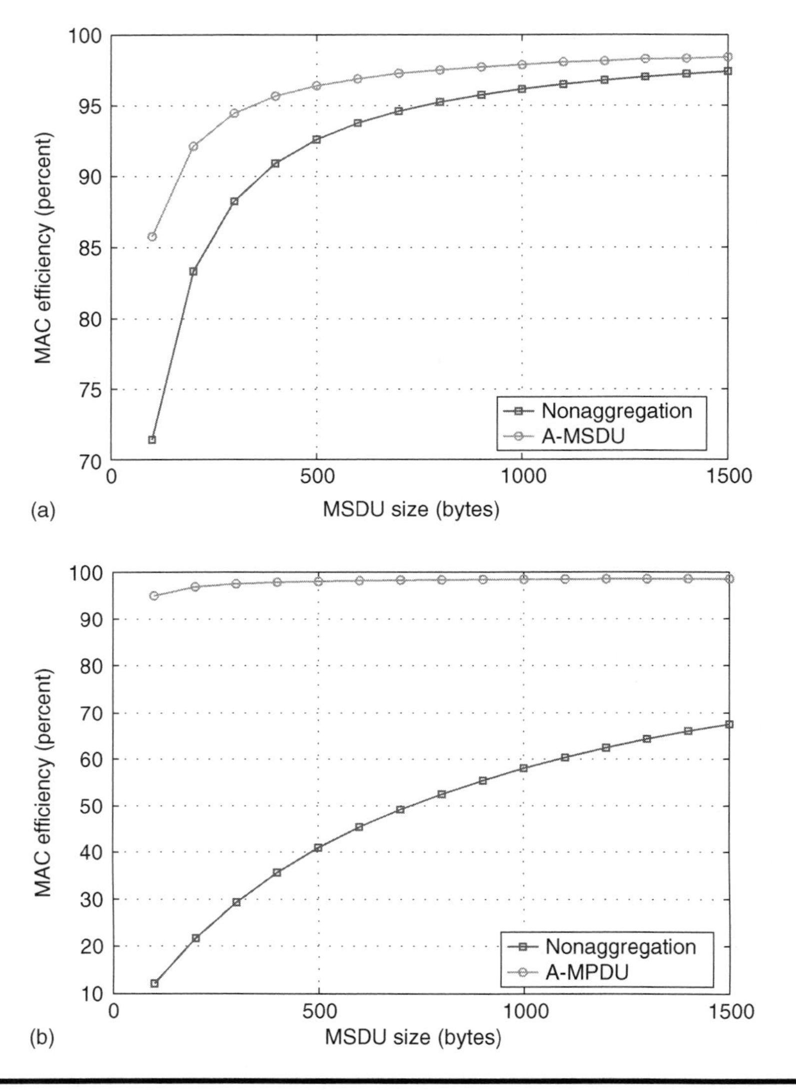

Figure 8.16 Performance of frame aggregation: a comparison. (a) A-MSDU, (b) A-MPDU: 2 x 2 MIMO, 144.4 Mbps PHY.

8.4.2.2 Block Acknowledgment

Exchange of A-MSDU or A-MPDU is made possible with a protocol called block acknowledgment (Block ACK) [34], which was initially introduced in 802.11e [1] for the purpose of efficiency improvement. The Block ACK provides the capability of acknowledging the reception of multiple packets (e.g., MSDUs or MPDUs) in a collective manner with a single acknowledgment message, rather than using one acknowledgment message per received packet. In conjunction with frame aggregation, the block acknowledgment feature can significantly improve the efficiency of the communications protocol between the transmitter and the receiver.

8.4.2.3 Reduced Interframe Space

Reduced interframe spacing (RIFS) is the newly defined type of interframe spacing in 802.11n, which can be used in place of SIFS to separate multiple transmissions from a single transmitter when no SIFS-separated response transmission is expected. Because RIFS is even shorter than SIFS, the use of RIFS can further reduce the channel idle time while still maintaining the normal protocol operation.

8.4.3 Reverse Direction Protocol

To improve the handling of bidirectional traffic generated by many interactive applications, a new feature called RDP has been accepted into the current version of 802.11n draft [2]. As its name suggests, this protocol allows the "piggybacking" of data along with an acknowledgment from the intended receiver back to the transmitter.

An illustration is provided in Figure 8.17a, wherein station 1 acquires a TXOP either through contention or allocation, and effectively shares it with receiver station 2. More specifically, station 1 first sends a PPDU burst to station 2, which contains not only data, but also a Block ACK request and a reverse direction grant. Station 2 then transmits a Block ACK as a response to the Block ACK request, and further sends back a response burst to station 1, consisting of one or multiple MPDUs and the associated Block ACK request. Station 1 finally regains the control of the TXOP and sends a Block ACK to station 2 to acknowledge the MPDUs transmitted by STA 2 in the response burst.

It has been demonstrated through analysis [35] and simulation [36] that RDP can significantly enhance the application-level capacity and QoS for TSs of a bidirectional nature. Figure 8.17b [36] illustrates the impact of RDP on the maximal number of simultaneous VoIP calls an 802.11n network can sustain. Both the 20 and 20/40 MHz hybrid channel bandwidth modes are simulated, whereas the MIMO configuration is set to the mandatory 2×2. Note that the 20/40 MHz hybrid channel bandwidth mode is another innovation accepted by the current 802.11n draft [2], which attempts to boost the capacity by using 40 MHz channel to carry data, although leaving control and management packets in 20 MHz channel for coexistence and backward compatibility purpose. As confirmed by Figure 8.17b, the adoption of RDP readily doubles the voice capacity of an 802.11n network, as it eradicates the need for the callee to obtain medium access once the caller acquires

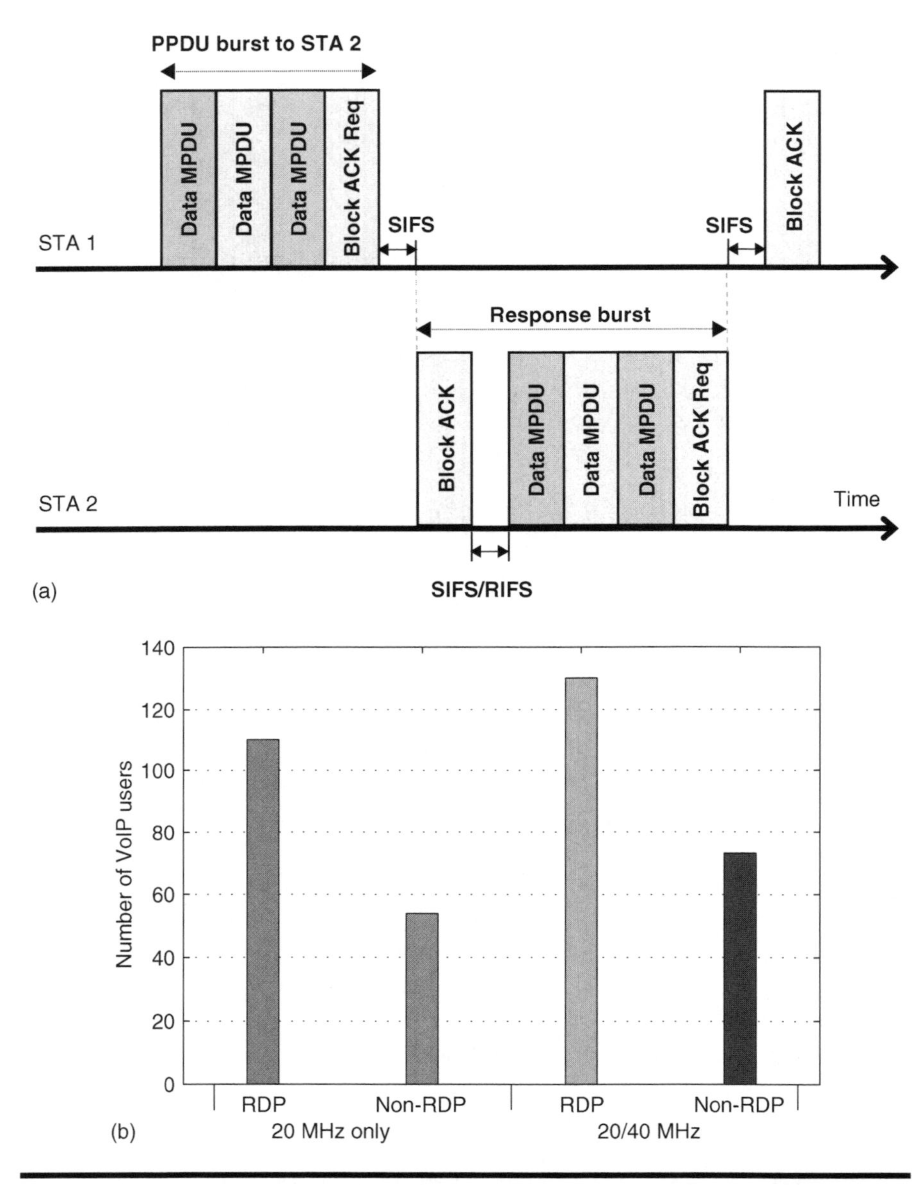

Figure 8.17 Reverse direction protocol. (a) Illustration of protocol operation, (b) impact on the number of concurrent VoIP calls.

the forward channel, and thus, significantly lowers the overhead associated with the channel access. RDP also proves to be helpful for TCP traffic in general, which generates an ACK for a certain number of received packets at transport layer, and therefore, would require a channel access on the reverse link.

8.5 QoS Mechanisms in IEEE 802.11s Mesh Network

8.5.1 Overview of IEEE 802.11s

To sustain the explosive growth of IEEE 802.11-compliant WLANs experienced in enterprise, public spaces, and residential applications, mesh networks have become important. They enable the interconnection of numerous APs in an efficient and cost-effective manner, especially in environments where PHY cabling is expensive or not permitted by building regulation (e.g., historical site). Mesh networks are also important in settings where telecommunication facilities must be installed quickly, as in disaster recovery or in combat, and in areas where a large coverage footprint is desirable, as in campus and municipal Wi-Fi networks. In general, a wireless mesh network may convey either traffic loads comparable to those carried by a single AP but spread out farther, or traffic aggregated by several APs and transported wirelessly. We refer to the latter as a "backbone" mesh network. The backbone of a mesh network generally consists of a collection of nodes that are either APs—each serving a collection of stations in a basic service set (BSS), or relay points. The nodes can forward traffic across the mesh and, when other functional differentiation is not relevant, they are simply referred to as MPs.

Mesh networks, which are deemed a special case of *ad hoc* networks, have gained immense attention in the research community from the very beginning, due to their tremendous economic and social impact and the abundant appealing research opportunities they generate [37,38]. Earlier literature has touched on many crucial aspects of mesh networking, ranging from novel MAC [39], innovative radio channel assignment [40] to efficient routing protocols [41].

In July 2004, the 802.11s task group was created to develop an amendment to the 802.11 standard for mesh networking [42]. One of the primary goals that 802.11s strives to achieve is to install sufficient capability to deliver proper QoS, especially in the backbone mesh network wherein the aggregated traffic load is anticipated to be enormous. Owing to the exacerbation of the hidden and exposed terminal problems in multihop networks [43], plus the frequent collisions and a significant underutilization of bandwidth, the current 802.11 MAC protocol, which is designed for a single-hop network, cannot form an efficient multihop network, let alone provide adequate QoS support in a backbone mesh environment [44].

To address these issues, the 802.11s task group has designed a mesh medium access coordination function (MCF) based on the 802.11e EDCA protocol [45,46], which deals with discovery, association, security, channel access control, topology learning, and forwarding. Several new MAC mechanisms have been introduced to facilitate congestion control, power saving, synchronization, and beacon collision avoidance. The task group has also contemplated the possibility of supporting multiple channel operations for stations equipped with single or multiple radios.

The MDA protocol, which is a reservation-based deterministic mechanism, has already been accepted into the current baseline draft of 802.11s [46]. Meanwhile, two multi-channel mesh medium access coordination protocols [47,48] were also presented and discussed in the 802.11s task group. The MDA and multi-channel MAC protocols are used here as an example to illustrate how the QoS can be strengthened in an 802.11s mesh environment. Detailed discussions of these protocols are presented in Sections 8.5.2 and 8.5.3, respectively.

8.5.2 Mesh Deterministic Access

MDA is an optional access method that allows a supporting MP to access the channel with lower contention rate than otherwise at selected times. Used in conjunction with distributed scheduling, the MDA protocol can provide enhanced QoS support in a large-scale mesh network.

An MP and its intended receiver exchange “request” and “reply” messages to set up an MDA TXOP (MDAOP), which has a predefined duration and start time. At the beginning of an MDAOP, the owner that initiates the MDAOP has the right to access the wireless channel using higher priority, as it uses a different set of EDCA [1] parameters (i.e., MDA_AIFS, MDA_CW_{min}, and MDA_CW_{max}). Nevertheless, as the priority is not an exclusive one, other MPs may still acquire the channel first, which will force the MDAOP holder to defer. The holder of MDAOP will attempt another channel access, when other transmissions end and the channel is sensed idle again.

To further reduce collision probability during MDA reservations, the MDAOP information is broadcast in beacon frames and gets rebroadcast by neighbor MPs again. Thus, the neighbors that are in direct transmission range or two hops away are informed of future transmissions. If the neighbor MPs are not involved in the MDAOP, they will refrain from channel access by resetting their network allocation vector (NAV) at the beginning of a neighboring MDAOP, thereby giving priority to the MDAOP owner.

In conjunction with distributed scheduling, MDA can offer a relatively predicable channel usage, thereby improving the QoS in the mesh network. Also, note that “a set of MDAOPs” may be set up for unicast transmissions from the source MP to the destination MP, which may be multiple hops away. An example for the use of the MDAOP set concept is to establish an MDAOP set for a single QoS flow, which can help ensure the QoS in an end-to-end manner. One drawback of MDA is that it requires proper synchronization among all the MPs involved in the MDAOP, which would incur additional protocol and implementation complexity.

8.5.3 Mesh Common Control Channel Protocol

IEEE 802.11 has been allocated multiple nonoverlapping frequency channels. More specifically, there are three channels available in the 2.4 GHz Industrial, Scientific and Medical (ISM) radio frequency (RF) band for 802.11b/g [5,7] and twelve in the 5 GHz Unlicensed National Information Infrastructure (U-NII) RF band for 802.11a [6]. Given the limited wireless spectrum and tremendous demand for wireless data, the possibility of dynamic frequency switching on a per packet basis has been extensively explored recently. Several multi-channel MAC protocols have already been proposed [39,49,50], all of which have interestingly pursued the common control channel (CCC) approach. As the name suggests, this solution approach dynamically selects one channel from the permissible pool for data communications, and signals the reservation of the chosen channel by performing a handshake in a separate dedicated control channel (CC) in either the time or frequency domain.

Two multi-channel MAC protocols [47,51] have been presented to IEEE 802.11s and have received significant interest from the group. Since these two CCC MAC protocols are proposed for mesh network, they will also be called CCC mesh MAC protocol (MMAC) in the ensuing discussion. Although proposed by two separate groups, the two protocols can be inherently integrated into one generic framework, which can find its application on MPs with an arbitrary number of radios.

- *Single radio case.* When an MP has only one radio, a protocol called common channel framework (CCF) [51] can be used. MPs use the handshake of request-to-switch (RTX) and clear-to-switch (CTX) control packets to negotiate the mesh traffic (MT) channels on which the subsequent data would be transmitted. As illustrated in Figure 8.18, a channel coordination window (CCW) defines a CCC to which all MPs should simultaneously tune their radio at certain intervals within each superframe and negotiate the usage of the channels. MPs that have completed the negotiation will switch their radio to the selected MT channel and sense it. If the wireless medium is detected as idle for a DIFS interval, data can then be transmitted. Unlike the RTS/CTS handshake that enables immediate reservation in the time domain, RTX/CTX handshake makes a reservation in the frequency domain. Also, it is worthwhile to note that all the MPs need to be synchronized so that they can tune into the CCW at the same time.
- *Multiple radio case.* Owing to its simplicity and low cost, single-radio MPs may still find their application in mesh network with light traffic. Nevertheless, it is expected that multiradio mesh devices will become necessary in large meshes to provide the transport capacity needed by the core mesh backbone. To coordinate access to multiple channels, these multiradio MPs can use a CCC protocol [47].

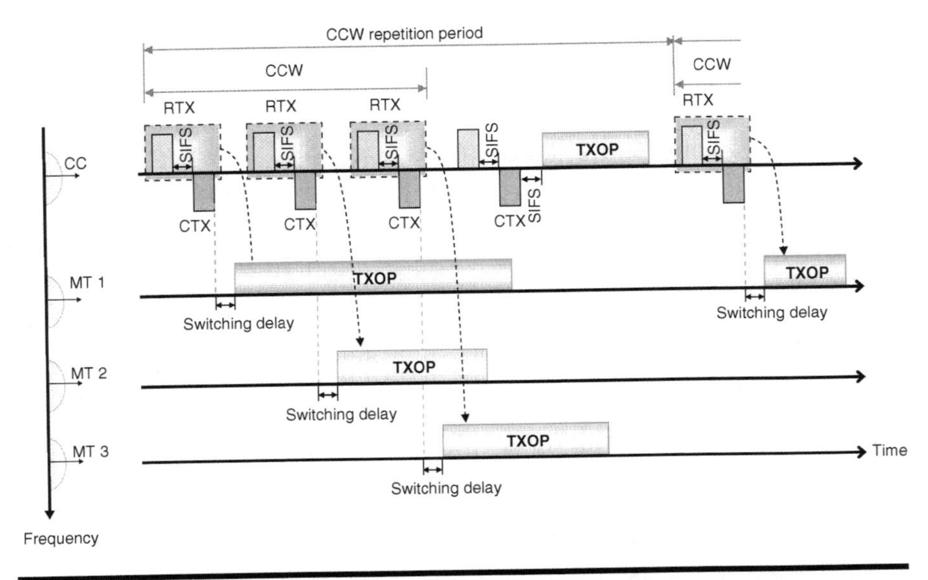

Figure 8.18 CCF MAC protocol: an illustration of three MT channels.

> The CCC protocol defines a CC, over which all mesh nodes will exchange control and management frames through a dedicated radio called the "control radio." The rest of the channels, called MT channels, are used to carry the mesh data traffic, which comprises traffic forwarded by MPs. Accordingly, the radios tuned to the MT channels are called "MT radios." Reservations of the various MT channels are made by exchanging control frames on the CC channel.

Figure 8.19 illustrates the concept of CC and MT channels as well as the basic channel access mechanism. In the illustrated example, the MPs have access to three MT channels and the CC. The CCC protocol observes the rules of EDCA for channel access and virtual carrier sensing [1]. As extensions of the legacy RTS and CTS messages, the mesh RTS (MRTS) and mesh CTS (MCTS) are exchanged on the CC to reserve an MT channel for the time it takes to transmit a TXOP. The particular MT channel selected for the transmission of the TXOP is indicated in a special field on the MRTS/MCTS. An MP keeps track of the time each MT channel has been reserved based on the value of the "duration" field of the MRTS/MCTS.

8.5.4 CCC MAC: Performance Evaluation

As confirmed by simulations in Refs 48 and 52, the CCC protocol framework can deliver the capacity and delay performance that meets the requirements demanded by the next-generation mesh backbones. Owing to space

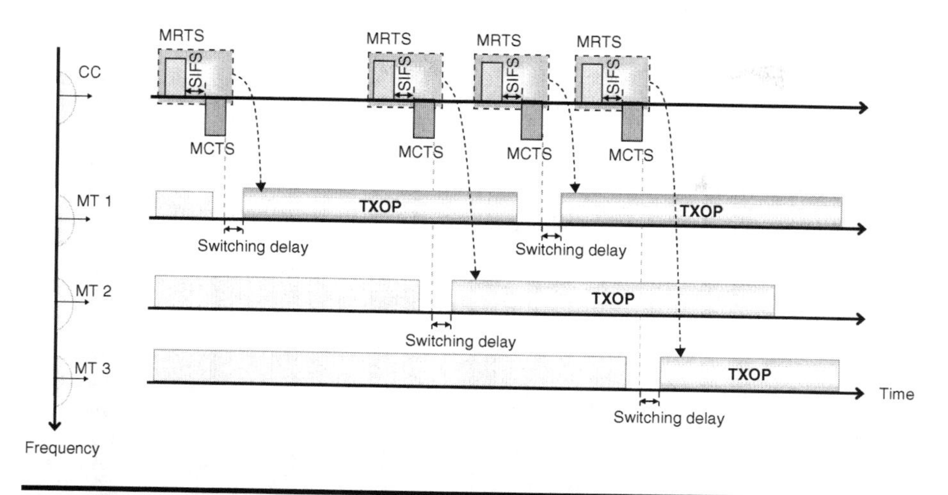

Figure 8.19 CCC MAC protocol: an illustration of three MT channels.

Table 8.8 Key MAC and Traffic Parameters

CW_{min}	CW_{max}	*AIFS*	*TXOP Size (Frames)*	*Payload (Bytes)*	*Interarrival Time*
31	1023	DIFS	1, 10, or 15	1500	Exponential

limitations, however, only some of the results for MPs with multiple radios are discussed hereafter in this section [52].

Performance was measured in OPNET by considering several independent TSs occurring along mesh links. The links were assumed to be situated sufficiently close so as to not permit channel reuse. It was further assumed that the MPs at the end of these links can hear all transmissions. Two major scenarios were examined, comprising four or eight independent TSs, respectively. Each MP is equipped with one control radio and one MT radio. Both radios operate in the 802.11a 5 GHz band. The control traffic is transmitted at 6 Mbps. The rate at which mesh data traffic is delivered varies, which could be either 24 or 54 Mbps in the simulation. MT is transmitted in TXOPs, the actual size of which also varies in the simulation scenarios. Table 8.8 presents the key MAC and traffic parameters.

8.5.4.1 Capacity

Figures 8.20a and 8.20b show the network capacity, achieved by four and eight TSs, respectively, for the scenarios specified earlier. The capacity is compared for different PHY transmission rates for mesh data traffic and different TXOP sizes.

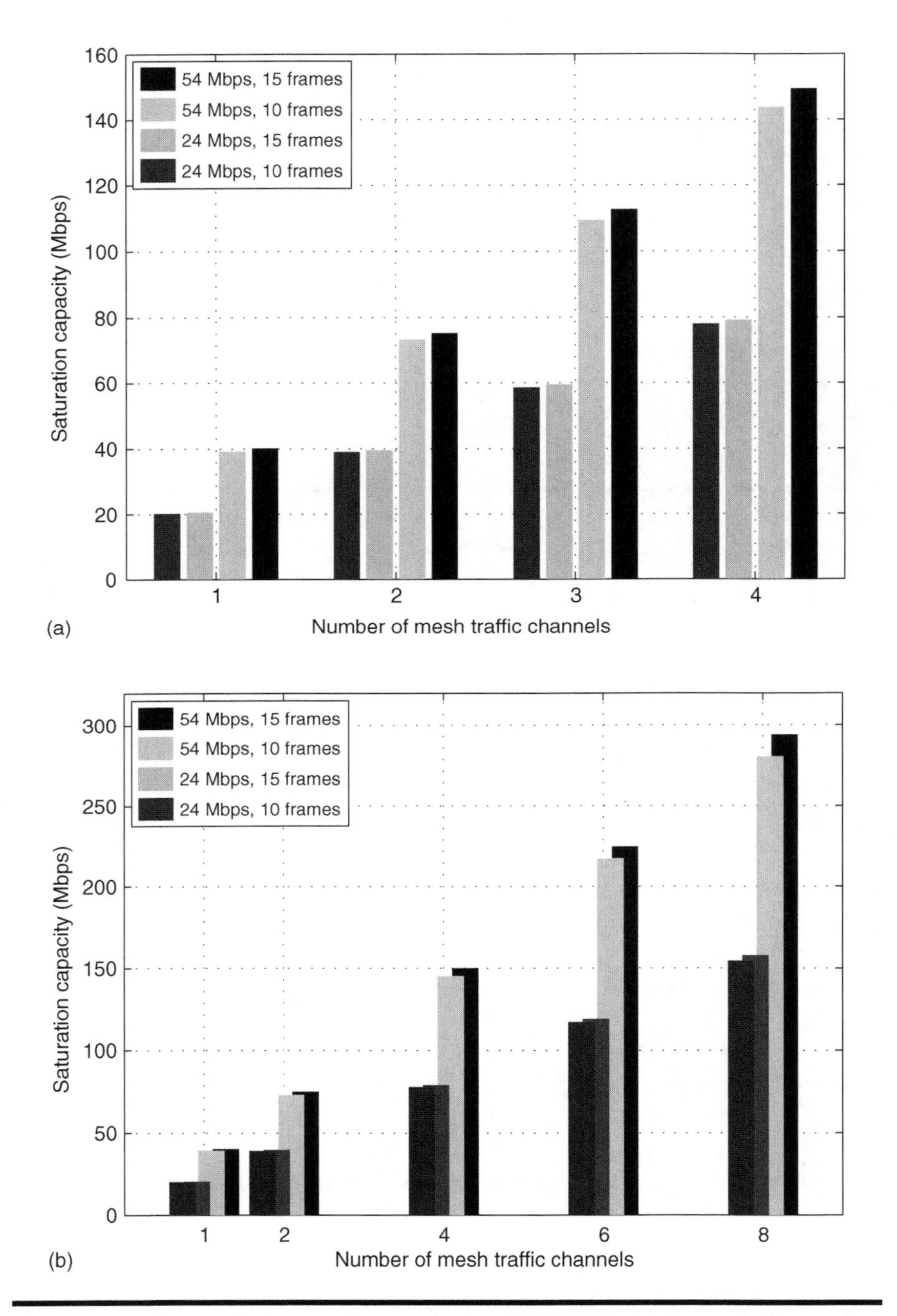

Figure 8.20 Saturation network capacity for CCC MAC. (a) Saturation capacity for four streams, (b) saturation capacity for eight streams.

The simulation results show that the network capacity (i.e., the aggregate throughput for all the independent TSs considered) increases linearly with the number of MT channels available. The maximum capacity is attained when the number of MT channels available is the same as the number of TSs. It reaches approximately 280 Mbps, when eight MT channels are deployed and the PHY transmission rate is 54 Mbps. The maximum traffic that can be carried by a wireless mesh using EDCA is approximately 40 Mbps for the same PHY-layer settings. This clearly shows that EDCA cannot be used in backbone mesh networks, which typically carry traffic from multiple BSSs.

The linear increase of capacity to the number of MT channels suggests that the CC is not a bottleneck, as saturation of the CC would have resulted in a diminishing rate of capacity increase. The concern about the CC becoming a bottleneck is further allayed by observing that capacity improves minimally when the TXOP size increases from 10 to 15 frames, which represents a 33.3 percent decrease in control traffic. Finally, we can see that the CC is not a bottleneck by examining the difference in the capacity attained with four and eight TSs, given the same ratio of MT channels to TSs. Although the control traffic in the latter case is twice the magnitude of the former, the capacity difference is negligible. Had the CC been a bottleneck, a lower capacity per TS and per channel would be observed in the latter case.

8.5.4.2 Delay

To examine the delay performance of the protocol, independent TSs were simulated at the mandatory data transmission rate of 24 Mbps. A number of TSs access MT channels, ranging from two to eight, through the CCC protocol. Two TXOP sizes are examined: 1 and 10 frames/TXOP. In lightly loaded meshes 1 frame/TXOP would be more common, carrying the equivalent of the traffic of a BSS, and used to extend the range of communication through multiple-hop transmissions. Longer TXOPs would be typical of mesh backbone networks conveying BSS traffic aggregated at different mesh APs.

A total traffic load of 15 Mbps was spread across eight TSs when TXOPs of a single frame were simulated. A total load of 19 Mbps was considered across eight TSs for TXOPs of 10 frames long. Note that a load of 19 Mbps approaches the saturation capacity of an 802.11a channel when EDCA is used. The channel access delay and queuing delay are compared with the same delays experienced when all MT radios share a single channel using EDCA. Figure 8.21 shows the distribution of channel access delay and queuing delay for 1 and 10 frames/TXOP, respectively. Table 8.9 summarizes the average delays for the two TXOP sizes considered.

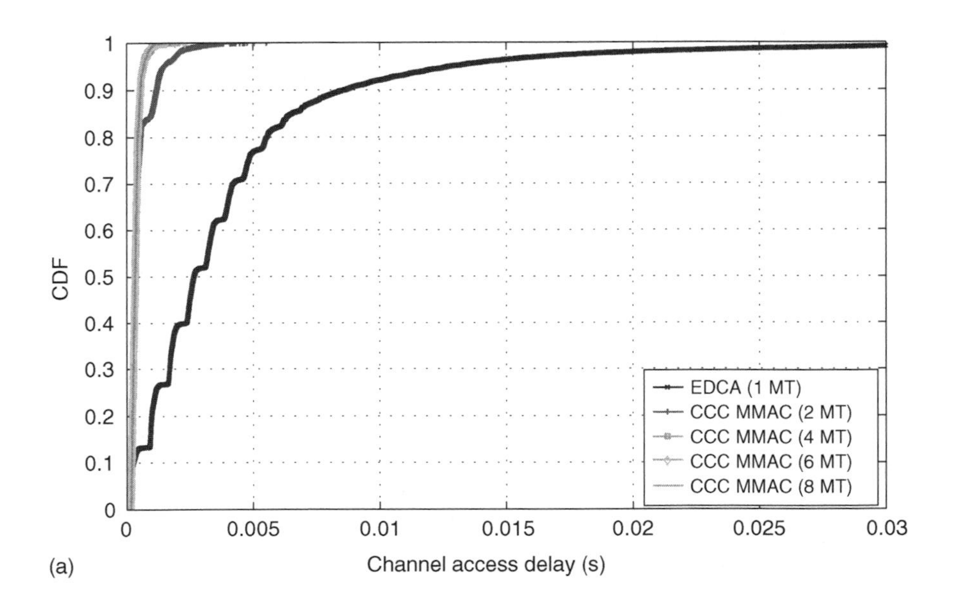

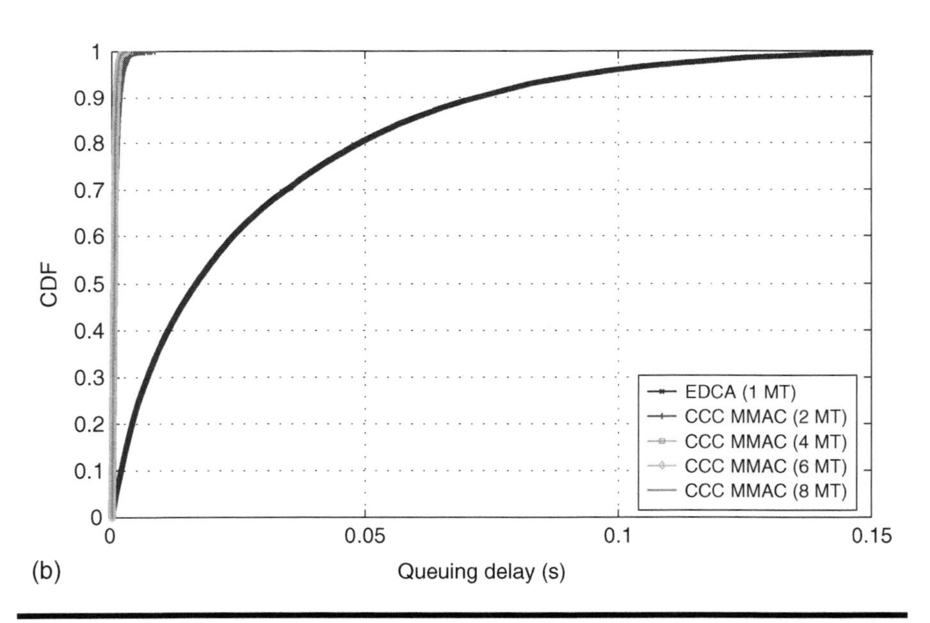

Figure 8.21 Delay performance for CCC MAC with a 24 Mbps data rate and eight TSs. (a) Channel access delay CDF for 1 frame/TXOP, (b) queuing delay CDF for 1 frame/TXOP, (c) channel access delay CDF for 10 frames/TXOP, (d) queuing delay CDF for 10 frames/TXOP.

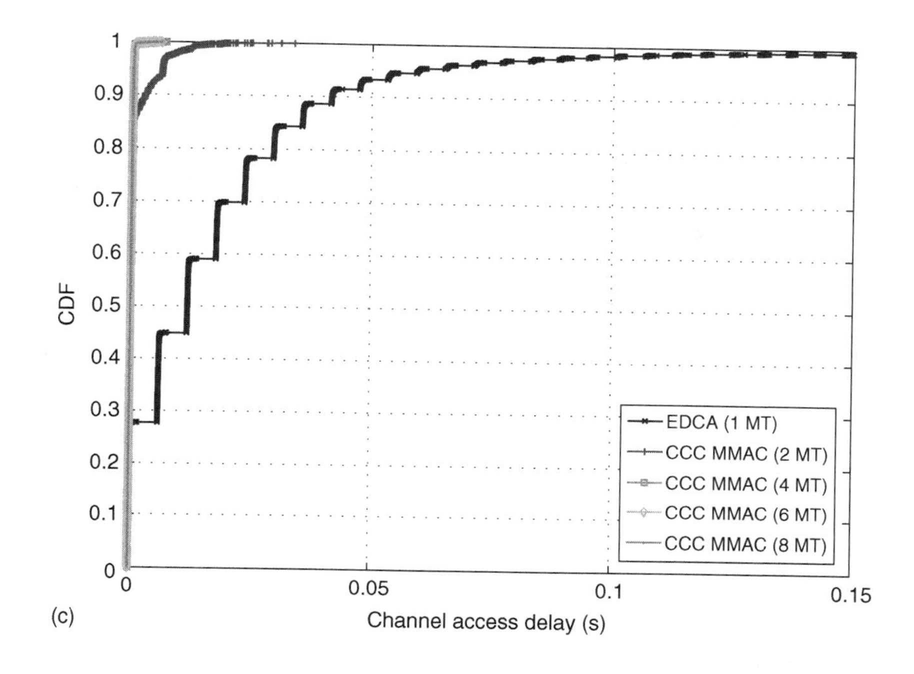

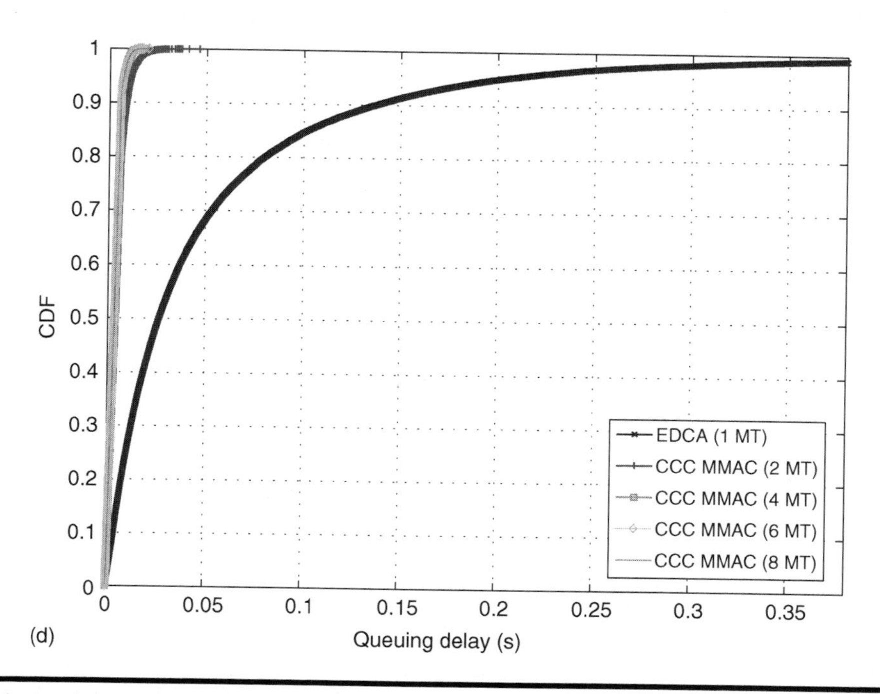

Figure 8.21 ***(Continued)***

Table 8.9 Summary of Per Hop Average Delay Statistics

Key Simulation Settings	*Scenario*	*Channel Access Delay (ms)*	*Queuing Delay (ms)*	*Total Delay (ms)*
1 frame/TXOP, 15 Mbps load, 8 TSs	EDCA	4.2	27.7	31.9
	CCC with 2 MT	0.5	0.7	1.2
	CCC with 4 MT	0.4	0.5	0.9
	CCC with 6 MT	0.4	0.5	0.9
	CCC with 8 MT	0.4	0.5	0.9
10 frames/TXOP, 19 Mbps load, 8 TSs	EDCA	18.2	53.4	71.6
	CCC with 2 MT	1.1	4.4	5.5
	CCC with 4 MT	0.3	3.5	3.8
	CCC with 6 MT	0.3	3.4	3.7
	CCC with 8 MT	0.3	3.4	3.7

The delays observed when EDCA is used as the MAC protocol are shorter when TXOPs consist of a single frame when compared to multiple frames, whereas CCC consistently results in significantly lower delays. When using mixed priorities, EDCA assures that high-priority portion of the traffic load delay performance is acceptable for QoS-sensitive applications if the TXOP consists of a single frame. When the entire traffic load has QoS requirements, however, the 31.9 ms average delay experienced per hop by a 15 Mbps traffic load with EDCA would be too long. Using CCC as the MAC protocol with just "two" MT channels would lead to acceptable delays, for example, 1.2 ms average delay per hop, for such a traffic load. For traffic loads in excess of 19 Mbps, EDCA is not even a feasible option.

As the TXOP size increases, simulations show that access and queuing delays experienced with EDCA grow rapidly, whereas CCC is not impacted negatively. For instance, using EDCA as the MAC protocol with a 19 Mbps traffic load would lead to an average delay of 71.6 ms/hop—18.2 ms of access delay and 53.4 ms of queuing delay, when using 10-frame TXOPs, as seen in Table 8.9. Both channel access and queuing delays, thus, improve significantly when CCC is employed as the MAC protocol. With four traffic channels, use of CCC reduces the observed per hop average delay to 3.8 ms—0.3 ms of access delay and 3.5 ms of queuing delay.

The improvement in delay jitter is even more dramatic. For 10 frames/TXOP case, as shown in Figures 8.21c and 8.21d, the 90 percentile of the queuing delay is 135.6 ms and of the access delay is 42.5 ms, when using EDCA. CCC with four MT channels reduces these statistics to 5.8 and 0.4 ms, respectively.

The preceding results show that EDCA cannot provide delay performance adequate for QoS applications in mesh networks that carry heavy

QoS-sensitive traffic. Backbone mesh networks can be described as such because they carry traffic aggregated by multiple APs. In contrast, CCC generates short delays per hop, adequate for all QoS applications. Because delay and jitter play a central role in determining the QoS for applications such as VoIP, a superior delay performance is essential for any mesh medium access coordination protocol to be relevant to mesh backbone networks.

8.6 Challenges and Future Perspectives

As new applications are constantly emerging, the continual evolution of the 802.11 standard will never come to completion. Indeed, with the amendment for "current" next-generation wireless LANs (i.e., IEEE 802.11n and 802.11s) still to be ratified, discussions on the next-generation wireless LAN [53,54] have recently been initiated within the IEEE 802.11 working group, which intends to further boost the PHY-layer transmission rate to well above 1 Gbps, enabling the system to carry uncompressed HDTV signals.

In addition, some existing IEEE 802 groups have also looked into the QoS design issue when nomadity or mobility has to be supported. More specifically, 802.11r [55] intends to help ensure QoS whereas mobile users roam from one AP to another. Meanwhile, IEEE 802.21 [56] attempts to establish a medium-independent handover framework, which strives to maintain the same QoS level even when the users move between heterogenous wireless networks (e.g., 802.11, 802.16 WiMAX, or 3GPP).

Moreover, given the fact that the 2.4 GHz band is a license-free spectrum, wireless LANs operating on this frequency have to coexist with IEEE 802.15.1 Bluetooth and even suffer interference from microwave ovens. As wireless access continuously gains momentum, the 2.4 GHz spectrum is anticipated to be saturated in the near future. Thus, the importance of proper coexistence and interference management cannot be underestimated. Without these features, basic 802.11 data transmission will be compromised, let alone the delivery of QoS-sensitive multimedia applications.

References

1. Part 11: Wireless LAN Medium Access Control (MAC) and Physical Layer (PHY) Specifications. Amendment 8: Medium Access Control (MAC) Quality of Service Enhancements, *IEEE Std 802.11e-2005*, November 2005.
2. Part 11: Wireless LAN Medium Access Control (MAC) and Physical Layer (PHY) Specifications. Draft Amendment for Enhancements for Higher Throughput, *IEEE P802.11n/D2.00*, February 2007.

3. Garg, S. and M. Kappes, Can I add a VoIP call? *Proceedings of IEEE International Conference on Communications (ICC'03)* (Anchorage, Alaska), May 2003.
4. Part 11: Wireless LAN Medium Access Control (MAC) and Physical Layer (PHY) Specifications, *ANSI/IEEE Std 802.11, 1999 Edition*, 1999.
5. Part 11: Wireless LAN Medium Access Control (MAC) and Physical Layer (PHY) Specifications: Higher-Speed Physical Layer Extension in the 2.4 GHz Band, *IEEE Std 802.11b-1999*, 1999.
6. Part 11: Wireless LAN Medium Access Control (MAC) and Physical Layer (PHY) Specifications: High-Speed Physical Layer in the 5 GHz Band, *IEEE Std 802.11a-1999*, 1999.
7. Part 11: Wireless LAN Medium Access Control (MAC) and Physical Layer (PHY) Specifications, Amendment 4: Further Higher Data Rate Extension in the 2.4 GHz Band, *IEEE Std 802.11g-2003*, 2003.
8. Stallings, W., *Data and Computer Communications, 7th Edition*. Prentice-Hall, New York, 2004.
9. Karn, P., MACA—A New Channel Access Method for Packet Radios, in *Proceedings of Amateur Radio 9th Computer Networking Conference*, London, Ontario, September 1990.
10. Ziouva, E. and T. Antonakopoulos, Efficient Voice Communications over IEEE 802.11 WLANs Using Improved PCF Procedures, *Proceedings of The Third International Network Conference (INC 2002)*, University of Plymouth, UK, July 2002.
11. Ni, Q., Performance analysis and enhancements for IEEE 802.11e wireless networks, *IEEE Network*, 19(4), 21–27, 2005.
12. Mangold, S., S. Choi, P. May, O. Klein, G. R. Hiertz, and L. Stibor, IEEE 802.11e Wireless LAN for Quality of Service, *Proceedings of European Wireless (EW2002) (invited paper)* (Florence, Italy), February 2002.
13. Mangold, S., G. Hiertz, and B. Walke, IEEE 802.11e Wireless LAN—Resource Sharing with Contention Based Medium Access, *Proceedings of IEEE PIMRC* (Beijing, China), September 2003.
14. Korakis, T. and L. Tassiulas, Providing quality of service guarantees in wireless lANs compliant with 802.11e, *Computer Networks*, 47(2), 239–255, 2005.
15. Fallah, Y. and H. Alnuweiri, A Controlled-Access Scheduling Mechanism for QoS Provisioning in IEEE 802.11e Wireless LANs, *Proceedings of the 1st ACM International Workshop on Quality of Service and Security in Wireless and Mobile Networks* (Montreal, Canada), October 2005.
16. Inan, I., F. Keceli, and E. Ayanoglu, An Adaptive Multimedia QoS Scheduler for 802.11e Wireless LANs, *Proceedings of IEEE International Conference on Communications* (Istanbul, Turkey), June 2006.
17. Ansel, P., Q. Ni, and T. Turletti, FHCF: An efficient scheduling scheme for IEEE 802.11e, *ACM/Kluwer Journal on Mobile Networks and Applications (MONET), Special Issue on Modeling and Optimization in Wireless and Mobile Networks*, 11(3), 391–403, 2006.
18. Part 3: Media Access Control (MAC) Bridges, *IEEE Std 802.1d-1998*, 1998.
19. Gu, D. and J. Zhang, A New Measurement-based Admission Control Method for IEEE 802.11 Wireless Local Area Networks, *Technical Report, Mitsubishi Electric Research Lab TR-2003-122*, www.merl.com/reports/docs/TR2003-122.pdf, October 2003.

20. Pong, D. and T. Moors, Call Admission Control for IEEE 802.11 Contention Access Mechanism, *Proceedings of IEEE GLOBECOM* (San Francisco, CA), December 2003.
21. Bellalta, B., M. Meo, and M. Oliver, Call Admission Control in IEEE 802.11e EDCA-based WLANs, *Technical Report, The 5th COST 290 Management Committee Meeting*, February 2006.
22. Wi-Fi Alliance, http://www.wi-fi.org/index.php.
23. Lindgren, A., A. Almquist, and O. Scheln, Quality of service schemes for IEEE 802.11 wireless LANs—an evaluation, *Special Issue of the Journal of Special Topics in Mobile Networking and Applications (MONET) on Performance Evaluation of QoS Architectures in Mobile Networks*, 8, 223–235, 2003.
24. Choi, S., J. Pavon, S. Shankar, and S. Mangold, IEEE 802.11e Contention-Based Channel Access (EDCF) Performance Evaluation, *Proceedings of IEEE International Conferences on Communications*, Anchorage, AK, May 2003.
25. Xu, K., Q. Wang, and H. Hassanein, Performance Analysis of Differentiated QoS Supported by IEEE 802.11e Enhanced Distributed Coordination Function (EDCF) in WLAN, *Proceedings of the IEEE Global Telecommunications Conference (GLOBECOM'03)* (San Francisco, CA), December 2003.
26. Robinson, J. W. and T. S. Randhawa, Saturation Throughput Analysis of IEEE 802.11e Enhanced Distributed Coordination Function, *IEEE Journals on Selected Areas in Communications*, 22, 917–928, 2004.
27. Kong, Z., D. Tsang, and B. Bensaou, Performance analysis of IEEE 802.11e contention-based channel access, *IEEE Journal on Selected Areas in Communications, Special Issue on Design, Implementation and Analysis of Communication Protocols*, 22(10), 2095–2106, 2004.
28. Tao, Z. and S. Panwar, An Analytical Model for the IEEE 802.11e Enhanced Distributed Coordination Function, *Proceedings of IEEE International Conference on Communications* (Paris, France), June 2004.
29. Tao, Z. and S. Panwar, Throughput and delay analysis for the IEEE 802.11e enhanced distributed channel access, *IEEE Transactions on Communications*, 54(4), 596–603, 2006.
30. Chou, C., K. Shin, and S. Shankar, Inter-Frame Space (IFS) Based Service Differentiation for IEEE 802.11 Wireless LANs, *Proceedings of IEEE Semiannual Vehicular Technology Conference—Fall (VTC'03)*, October 2003.
31. Abraham, S., A. Meylan, and S. Nanda, 802.11n MAC Design and System Performance, *Proceedings of IEEE International Conference on Communications* (Seoul, Korea), May 2005.
32. Gast, M., *802.11 Wireless Networks: The Definitive Guide, 2nd Edition*. O'Reilly, 2005.
33. Xiao, Y. and J. Rosdahl, Throughput and delay limits of IEEE 802.11, *IEEE Communications Letters*, 6, 355–357, 2002.
34. Lee, I., S. Yoon, and S. Park, Throughput Analysis of IEEE 802.11e Wireless LANs and Efficient Block Ack Mechanism, *Proceedings of IEEE International Symposium on Communications and Information Technology (ISCIT'04)* (Sapporo, Japan), October 2004.
35. Liu, C. and A. P. Stephens, An Analytic Model for Infrastructure WLAN Capacity with Bidirectional Frame Aggregation, *Proceedings of IEEE Wireless*

Communications and Networking Conference (WCNC'05) (New Orleans, LA), March 2005.
36. Akhmetov, D., 802.11N: Performance Results of Reverse Direction Data Flow, *Proceedings of IEEE 17th International Symposium on Personal, Indoor and Mobile Radio Communications (PIMRC'06)* (Helsinki, Finland), September 2006.
37. Akyildiz, I. F. and X. Wang, A Survey on Wireless Mesh Networks, *IEEE Communications Magazine*, 43(9), S23–S30, September 2005.
38. Lee, M. J., J. Zheng, Y. Ko, and D. M. Shrestha, Emerging standards for wireless mesh technology, *IEEE Wireless Communications*, 13(2), 56–63, 2006.
39. Kyasanur, P., J. Padhye, and P. Bahl, A Study of an 802.11-Like Control Channel-Based MAC, *Proceedings of IEEE Broadnets* (Boston, MA), October 2005.
40. Wu, H., F. Yang, K. Tan, J. Chen, Q. Zhang, and Z. Zhang, Distributed channel assignment and routing in multiradio multichannel multihop wireless networks, *IEEE Journal on Selected Areas in Communications*, 24(11), 1972–1983, 2006.
41. Alicherry, M., R. Bhatia, and L. Li, Joint channel assignment and routing for throughput optimization in multiradio wireless mesh networks, *IEEE Journal on Selected Areas in Communications*, 24(11), 1960–1971, 2006.
42. D. E. 3rd, IEEE P802.11s Call for Proposals, *IEEE 802.11s document, DCN 802.11-04/1430r12*, http://www.802wirelessworld.com/index.jsp, January 2005.
43. Xu, S. and T. Saadawi, Revealing the problems with 802.11 Medium Access Control Protocol in multi-hop wireless *ad hoc* networks, *Computer Networks: The International Journal of Computer and Telecommunications Networking*, 38, 531–548, 2002.
44. Zhao, R., B. Walke, and G. R. Hiertz, An efficient IEEE 802.11 ESS mesh network supporting quality-of-service, *IEEE Journal on Selected Areas in Communications*, 24(11), 2005–2017, 2006.
45. Hiertz, G., S. Max, E. Weiss, L. Berlemann, D. Denteneer, and S. Mangold, Mesh Technology Enabling Ubiquitous Wireless Networks, *Proceedings of the 2nd Annual International Wireless Internet Conference* (Boston, MA), August, 2006.
46. Part 11: Wireless LAN Medium Access Control (MAC) and Physical Layer (PHY) Specifications. Draft Amendment for ESS Mesh Networking, *IEEE P802.11s/D1.02*, March 2007.
47. Benveniste, M., CCC MAC Protocol Framework and Optional Features, *IEEE 802.11s document, DCN 802.11-05/0880r0*, http://www.802wirelessworld.com/index.jsp, September 2005.
48. Taori, R. et al., 802.11 TGs MAC Enhancement Proposal, *IEEE 802.11s Document, DCN 802.11-05/0608r1*, http://www.802wirelessworld.com/index.jsp, July 2005.
49. Kyasanur, P., J. Padhye, and P. Bahl, A New Multi-Channel MAC Protocol with On-Demand Channel Assignment for Multi-Hop Mobile *Ad Hoc* Networks, *Proceedings of IEEE International Symposium on Parallel Architectures, Algorithms and Networks (ISPAN)* (Dallas, TX), December 2000.
50. Hass, Z. J. and J. Deng, Dual busy tone multiple access (DBTMA)—a multiple access control for *ad hoc* networks, *IEEE Transactions on Communications*, 50(6), 975–985, 2002.
51. Aboul-Magd., O et al., Joint SEE-Mesh/Wi-Mesh Proposal to 802.11 TGs, *IEEE 802.11s Document, DCN 802.11-06/0328r0*, http://www.802wirelessworld.com/index.jsp, February 2006.

52. Benveniste, M. and Z. Tao, Performance Evaluation of an MMAC Protocol for IEEE 802.11s Mesh Network, *Proceedings of IEEE Sarnoff Symposium* (Princeton, NJ), May 2006.
53. Venkatesan et al., Audio/Video Streaming over 802.11, *IEEE 802.11wng Document, DCN 802.11-07/0400r1*, http://www.802wirelessworld.com/index.jsp, March 2007.
54. Engwer, D., 802.11 Looking Ahead to the Future, *IEEE 802.11wng Document, DCN 802.11-07/0412r1*, http://www.802wirelessworld.com/index.jsp, March 2007.
55. Bangolae, S., C. Bell, and E. Qi, Performance Study of Fast BSS Transition Using IEEE 802.11r, *Proceeding of the International Conference on Communications and Mobile Computing* (Vancouver, Canada), July 2006.
56. IEEE 802.21 official website, http://www.ieee802.org/21/.

Chapter 9

Peer-Assisted Video Streaming over WLANs

Danjue Li, Chen-Nee Chuah, Gene Cheung, and S.J. Ben Yoo

Contents

9.1 Introduction 306
9.2 Why Peer-Assisted Streaming Schemes 308
    9.2.1 Challenges in Streaming Video Content over IEEE 802.11 WLANs 308
    9.2.2 Prior Protection-Based Approaches 308
    9.2.3 Enhancing Streaming Quality by Exploiting Peer Resources 309
9.3 Peer-Assisted Video Streaming 311
    9.3.1 Content Discovery 312
    9.3.2 Peer Mobility Tracking 314
    9.3.3 Peer Selection 315
    9.3.4 Data Selection: Rate-Distortion Optimization 317
9.4 Enhancing Streaming Quality Using Caching 319
    9.4.1 Resource-Based Caching Algorithm 319
    9.4.2 Popularity-Based Caching Algorithm 320
    9.4.3 Prefix Caching Algorithm 320
    9.4.4 Selective Partial Caching Algorithm 321
    9.4.5 Distortion Minimizing Smart Caching Algorithm 321

9.5 Energy Efficiency and Awareness....................................322
9.5.1 Energy-Saving Strategies 323
9.5.2 Energy-Aware Peer Selection....................................325
9.6 Open Issues 327
9.6.1 Dynamic Wireless Channel Quality 327
9.6.2 Traffic with Different Quality-of-Service Requirements 327
9.6.3 Incentive Mechanism for Node Cooperation....................................327
9.6.4 Security Issues 328
9.7 Summary 328
References 329

In recent years, there has been an increasing demand to deliver high-quality, high-bandwidth video streaming services over IEEE 802.11 wireless local area networks (WLANs). However, streaming over wireless networks presents a number of unique challenges compared to delivery over wired networks. In particular, the end-to-end perceived video quality can fluctuate vastly due to the higher-channel bit error rate (BER), fading, interference, and end-host mobility, among other reasons. This chapter examines the state-of-the-art in supporting high-quality video streaming services over IEEE 802.11 wireless networks and, particularly, focuses on introducing the latest advances in video streaming designs that use approaches assisted by peer-to-peer (P2P) concepts. The problem is first approached by evaluating other existing methods of streaming video content over IEEE 802.11 WLAN. Next, the design details for peer-assisted streaming schemes are explored, which entail content discovery, mobility tracking, peer selection, node caching, and energy efficiency. The chapter concludes by identifying some open research issues in the implementation of the peer-assisted streaming schemes.

9.1 Introduction

Advances in wireless communication technologies and mobile devices have made ubiquitous Internet access a reality in an increasing number of areas. IEEE 802.11-based WLANs [1] have emerged as an attractive solution for providing network connectivity in businesses, educational institutions, and public areas. Although current uses of WLANs are mainly for generic data transmission, the desire for supporting delay-sensitive applications such as Voice-over-Internet Protocol (VoIP) and video streaming is a natural progression as these applications gain in popularity on the Internet as a whole.

In contrast to general datafile transmission, video communication is generally more tolerant to packet losses but has a more stringent delay requirement. Any packet that arrives later than its delivery deadline will be useless for decoding and be equivalent to a packet loss. Interdependency among video data can cause error propagation and further degrade the video quality. Furthermore, compared to wired links, wireless links are more error-prone and unpredictable. End-host mobility, time-varying channel fading, interference, high BER, etc., can all cause the end-to-end perceived video quality to fluctuate greatly. Thus, the use of WLANs to deliver high-quality video streaming services requires a robust streaming scheme that can cope with the aforementioned challenges.

Earlier literature contains many such designs. Pure protection-based streaming schemes [2–5] focus on using forward error correction (FEC) [6] or automatic repeat request (ARQ) [7] to help streaming clients recover from quality degradation caused by packet losses. Combined with cross-layer optimization mechanisms, these schemes can effectively adapt to wireless channel quality variation. Taking a different approach, peer-assisted video streaming schemes were conceived by observing that (i) part of the multimedia content desired by a streaming client already resides on peers that are located in the same WLAN, and (ii) connections to these peers typically have shorter delay and better performance than connection to a remote video server. They focus on combining the advantages of two communication modes offered by IEEE 802.11 networks (i.e., infrastructure mode and *ad hoc* mode) and leveraging source diversity (i.e., media server plus nearby peers) to enhance streaming performance. Compared to the protection-based streaming schemes, peer-assisted video streaming schemes can better utilize both wired and wireless bandwidth. By distributing the streaming load across the media server and neighboring peers, they can effectively prevent potential server congestion problems caused by the appearance of a flash crowd as well. We also note that in everyday operations, there is an increasing tendency that streaming clients store information temporarily in their disks for later replay, which makes the peer-assisted approach practical and promising. This chapter mainly focuses on introducing peer-assisted streaming schemes. Note that in real implementations, peer-assisted approaches can be jointly used with certain protection schemes to further improve streaming performance.

The rest of the chapter is organized as follows. Section 9.2 reviews some fundamentals of protection-based schemes for streaming video content through IEEE 802.11 WLAN to motivate the design of peer-assisted video streaming schemes. Section 9.3 introduces peer-assisted streaming schemes and discusses some practical design issues, which include content discovery, mobility tracking, peer selection, and data selection. Section 9.4 explores how caching and local retransmission can be used in the peer-assisted streaming scheme design to further improve streaming performance. Because most wireless streaming clients are battery-powered, energy-related issues, such as improving energy efficiency and performing energy-aware peer selection (EPS), are investigated in Section 9.5. Section 9.6 concludes by discussing some open issues pertaining to peer-assisted approaches.

9.2 Why Peer-Assisted Streaming Schemes

9.2.1 Challenges in Streaming Video Content over IEEE 802.11 WLANs

The various IEEE 802.11 protocols provide a best-effort service to transport datagrams from the source to the destination without making any guarantees on the end-to-end performance for an individual packet. Neither do they make any promises about the variation of the packet delay, bandwidth, and losses within a packet stream. Such best-effort service is effective enough to support file-transfer applications on top of a reliable transport protocol such as Transmission Control Protocol (TCP) due to their tolerances to large delays and jitter. However, its lack of any quality-of-service (QoS) guarantees makes it extremely challenging to support delay-sensitive streaming applications.

The wireless nature of the underlying network creates additional difficulties. Portable devices in the network often have limited bandwidth, storage space, and are battery-powered. When they are served, another set of problems arise. These problems include how to scale streaming rates (SRs) to fit the bandwidth limitations of each node, cope with node mobility, and provide energy-efficient service to allow every node to participate for as long as possible.

9.2.2 Prior Protection-Based Approaches

Protection-based approaches use FEC [6] or ARQ [7] to recover from quality degradation caused by packet losses. Most of the early protection-based multimedia streaming schemes concentrated on only optimizing the application layer in the protocol stack. Recent results indicate that local optimization of individual layers may not lead to global optimization. By looking at optimization across multiple layers, cross-layer optimization mechanisms can be combined with protection schemes such as FEC and ARQ to help the system adapt better to variations in the underlying network.

As an error control mechanism, ARQ makes use of acknowledgments (i.e., messages sent by the receiver to indicate successful receptions of packets) and time-outs (i.e., a reasonable amount of time after sending the data frame) to achieve reliable data transmission. When the transmitter does not receive the acknowledgment before the time-out occurs, it retransmits the frame until it is either correctly received by the receiver or the error persists beyond a predetermined number of retransmissions.

Because the transmitter only retransmits unsuccessful packets, ARQ is inherently channel-adaptive. On good channels, ARQ is an attractive approach to perform error control, whereas on poor channels it may introduce significant delays due to the round-trip delay of the retransmission request and its response. These delays significantly limit the applicability of ARQ to streaming applications.

FEC can improve the error resilience of a streaming scheme without incurring long delays, which represents an obvious advantage over the ARQ approach for real-time applications. The basic idea of FEC is to send redundant information along with the raw packet stream. At the cost of increasing the transmission rate of the packet stream, redundant information can be used to reconstruct lost packets at the receiver. For streaming applications, FEC can be performed at two levels: the bit level and the packet level. These two FEC schemes can also be applied jointly to the encoded video bitstream to achieve even higher error resilience [8].

Implementing FEC requires a certain portion of bandwidth always to be allocated ahead of time to send redundant information. When a client has limited bandwidth, one has to carefully evaluate the trade-off between error resilience and the streaming data rate. In a network setting, FEC techniques can be used by themselves, or in conjunction with the aforementioned ARQ techniques. A hybrid FEC/ARQ scheme [9] that combines the advantages of both can offer a better performance, especially for the time-varying wireless channels.

Because network bandwidth availability is constantly affected by traffic flows, adjusting FEC parameters and ARQ intervals on the fly, based on network feedback, can help to avoid overloading the network and improve bandwidth utilization as well. In contrast to the traditional approach, where each layer in the protocol stack is independently designed, cross-layer design aims at performing joint optimization across the layers, allowing more efficient resource utilization. It can encompass the entire protocol stack or only consider joint optimization of some subset of layers, such as the physical and Media Access Control (MAC) layers and the application and network layers. In the example illustrated in Figure 9.1, the three layers—the application layer, physical layer, and MAC layer—are jointly considered in the optimization.

9.2.3 Enhancing Streaming Quality by Exploiting Peer Resources

Although protection-based approaches have demonstrated their efficacy in improving streaming quality, system performance still deteriorates rapidly as the number of clients increases due to the current "server–client" service model. Because this simple server–client model overlooks the possibility

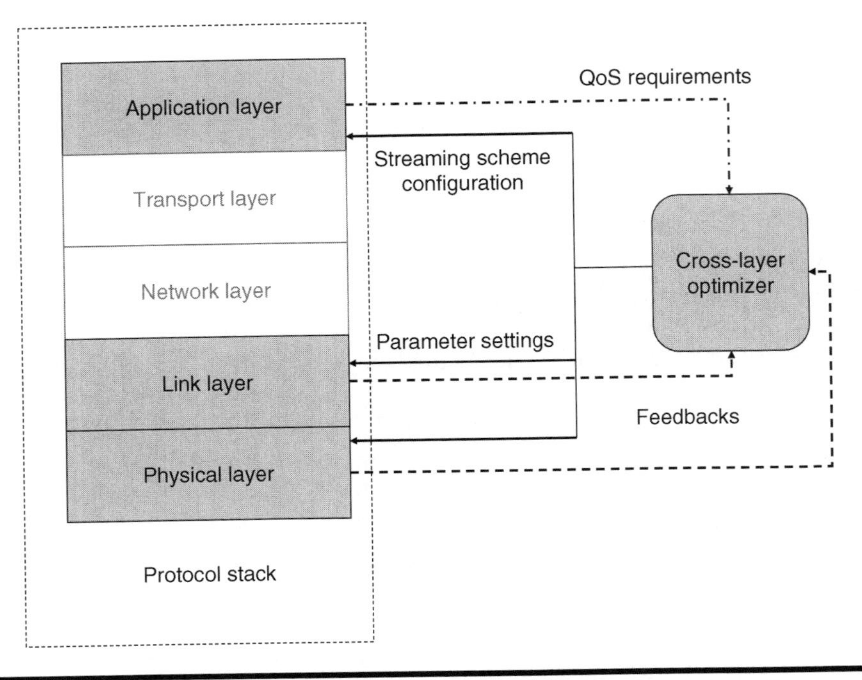

Figure 9.1 An example of cross-layer optimization architecture for video streaming over WLAN.

that the desired video content may already exist among peers located in the same WLAN, each client always uses the media server as its sole source. When a flash crowd appears, the server can be easily overloaded. In addition, in most existing WLAN streaming scenarios (as shown in Figure 9.2), the media server is always remotely accessed via the Internet. The quality of the connection traversing the Internet can frequently degrade due to network congestion and link failure [10]. Compared to the connection to the remote media server, connections to peers usually have shorter delay and potentially higher bandwidth. Under the assumption that streaming clients are cooperative, "peer-assisted approaches" explore the possibility that the desired video content may exist among peers and make those nearby resources accessible to video subscribers. In the case when the server connection bandwidth is smaller than the local area network (LAN) bandwidth, such peer-assisted streaming schemes can obtain extra bandwidth from peers and the aggregated SR will increase. Even in the case when server connection bandwidth is large enough, peer-assisted streaming schemes can still improve transmission quality by leveraging source/path diversity and using peer connections to shorten the transmission delay.

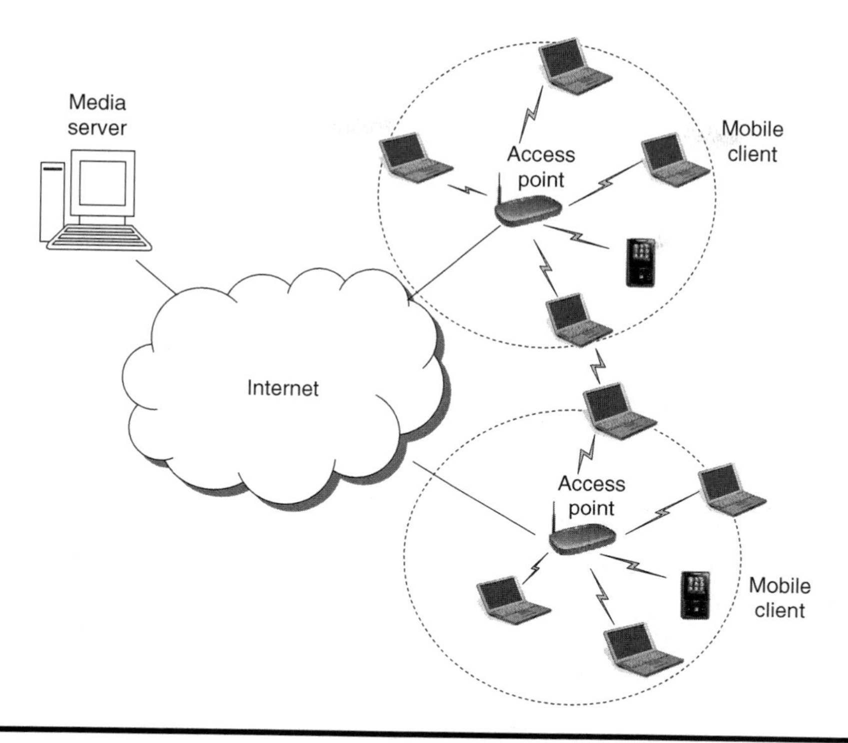

Figure 9.2 Video streaming over IEEE 802.11 wireless networks.

9.3 Peer-Assisted Video Streaming

In Section 9.2, it was seen that peer-assisted video streaming provides communication among peers to allow a streaming client to simultaneously stream from both a media server and its peers. By exploiting shorter peer connections and leveraging source/path diversity, it can achieve better streaming performance. In support of such services, the streaming system should be able to provide the following mechanisms:

- *Content discovery.* In scenarios where peer-assisted approaches are used, available content is normally scattered in the WLAN among different peers. To exploit these peer resources, the system needs to first find out which peers are potential senders. This process is called content discovery.
- *Peer mobility tracking.* Because the reliability of wireless connections to these peers depends heavily on the node mobility and physical channel characteristics, peer mobility tracking elaborates on how a streaming system can track peer mobility and quantify its impact.

- *Peer selection.* On gathering peer information, the system should then decide how to select peers from which it will request information. Peer selection refers to the process of determining whether a peer should be selected as a content provider, and if selected, how frequently the client should request information from it.
- *Data selection.* Besides the preceding three tasks, the system should also be able to schedule data requests/(re)transmissions efficiently to optimize the streaming performance. By performing data selection, the system decides an optimal scheduling scheme for data transmissions, that is, which data packet to send and how to send it.

We will examine these four issues in detail in the following sections.

9.3.1 Content Discovery

In general, the remote media server stores a complete copy of the desired video file, whereas peers may only have a part of it. This assumption about partial content availability at peers is reasonable because peers may delete a part of the content after viewing due to storage limitations or may have only downloaded the beginning portions of the video due to a lack of interest or early exit in their earlier streaming sessions. Performing content discovery is necessary for locating the desired video content among peers.

There are different approaches to perform content discovery, depending on which entity in the system is responsible for sending/receiving control messages. If we assume a completely distributed system, where each streaming client identifies its potential content providers and schedules its streaming process independent of others, then the content discovery will be initiated by the requesting streaming client. By sending requests to peers in the neighborhood either through the access point (AP) or through direct *ad hoc* connections, the client expects replies from willing peers to indicate their content availability and their maximum allowable SR. It then forms a table to record this information and track updates. In this client-centric approach, a distributed hash table (DHT) [11] is often used by the client to help accomplish content discovery. Taking a different approach, a server-centric relies on the server to locate peers in the same coverage area that can become potential senders, distribute the streaming load among itself and the willing peers, and schedule the streaming process. Another efficient alternative is to use a streaming proxy. By collocating a streaming proxy with the AP to coordinate the peer-assisted streaming process, one can avoid synchronization problems among multiple senders and can more easily adapt to instantaneous changes in the network, which represents an advantage over the server-centric approach. Meanwhile, with the coordination of the proxy, the system can also relieve quality degradation caused by

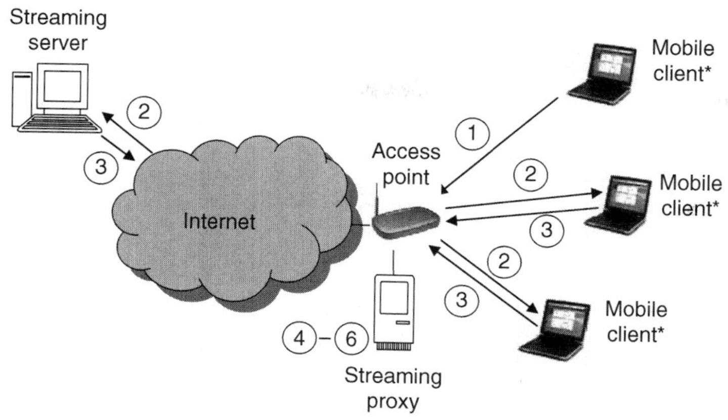

* Equipped with IEEE 802.11 interface cards

(a)

- Step 1—Client sends a SREQ (VFN, UID, SR) message to the proxy.
- Step 2—Proxy sets a time-out t_{SRTO} and broadcasts an MREQ (VFN, UID, SR) message to the media server and all the mobile users in the WLAN.
- Step 3—Server replies a SACK (RDP, AB) message once the request is authorized. Willing peers send a PACK (CL, UID, AB) message back to the proxy.
- Step 4—Proxy builds up a content table to save all the replies. When the time-out t_{SRTO} expires, proxy builds up a joint sender group based on the replies and mobility-tracking results.
- Step 5—Proxy requests data from the sender group on behalf of the client and forwards them to the client on receiving the data.
- Step 6—Peer sends out the PUPD (CL, UID, AB) message to the proxy if the available video content changes over time. Proxy updates content table once it gets a PUPD (CL, UID, AB) message or detects the peer departure.

(b)

Figure 9.3 Peer-assisted video streaming with the coordination of a proxy.

contention that is often seen in a client-centric approach. For these reasons, we will use the proxy-centric approach to discuss how to address the four design issues in constructing a peer-assisted streaming scheme.

In a proxy-assisted setting, the content discovery process proceeds as the following (as illustrated in Figure 9.3). Once the proxy gets the "streaming request" (SREQ) message from a client, it will initiate a new streaming session for this client and randomly generate an unused integer number as the streaming session ID (SSID). Then the proxy will start a timer t_{SRTO} and broadcast a "media request" (MREQ) message to the remote media server and all the peers in its transmission range, including the desired video file name (VFN), client's user ID (UID), and the preferred SR. The request will be rebroadcasted if no reply arrives at the proxy before t_{SRTO} expires. If the proxy gets no reply before t_{SRTO} expires, it will send a "system busy"

Table 9.1 Content Table

SSId	*UID*	*Content Portions*	*Allocated Bandwidth*
	Server	[0 s, 60 s)	Variable
12	8088	[0 s, 5 s)	100 kbps
	8092	[50 s, 60 s)	200 kbps

message to the client. The client will wait for a random time before it sends another SREQ to the proxy. When peers receive the request from the proxy, they will search their local archives to see if they have the desired video content. Any peer that has kept full or partial copies of the desired content and is willing to share will reply to the proxy with its UID, a content list (CL) showing which portion of the requested video content it has, and the allocated bandwidth (AB) for this session. Meanwhile, the media server will also reply with its content information and AB after authorizing the request. After t_{SRTO} expires, the proxy will construct a content table as shown in Table 9.1. To keep the information in Table 9.1 up-to-date, the proxy will send out a periodical probing beacon to detect the existence of peers. At the same time, whenever the available content for VFN changes in a peer, a "peer update" (PUPD) message will be automatically sent to update the entry in the content table. Once a peer leaves the WLAN, the proxy will delete the corresponding entry from the content table.

9.3.2 Peer Mobility Tracking

The mobility of peers makes membership of the candidate sender group rather dynamic. This can potentially lead to a high variation of the video quality perceived by the client. To smooth the video quality variation, the system intends to enlist the most stable nodes into the joint sender group. There are different ways to quantify the mobility of a given peer. One way to characterize it is to use the session duration and revisit interval metrics [12]. The session duration refers to the amount of time that a user stays associated with an AP (and hence served by our proxy) before moving to another AP or leaving the network, whereas the revisit interval refers to the amount of time before the next visit of the peer. Longer session durations and shorter revisit intervals represent lower mobility. By keeping track of these two metrics, one can infer how "stable" a peer may be as a potential video source.

With a proxy-centric approach, the proxy tracks peer mobility by recording the association/disassociation behaviors for each peer in the WLAN, as illustrated in Table 9.2. When the proxy gets a streaming request from the

Table 9.2 Mobility Tracking Table

UID	Association Time (a.m.)	Disassociation Time (a.m.)
8088	10:10:10	10:23:19
	10:39:10	10:53:19
	10:59:10	11:08:09
8092	10:10:10	10:43:10
	10:45:10	11:23:19

```
1   if The proxy detects that peer j joins the WLAN at time t_a
2           if peer j has not visited WLAN in the past, i.e., T_j^a ≡ Ø
3                   T_j^a ← dummy value
4           else
5                   T_j^a ← {T_j^a, t_a}
6   if The proxy detects that peer j leaves the WLAN at time t_b
7           if peer j has not visited WLAN in the past, i.e., T_j^d ≡ Ø
8                   T_j^d ← dummy value
9           else
10                  T_j^d ← {T_j^d, t_b}
11
12  ℜ_j ← T_j^d − T_j^a, ℑ_j ← T_j^a(k) − T_j^d(k − 1)
13  if The proxy needs to construct a joint sender group
14          for Each peer listed in the content table
15                  m_j ← E{R_j} / (E{V_j}(C_1σ²(R_j) + C_2σ²(V_j)))
```

Figure 9.4 Pseudocode for generating mobility factor for mobile user.

client, it will use the information in Table 9.2 to generate a mobility factor m_j for each peer, as illustrated by Figure 9.4. T_j^a is the set of association times for peer j, T_j^d the set of dissociation times for peer j, $\Re_j$ the set of session duration times R_j for peer j, $\Im_j$ the set of revisit intervals V_j for peer j, and C_1 and C_2 are weighting coefficients. A larger m_j value represents higher stability.

9.3.3 Peer Selection

On completion of "content discovery" and "mobility tracking," the system will construct a sender group consisting of the media server and peers. To construct a stable sender group, the criteria consist of the available content, the maximum SR allowed by the current network status and the node restriction, and the mobility level. Peer with more available content, higher SR, and lower mobility level is given a higher priority to be selected. A comprehensive peer selection algorithm carefully weighs these three criteria to determine whether or not a peer should be selected, incurring long processing time. Because the start-up delay is a concern for streaming applications, one simple but timely way to construct a sender group is to pick peers starting with the most stable (i.e., having the highest mobility factor and the largest content segment from the current streaming point) until the aggregated bandwidth from peers and the server reaches the link capacity between the proxy and the client. To maintain the stability of the sender group and minimize the impact of sudden peer departure, the system keeps the record of a few additional peers. By monitoring the sender group, the system can switch to the backup peer connections once it detects signs that indicate a potential peer departure.

Scheduling the streaming process among the peers/media server in the sender group is the next task to accomplish in "peer selection." The key design issue here is to determine how frequently each sender in the group should send out its video content. Because the SR from a given sender is jointly determined by the underlying network status as well as the resource allocated for streaming by the sender, one can determine the rate for each sender by taking into account the available network bandwidth, channel characteristics, and a prespecified, fixed level of FEC to minimize the probability of packet loss. A sender that can allocate more bandwidth for streaming and has a better connection to the client will be selected to stream data more often.

The concept of "asynchronous clocks" [13] has been proposed for this matter. The idea is to set up a clock j for each sender–receiver pair j.* This clock j wakes up at regular intervals of Δ_j. Δ_j is inversely proportional to the channel quality and the allocated SR, where the channel quality is quantitatively measured using the TCP-friendly rate control (TFRC) [14], that is,

$$\Omega_j = \frac{L}{\mu_j\sqrt{2\alpha_j/3} + t_{\mathrm{RTO}}\left(3\sqrt{3\alpha_j/8}\right)\alpha_j\left(1 + 32\alpha_j^2\right)} \tag{9.1}$$

where L is the packet size, μ_j the round-trip time, α_j the loss event rate perceived by the receiver for the connection to the sender j, and t_{RTO} the TCP retransmission time-out value. According to Floyd et al. [14], it is reasonable and practical to estimate t_{RTO} by using $t_{\mathrm{RTO}} = 4\mu_j$.

Once the clock j wakes up, it signals that a transmission opportunity is immediately granted to request a packet from its associated sender. After the request is initiated, the clock j will be reset to wake up after another Δ_j.

* If a proxy is used to coordinate the streaming proxy, these clocks will be set up at the proxy. In receiver-driven scenarios they will be set up at each receiver instead.

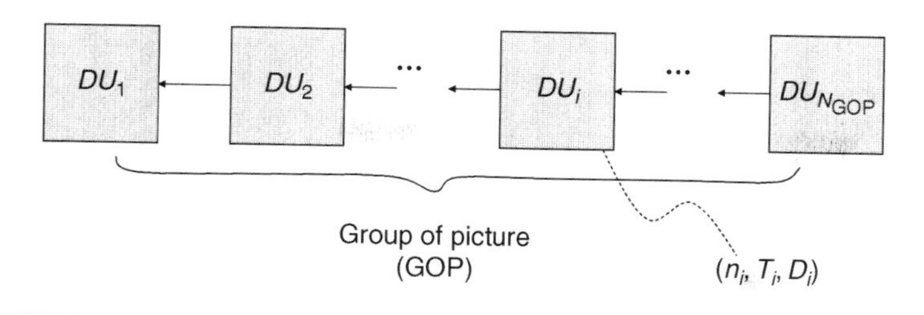

Figure 9.5 Linear DAG.

9.3.4 Data Selection: Rate-Distortion Optimization

A compressed video representation is normally assembled into packets before being transmitted over the network. Because the video file is predictively encoded, packets in the bitstream will be dependent on one another. Such dependency can be modeled using a directed acyclic graph (DAG) [15], where each DU_i represents a frame i and a directed edge represents the dependency relation between them. Parameters associated with each DU_i include the size n_i, the decoding time T_i, and the distortion reduction D_i. The size n_i is the size of DU_i measured in number of Real-Time Transport Protocol (RTP) packets [27]. The decoding time T_i is the time by which DU_i must arrive at the client. D_i is the distortion reduction if DU_i arrives at the client on time and is successfully decoded. Figure 9.5 shows an example of a DAG graph where linear interdependency exists among packets.

The main task of "data selection" is to determine which DU_i should be requested from a selected sender for (re)transmission. Traditional streaming systems normally stream the packetized media in a fixed sequence according to the presentation time. Instead of streaming packets in order, a rate-distortion optimized (RaDiO) streaming scheme [15] allows each packet to be transmitted according to a transmission policy that minimizes the expected end-to-end distortion of the entire presentation subject to a transmission rate constraint and other possible constraints. Specifically, it will address the following scheduling issue: given a set of interdependent video packets with their respective deadlines, a set of scheduling slots (also called transmission opportunities), and a set of streaming rates constrained by network status, which packet should be sent in each slot to yield the lowest distortion perceived by the receiver? The solution to this packet scheduling problem is obtained by minimizing the distortion perceived by the receiver, that is,

$$D(\pi) = D_0 - \sum_{i=1}^{N} D_i \prod_{l \leq i} q_l(\pi_l) \tag{9.2}$$

although taking into account the dependence relationships of packets in addition to their different delivery deadlines and basic importance [15]. In Equation 9.2, $l \leq i$ denotes the set of DU_is that precede or are equal to DU_i in the DAG. D_0 is the overall expected distortion for the group given no packet is received; D_i the distortion reduction if DU_i is successfully decoded; and $q_l(\pi_l)$ the timely arrival probability under the transmission policy π_l. As Chou and Miao [15] have shown in their work, the optimal DU_i for (re)transmission is the one with the largest $\lambda_i = \lambda'_i S_i / n_i$, where λ'_i is the increase in successful-delivery likelihood, given one transmission is sent at the optimization instant and S_i is the data sensitivity. λ'_i and S_i can be defined as the following:*

$$\lambda'_i = q_i(\pi_{i,1}) - q_i(\pi_{i,0}) \tag{9.3}$$

$$S_i = \sum_{k \geq i} D_k \prod_{\substack{l \leq k \\ l \neq i}} q_l(\pi_l) \tag{9.4}$$

where $\pi_{i,1}$ is the transmission policy of DU_i, given one more transmission request is sent to j at time t_o, and $\pi_{i,0}$ the policy of DU_i, given no request is sent out at time t_o. Here transmission policy corresponds to a sequence of transmission schedules, which dictates when and how the packet should be requested or transmitted.

When being jointly implemented with protection schemes (FEC and ARQ), and even cross-layer approaches, RaDiO-based data selection can use network bandwidth more effectively and also lead to better streaming performance.

Figure 9.6 compares the performance of the peer-assisted streaming system with that of a single-source streaming system. Hereafter, we use two 300 frame Class-B† standard video sequences, "Foreman" and "Container" to drive the simulation, which are both encoded at 120 kbps as shown in Figure 9.5 using H.263 version 2 at Quarter Common Intermediate Format (QCIF), with 30 frames per second and a 1/25 I-frame frequency. To get rid of the randomness of simulation results, we repeat the simulations 50 times with different random speeds, and plot the cumulative distribution function (CDF) of the across-run average peak signal-to-noise ratio (PSNR) (computed across multiple runs for each frame) shown in Figure 9.6. Figure 9.6 shows that peer-assisted streaming schemes achieve better performance

* These two equations represent only the simplified version of the solution to the RaDiO framework presented in Ref. 15. The complete version, involving a complete Markov process, is much more complicated.

† Class-B sequences normally have medium spatial detail and low amount of movement or vice versa.

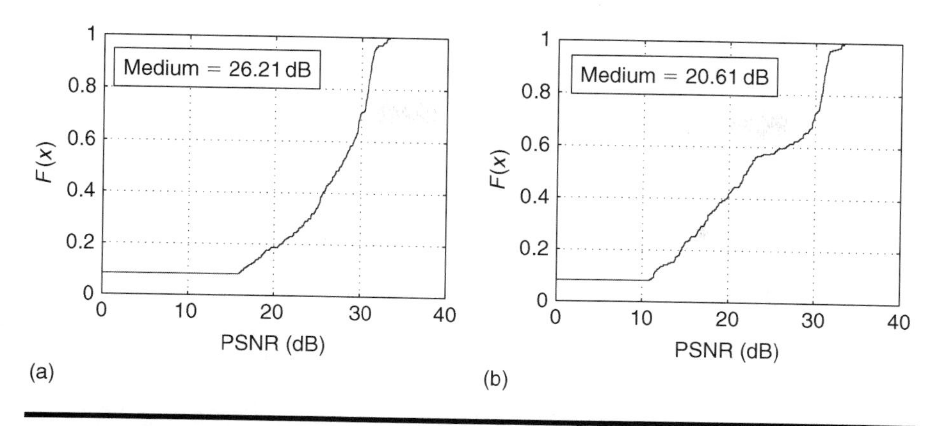

Figure 9.6 CDF of the across-run average PSNR (Foreman). (a) CDF for PSNR w/no cache w/multiple senders, (b) CDF for PSNR w/no cache w/single sender.

than schemes that use only the media server by effectively leveraging the content available at peers.

9.4 Enhancing Streaming Quality Using Caching

All the earlier discussions are based on the assumption that caching is not allowed during the streaming process. However, we notice that a wireless channel is much more error-prone compared to a wireline channel due to high BER, contention, and end-host mobility. Local caching and retransmission can be performed to improve wireless channel throughput by concealing wireless losses from end users. Caching can be performed either at peers or at the proxy, if one is used to coordinate the streaming process. Owing to memory limitations, storage limitations, and power constraints of the peers, caching usually takes place at the proxy. Because the cache size is finite, contention will take place when a packet arrives at a full cache. To solve the cache-contention problem, a comprehensive cache management is needed. Here, some popular cache management strategies that can be applied to the streaming scenario illustrated in Figure 9.3 are reviewed.

9.4.1 Resource-Based Caching Algorithm

Tewari et al. [16] defined a new disk-based caching policy called the Resource-Based Caching (RBC) Algorithm, which considers bandwidth as well as storage capacity constraints and caches a mixture of intervals and full files that have the greatest caching gain. As an improvement over the RBC, Almeida et al. [17] proposed a pooled RBC policy to allow sharing

AB of cached files. Whenever a new full file is added to the cache, its bandwidth allocation is added to the bandwidth pool. When a request for a cached file arrives, a new stream from the bandwidth pool is assigned (if possible) to that request. When it finishes delivering the file, the bandwidth is returned to the pool and can be assigned to a new request, even if that request is for a different cached file.

9.4.2 Popularity-Based Caching Algorithm

Rejaie et al. [18] defined a popularity-based caching (PBC) policy to simply cache the media files or partial files that are estimated to have the highest access frequency at the current time. Unlike replacement algorithms for Web caching that make a binary replacement decision, that is, pages are cached and flushed in their entirety, the PBC policy for streaming media cannot only make multivalued replacement decisions but also perform replacement with different granularities. Consider scenarios where layered-encoded streams are used to cope with the heterogeneity of end hosts and underlying networks. When a PBC policy is implemented, as the popularity of a cached stream decreases, its quality is gradually reduced by dropping layers/segments with the lowest popularity till it is completely flushed out.

9.4.3 Prefix Caching Algorithm

In a complementary work, Sen et al. [19] proposed a prefix caching strategy that caches the initial frames of the video stream to reduce the client start-up delay and decrease variation in quality. When the prefix caching strategy is implemented, on receiving a request for the stream, a multimedia proxy initiates transmission to the client, and then simultaneously streams the initial frames from its cache and fetches the remaining frames from the server. By storing the first few seconds of popular video clips, a multimedia proxy shields users from the delay, throughput, and loss properties of the path between the server and the proxy, resulting in a more stable streaming quality.

The size of the video prefix is jointly decided by the channel quality of the connection between the server and the proxy and the target client playback delay. It should be large enough to relieve the adverse impact of delivery delay and jitters, allow retransmission of lost packets, and absorb the variation of available bandwidth. In general, 1 or 2 s of buffered prefix can be used to handle the most common packet loss cases [19]. This caching algorithm can be applied to both constant bit rate (CBR) and variable bit rate (VBR) streams. By performing workahead smoothing into the client playback buffer to reduce the network resource requirements, it is especially beneficial for delivering VBR streams exhibiting significant burstiness.

9.4.4 Selective Partial Caching Algorithm

Selective partial caching [20,21] has also been introduced to accommodate variations in media display rate at the client end. Different from Prefix Caching Algorithm, the proxy implementing Selective Partial Caching Algorithm does not cache the frames immediately following the initial segment. Instead, it selects intermediate frames for caching based on the knowledge of user buffer size and video stream properties. It tries to give the maximum benefit to the user, in terms of increasing the robustness of entire video stream against network congestion, although not violating the user buffer size limit. For example, to cache an Motion Picture Experts Group (MPEG)–encoded stream using the Selective Partial Caching Algorithm, one can place as many of the I-frames as possible in the cache to facilitate more efficient playback at an arbitrary frame position, reduce bandwidth demands, and provide functionalities such as rewinding and fast forward.

9.4.5 Distortion Minimizing Smart Caching Algorithm

Distortion Minimizing Smart Caching (DMSC) [13] combines the design ideas of RBC and selective caching to determine which video frames should be cached to optimally improve the quality at the client end. DMSC accomplished this using the rate-distortion optimization framework. The basic idea of DMSC is to use the distortion reduction contributed by the successful delivery of DU_i to denote its importance and cache data that will maximize the distortion reduction at the client under a certain cache size constraint. Here, we use a streaming system, as illustrated in Figure 9.3, as an example to explain the DMSC Algorithm. Let t be the time when the cache contention happens. At time t, the relative importance I_i of DU_i is measured by λ_i and its time to live (TTL_i):

$$I_i = \min\left\{\lambda_i \cdot \left\lfloor \frac{\mathrm{TTL}_i}{t_o - t + \mu_P} \right\rfloor, \lambda_i\right\} \tag{9.5}$$

where TTL_i is given by $\mathrm{TTL}_i = T_i - t$, t_o is the first transmission opportunity appearing after t, μ_P the measured mean round-trip time for the connection between the proxy and the client, and λ_i the benefit of transmitting DU_i. The smaller the I_i, the less important the DU_i will be. Equation 9.5 shows that DU_i that cannot arrive at the client by its play-out deadline T_i will be the least important. With the relative importance defined by Equation 1.5, if the incoming packet is more important than some packets in the cache, we need to flush the least important packet from the cache to make room for the incoming packet. If the incoming packet is less important than any packet in the cache, the proxy will choose to drop the incoming packet.

Table 9.3 Effect of Different Caching Strategies on the System Performance When Peer-Assisted Streaming Is Supported

Sequence (dB)	*Foreman*	*Container*
PSNR w/o loss	30.71	34.32
PSNR w/loss w/DMSC	29.16	32.40
PSNR w/loss w/SCSF	25.56	30.45
PSNR w/loss w/NC	24.45	29.77

Note that by implementing DMSC in the system as shown in Figure 9.3, there is no packet forwarding anymore. Instead, packets arriving in the proxy will be cached/dropped based on their importance, and the proxy cache will operate as a secondary sender, which can actively send out packets to the client under the regulation of its associated asynchronous clock.

To evaluate how the introduction of cache affects the streaming quality, we simulate three schemes—DMSC, Simple Caching Simple Fetch (SCSF), and No Caching (NC)—in a peer-assisted streaming system. Here, SCSF refers to the scheme that allows packets to be cached in the proxy only if its cache is not full. Otherwise, the proxy will directly forward the incoming data to the client. When the packet to be retransmitted is in the cache, the proxy will fetch it from its cache instead of requesting it from the joint sender group (i.e., local retransmission). Again, we repeat the experiments 40 times each, an average the results over 40 runs to get the across-run average PSNR perceived by the client. Table 9.3 shows that using limited cache helps to improve the performance for the peer-assisted streaming scheme. With the same cache capacity, the DMSC strategy can provide better system performance compared to the SCSF strategy.

9.5 Energy Efficiency and Awareness

Because peers in the system might be battery-powered, this section focuses on addressing energy-related issues such as how to provide energy efficiency and how to distribute streaming loads better among peers to prevent premature draining of low-power nodes. Toward this end, we first explore how to save energy by exploiting the power-saving mode of wireless devices. We then introduce an EPS scheme to help maximize the life of peers in the network. Here, we are mainly concerned with the energy consumed by the wireless network interface cards (WNICs), which is determined by how much information is sent or received and how long the WNIC is activated.

9.5.1 Energy-Saving Strategies

Several studies address the energy problem at various protocol layers [13,22–24]. Ramos et al. [22] proposed a link-layer adaptation scheme to dynamically adapt the fragmentation threshold, transmission power, and retry limit according to channel conditions. Tamai et al. [24] proposed an energy-aware video streaming system based on battery capacity, desired playback duration, and relative importance of video segments. Taking a different approach, Chandra and Vahdat [23] and Li et al. [13] explored ways to transmit data packets in a predictable fashion to allow the clients to periodically transition the WNIC to a low-power sleep state. This type of approach is widely used to conserve energy for handheld devices due to its simplicity. Here, we explain details on how to use this idea in the context of peer-assisted streaming illustrated in Figure 9.3.

Figure 9.7a illustrates the phases in one transmission cycle of sender $j \in \{1,\ldots,M-1\}$ without any energy-saving strategies. Here a transmission cycle refers to the time period between two adjacent request arrivals on sender j. The various tasks performed by the MAC protocol in one transmission cycle will correspond to different radio modes, namely, "receive," "transmit," "idle," or "sleep" modes. Although power consumption for each mode might vary for different WNICs, all of them share the characteristic property that the transmit, receive, and idle modes always consume a similarly large amount of power, whereas sleep mode consumes similarly less amount of power [25]. Ignoring the energy consumed for channel

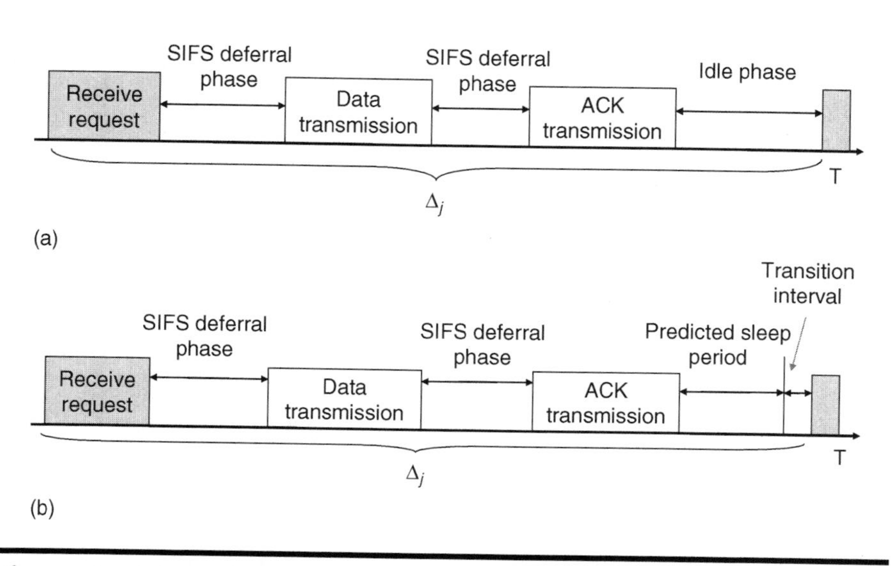

Figure 9.7 A transmission cycle of a selected sender *j*. (a) Without power-saving strategy, (b) with power-saving strategy.

sensing, the total energy consumed by sender j during one transmission cycle will include the energy spent to receive request and acknowledgment, e_j^r; the energy spent transmitting requested data packets, e_j^t; and the energy drainage during the idle mode, e_j^i, that is, $e_j = e_j^t + e_j^r + e_j^i$.

To compute e_j^t, e_j^r, and e_j^i, we first define the following variables. Let ϵ_j be the BER perceived by sender j, R the physical-layer data transmission rate of the WNIC, b_r the request packet size, b_a the acknowledgment packet size, b the average RTP packet size, n the number of RTP packets requested each time from sender j, and T_{SIFS} be the length of short interframe space (SIFS). Let P_t, P_r, P_i, and P_s the power of the WNIC under transmit, receive, idle, and sleep modes, respectively. We define Δ_j as the period to request packets from sender j. We can now compute e_j^t, e_j^r, and e_j^i as

$$e_j^t = \frac{n \cdot b}{R} \cdot \frac{1}{(1-\epsilon_j)^b} \cdot P_t \tag{9.6}$$

$$e_j^r = \frac{b_r + n \cdot b_a}{R} \cdot P_r \tag{9.7}$$

$$e_j^i = \left(\Delta_j - \frac{n \cdot b/R}{(1-\epsilon_j)^b} - \frac{n \cdot b_a + b_r}{R}\right) \cdot P_i \tag{9.8}$$

From Figure 9.7a, we can see that a WNIC stays idle for most of its transmission cycle, which contributes to a significant part of total energy consumption. If a power-saving mechanism is supported and the WNIC can switch among different communication modes for arbitrary durations of time, transitioning to low-power mode (i.e., sleep mode) when the WNIC is not sending/receiving data can result in large energy savings. Choosing sleep time, T_j^s, for sender j is a trade-off between saving power and increasing the packet loss rate: with a smaller T_j^s, the WINC spends lots of time in idle state, potentially missing out on further energy savings; on the contrary, with a larger T_j^s, the WINC might be asleep while a packet was being delivered, potentially missing a packet. From the earlier discussions on peer selection, we know that a selected sender j only receives requests and sends data when its associated asynchronous clock wakes up. By taking advantage of the asynchronous clock settings, a power-conserving strategy is proposed as illustrated in Figure 9.7b. When an asynchronous clock wakes up, the proxy selects a sender j from the joint sender group and generates a request with the clock period information, Δ_j piggybacked. On receiving a new request, the sender strips off the clock period information. Because packets might experience different delays before arriving in sender j, one can use a sliding window with size equal to W requests to smooth out the effect of delay jitter and improve the estimation accuracy of T_j^s:

$$T_j^s = \left(\Delta_j - \frac{n \cdot b/R}{(1-\epsilon_j)^b} - 2n \cdot T_{\text{SIFS}} - \frac{n \cdot b_a + b_r}{R} - T_j^t + \frac{\sum_{k=2}^{W} \left| I_j^{(k)} - I_j^{(k-1)} \right|}{W-1} \cdot \frac{I_j^{(W)} - I_j^{(W-1)}}{\left| I_j^{(W)} - I_j^{(W-1)} \right|} \right) \cdot \alpha \tag{9.9}$$

where $I_j^{(k)}$ is the packet arrival time at sender j and T_j^t the WNIC waking-up time. α is the "estimation adjust factor," which scales from 0 to 1, as a measure of how conservative we choose the sleep period. A smaller estimation adjust factor means a more conservative estimation. Because the WNIC will be placed into sleep mode for T_j^s instead of staying idle, the total energy consumption e_j will be adjusted to be $e_j = e_j^t + e_j^r + e_j^i - T_j^s(P_i - P_s)$.

9.5.2 Energy-Aware Peer Selection

As mentioned earlier, most peers rely on batteries for energy, whereas streaming video usually tends to be a lengthy process that can consume a significant amount of that energy. Because the client streams different portions of video content from different peers, unexpected power depletion in some peers can cause peer failure and greatly degrade the received video quality. Therefore, it is important to design an EPS scheme to avoid completely draining low-power nodes although still providing satisfying video quality.

Previous discussions have explained how peer selection can be performed using asynchronous clocks whose periods depend on the underlying channel quality and the bandwidth by senders allocated for the streaming process. By integrating energy awareness into the peer selection scheme design, now the period of each clock will also depend on the energy level of sender j, E_j. Toward this goal, the proxy needs to track the energy level of each sender based on its updated energy information, and decide whether the remaining energy is higher than its preset threshold, which would be, for example, 10 percent of its initial energy level. It will remove those senders that have energy lower than the threshold. Then for the remaining senders, by jointly considering maximum SR R_j and E_j, the period of asynchronous clock, which regulates the connection between sender j and the proxy, will be set as

$$\Delta_j^e = \begin{cases} \max\left\{ \frac{b}{\frac{R_j \cdot E_j}{\sum_{l=1}^{M-1} R_l \cdot E_l} \cdot (R_0 - R_S)}, \frac{b}{R_j} \right\} & j = 1, 2, \ldots, M-1 \\ \frac{b}{R_S}, & j = S \end{cases} \tag{9.10}$$

Table 9.4 Performance Comparison

Video Sequence	*Foreman*	*Container*
PSNR w/o loss (dB)	30.71	34.32
PSNR w/loss w/EPS (dB)	25.88	31.97
PSNR w/loss w/o EPS (dB)	24.12	30.32
Consumed energy w/EPS (J)	20.01	22.65
Consumed energy w/o EPS (J)	35.28	33.55

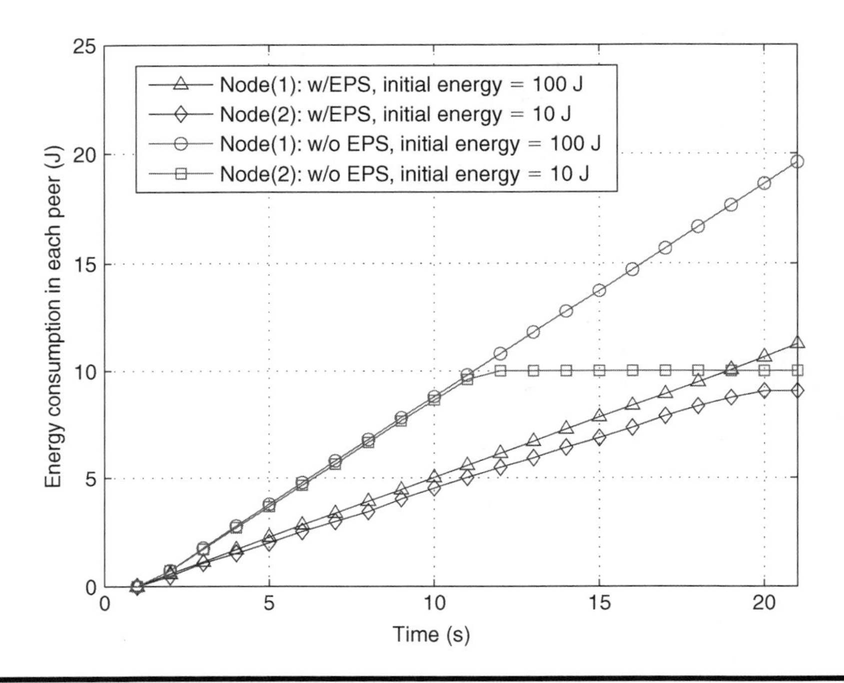

Figure 9.8 Cumulative energy consumption on each peer (Foreman).

which shows that the amount of information streamed from j will be proportional to its R_j and E_j. A sender with better connection and higher energy will get more opportunities to be queried by the proxy to send out packets. Table 9.4 compares the streaming performance of the peer-assisted streaming system that uses EPS with that of a system without using EPS, whereas Figure 9.8 shows the energy consumption on each peer throughout the simulation. We can see that the available energy of peer 2 is not enough for it to participate throughout a simulation run. By using energy-aware scheduling, the peer-assisted streaming system can achieve better performance by quickly reacting to quality degradation caused by a dying peer connection.

9.6 Open Issues

9.6.1 Dynamic Wireless Channel Quality

Compared to their wired counterparts, wireless networks typically offer lower overall channel quality as well as greater channel quality variation. Several effects occur in addition to the attenuation caused by the distance between the sender and receiver, including shadowing, reflection, scattering, and diffraction. The results of these effects can lead to severe radio channel impairments and result in wireless channel quality that is time, location, and frequency dependent. Meanwhile, the interference caused by transmitting in overlapping wireless channels can result in much higher error rates (e.g., 10^{-4} instead of 10^{-10} for fiber optics). Although we can integrate various models into the analysis to reflect the impacts of these effects on the system performance and design streaming schemes to be robust, ensuring high-quality services under dynamic wireless channel quality in a real deployment environment is still challenging.

9.6.2 Traffic with Different Quality-of-Service Requirements

In a practical system, video streaming traffic must share network resources with other types of traffic. These traffic types, such as Web traffic, voice traffic, and text messaging, have different QoS requirements from video streaming traffic in terms of bandwidth, delay, and packet loss. IEEE 802.11e since late 2005 has been approved as a standard to provide better QoS for applications supported in 802.11 WiFi networks. By classifying and prioritizing traffic according to their QoS requirements, the IEEE 802.11e standard [26] is considered of critical importance for delay-sensitive applications, such as voice over wireless IP and video streaming. However, currently the standard only offers a centralized approach for a single-hop scenario. When we extend our scope to consider multihop 802.11 networks, ensuring good streaming quality without sacrificing the performance of other traffic still remains an open question.

9.6.3 Incentive Mechanism for Node Cooperation

Like any other system that leverages peer resources, the success of the streaming system that we have proposed in this chapter depends on the cooperative resource contributions from individual peers. However, past studies have shown that it is not rare to find that noncooperative peers (also known as free riders) are prevalent in current P2P systems. Although various incentive mechanisms have been proposed in the past to encourage cooperative behavior among peers for P2P file sharing and streaming over wired networks, how to design an efficient incentive mechanism for P2P streaming over wireless networks remains to be addressed.

9.6.4 Security Issues

To exploit peer resources to enhance streaming quality over wireless networks, there are two types of security issues needed to be solved. The first type stems from the fact that radio interfaces make eavesdropping much easier in wireless networks than wired or fiber-optic technologies. The second appears in the context of peer-assisted streaming. Like all existing P2P applications, peer-assisted streaming schemes are based on the assumption that all peers are trustworthy and they will not attack their peers. However, in reality, many P2P networks are under constant attacks such as poisoning attacks (e.g., providing files whose contents are different from the description), polluting attacks (e.g., inserting bad chunks/packets into an otherwise valid file on the network), and insertion of viruses by people with a variety of motives. Addressing these security issues is still open to further exploration.

9.7 Summary

In this chapter, we learned how to exploit peer resources to design peer-assisted video streaming schemes to support high-quality video streaming service over 802.11 WLANs.

We started the chapter by reviewing some fundamentals of streaming video content over IEEE 802.11 WLANs and existing approaches for improving streaming performance. Then we looked at the design of peer-assisted streaming schemes, which entails four parts: content discovery, mobility tracking, peer selection, and data selection. Content discovery addresses the issue of finding out peers that can become potential senders. Because the reliability of wireless connections to these peers heavily depends on the node mobility and physical channel characteristics, mobility tracking elaborates on how a streaming system can track peer mobility and quantify its impact. Peer selection and data selection explain how to use request data from peers that can provide video content. Although we later focus solely on examples where a proxy works as a streaming coordinator, the design principles are general enough to be applied to other scenarios as well.

In this chapter, we also explored how caching can enhance streaming performance. Several caching strategies were introduced including RBC Algorithm, PBC Algorithm, Prefix Caching Algorithm, Selective Partial Caching Algorithm, and DMSC Algorithm. We should keep in mind that some of the caching strategies were originally designed for wired networks. They assume fixed network topology, and normally cause high computation and communication overhead. These assumptions might not be valid anymore in wireless networks where caching nodes are mobile hosts with scarce bandwidth resource and limited memory, storage, and power.

At the end of the chapter, we examined two energy-related issues. The first issue is how to improve the energy efficiency of mobile streaming peers. We learned that by periodically transitioning sender WNICs into a low-energy consuming sleep mode, peers can save energy while maintaining good streaming quality, and as a result, they can participate in the streaming process for a longer period of time. The second is how to prevent sudden peer departure caused by premature energy draining. We revisited the problem of peer selection by jointly considering network conditions and energy levels of each peer node. We then showed that the EPS scheme will stream more information from high-power peers, and consequently, it can prevent the premature draining of low-power nodes. It can also avoid sudden peer failure by monitoring node energy level and gracefully switching to new peer nodes once it detects energy levels dropping lower than a preset threshold.

References

1. IEEE Std. 802.11, IEEE 802.11b/d3.0 Standard for Wireless Medium Access Control (MAC) and Physical Layer (PHY) Specification, August 1999.
2. Qiao, D. and K. G. Shin, A Two-Step Adaptive Error Recovery Scheme for Video Transmission over Wireless Networks, *IEEE Conference on Computer Communications (INFOCOM)*, Tel Aviv, Israel, June 2000.
3. Xu, X., M. van der Schaar, S. Krishnamachari, S. Choi, and Y. Wang, Adaptive Error Control for Fine-Granular-Scalability Video Coding over IEEE 802.11 Wireless LANs, *IEEE International Conference on Multimedia & Expo (ICME)*, Baltimore, MD, July 2003.
4. Krishnamachari, S., M. van der Schaar, S. Choi, and X. Xu, Video Streaming over Wireless LANs: A Cross-Layer Approach, *Proceedings of Packet Video Workshop*, Nantes, France, April 2003.
5. Majumdar, A., D. G. Sachs, I. V. Kozintsev, K. Ramchandran, and M. M. Yeung, Multicast and unicast real-time video streaming over wireless LANs, *IEEE Transactions on Circuits and Systems for Video Technology*, 12(6), 524–534, 2002.
6. Bolot, J., S. Fesse-Parisis, and D. Towsley, Adaptive FEC-Based Error Control for Interactive Audio on the Internet, *IEEE Conference on Computer Communications (INFOCOM)*, New York, March 1999.
7. Podolsky, M., S. McCanne, and M. Vetterli, Soft ARQ for Layered Streaming Media, Technical Report UCB/CSD-98-1024, UC Berkeley, CA, November 1998.
8. Shan, Y., S. Yi, S. Kalyanaraman, and J. W. Woods, Two-Stage FEC Scheme for Scalable Video Transmission over Wireless Networks, *SPIE Communications/ITCom, Multimedia Systems and Applications*, Boston, MA, October 2005.
9. Sachs, D. G., I. Kozintsev, M. Yeung, and D. L. Jones, Hybrid ARQ for Robust Video Streaming over Wireless LANs, *Proceedings of Information Technology: Coding and Computing*, Las Vegas, NV, USA, April 2001.

10. Cheung, G., C. Chuah, and D. Li, Optimizing Video Streaming against Transient Failures and Routing Instability, *IEEE International Conference on Communications (ICC)*, Paris, France, June 2004.
11. Liang, J. and K. Nahrstedt, RandPeer: Membership Management for QoS Sensitive Peer-to-Peer Applications, *IEEE Conference on Computer Communications (INFOCOM)*, Barcelona, Catalonia, Spain, April 2006.
12. Balazinska, M. and P. Castro, Characterizing Mobility and Network Usage in a Corporate Wireless Local-Area Network, *First International Conference on Mobile Systems, Applications, and Services (MobiSys)*, San Francisco, CA, May 2003.
13. D. Li, C. Chuah, G. Cheung, and S. J. Ben Yoo, MUVIS: multi-source video streaming server over WLANs, *IEEE/KICS Journal of Communications and Networks—Special Issue on Towards the Next Generation Mobile Communications*, 7(2), 144–156, 2005.
14. Floyd, S., M. Handley, J. Padhye, and J. Widmer, Equation-Based Congestion Control for Unicast Applications, *ACM SIGCOMM*, Stockholm, Sweden, August 2000.
15. Chou, P. and Z. Miao, Rate-Distortion Optimized Streaming of Packetized Media, Microsoft Research Technical Report MSR-TR-2001-35, February 2001.
16. Tewari, R., H. Vin, A. Dan, and D. Sitaram, Caching in Bandwidth and Space Constrained Hierarchical Hyper-Servers, Technical Report CS-TR-96-30, Department of Computer Sciences, University of Texas, Austin, TX, January 1997.
17. Almeida, J., D. Eager, and M. Vernon, A hybrid caching strategy for streaming media files, *Proceedings of SPIE/ACM Conference on Multimedia Computing and Networking*, 4312, 200–212, 2001.
18. Rejaie, R., H. Yu, M. Handley, and D. Estrin, Multimedia Proxy Caching Mechanism for Quality Adaptive Streaming Applications in the Internet, *IEEE Conference on Computer Communications (INFOCOM)*, Tel Aviv, Israel, June 2000.
19. Sen, S., J. Rexford, and D. Towsley, Proxy Prefix Caching Multimedia Streams, *IEEE Conference on Computer Communications (INFOCOM)*, New York, March 1999.
20. Wang, Y., Z. L. Zhang, D. Du, and D. Su, A Network Conscious Approach to End-to-End Video Delivery over Wide Area Networks Using Proxy Servers, *IEEE Conference on Computer Communications (INFOCOM)*, New York, April 1999.
21. Miao, Z. and A. Ortega, Proxy Caching for Efficient Video Services over the Internet, Proceeding of Packet Video Workshop, New York, April 1999.
22. Ramos, N., D. Panigrahi, and S. Dey, Energy-Efficient Link Adaptations in IEEE 802.11b Wireless LAN, *IASTED International Conference on Wireless and Optical Communications*, Banff, Alberta, July 2003.
23. Chandra, S. and A. Vahdat, Application-Specific Network Management for Energy-Aware Streaming of Popular Multimedia Formats, *Proceedings of the USENIX Annual Technical Conference*, Monterey, CA, June 2002.
24. Tamai, M., T. Sun, K. Yasumoto, N. Shibata, and M. Ito, Energy-Aware Video Streaming with QoS Control for Portable Computing Devices, ACM NOSSDAV, Kinsale, County Cork, Ireland, June 2004.

25. Jiao, Y. and A. R. Hurson, Adaptive Power Management for Mobile Agent-Based Information Retrieval, *Proceeding of Advanced Information Networking and Application*, Tamkang University, Taiwan, March 2005.
26. IEEE Std. 802.11eD13.0, IEEE Computer Society, Part 11: Wireless Medium Access Control (MAC) and Physical Layer (PHY) Specifications: Medium Access Control (MAC) Enhancements for Quality of Service (QoS), January 2005.
27. Schulzrine, H., S. Casner, R. Frederick, and V. Jacobson, RTP: A Transport Protocol for Real-Time Application, IETF RFC 1889, January, 1996.

Chapter 10

Multimedia Services over Broadband Wireless LAN

Jianhua He, Zuoyin Tang, Yan Zhang, and Zongkai Yang

Contents

10.1 Introduction....335
10.1.1 Technical and Applications Trends....335
10.1.2 Wireless LAN....336
10.1.2.1 PHY Layer....336
10.1.2.2 MAC Protocols....337
10.1.3 Multimedia Application Requirements and Standards....337
10.1.3.1 Moving Picture Experts Group-4 and H.264....338
10.1.3.2 JPEG2000....339
10.2 Research Challenges....339
10.2.1 Link Capacity and Adaptation....340
10.2.2 MAC QoS Scheme....340
10.2.3 End-to-End QoS Scheme....341
10.2.4 Cross-Layer Design and Adaptation....341
10.2.5 Mobility Support....342
10.3 IEEE 802.11 QoS Support....343
10.3.1 Physical-Layer Enhancement....343
10.3.2 WLAN QoS Mechanism....343

10.3.3 WLAN QoS Algorithms 344
10.3.3.1 Scheduling 344
10.3.3.2 Admission Control and Reservation 345
10.4 End-to-End QoS Support..347
10.4.1 IntServ and DiffServ Internet QoS Architectures..347
10.4.2 DiffServ-Based End-to-End QoS Support 348
10.4.3 IntServ-Based End-to-End QoS Support..348
10.5 Cross-Layer Optimization and Adaption 348
10.5.1 Solutions for Cross-Layer Optimization..349
10.5.2 Adaptation Architecture..350
10.6 Case Study of Media-Oriented Applications over WLAN..350
10.6.1 Image Transmission..351
10.6.2 Voice-over-IP..352
10.6.3 Video Streaming..352
10.7 Seamless Multimedia QoS Support..354
10.7.1 WLAN Roaming..354
10.7.2 Interworking with Cellular Networks..354
10.8 Open Issues..356
10.8.1 Systematic Study..356
10.8.2 Multimedia Support over New MAC 356
10.8.3 Multimedia over Multihop WLAN 357
10.9 Conclusion 357
References..357

The ultimate goal of future-generation wireless communications is to provide ubiquitous seamless connections between mobile terminals so that users can enjoy high-quality multimedia services anytime anywhere. Developed as a simple and cost-effective wireless technology for best-effort services, Institute of Electrical and Electronics Engineers (IEEE) 802.11–based wireless local area networks (LANs) have gained popularity at an unprecedented rate [1]. It is inevitable to support multimedia applications over 802.11 wireless LANs. However, 802.11 Media Access Control (MAC) protocol has been designed primarily for best-effort (BE) data services. Owing to the lack of built-in quality-of-service (QoS) support, IEEE 802.11 experiences serious challenges in meeting the demands of multimedia services and applications, which normally have strict QoS requirement.

IEEE 802.11 working group amends the 802.11 standard and enhances the QoS capabilities in the new standard 802.11e [2]. A new high-speed physical (PHY) layer named IEEE 802.11n is also under development, which sets up a throughput requirement above 100 Mbps at the MAC layer. In addition, significant advances are made on multimedia standards and signaling protocols for efficient multimedia compression and delivery over general wireless packet networks. Providing multimedia services to mobiles and fixed users through wireless LAN can be a reality with the development of the high-speed PHY layer and QoS-based MAC layer and the advances of multimedia standards. However, how to efficiently combine these techniques together and provide end-to-end QoS guarantee for WLAN multimedia services is still very challenging and needs further study.

In this chapter, the state of the art for supporting multimedia over broadband WLAN is examined. We identify the research challenges of supporting multimedia over WLAN. Current approaches proposed to address these challenges are then presented. The chapter concludes by identifying some open issues for the delivery of multimedia over broadband WLAN.

10.1 Introduction

10.1.1 Technical and Applications Trends

In the past 20 years, wireless communication has evolved at an amazing speed. General Packet Radio Service (GPRS) has been deployed to give Internet protocol (IP) connectivity to Global System for Mobile (GSM) users; third-generation (3G) cellular systems, such as Universal Mobile Telecommunications System (UMTS) and CDMA2000 (code division multiple access), are enhanced for mobile Internet solutions [3,4]. In parallel with the evolution of cellular systems, a number of other access technologies such as wireless LAN have emerged. A wireless LAN gives users the mobility to move around within a broad coverage area and still be connected to the network. For the home user, wireless has become popular due to ease of installation and freedom of location with the gaining popularity of laptops. Public businesses such as coffee shops or malls have begun to offer wireless access to their customers. Wireless LAN hot spots experienced a rapid growth across the entire world, which attests to the growing importance of broadband wireless for a variety of enterprise and consumer applications.

The rapid growth of communication systems has radically changed the ways of conducting business and consumer habits. Cellular- and WLAN-networked devices are increasingly used for a new range of audio- and video-based entertainment and media applications such as Voice-over-IP (VoIP), video and audio downloads and playback, mobile TV and interactive games, specifically targeted to the wireless market. The current wireless communication system needs to offer not just simple voice telephone calls, but also rich multimedia services. This has driven research and development to multimedia services over wireless systems—in particular, the next-generation cellular system and wireless LAN—to achieve cost-effective broadband wireless access and ubiquitous multimedia services. Significant advances have been achieved on the network and signaling protocols, network specifications, and multimedia standards (such as JPEG2000 for image coding and H.264/AVC for video coding) [5,6].

In the new multimedia standards, higher compression efficiency and flexibility can be achieved. In addition, a set of error resiliency techniques was introduced. However, these techniques are insufficient for multimedia transmission over wireless networks because the resource management and protection strategies available in the lower layers (PHY and MAC) are not optimized explicitly considering the specific characteristics of multimedia applications. This section gives a brief overview of wireless LAN techniques and multimedia standards, with an emphasis on the techniques pertaining to multimedia transmission from the individual layers. The research challenges and proposed solutions are discussed in later sections.

10.1.2 Wireless LAN

Today, most of the WLAN cards are implemented based on the various versions of IEEE 802.11 (Wi-Fi). An alternative ATM-like 5 GHz standardized technology, high performance radio LAN (HIPERLAN), was developed in Europe with strong QoS capabilities [7]. But it has not been a success in the market so far. Therefore, we will focus on the IEEE 802.11–based wireless LANs.

10.1.2.1 PHY Layer

IEEE 802.11 denotes a set of wireless LAN standards developed by the IEEE 802.11 working group. The 802.11 family currently includes six over-the-air modulation techniques, all of which use the same protocol. The most popular techniques are those defined by the a, b, and g amendments to the original standard. Many new features at both MAC and PHY layers for high-speed applications are defined by 802.11n. The first widely accepted wireless networking standard was 802.11b, followed by 802.11a and 802.11g. As 802.11b and 802.11g standards use the 2.40 GHz frequency band, 802.11b and 802.11g equipments can incur interference from microwave ovens, Bluetooth devices, and other appliances using the same band. The 802.11a standard uses the 5 GHz band, and is therefore, not affected by products operating on the 2.4 GHz band.

802.11 specifies multiple transmission rates in the PHY layer that are achieved by adaptive modulation and coding (AMC) technique to maximize throughput, which is the trend in the emerging access networks [1]. AMC adaptively changes the level of modulation—binary phase shift key (BPSK), quaternary PSK (QPSK), 8-PSK, 16-quadrature amplitude modulation (QAM), and so on—and amount of redundancy for an error correction code. A higher level of modulation with no error correction code can be used by users with good signal quality to achieve higher bandwidth. Typically, the average bit error rate (BER) requirement is set in advance, depending on the class of application. Then, AMC is applied to ensure the QoS, which is evaluated directly by the signal-to-interference ratio (SIR) or indirectly measured by BER.

10.1.2.2 MAC Protocols

The IEEE 802.11 MAC provides a shared access to the wireless channel and supports two medium access protocols: contention-based Distributed Coordination Function (DCF) and optional Point Coordination Function (PCF) [1]. DCF is used as a basis for PCF. When PCF is enabled, the wireless channel is divided into superframes. Each superframe consists of a contention-free period (CFP) for PCF and a contention period (CP) for DCF. At the beginning of CFP, the point coordinator usually located in the access point (AP) contends for access to the wireless channel. Once it acquires the channel, it cyclically polls high-priority stations and grants them the privilege of transmitting. PCF starts with a beacon frame, which is a management frame that maintains the synchronization between stations and delivers timing-related parameters. The priority of PCF over DCF is guaranteed with PIFS being smaller than DCF interframe spaces (DIFS). This ensures that other stations will not attempt to access the channel.

DCF employs a Carrier Sense Multiple Access with Collision Avoidance (CSMA/CA) as the access method [1]. A truncated binary exponential back off scheme is used in the access method. Before initiating a transmission, each station is required to sense the medium. If the medium is busy, the station defers its transmission and initiates a back off timer. The back off timer is randomly selected between 0 and contention window (CW). Once the station detects that the medium has been free for a duration of DIFS, decrement of the back off counter begins as long as the channel is idle. As the back off timer expires and the medium is still free, the station begins to transmit. If an acknowledgment is not received within a timeout period, the transmitted packet is inferred to be lost due to either packet collision or corruption. Then the mentioned back off procedure repeats to retransmit the packet. The size of the CW is doubled until it reaches the CW_{max} value. If the number of retransmissions for a packet reaches the maximal allowed retries, the packet is discarded. If a packet is successfully transmitted, back off procedure is initiated for new packets with the CW reset to CW_{min}.

10.1.3 Multimedia Application Requirements and Standards

Media content such as voice and video differs from other types of data in at least three ways that are relevant to wireless communication.

- Many media applications are highly sensitive to delay.
- Media can tolerate a certain measure of packet loss while remaining useful.
- Streaming media consists of a sequence of frames, each with a different delivery deadline.

Therefore, delivery of real-time (RT) multimedia traffic typically has QoS requirements, for example, bandwidth, delay, and error requirements [8]. Video transmissions, for example, require a large amount of bandwidth. Throughput requirements range from 300 to 800 kbps for videoconferencing applications to 19.2–1500 Mbps for high-definition television (HDTV) to transmit high-quality noncompressed video frames integrated with high-quality audio. RT video has also strict delay constraints (e.g., 1 s) because RT video must be played out continuously. If the video packet does not arrive in a timely manner, the playout process will pause, which is annoying to the human eye. Video applications typically impose upper limits on BER (e.g., 1 percent). Because video compression techniques seek to reduce redundancies between successive frames, too many bit errors would seriously degrade the video presentation quality.

10.1.3.1 Moving Picture Experts Group-4 and H.264

Moving Picture Experts Group (MPEG)-4 is a standard used primarily to compress audio and visual (AV) digital data [5]. Introduced in late 1998, it is the designation for a group of audio and video coding standards and related technology agreed on by the International Organization for Standardization (ISO)/International Electrotechnical Commission (IEC) MPEG. The uses of the MPEG-4 standard are Web (streaming media) and compact disk (CD) distribution, conversation (videophone), and broadcast television, all of which benefit from compressing the AV stream. MPEG-4 absorbs many of the features of MPEG-1 and MPEG-2 and other related standards, adding new features such as support for three-dimensional (3-D) rendering, object-oriented composite files, and support for externally specified digital rights management and various types of interactivity. MPEG-4 is still a developing standard and is divided into a number of developing parts. The key parts to be aware of are MPEG-4 part 2 (MPEG-4 SP/ASP) and MPEG-4 part 10 (MPEG-4 AVC/H.264).

H.264 consists of two conceptually different layers [6]. First, the video coding layer (VCL) contains the specification of the core video compression engines that achieve basic functions such as motion compensation, transform coding of coefficients, and entropy coding. This layer is transport-unaware, and its highest data structure is the video slice, a collection of coded macroblocks (MBs) in scan order. Second, the network abstraction layer (NAL) is responsible for the encapsulation of the coded slices into transport entities of the network.

The NAL defines an interface between the video codec and the transport world. It operates on NAL units (NALUs) that improve transport abilities over almost all existing networks. An NALU consists of a 1 byte header and a bit string that represents, in fact, the bits constituting the MBs of a slice. The header byte consists of an error flag, a disposable NALU flag, and the NALU type. Finally, the NAL provides a means to transport high-level syntax (i.e., syntax assigned to more than one slice, e.g., to a picture or group of pictures) to an entire sequence.

The H.264 standard introduces a set of error resilience techniques such as slice structure, data partitioning (DP), and flexible macroblock ordering (FMO) [6]. Among these techniques, DP is an effective application-level framing technique that divides the compressed data into separate units of different importance. Generally, all symbols of MBs are coded together in a single bit string that forms a slice. However, DP creates more than 1 bit string (partition) per slice, and allocates all symbols of a slice into an individual partition with a close semantic relationship.

10.1.3.2 JPEG2000

JPEG2000 is a wavelet-based image compression standard [5]. It was created by the Joint Photographic Experts Group committee with the intention of superseding their original discrete cosine transform-based JPEG standard. JPEG2000 can operate at higher compression ratios without generating the characteristic artifacts of the JPEG standard. It also allows more sophisticated progressive downloads.

There are several advantages of JPEG2000 over the ordinary Joint Photographic Experts Group (JPEG) standard. In addition to the superior compression performance, multiple resolution representation, and random codestream access and processing, JPEG2000 has capabilities of progressive transmission, error resilience, and flexible file format. JPEG2000 is robust to bit errors introduced by noisy communication channels. This is accomplished by the inclusion of resynchronization markers, the coding of data in relatively small independent blocks, and the provision of mechanisms to detect and conceal errors within each block. With flexible file format, handling of color-space information, metadata, and interactivity in networked applications are allowed.

10.2 Research Challenges

IEEE 802.11 wireless LAN gained global acceptance and popularity in wireless computer networking and is anticipated to continue being the preferred standard for supporting wireless LAN's high-speed multimedia applications. Multimedia delivery over wireless LAN faces numerous challenges, among them are high BERs, limited and unpredictably varying bit rates, lack of MAC and end-to-end QoS support schemes, and cross-layer optimization. The research challenges are identified and analyzed in this section.

10.2.1 Link Capacity and Adaptation

Wireless networks differ from the wired networks, for example,

- Wireless channels are inherently more lossy than wired channels, with packet loss rates that can be orders of magnitude higher than their wired counterparts.
- Wireless is a broadcast medium, and although multiple receivers can listen on a single transmission, they each have a different quality of reception.
- Wireless links have time-varying capacities due to channel fading and interference. The speed on most wireless LANs (typically 1–54 Mbps) is far slower than even the slowest common wired networks (100 Mbps up to several gigabit per second).

The limited, dynamic, and unpredictable link capacity makes it difficult to support a large number of high-speed multimedia applications concurrently over wireless LANs. In addition, as described in the earlier section, link adaptation is required to adaptively choose modulation and forward error correction (FEC) in 802.11 PHY layer to achieve better link utilization. The average BER requirement is set beforehand, depending on the application. Then adaptive modulation and FEC are applied to ensure the QoS. However, the rate adaptation and signaling mechanisms are intentionally left open by 802.11. It becomes more challenging to design efficient link adaptation algorithms with the limited link bandwidth to ensure diverse QoS required by multimedia applications.

10.2.2 MAC QoS Scheme

In WLANS, the MAC protocol is the key component that provides the efficiency in sharing the common radio channel, while satisfying the QoS requirements of various multimedia traffic. However, frames in DCF, the basic access method in the legacy 802.11 MAC-layer protocol, do not have priorities and there is no other mechanism to enforce a guaranteed bandwidth and access delay bound. As a result, RT multimedia applications such as voice or live video transmissions may suffer from unacceptable packet loss or delay with this protocol. The second access mode of the IEEE 802.11 MAC-layer protocol, PCF, is designed for delay-bounded services. However, it is centralized and can only be used in the network of infrastructure mode. In addition, the loose specification of PCF leaves many issues unsolved [9].

- No access control policy is included in the protocol.
- PCF may experience substantial delay at low load, because stations must wait for polling, even in an otherwise idle system.

- Because the AP needs to contend for the channel using DCF at the beginning of a CFP, the period of contention-free polling may vary and be out of control.
- With a large number of interactive streams to be polled, the applications using DCF contention can be harmed.
- PCF is a centralized approach, which will make the system run into risks if the point coordinator does not work.

Therefore, the legacy 802.11 MAC-layer protocol cannot provide necessary QoS at the link level for end-to-end QoS requirements for multimedia applications.

10.2.3 End-to-End QoS Scheme

IEEE 802.11 uses a shared medium and provides differentiated control of access to the medium to handle data transfers with QoS requirements. The IEEE 802.11 LAN will become part of a larger network providing end-to-end QoS delivery or to function as an independent network providing transport on a per-link basis. It is required to integrate IEEE 802.11 protocols with other end-to-end QoS delivery mechanism for multimedia applications. Considering the unreliability and bandwidth fluctuations of wireless channels and congestion in Internet links, the end-to-end QoS schemes including the components of priority setting and mapping, admission control and resource reservation is a big challenge for wireless LAN multimedia applications.

10.2.4 Cross-Layer Design and Adaptation

Classical networking technology has effectively separated communications issues from end-user applications through the abstractions of the Open Systems Interconnection (OSI) framework [10,11]. This model has successfully defined protocols, by which developers could focus their work at a level appropriate to their development needs, for example, PHY, data link, network, transport, and application layers. This architectural concept has been enormously successful and implemented almost ubiquitously. However, in a time-varying wireless environment, applications with the OSI model may experience very poor performance. Consider the case of multimedia transmission over broadband wireless LAN. It is clear that there are some strategies available for video source control (e.g., rate control and error concealment) in the application layer, reliability and congestion control in the transport layer, QoS support in the network layer, and reliability and channel access control in the MAC layer (e.g., random channel access parameters, scheduling and transmission opportunities control, and

AMC). And there are dependencies between some of the strategies. It is almost impossible to optimize the application's performance by determining the strategies from the point of individual layers without knowledge from other layers in the time-varying wireless environment. An interesting approach to address this problem is using cross-layer design and optimization technique.

However, finding the optimal solution to the earlier-mentioned cross-layer optimization problem is very challenging. Even considering the cross-layer optimization problem in the PHY and MAC layers of IEEE 802.11e, it is difficult to derive analytical expressions for the throughput, delay, and power as functions of channel conditions, because these functions are nondeterministic and nonlinear [12]. Only the worst case or average values can be determined. The dependencies between some of the strategies in the PHY, MAC, and application layers make the cross-layer optimization problem more difficult.

In addition to the necessity of cross-layer design, adaptation is an important technique to achieve good application performance [11]. Most current video applications are insensitive to changing network conditions. In a time-varying wireless environment, however, video applications must be robust and adaptive in the presence of unreliability and bandwidth fluctuations. The wireless channel conditions, the wired links, and multimedia content characteristics may change continuously. In addition, mobility and handoff will also cause bandwidth fluctuations. Without constant information update and network or application adaptations, the multimedia application performances will degrade significantly.

10.2.5 Mobility Support

Currently, 802.11 WLAN service is available for low-mobility devices in isolated hot spots with coverage from dozens of meters up to a few hundred meters. 802.11 also supports mobile stations within an 802.11-extended service set to roam among multiple APs. In addition to roaming and horizontal handoff among 802.11 WLANs, supporting QoS anytime anywhere, and by any media requires seamless vertical handoffs between WLAN and other different wireless networks such as GPRS, 3G, and Bluetooth networks.

However, with the decreasing size of cells in next-generation multimedia-enabled wireless networks, the number of handoffs during a call's lifetime increases. When a handoff takes place, the available bandwidth may vary drastically (e.g., from a few megabits per second to a few kilobits per second). And the AP or base station may not have enough unused radio resource to meet the demand of a newly joined mobile host. Consequently, seamless mobility support will pose a serious problem for RT multimedia transmission over wireless LANs.

10.3 IEEE 802.11 QoS Support

10.3.1 Physical-Layer Enhancement

To provide QoS support for high-speed multimedia applications, IEEE formed a new 802.11 Task Group (TGn) to develop a new amendment to the 802.11 standard. The real data throughput is estimated to reach a theoretical 540 Mbps. 802.11n builds on earlier 802.11 standards by adding multiple-input multiple-output (MIMO) [13,14]. MIMO uses multiple transmitter and receiver antennas to allow for increased data throughput through spatial multiplexing and increased range by exploiting the spatial diversity, perhaps through coding schemes such as Alamouti coding [14].

Link adaptation is another mechanism of improving link-layer capacity. Existing algorithms can be classified into three categories: signal-to-noise ratio (SNR)-based, packet retransmission-based, and hybrid [15–17]. In the SNR-based link adaptation algorithms, received signal strength (RSS) is used as an indication of link quality [15]. Transmission rate is selected based on the average or instantaneous RSS information from a predetermined SNR rate table. Normally, the RSS information is collected by the transmitting station from earlier frames received from the receiving station. Receiver base rate fallback (RBRF) is a typical example of such algorithms [16]. In the packet retransmission-based link adaptation algorithm, the transmitting station counts the outcome (either successful or failed) of each transmission attempt. Based on the packet transmissions (losses) history, the transmitting rate can be adaptively raised by a level or fallback or be kept. Auto rate fallback (ARF) is the first documented bit-rate selection algorithm [17]. The ARF algorithm also performs well in situations where link conditions change on the order of tens of packets.

10.3.2 WLAN QoS Mechanism

As we explained in the earlier sections, the legacy IEEE 802.11 MAC can't provide efficient QoS support for multimedia applications. The need for a better access mechanism to support service differentiation and QoS support has led TGe of IEEE 802.11 to develop 802.11e standard. The 802.11e standard introduces the hybrid coordination function (HCF) that concurrently uses a contention-based mechanism enhanced distributed channel access (EDCA), and a polling-based mechanism, HCF-controlled channel access (HCCA) [2].

The EDCA mechanism provides differentiated and distributed access to the channel using eight different user priorities (UPs), which are mapped to four access categories (ACs) [2]. For each AC, an enhanced variant of the DCF, called an EDCA function (EDCAF), contends for transmission opportunities (TXOPs) using a set of distinct EDCA parameters, including arbitration interframe space (AIFS) instead of DIFS in DCF and a pair of CW_{min} and CW_{max}.

The HCCA mechanism is used for contention-free transfer. It uses a QoS-aware centralized coordinator, called a hybrid coordinator (HC), and operates under rules that are different from the point coordinator (PC) of the PCF [2]. The HC is collocated with the QoS-enabled AP and uses the HC's higher priority of access to the channel to initiate frame exchange sequences and to allocate TXOPs to itself and other stations to provide limited-duration controlled access phase (CAP) for contention-free transfer of QoS data. The HC traffic delivery and TXOP allocation may be scheduled during the CP and any locally generated CFP (generated optionally by the HC) to meet the QoS requirements of a particular traffic stream. TXOP allocations and contention-free transfers of QoS traffic can be based on the HC's knowledge of the amounts of pending traffic belonging to different traffic streams and are subject to QoS policies.

10.3.3 WLAN QoS Algorithms

IEEE 802.11e standard specifies both channel access differentiation mechanism by EDCA and delay-bounded channel access mechanism by HCCA. However, those mechanisms cannot guarantee QoS for multimedia applications on their own. Providing QoS guarantees requires extra functionalities from the network devices beyond MAC layer QoS mechanisms. These functionalities include classifications, queuing and scheduling, buffer management, and resource reservation. Wireless LAN AP and stations are parts of devices in the end-to-end network path, and are required to implement the earlier-mentioned functionalities for QoS support as well. In the following subsections, several algorithms implementing the functionalities are introduced.

10.3.3.1 Scheduling

A number of scheduling algorithms have been proposed for network-layer packet handling [8]. They can be used directly to schedule the packets in the IEEE 802.11e MAC-layer buffer. These algorithms seek to allocate link bandwidth among the various classes of traffic in *a priori* or fair manner, and provide statistical, aggregate, or per-flow guarantees on network parameters such as delay, jitter, and packet loss. Most scheduling algorithms can be divided into two classes [8].

- Work-conserving scheduler, which is not idle when there is a packet to transmit in any of its queues.
- Non-work-conserving scheduler, which may choose to remain idle even if there is a packet waiting to be served. This type of scheduler may be useful in reducing burstiness of the traffic or in providing a strict guarantee.

In the 802.11e standard, a simple reference scheduling algorithm, referred to as TGe scheduler, is proposed for controlling the medium access [2]. The HC first decides on a service interval (SI), the highest submultiple value of the beacon interval, which is smaller than the minimum of the maximum SI (MSI) of all admitted flows. According to the SI, aggregate HCCA TXOPs allocated to the station are calculated. The transmission duration TD_i for station i with n admitted traffic streams is computed by [2]

$$TD_i = \sum_{j=1}^{n} \max\left(\frac{N_j L_j}{R_j} + O,\ \frac{M_j}{R_j} + O\right) \tag{10.1}$$

where R_j denotes the minimum PHY rate and ρ_j the mean data rate requirement for the j stream, $N_j \rho_j = (SI)/L_j$, and O the total MAC/PHY overhead in the frame exchange sequences. The scheduler issues polls to the QSTAs consecutively in a round-robin manner.

Inan et al. [18] proposed an application-aware adaptive HCCA scheduler, which uses the earliest deadline first (EDF)–scheduling discipline. The EDF approach makes the polling order adaptive according to the calculated deadlines of associated traffic profiles. The multimedia traffic is scheduled in the sense that each QoS station has a distinct SI. The SI and the TXOP of the QoS station are adapted according to the traffic characteristics as well as instantaneous buffer occupancy information in RT.

10.3.3.2 Admission Control and Reservation

An IEEE 802.11 network may use admission control to administer policy or regulate the available bandwidth resources [2]. Admission control is also required when a station desires guarantee on the amount of time it can access the channel. The HC can be used to administer admission control in the network. Admission control, in general, depends on vendors' implementation of the scheduler, available channel capacity, link conditions, retransmission limits, and the scheduling requirements of a given stream. All of these criteria affect the admissibility of a given stream. As the QoS facility supports two access mechanisms, there are two distinct admission control mechanisms: one for contention-based access and the other for controlled access.

Similar to DCF, EDCA is very likely to be the dominant channel access mechanism in WLANs because it is a distributed MAC scheme and easy to implement. In the past few years, there has been a lot of research work focusing on admission control in EDCA. Basically, the existing EDCA admission control schemes can be classified into two categories: measurement- and model-based [19]. In measurement-based schemes, admission control decisions are based on continuously measured network conditions such as throughput and delay. In contrast, the model-based schemes construct certain performance metrics to evaluate the status of the network.

For the controlled-based access, a reference design of a simple scheduler and admission control is proposed in the 802.11e standard [2]. This scheduler takes the QoS requirements of flows into account and allocates TXOP duration to stations on the basis of the mean sending rate and mean packet size. The QoS requirements include mean data rate, nominal MAC service data unit (MSDU) size, and maximum SI or delay bound. It is shown that the reference design is efficient to the flows with constant bit rate (CBR) characteristic. But it does not take the rate and packet size fluctuation of variable bit rate (VBR) traffic into account. The packet loss probability of VBR traffic may be large.

Gao et al. [20] proposed a modified admission control algorithm, named PHY rate-based admission control (PRBAC), which considers PHY rate variance due to station mobility and wireless channel characteristics. The basic idea is to use the long-term average PHY rates for admission control, and

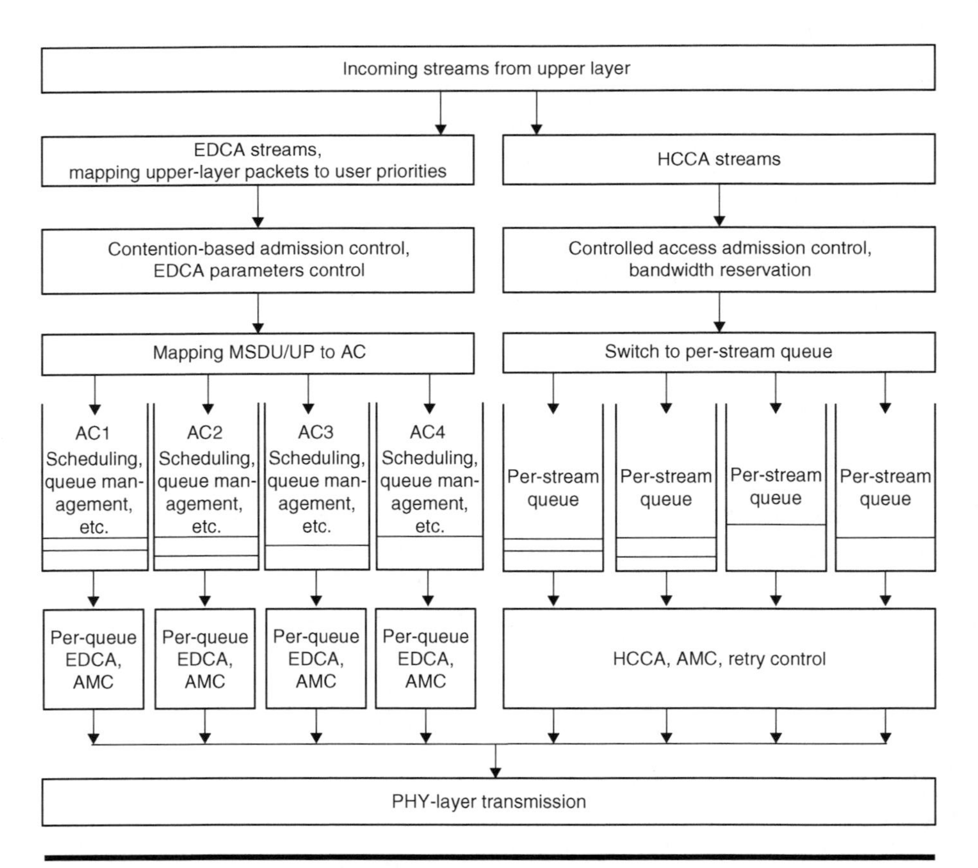

Figure 10.1 IEEE 802.11 MAC QoS schemes.

at the same time use the instantaneous PHY rates to distribute TXOPs to individual stations. In this way, the algorithm can admit more flows than the reference scheme. An admission control scheme for VBR traffic was further proposed in Ref. 21.

Van et al. [22] proposed an optimized and scalable HCCA-based admission control for delay-constrained video streaming applications that leads to a larger number of stations being simultaneously admitted. The scheduling problem is formulated as a linear optimization task that enables to obtain its optimal solution by the simplex algorithm. Performance of the scheduler based on linear programming is evaluated for video calls with VBR traffic sources.

A reference implementation model is presented in Figure 10.1, in which both 802.11e MAC QoS mechanisms (EDCA and HCCA mechanisms) and MAC QoS schemes (admission control, scheduling, priority mapping, etc.) are included.

10.4 End-to-End QoS Support

IEEE 802.11 WLANs have been successfully applied as the last-mile technology in the increasingly pervasive computing environments where wireless and mobile users access Internet services via the AP. With more and more RT multimedia applications subscribed to by mobile users, there is an immediate demand for end-to-end QoS guarantee to be provided in wired-cum-WLAN heterogeneous networks. In this section, two end-to-end QoS architectures are presented.

10.4.1 IntServ and DiffServ Internet QoS Architectures

Two service architectures are adopted by the Internet Engineering Task Force (IETF): integrated services (IntServ) and differentiated services (DiffServ), for providing per-flow and aggregated-flow service guarantees, respectively [23,24]. The IntServ framework introduces service classes with different traffic characteristics to match the application of QoS requirements. Traffic from these service classes is treated differently at the routers with the aid of classifiers, queues, schedulers, and buffer-management schemes. An application in the IntServ environment uses a Resource Reservation Protocol (RSVP) [25] to signal and reserve the appropriate resources at the intermediate routers along the path from its source to destination(s).

The IntServ service model, in spite of its flexibility and per-flow QoS provisioning, has not been successfully deployed on the public Internet [8]. It may still be used in small-scale networks and within a customer network, but it is not suitable for the core of the Internet because of its inability to respond to traffic changes and lack of scalability. To address the problems of IntServ, the IETF decided to look for a more scalable alternative and developed the DiffServ architecture. DiffServ is based on resource provisioning rather than resource signaling and reservation as in IntServ [24]. The DiffServ handles traffic aggregates instead of microflows. As a result, on a per-flow basis, DiffServ provides qualitative instead of quantitative QoS guarantees provided by IntServ. The DiffServ standards define forwarding treatments, not end-to-end services such as the guaranteed and controlled load services in IntServ.

10.4.2 DiffServ-Based End-to-End QoS Support

Currently, there are some approaches that focus on the interoperability between IEEE 802.11 and DiffServ or IntServ. These approaches explore the possibility of protocol interoperability at the AP. Integrating DiffServ with IEEE 802.11 WLAN has been studied in Ref. 26. A collaborative end-to-end QoS architecture between DiffServ and IEEE 802.11e across wired wide area network (WAN), wired LAN, and WLAN was proposed based on DiffServ, IEEE 802.1D/Q, and 802.11e [26]. The user priority values are carried by 802.3 MAC frames via the 802.1Q virtual LAN (VLAN) tag. These user priorities are forwarded through 802.1D MAC bridge to 802.1e MAC and used by Enhanced Distributed Coordination Function (EDCF) to differentiate flows. DiffServ code point (DSCP) values, defined in the DiffServ (DS) field, and traffic category identification (TCID), defined in 802.1e are mapped with two methods: the IP packets are encapsulated and placed in priority queues without preemption, or IP packets are classified and shaped according to the priority of the DSCP values before being forwarded to 802.1e priority queues.

10.4.3 IntServ-Based End-to-End QoS Support

For the wired side, RSVP has been widely accepted as a flow reservation scheme in IEEE 802 style LANs. Li and Prabhakaran [27] investigate the integration of RSVP and a flow reservation scheme for the support of IntServ in support of heterogeneous wired-cum-wireless networks. A local bandwidth estimation and reservation scheme called "WRESV" is proposed for WLAN. In addition, the signaling and procedures for integrating RSVP and WRESV are designed. Message mappings at the AP are implemented by cross-layer interaction, and user priorities are mapped to 802.11 MAC priorities with 802.1p. Because WRESV can work with most of the existing MAC schedulers such as DCF, EDCF, and distributed fair scheduling (DFS), this integration scheme is general.

10.5 Cross-Layer Optimization and Adaption

Depending on the multimedia application, wireless infrastructure, and flexibility of the 802.11e standards, different approaches can lead to optimal performance. This section presents the cross-layer design and adaptation architectures.

10.5.1 Solutions for Cross-Layer Optimization

A classification of the possible solutions for cross-layer optimization is proposed in Ref. 12 based on the order in which cross-layer optimization is performed, to gain further insights into the principles.

- *Application-centric approach.* The application (APP) layer optimizes the lower-layer parameters one at a time in a bottom-up (starting from the PHY) or top-down manner, based on its requirements. However, this approach is not always efficient. The APP operates at slower timescales and coarser data granularities than the lower layers, and is not able to instantaneously adapt their performance to achieve an optimal performance.
- *MAC-centric approach.* The application layer passes its traffic information and requirements to the MAC that decides which application-layer packets or flows should be transmitted and at what QoS level. The MAC also decides the PHY-layer parameters based on the available channel information.
- *Integrated approach.* In this approach strategies are determined jointly. Exhaustively trying all the possible strategies and their parameters to choose the composite strategy leading to the best quality performance is impractical due to the associated complexity.

A cross-layer design problem is formulated as an optimization with the objective of selecting a joint strategy across PHY, MAC, and application layers for one-hop wireless LAN [12]. In the PHY layer, the strategies can represent the various modulation and coding schemes. In the MAC layer, the strategies correspond to different packetization, automatic repeat request (ARQ), scheduling, and admission control mechanisms. The strategies in the application layer correspond to adaptation of video compression parameters, frame selection, error concealment or recovery, packetization, traffic shaping, traffic prioritization, scheduling, ARQ, and FEC mechanisms. The cross-layer optimization problem is to find the optimal composite strategy over the strategy combinations subject to the wireless station and overall system constraints. Cooperation among the applications, MAC, and PHY layers in selecting the optimal PHY modulation strategy is shown resulting in the highest multimedia quality.

Ksentini et al. [28] establish another method of solving the cross-layer adaptation and optimization problem. First, formal procedures are used to establish optimal initialization, grouping of strategies at different stages (i.e., which strategies should be optimized jointly), and ordering (i.e., which strategies should be optimized first) to perform the cross-layer adaptation and optimization; second, different practical considerations (e.g., buffer sizes, ability to change retry limits or modulation strategies at the packet level) for the deployed 802.11e QoS MAC are taken into account; third, reduce the space of possible solutions and find the best strategy. They also present an example to investigate the dependencies among these strategies.

10.5.2 Adaptation Architecture

Owing to the mobility, burst multimedia traffics, dynamic wireless channels, and Internet links, adaptation is critical to the implementation of cross-layer optimization. To address this issue, an adaptation framework for scalable video transport over broadband wireless networks has been proposed [11]. In the adaptive framework, there are three basic components: (1) scalable video representations, each of which has its own specified QoS requirements; (2) network-aware end systems, which are aware of network status and can adapt the video streams accordingly; and (3) adaptive services, with which the networks support the adaptive QoS required by scalable video representations. Under this framework, as wireless channel conditions change, wireless stations and network elements can scale the video streams and transport the scaled video streams to receivers with a smooth change in perceptual quality.

Although the adaptation framework was originally proposed for video applications over general wireless networks, it can be directly extended for the purpose of delivering multimedia applications over IEEE 802.11 WLANs. A modified adaptation framework for IEEE 802.11 WLAN multimedia applications is presented in Figure 10.2. In the fixed stations and 802.11 wireless stations, raw multimedia (video, audio, image, etc.) traffics are encoded–decoded with scalable codecs. The encoding procedure is assisted by the network (and wireless channel) monitor for encoding parameters control (such as rate control, error recovery, or concealment). The network monitor module is also responsible for coordinating with the network transport module on congestion and reliability control. Wireless channel monitor module will help in the control of 802.11 MAC QoS schemes, AMC, and ARQ schemes. The direction of multimedia streams can be either from fixed network stations to 802.11 wireless stations or in a reverse direction.

10.6 Case Study of Media-Oriented Applications over WLAN

The concepts of cross-layer design and adaptation have gained research interests for multimedia transmission over wireless LANs. In this section,

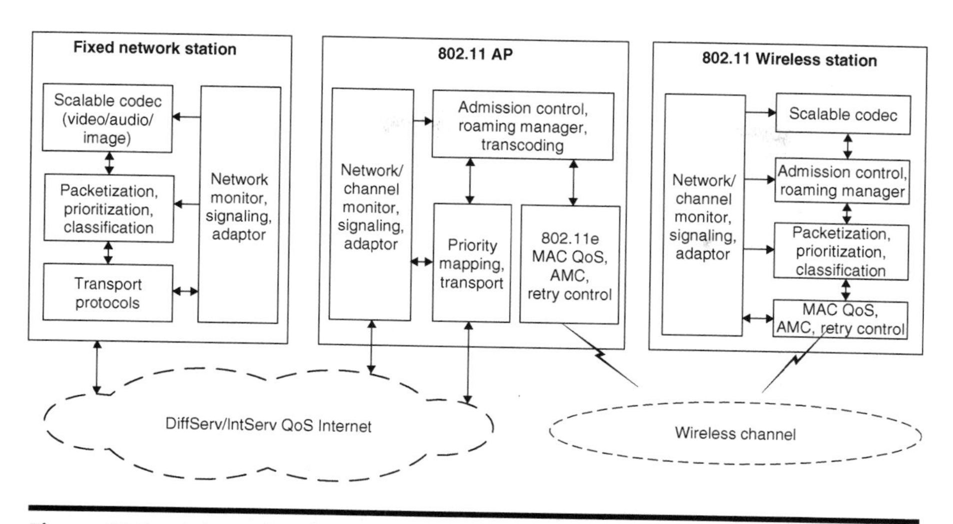

Figure 10.2 Adaptation framework for multimedia applications over IEEE 802.11 WLAN.

some applications of cross-layer and adaptation into different types of multimedia transmissions are presented, although not all of the strategies available in the considered layers are taken into account in the optimization.

10.6.1 Image Transmission

Frescura et al. [29] propose a MAC-centric cross-layer JPEG2000 transmission scheme. The performance of error resilience tools in JPEG2000 is first analytically modeled, which is validated by simulation results. The analytical model is utilized by the AP for designing efficient unequally error protection schemes for JPEG2000 transmission. A utility function is defined to make trade-off between the image quality and the cost for transmitting the image over wireless channel.

He et al. [30] propose another MAC-centric cross-layer transmission scheme for JPEG2000 and motion JPEG2000 codestream in an 802.11 WLAN environment. JPEG2000 and MJPEG2000 codestreams are transmitted using unreliable protocols such as User Datagram Protocol–Real-Time Transport Protocol (UDP-RTP) with a protection scheme. For the transmitted packets, a table contains the values which are absolute or relative to a measure of the sensitivity. Through this information, unequal error protection (UEP), different QoS policies, and intelligent retransmission techniques can be achieved. The proposed technique may be used in a wide range of applications, including point-to-multipoint digital slide projection and wireless video-surveillance systems.

10.6.2 Voice-over-IP

Yu et al. [31] propose a MAC layer priority queuing-based approach to enhance the VoIP performance over 802.11 WLAN. Two queues are implemented along with strict priorities on top of the 802.11 MAC controller. The proposed scheme is seen to be remarkably effective for VoIP service in coexistence with non-RT traffic, by the use of flow-control mechanism of the Transmission Control Protocol (TCP).

Wan and Du [32] propose another MAC-layer polling-based scheme for VoIP over 802.11 WLAN. The multimedia packets access delay incurred in the AP is analyzed based on a prioritized M/G/1 analysis model. An improved PCF polling-scheduling algorithm is developed for improving the WLAN's multimedia traffic application by reducing the access delay.

Wu et al. [33] propose a MAC-layer-based software solution, called layer 2.5 SoftMAC for VoIP services over commercial IEEE 802.11 MAC networks. SoftMAC resides between the IEEE 802.11 MAC layer and the IP layer to coordinate the RT multimedia and BE data-packet transmission. In the scheme, distributed admission control is used to regulate the load of RT traffic; rate control is used to minimize the impact of BE traffic on real-time one; and nonpreemptive priority queuing is used to provide high-priority service to VoIP traffic. The effectiveness of their proposed solution is demonstrated.

Narbutt and Davis [34] investigated the relationship between resource utilization and the quality of VoIP calls transmitted over in the wireless LAN. Through experimentation with various codecs and packetization schemes, it was found that as the load (number of calls) reaches the available capacity level, packet delays and jitter increase dramatically resulting in degraded call quality. In addition, a method of using MAC bandwidth to assess the VoIP call quality on 802.11 WLANs is proposed.

10.6.3 Video Streaming

Van der Schaar et al. [35] proposed an application-MAC-PHY adaptation-based cross-layer admission control for delay-constrained video streaming applications with IEEE 802.11e HCCA mechanism. Given the allocated TXOPs, each station deploys an optimized scheduling and UEP strategy that is facilitated by the fine-grain layering provided by the scalable bit-stream. This cross-layer strategy enables graceful quality degradation even when the channel conditions or the video sequence characteristics change.

Ksentini et al. [28] proposed an 802.11e QoS mechanism-based application and MAC cross-layer transmission scheme for H.264 video streams over IEEE 802.11 WLAN. The proposed cross-layer architecture relies on a DP technique at the application layer and an appropriate QoS mapping at the 802.11e-based MAC layer. By employing DP, the H.264 encoder partitions the compressed data in separate units of different importance. Based on the QoS requirements of these different partitions, a marking algorithm is specified at the MAC layer, associating each partition with an AC provided by 802.11e EDCA. The application layer can pass its streams along with their requirements to protect the most important H.264 information, which guarantees low degradation of received H.264 stream.

Wong et al. [36] proposed a classification-based MAC and application-layer cross-layer strategy for wireless multimedia transmission over 802.11 wireless LAN. Both content- and channel-related features are used to select a joint application-MAC strategy from the strategies available at the various layers. Retry limit is used as the MAC-layer strategy. Preliminary results indicate that considerable improvements can be obtained.

Tan et al. [37] proposed a PHY and application-layer-based cross-layer end-to-end joint source/channel coding protocol for MPEG-4 video over 802.11b WLAN. The idea of receiver-driven layered multicast has been extended to handle packet loss, by adding receiver-driven UEP. The protocol can efficiently recover from rate reduction of heavy burst loss, although having good TCP-friendliness under low-loss cases.

The left video streaming schemes used MAC-layer retry limit–based MAC-application-layer cross-layer designs. Wong et al. [38] provide an adaptive DCF for video transmission over 802.11 WLAN. Adaptive retransmission limit is used for video packets with different priorities. The actual arrival rates of video flow packets are also considered. Li and van der Schaar [39] use a similar idea in designing error protection method for wireless video. The retry limit settings of the MAC layer is optimized to minimize the overall packet losses and maximize video quality. Simulations show that the proposed cross-layer protection method can result in better video quality. Lu et al. [40] propose a time stamp-based content-aware adaptive retry (CAR) mechanism for MPEG video streaming over 802.11 WLANs. In the scheme, MAC dynamically determines whether to send or discard a packet based on its retransmission deadline. The retransmission deadline is assigned to each packet according to its temporal relationship and error propagation characteristics with respect to other video packets in the same group of pictures. Simulation results show CAR significantly improves video quality and saves channel bandwidth. Ni et al. [41] develop a bandwidth adaptive technique for media streaming over IEEE 802.11 networks. An RT detection of wireless bandwidth change is designed; streaming media is dynamically adapted to optimize the quality; coordination and performance optimization are performed among media streaming server, wireless AP points, and wireless clients. Bucciol et al. [42] developed a cross-layer ARQ algorithm for video streaming over 802.11 wireless networks. The algorithm combines application-level information about the perceptual and temporal importance of each packet into a single priority value. The priority value is then used to drive packet selection at each retransmission opportunity.

H.264 video streaming based on the proposed technique has been simulated. Results show that the proposed method consistently outperforms the standard MAC-layer 802.11 retransmission scheme, delivering more than 1.5 dB peak signal-to-noise ratio (PSNR) gains using approximately half of the retransmission bandwidth.

10.7 Seamless Multimedia QoS Support

Recent efforts have been made to extend 802.11 WLANs into outdoor cellular networks to provide complete mobile broadband service with ubiquitous coverage and high-speed connectivity. The fluctuations of bandwidths during handoffs and unpredictability of terminal mobilities pose big challenges on the QoS support for multimedia applications over wireless LAN. This section presents some approaches on providing seamless multimedia QoS support between wireless LAN and other wireless networks.

10.7.1 WLAN Roaming

Samprakou et al. [43] present a fast IP handoff method for 802.11 multimedia applications, which quickly restores IP connectivity for mobile clients. Brickley et al. [44] present a cell breathing–based technique to optimize the received quality for delay sensitive RT multimedia applications in 802.11b WLAN, by balancing the load between various APs, although changing the coverage area of each AP. Raghavan and Zeng [45] propose an adaptive scheme based on the RSS for improving the quality of the multimedia over 802.11 WLAN. Mobility information from individual users and that combined from all wireless users are used to reduce the inter-AP communication and thus, provide quality improvement with a reduced overhead. QoS for multimedia traffic is improved with this scheme.

10.7.2 Interworking with Cellular Networks

Refs 46 and 47 investigate QoS interworking of GPRS and WLANs. Chen and Lin [46] present a gateway approach to the integration of GPRS and WLANs. The proposed architecture leverages mobile IP as the mobility management protocol over WLANs. The interworking between GPRS and WLANs is achieved by a gateway that resides on the border of GPRS and WLAN systems. Empirical experiments with multimedia applications were conducted to analyze the performance. Indulska and Balasubramaniam [47] describe a generic, context-aware handover solution for multimedia applications and illustrate how this handover works for redirection of communications between WLANs and GPRS or UMTS networks.

QoS interworking of WLANs and 3G wireless networks is investigated in Refs 48–52. Yu and Krishnamurthy [48] propose a joint-session admission-control scheme to optimize the utilization of radio resources in 802.11e and Wideband Code Division Multiple Access (WCDMA) network. The concept of effective bandwidth is used in the CDMA network to derive the unified radio resource usage. Lee et al. [49] propose a heterogeneous RSVP extension mechanism called as Heterogeneous Mobile RSVP (HeMRSVP), which allows mobile hosts to reach the required QoS service continuity while roaming across UMTS and WLAN networks. Numerical results show that the HeMRSVP achieves significant performance improvement. Wu et al. [50] analyze the delay associated with vertical handoff using Session Initiation Protocol (SIP) in the WLAN-UMTS Internetwork. Analytical results show that WLAN-to-UMTS handoff incurs unacceptable delay for supporting RT multimedia services, and is mainly due to transmission of SIP signaling messages over erroneous and bandwidth-limited wireless links. In contrast, UMTS-to-WLAN handoff experiences much less delay, mainly contributed by the processing delay of signaling messages at the WLAN gateways and servers. Although the former case requires the deployment of soft handoff techniques to reduce the delay, faster servers and more efficient host configuration mechanisms can do the job in the latter case. Salkintzis et al. [51] evaluated the conditions and restrictions under which seamless multimedia QoS provision across UMTS and IEEE 802.11e WLANs is feasible. Results indicate that WLANs can support seamless continuity of video sessions for a relatively small number of UMTS subscribers, depending on the applied WLAN policy, access parameters, and QoS requirements. The 3rd Generation Partnership Project (3GPP) has specified the IP multimedia subsystem (IMS) for the provisioning of multimedia services in UMTS Release 5 and later. Interconnection at the service layer between 3GPP and LAN networks requires interworking between IMS and WLAN functionalities. Marquez et al. [52] analyzed how the interconnection of 3GPP and WLAN networks may be performed to support different levels of service interconnection. Special attention is paid to the interconnection at the session negotiation level, using SIP (the base protocol of the IMS) to provide session negotiation with QoS and authentication, authorization, and accounting (AAA) support.

QoS interworking of WLAN and general cellular networks is studied in Refs 53–55. Saravanan et al. [53] proposed a priority-based service interworking architecture with hybrid coupling for address resource management and call admission control in cellular or WLAN interworking. Better QoS is provided to high-prioritized voice traffic. Ruuska and Prokkola [54] developed a new approach to end-to-end QoS in wireless heterogeneous networks. DiffServ is found to be more appropriate to support widely roaming users compared to IntServ. Schorr et al. [55] present a system that adapts media streams to the respective characteristics of different (wireless) access networks. An interface selection subsystem monitors characteristics of the network interfaces. A service subsystem initiates media stream adaptation to match network resource availability and provider constraints. A media subsystem transmits the media streams and applies appropriate error correction mechanisms (e.g., FEC and ARQ) depending on network characteristics.

10.8 Open Issues

Although significant research efforts have been made toward ubiquitous high-speed multimedia applications over broadband wireless LANs, the problems of seamless QoS and mobility support are not completely solved. This section identifies several open issues.

10.8.1 Systematic Study

As described in the previous sections, there are various approaches for multimedia support over wireless LAN. However, in most of the approaches, only a part of combinations of strategies available in the PHY, MAC, transport, and application layers of the OSI model are considered. Considering the architectures of cross-layer design and adaptation, systematic study will be required to optimize the performance of multimedia applications over broadband wireless LAN. In addition, the multimedia application and wireless network performances as functions of the available strategies should be thoroughly analyzed.

10.8.2 Multimedia Support over New MAC

The limited wireless network capacity of wireless LAN poses big challenges on supporting a large number of high-quality multimedia applications. In addition, for the wireless LANs operating in the 2.4 GHz frequency bands, there can be large interference from other wireless networks or electronic devices. In parallel with the standard efforts of developing high-speed IEEE 802.11n PHY amendment, there are research interests in utilizing the available multichannels and dynamic frequency selections for capacity improvement and interference resistance. Currently, there are few research works on how to support QoS for multimedia applications over wireless LAN MACs operating with multichannels. In contrast, opportunistic spectrum access received large research interests. The next-generation wireless LAN MAC protocols may be required to detect and use the opportunistic radio spectrum and cooperate with other opportunistic spectrum access systems. With the new MAC protocols, the QoS mechanisms, end-to-end QoS support, cross-layer adaptation, and optimization for multimedia applications over broadband wireless LANs will require new design considerations.

10.8.3 Multimedia over Multihop WLAN

Multihop networking and relaying have been proved to be effective in improving wireless link and network reliability and increasing network coverage and capacity. However, for the multimedia applications delivered over multihop wireless LAN, the frequent network topology changes and the distributed operations of mobile terminals make the QoS support problem more challenging. Although there are many research efforts on the multimedia support over multihop wireless LANs, to the best of our knowledge, there is no efficient QoS scheme proposed for IEEE 802.11 standards-based multihop wireless LANs. Further studies will be required for the QoS support problem.

10.9 Conclusion

This chapter investigates the research issues of transporting multimedia applications over broadband wireless LANs. The dynamic wireless channel conditions, network bandwidth fluctuations, and insufficient QoS support schemes make the issues very challenging. After introducing the basic features of wireless LAN and multimedia standards, we identified a number of research challenges. Research efforts and existing approaches to resolve the issues are discussed, with focus on the wireless LAN QoS mechanisms and algorithms, end-to-end QoS support, cross-layer adaptation, and optimization. Case studies of media-type oriented applications over wireless LANs and seamless QoS support are also introduced. The final section highlights open issues and future research directions.

References

1. IEEE Std. 802.11, Wireless LAN Medium Access Control (MAC) and Physical Layer (PHY) Specifications, 1999.
2. IEEE Std. 802.11e, Wireless LAN Medium Access Control (MAC) Enhancements for Quality of Service (QoS), 2005.
3. Richardson, K. W. UTMS overview, *IEEE Electron. Control Eng. J.*, 12, 93–100, 2000.
4. CDMA2000 High Rate Broadcast-Multicast Packet Data Air Interface Specification. 3GPP2, 2005.
5. Moccagatta, I., S. Soudagar, J. Liang, and H. Chen, Error-resilient coding in JPEG-2000 and MPEG-4, *IEEE J. Select. Areas Commun.*, 18(6), 899–914, 2000.
6. ITU-T Rec. H.264/ISO/IEC 14496-10 AVC, JVT-G050, Joint Video Team (JVT) ISO/IEC MPEG, ITU-T VCEG, Final Draft, International Standard on Joint Video Specification, 2003.
7. Doufexi, A., S. Armour, M. Butler, A. Nix, D. Bull, J. McGeehan, and P. Karlsson, A comparison of the HIPERLAN/2 and IEEE 802.11a wireless LAN standards, *IEEE Commun. Mag.*, 40, 172–180, 2002.

8. El-Gendy, M., A. Bose, and K. G. Shin, Evolution of the Internet QoS and support for soft real-time applications, *Proc. IEEE*, 91(7), 1086–1104, 2003.
9. Ferre, P., A. Doufexi, A. Nix, and D. Bull, Throughput analysis of IEEE 802.11 and IEEE 802.11e MAC, *Proc. IEEE WCNC*, 2, 783–788, 2004.
10. Caripe, W., G. Cybenko, K. Moizumi, and R. Gray, Network awareness and mobile agent systems, *IEEE Commun. Mag.*, 36, 44–49, 1998.
11. Wu, D., Y. Hou, and Y. Zhang, Scalable video coding and transport over broadband wireless networks, *Proc. IEEE*, 89(1), 6–20, 2001.
12. Van der Schaar, M. and S. Shankar, Cross-layer wireless multimedia transmission: challenges, principles, and new paradigms, *IEEE Wireless Commun.*, 12, 50–58, 2005.
13. Tarokh, V., A. Naguib, N. Seshadri, and A. R. Calderbank, Space-time codes for high data rate wireless communication: performance criterion and code construction, *IEEE Trans. Inf. Theory*, 44, 744–765, 1998.
14. Alamouti, S., Space block coding: a simple transmitter diversity technique for wireless communications, *IEEE J. Select. Areas Commun.*, 16, 1451–1458, 1998.
15. Pavon, J. and S. Choi, Link adaptation strategy for IEEE 802.11 WLAN via received signal strength measurement, *Proc. IEEE ICC*, 2, 1108–1113, 2003.
16. Qiao, D., S. Choi, and K. Shin, Goodput analysis and link adaptation for IEEE 802.11a wireless LANs, *IEEE Trans. Mobile Comp.*, 1, 278–292, 2002.
17. Kamerman, A. and L. Monteban, WaveLAN-II: a high-performance wireless LAN for the unlicensed band, *ATT Bell Lab. Tech. J.*, 2, 118–133, 1997.
18. Inan, I., F. Keceli, and E. Ayanoglu, An adaptive multimedia QoS scheduler for 802.11e wireless LANs, *Proc. IEEE ICC*, 11, 5263–5270, 2006.
19. Gao, D., J. Cai, and K. Ngan, Admission control in IEEE 802.11e wireless LANs, *IEEE Network*, 19, 6–13, 2005.
20. Gao, D., J. Cai, and L. Zhang, Physical rate-based admission control for HCCA in IEEE 802.11e WLANs, *Proc. IEEE AINA*, Taiwan, March 2005.
21. Fan, W. F., D. Y. Gao, D. H. K. Tsang, and B. Bensaou, Admission control for variable bit rate traffic in IEEE 802.11e WLANs, *Proc. IEEE LANMAN04*, Mill Valley, CA, 2004.
22. Van der Schaar, M., Y. Andreopoulos, and H. Zhiping, Optimized scalable video streaming over IEEE 802.11 a/e HCCA wireless networks under delay constraints, *IEEE Trans. Mobile Comp.*, 5, 755–768, 2006.
23. Braden, R., D. Clark, and S. Shenker, Integrated services in the Internet architecture: an overview, *IETF, RFC 1633*, June 1994.
24. Blake, S., D. Black, M. Carlson, E. Davis, Z. Wang, and W. Weiss, An architecture for differentiated services, *IETF, RFC 2475*, December 1998.
25. Braden, B., L. Zhang, S. Berson, S. Herzog, and S. Jamin, Resource reservation protocol (RSVP)—version 1 functional specification, *IETF, RFC 2205*, September 1997.
26. Park, S.-Y., K. Kim, D. C. Kim, S. Choi, and S. Hong, Collaborative QoS architecture between DiffServ and 802.11e wireless LAN, *Proc. IEEE VTC'03-Spring*, Jeju, Korea, April 2003.
27. Li, M. and B. Prabhakaran, MAC layer admission control and priority re-allocation for handling QoS guarantees in non-cooperative wireless LANs, *ACM/Springer MONET*, October 2005.
28. Ksentini, A., M. Naimi, and A. Gueroui, Toward an improvement of H.264 video transmission over IEEE 802.11e through a cross-layer architecture, *IEEE Commun. Mag.*, 44, 107–114, 2006.

29. Frescura, F., M. Giorni, C. Feci, and S. Cacopardi, JPEG2000 and MJPEG2000 transmission in 802.11 wireless local area networks, *IEEE Trans. Consumer Electron.*, 49, 861–871, 2003.
30. He, J., Z. Yang, D. Yang, Z. Tang, J. Hu, and C. T. Chou, Analysis and representation of statistical performance of JPEG2000 encoded image over wireless channels, *Proc. IEEE Int. Conf. Electron., Circuits Syst.*, 3, 1051–1054, 2002.
31. Yu, J., S. Choi, and J. Lee, Enhancement of VoIP over IEEE 802.11 WLAN via dual queue strategy, *Proc. IEEE ICC*, 6, 3706–3711, 2004.
32. Wan, Q. and M. Du, Queueing analysis and delay mitigation in the access point of VoWLAN, *IEEE Intl. Symp. Commun. Inf. Tech.*, 2, 1160–1163, 2005.
33. Wu, H., Y. Liu, Q. Zhang, and Z. Zhang, SoftMAC: layer 2.5 collaborative MAC for multimedia support in multihop wireless networks, *IEEE Trans. Mobile Comp.*, 6, 12–25, 2007.
34. Narbutt, M. and M. Davis, Gauging VoIP call quality from 802.11 WLAN resource usage, *Proceedings of the IEEE WoWMoM*, Niagara-Falls, Canada, 2006.
35. Van der Schaar, M., Y. Andreopoulos, and Z. Hu, Optimized scalable video streaming over IEEE 802.11 a/e HCCA wireless networks under delay constraints, *IEEE Trans. Mobile Comp.*, 5, 755–768, 2006.
36. Wong, R., M. Schaar, and D. Turaga, Optimized wireless video transmission using classification, *Proceedings of the IEEE ICME*, Amsterdam, The Netherlands, 2005.
37. Tan, K., Q. Zhang, and W. Zhu, An end-to-end rate control protocol for multimedia streaming in wired-cum-wireless environments, *Proc. Int. Symp. Circuits Syst.*, 836–839, 2004.
38. Wong, R., N. Shankar, and M. Schaar, Integrated application MAC modeling for cross-layer optimized wireless video, *Proc. IEEE ICC*, 2, 1271–1275, 2005.
39. Li, Q. and M. van der Schaar, Error protection of video over wireless local area networks through real-time retry limit adaptation, *Proc. IEEE ICASSP*, 5, 993–996, 2004.
40. Lu, M., P. P. Steenkiste, and T. Chen, Video streaming over 802.11 WLAN with content-aware adaptive retry, *Proc. IEEE ICME*, 1, 723–726, 2005.
41. Ni, Y., Y. Zhu, and H. H. Li, Bandwidth adaptive multimedia streaming for PDA applications over WLAN environment, *Proc. IEEE Int. Symp. Ind. Electron.*, 609–613, 2004.
42. Bucciol, P. P., G. Davini, E. Masala, E. Filippi, and J. C. De Martin, Cross-layer perceptual ARQ for H.264 video streaming over 802.11 wireless networks, *Proc. IEEE GLOBECOM*, 5, 3027–3031, 2004.
43. Samprakou, I., C. Bouras, and T. Karoubalis, Fast IP handoff support for VoIP and multimedia applications in 802.11 WLANs, *Proc. IEEE WoWMoM*, 1, 332–337, 2005.
44. Brickley, O., S. Rea, and D. Pesch, Load balancing for QoS optimisation in wireless LANs utilising advanced cell breathing techniques, *Proc. IEEE VTC-Spring*, 3, 2105–2109, 2005.
45. Raghavan, M. and Q. Zeng, Adaptive quality improvement scheme (AQIS) with composite mobility prediction for multimedia traffic in WLAN handoff, *Proc. Symp. Wireless Telecommun.*, 1, 1–7, 2006.
46. Chen, J. and H. Lin, A gateway approach to mobility integration of GPRS and wireless LANs, *IEEE Wireless Commun.*, 12, 86–95, 2005.
47. Indulska, J. and S. Balasubramaniam, Context-aware vertical handovers between WLAN and 3G networks, *Proc. IEEE VTC-Spring*, 5, 3019–3023, 2004.

48. Yu, F. and V. Krishnamurthy, Optimal joint session admission control in integrated WLAN and CDMA cellular networks with vertical handoff, *IEEE Tran. Mobile Comp.*, 6, 126–139, 2007.
49. Lee, G., L. Li, and W. Chien, Heterogeneous RSVP extension for end-to-end QoS Support in UMTS/WLAN interworking systems, *Proc. Int. Conf. PDCAT*, 1, 170–175, 2006.
50. Wu, W., N. Banerjee, K. Basu, and S. K. Das, SIP-based vertical handoff between WWANs and WLANs, *IEEE Wireless Commun.*, 12, 66–72, 2005.
51. Salkintzis, A., D. Skyrianoglou, and N. Passas, Seamless multimedia QoS across UMTS and WLANs, *Proc. IEEE VTC*, 4, 2284–2288, 2005.
52. Marquez, F. G., M. G. Rodriguez, T. R. Valladares, T. de Miguel, and L. A. Galindo, Interworking of IP multimedia core networks between 3GPP and WLAN, *IEEE Wireless Commun.*, 12, 58–65, 2005.
53. Saravanan, I., G. Sivaradje, and P. P. Dananjayan, QoS provisioning for cellular/WLAN interworking, *Proceedings of IFIP International Conference on Wireless and Optical Communication Networks*, Bangalore, India, 2006.
54. Ruuska, P. P. and J. Prokkola, Supporting IP end-to-end QoS architectures at vertical handovers, *Proceedings of the IEEE AINA'06*, Vienna, Austria, 2006.
55. Schorr, A., A. Kassler, and G. Petrovic, Adaptive media streaming in heterogeneous wireless networks, *Proc. IEEE Workshop Multimedia Signal Process.*, 1, 506–509, 2004.

Chapter 11

Improving User-Perceived Quality for Video Streaming over WLAN

Nikki Cranley and Gabriel-Miro Muntean

Contents

11.1 Introduction 363
11.2 Wireless Local Area Networks 364
  11.2.1 Overview 364
  11.2.2 IEEE 802.11 Family 365
    11.2.2.1 IEEE 802.11b 366
    11.2.2.2 IEEE 802.11a 367
    11.2.2.3 IEEE 802.11g 367
    11.2.2.4 Distance Range and Transmission Rate Comparisons 368
  11.2.3 IEEE 802.11e: MAC Enhancement for QoS 368
    11.2.3.1 Enhanced Distributed Coordination Access 369
11.3 Adaptive Multimedia Streaming Systems and Solutions 371
  11.3.1 Multimedia Streaming: Overview 371
  11.3.2 Adaptive Multimedia Streaming: Principle 372

11.3.3 Adaptive Multimedia Streaming: Solutions 373
11.3.3.1 TCP-Friendly Rate Control and TFRC Protocol 375
11.3.3.2 Enhanced Loss Delay Adjustment Algorithm 376
11.3.4 Adaptive Multimedia Streaming: Discussion 376
11.4 Multimedia Streaming User-Perceived Quality 379
11.4.1 QoS-Related Factors Affecting Multimedia Streaming 379
11.4.1.1 Heterogeneity 379
11.4.1.2 Congestion 379
11.4.1.3 Bandwidth Fluctuations 379
11.4.1.4 Delay 380
11.4.1.5 Jitter 380
11.4.1.6 Loss 381
11.4.1.7 Noise 381
11.4.2 Objective Video Quality Metrics 381
11.4.2.1 Mathematical Metrics 382
11.4.2.2 Model-Based Metrics 383
11.5 Optimum Adaptation Trajectory-Based Quality-Oriented Adaptation Scheme 384
11.5.1 User Perception in Adaptive Video Streaming 384
11.5.2 Quality-Oriented Adaptation Scheme Principle 384
11.5.3 Optimum Adaptation Trajectory Principle 386
11.5.3.1 Test Methodology 387
11.5.3.2 Test Sequence Preparation 389
11.5.3.3 Results 390
11.5.3.4 Discussion 395
11.6 Quality-Oriented Multimedia Streaming Evaluation 395
11.6.1 Simulation Setup 396
11.6.2 Multimedia Clips 397
11.6.3 Simulation Models 397
11.6.4 Testing Scenarios 397
11.6.5 Testing Results 398
11.7 Conclusions 401
References 403

Multimedia streaming over wireless networks is becoming increasingly popular. Adaptive solutions are proposed to compensate for high fluctuations in the available bandwidth to increase streaming quality. In this chapter we characterize major wireless local area network (WLAN) technologies and discuss the limitations of existing adaptive solutions during wireless streaming. We then show that streaming quality increases by using the quality-oriented adaptation scheme (QOAS), which explicitly considers user-perceived quality in the adaptation process. Further improvement is achieved by performing the rate adaptation according to an optimal adaptation trajectory (OAT) through the set of possible encoding configurations to obtain the same average bit rate.

11.1 Introduction

There is an increasing demand for streaming video applications over both wired and wireless Internet Protocol (IP) networks. Unlike wired broadband connectivity that offers large bandwidth to video streaming applications, these applications are put under pressure when used over bandwidth-limited WLANs and consequently may offer services of lower quality to their users. To address this issue, significant efforts have been made to develop compression-efficient encoding standards such as MPEG-4 and H.264. Apart from WLANs limited bandwidth, they are also susceptible to fluctuating bandwidth and time-varying delays making it challenging to support the delivery of high-quality video streaming under these conditions. Adaptive video streaming delivery solutions have been developed and achieve good streaming quality, but very few explicitly consider end-user-perceived quality during the adaptation process.

In general, video quality adaptation and video quality evaluation are considered distinct activities. Recent work in multimedia streaming adaptation has seen the integration of objective video quality metrics to evaluate the end-user-perceived quality as a key component of the adaptation process. However, many of these objective metrics have been found to be poorly designed to measure the fluctuations in video quality caused by time-varying network conditions. Moreover many objective video quality metrics have been found to have a poor correlation to the human visual system. This chapter provides an overview of WLAN solutions and describes main challenges for multimedia streaming over WLAN networks. The limitations of existing adaptive wireless streaming solutions are also discussed. The chapter shows how knowledge of video quality perception can be integrated and used to optimally adapt the video quality so as to maximize end-user-perceived quality. In this context we demonstrate that significant improvements in terms of both end-user-perceived quality and bandwidth efficiency for wireless video streaming can be achieved. This chapter also provides a review of objective video quality metrics and subjective video-testing methodologies.

Through extensive subjective testing it has been found that an OAT, through the set of possible video encoding configurations, exists. The OAT indicates how the video quality should be adapted in terms of the encoding configuration parameters to maximize the end-user-perceived quality. The OAT varies according to the time-varying characteristics of the video content. For example, when the bit rate of the video stream must be reduced to achieve a target transmission bit rate, for high-action video clips the resolution should be reduced and the frame rate preserved because users cannot assimilate all the spatial detail but are more sensitive to continuity of motion within the scene. In contrast, for low action clips the frame rate can be reduced because there is low temporal activity within the scene and continuity of motion can be preserved in spite of a reduced frame rate; however, users are more sensitive to the loss of spatial information (SI) for this content type.

The QOAS considers end-user-perceived quality estimates during adaptive multimedia streaming. Based on feedback, QOAS performs bit rate adaptation of the video stream and achieves increased streaming quality in highly variable and bandwidth-constrained network delivery conditions. The QOAS uses the OAT to determine how the video quality should be adapted to achieve the target transmission bit rate. Experimental results illustrate that using a two-dimensional adaptation strategy based on the OAT outperforms one-dimensional adaptation schemes that affect either multimedia stream's resolution or its frame rate, providing better user-perceived quality. Simulation-based test results are presented showing how by using QOAS for video streaming loss rate decreases, average throughput increases, and the user-perceived quality improves. When the OAT is used in conjunction with QOAS, end-user-perceived quality is further improved during the delivery of the video stream over wireless networks.

This chapter describes methods to provide end-user-perceived quality of service (QoS) for video streaming applications over IEEE 802.11b. The remainder of this chapter is structured as follows. Section 11.2 describes briefly the WLAN family of standards. Section 11.3 provides an overview of adaptive multimedia streaming applications and the challenges faced by systems providing such services. Section 11.4 describes the factors that affect end-user-perceived quality. The objective video quality metrics that can be used to provide a measure of distortion of the multimedia stream is described. Section 11.5 describes an adaptive video streaming system that provides optimized delivery of multimedia services over IEEE 802.11b WLAN and that explicitly considers end-user-perceived quality during the adaptation process. Section 11.6 presents simulation results that demonstrate the superior performance of the optimal adaptation trajectory-based-quality-oriented adaptation scheme (OAT-QOAS) adaptive system over two widely used multimedia adaptation systems, namely, Loss Delay Adjustment Algorithm (LDA+) and TCP-Friendly Rate Control (TFRC), and nonadaptive multimedia streaming.

11.2 Wireless Local Area Networks

11.2.1 Overview

The proliferation of wireless communications is affecting the way modern technology users perceive the traditional "wired world." The desire for systems that offer improved flexibility and mobility has stimulated people into considering a whole new breed of wireless technologies as viable alternatives to the traditional wired networks. Wireless technologies also offer the potential of low deployment costs and broadband bandwidth capability. However, an increased interest in the area has resulted in a knock on necessity to develop new and improved standards to cope with the consumer's growing needs. As various bodies develop these new standards, it is important that at all times the spectrum allocation process is carefully monitored and controlled. Loose regulation may not only accelerate the expansion of the WLAN market but also create interference problems; in contrast, strict regulation could allocate the spectrum well but might impede market development. Section 11.2.2 provides a detailed presentation of WiFi—the IEEE 802.11 family. From a long list of existing wireless technologies—WiMax IEEE 802.16, Bluetooth, UltraWideBand, and HIPERLAN2—WiFi is discussed because it is the most popular wireless technology for multimedia delivery.

11.2.2 IEEE 802.11 Family

Before the creation of the IEEE 802.11, very few wireless standards existed and those that did exist were not widely implemented as they offered very few practical benefits. However, in 1985 the Federal Communications Commission (FCC) of the United States authorized the industrial, scientific and medical (ISM) frequency bands. These three ISM bands accelerated the development of WLANs because vendors no longer needed to apply for frequency licenses for their products.

Today, IEEE 802.11 is a standard commonly referred as "wireless Ethernet" and branded as "WiFi" meaning WIreless FIdelity. The purpose of this technology is to enable high-speed WLAN access.

In 1989, the Institute of Electrical and Electronic Engineers (IEEE) 802.11 Working Group began elaborating on the WLAN Media Access Control (MAC) and physical layer (PHY) specifications. The IEEE finalized the initial standard for WLAN (called IEEE 802.11) in June 1997. This standard specified a 2.4 GHz operating frequency with data rates of 1 and 2 Mbps. It was initially developed to be a simple and cost-effective technology, which aimed to replace existing wired networks as well as offering other potential services that was not possible with earlier technologies.

The original 802.11 standard has now been replaced by a family of standards (e.g., 802.11a, 802.11b, and 802.11g) and is the most popular group of wireless standards. It offers users solutions for wireless network connectivity at varying rates over short to medium distances. The standards cover the PHY and MAC layers only, but these can be implemented in conjunction with existing IEEE protocols such as 802.2 for the logical link control layer. Enhancements or extensions to the basic functionality were developed under alphabetically named task groups.

IEEE 802.11 is designed for best-effort (BE) services only. The lack of a built-in mechanism for support of real-time services makes it very difficult to provide QoS guarantees for throughput- and delay-sensitive multimedia applications. However, it is hoped that the recently standardized 802.11e specification will rectify this situation to a certain extent.

There are two possible network configurations specified in the IEEE documentation for using 802.11: *ad hoc* and infrastructure.

- *Ad hoc* networks do not require a base station. 802.11 wireless-enabled devices may search for other devices within range to form a network. Devices may also search for target nodes that are out of range by flooding the network with broadcasts that are forwarded by each node.
- Infrastructure configuration involves the usage of a base station that acts as a central point between two or more wireless devices. The base station controls all traffic and communications on the network; thus all devices that wish to use the network must be able to access it. Multiple base stations can be linked together by a wired network and thus they cover a wider area where mobile devices can communicate with them.

A brief summary of the standards proposed as part of the 802.11 family is presented next. 802.11a, 802.11b, and 802.11g offer improvements at PHY, whereas 802.11e covers MAC layer.

11.2.2.1 IEEE 802.11b

Within two years of the introduction of 802.11, the initial standard was revised and on September 16, 1999, the 802.11b standard was introduced. The new standard promised interoperability among products from different vendors and compatibility with legacy 802.11 products. It is often referred to as Wi-Fi, a term promulgated by the Wireless Ethernet Compatibility Alliance (WECA). Products certified as Wi-Fi by WECA are interoperable with one another even if they are from different manufacturers.

Like its predecessor, 802.11b operates in the 2.4 GHz ISM band. 802.11b is an enhancement of the initial 802.11, and it includes transmission rates of 5.5 and 11 Mbps in addition to the 1 and 2 Mbps transmission rates of the initial standard. However, the actual maximum data rate for a user is 6 Mbps [1]. The higher transmission rates were provided through complementary code keying (CCK), a modulation technique that makes efficient use of the radio spectrum.

Although IEEE 802.11b successfully promoted the introduction of WLANs into the marketplace and is still the most popular form of WLAN, it has some drawbacks. The standard introduces problems in the area of interference because of its operation in the 2.4 GHz ISM band. This band is also utilized by medical equipment, household appliances (e.g., cordless telephones and microwaves) as well as newer technologies such as Bluetooth—all of which can cause major interference problems.

11.2.2.2 IEEE 802.11a

With the aim of improving the problem of radio interference in 802.11b, the IEEE decided to create an additional 802.11 standard that would make use of an alternative frequency band. The 802.11a standard was therefore introduced in September 1999. It operates in the 5 GHz band and therefore offers much less potential for radio frequency (RF) interference than other 802.11 standards (e.g., 802.11b and 802.11g) that utilize 2.4 GHz frequencies. 802.11a-based products became available in late 2001.

The standard allows high-speed data transfers of 6, 9, 12, 18, 24, 36, 48, and 54 Mbps, between computer systems and various peripheral devices at speeds approaching that of fast-wired local area networks. The standard makes use of orthogonal frequency division multiplexing (OFDM) to offer the maximum standardized data rate of 54 Mbps. OFDM has been selected as the modulation scheme for the standard due to its good performance on highly dispersive channels [2]. The standard also offers up to 12 nonoverlapping channels (as opposed to three with 802.11b).

With high data rates and relatively low interference, 802.11a is suited to support multimedia applications and densely populated user environments. This makes 802.11a an excellent long-term solution for satisfying current and future requirements [3–6].

However, 802.11a does have a number of disadvantages. As it operates in the 5 GHz band it is incompatible with the previous 802.11 and 802.11b standards that operate in the 2.4 GHz band. This means that companies or individuals wishing to upgrade from the previous standards to benefit from improved transfer speeds and reduced interference must replace all existing products with the new 802.11a products. In addition to this, the 5 GHz band in which the standard operates is not license-free in every country.

11.2.2.3 IEEE 802.11g

As a result of the problems with 802.11a it became desirable to create a standard that offered improved data transmission rates although still maintaining backward compatibility with the existing 802.11 and 802.11b products. For this reason, the IEEE proposed the 802.11g standard in November 2001. In June 2003, it was formally ratified by the IEEE.

IEEE 802.11g is a "mixture" of 802.11a and 802.11b. It can reach 54 Mbps with the OFDM technique and uses the 2.4 GHz ISM band. The standard specifies transmission rates of 6, 12, 24, 36, 48, and 54 Mbps [7], similar to 802.11a. Note that the highest bit rates can be obtained only in a short range of distance.

Because 802.11g uses the same spectrum between 2.4 and 2.4835 GHz and is inherently backward compatible with 802.11b, it is attracting more attention from industry than the earlier standardized 802.11a. There has been much debate over the use of 802.11g versus 802.11a for satisfying needs for higher-performance WLAN applications, but it seems that the IEEE 802.11g has started to dominate the high-performance WLAN market and the situation will not change in the near future.

11.2.2.4 Distance Range and Transmission Rate Comparisons

The range of an 802.11 access point (AP) can be around 100 m or more. Maximum range estimates are higher and claims of up to 10 km are made when optimization is introduced [8]. However, the distance from the AP or other device will affect the bit-rate. As distance increases, the maximum bit rate will decline. This is caused by noise due to the nature of a wireless interface over the air. Wireless links are more prone to noise and interference than their wired counterparts. The bit rate per user also declines when more users join an AP. For indoor networks, Rayleigh fading is another effect to be considered.

Table 11.1 presents a summary of 802.11 standard versions' characteristics. Because 802.11a, 802.11g, and 802.11b are the most used standards, Table 11.2 gives the range of distance of these versions depending on the given bit rates.

11.2.3 IEEE 802.11e: MAC Enhancement for QoS

Without QoS support, the discussed versions of the 802.11 standard (that offer improvements at the PHY layer) do not optimize the transmission of voice and video. There is no effective mechanism to prioritize traffic within 802.11. As a result, the 802.11e task group defined the 802.11 MAC

Table 11.1 Characteristics of the IEEE 802.11 Standards

Standard	*Frequency Band (GHz)*	*Transmission Rates (Mbps)*	*Throughput (Mbps)*
802.11	2.4–2.48	1, 2	1
802.11a	5.15–5.35, 5.72–5.82	6, 9, 12, 18, 24, 36, 48, 54	25
802.11b	2.4–2.48	1, 2, 5.5, 11	6
802.11g	2.4–2.48	6, 9, 12, 18, 24, 36, 48, 54	8–22

Table 11.2 Transmission Rates and Range for IEEE 802.11a, g, and b

Tx Rate (Mbps)	*802.11a (40 mW 6 dB Gain Diversity Patch Antenna) Range*	*802.11g (30 mW 2.2 dB Gain Diversity Dipole Antenna) Range*	*802.11b (100 mW 2.2 dB Gain Diversity Dipole Antenna) Range*
54	45 ft (13 m)	90 ft (27 m)	–
48	50 ft (15 m)	95 ft (29 m)	–
36	65 ft (19 m)	100 ft (30 m)	–
24	85 ft (26 m)	140 ft (42 m)	–
18	110 ft (33 m)	180 ft (54 m)	–
12	130 ft (39 m)	210 ft (64 m)	–
11	–	160 ft (48 m)	160 ft (48 m)
9	150 ft (45 m)	250 ft (76 m)	–
6	165 ft (50 m)	300 ft (91 m)	–
5.5	–	220 ft (67 m)	220 ft (67 m)
2	–	270 ft (82 m)	270 ft (82 m)
1	–	410 ft (124 m)	410 ft (124 m)

(medium access layer) enhancements to improve QoS for better support of audio and video applications.

Therefore, 802.11e version provides a MAC-layer enhancement. 802.11e adds an extra MAC function called the hybrid coordination function (HCF), made up of the enhanced distributed coordination access (EDCA) and the HCF-controlled channel access (HCCA) mechanism.

Looking at the EDCA first, each node has four queues that allow traffic to be split based on priority. The size of the interframe spacing (IFS) used depends on the priority of the queue. For the HCCA part, frames are transmitted using the HCCA mechanism. HCCA sets up virtual connections and negotiates QoS before transmission.

Other new features introduced by 802.11e include a more efficient way of doing block acknowledgement of packets and allowing nodes to communicate directly without going through the AP in infrastructure mode.

11.2.3.1 Enhanced Distributed Coordination Access

A significant limitation of IEEE 802.11b is its inability to enable QoS or take into consideration the characteristics and performance requirements of the traffic. In terms of access methods, IEEE 802.11 standard makes it mandatory that all stations implement the Distributed Coordination Function (DCF), a form of Carrier Sense Multiple Access with Collision Avoidance (CSMA/CA), and optional that they implement the Point Coordination Function (PCF), which enables the transmission of time-sensitive information. DCF provides channel access with equal probabilities to all stations

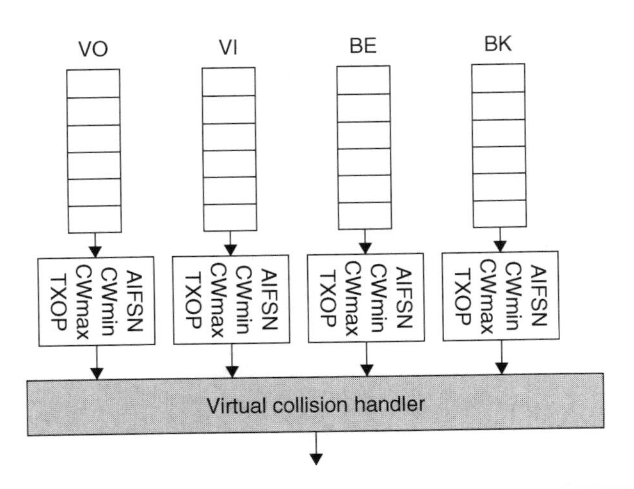

Figure 11.1 IEEE 802.11e access categories.

contending for the channel access in a distributed manner regardless of the requirements of the traffic. The IEEE 802.11e QoS MAC enhancement standard enables traffic differentiation by allowing up to four different transmission queues with different access priorities [9]. This allows the AP to provide differentiated service to various application types enabling them to meet their target QoS requirements.

EDCA is designed to provide differentiated, distributed channel accesses. EDCA can be used to provide eight different levels of priority (from 0 to 7) by enhancing the DCF. EDCA is not a separate coordination function. Rather, it is a part of a single coordination function, called the hybrid controller (HC) of the 802.11e MAC. The HC combines the aspects of both DCF and PCF. The 802.11e standard defines four AC queues into which different traffic streams can be directed: voice (VO), video (VI), BE, and background (BK) as shown in Figure 11.1. Each frame arriving at the MAC with a priority is mapped into a particular AC.

Each AC is configured with the EDCA parameters: AIFS[AC], CWmin[AC], CWmax[AC], and TXOP[AC]. The duration of AIFS[AC] is determined by the AIFSN[AC]. In 802.11b, the duration of DIFS had an AIFSN[AC] of at least 2. In 802.11e, the duration of AIFS[AC] is determined by SIFS + AIFSN[AC] $*$ TS. The smaller the AIFSN[AC] the higher the medium access priority. The back off period of each AC is chosen according to a uniform distribution over the interval [0, CW[AC]]. The CW size is initially assigned CWmin value, which is doubled when transmission fails up to the maximum value defined by CWmax.

Each AC behaves as a single enhanced DCF-contending entity where each AC has its own EDCA parameters and maintains its own back off counter (BC). When two or more competing ACs finish the back off process at the same time, the collision is handled in a virtual manner. The frame from the highest priority AC is chosen and transmitted whereas the lower priority ACs perform a back off with increased CW values. The EDCA parameters can be used to differentiate the channel access among different priority traffic. Smaller AIFSN and CWmin values reduce the channel access delay and provide a greater capacity share for the AC. However, using smaller values of CWmin increases the probability of collisions. The EDCA parameters are announced by the AP via beacon frames and can be dynamically adapted to meet the traffic requirements and network load conditions.

The IEEE 802.11e standard also defines a transmission opportunity (TXOP) as the interval of time during which an AC has the right to initiate transmissions without having to recontend for access. During an EDCA TXOP, an AC is allowed to transmit multiple MAC protocol data units from the same AC with an SIFS time gap between an ACK and the subsequent frame transmission [10]. The duration of the TXOP is determined by the value of the TXOP limit parameter. The TXOP limit parameter is an integer value in the range (0, 255) and gives the duration of the TXOP interval in units of 32 μs. If the calculated TXOP duration requested is not a factor of 32 μs, the value is rounded up to the next higher integer that is a factor of 32 μs. The maximum allowable TXOP limit is 8160 μs with a default value of 3008 μs.

11.3 Adaptive Multimedia Streaming Systems and Solutions

11.3.1 Multimedia Streaming: Overview

Streaming is a server–client technology that allows multimedia data to be transmitted and consumed. Streaming applications include E-learning, videoconferencing, video on demand, etc. The main goal of streaming is that the stream should arrive and play out continuously without interruption. In general, streaming involves sending multimedia (e.g., audio and video) from a server to a client over a packet-based network such as the Internet. There are two main types of streaming: progressive streaming and real-time streaming.

Progressive streaming is often called progressive download. In progressive streaming, a compressed video file is transferred to the hard disk of the client progressively. This streaming method is used typically when the movie size is relatively short (i.e., less than three minutes), for example, movie trailers, short movie clips, and advertisements. However, depending on the format of the video, some progressive files require that the entire movie be downloaded before it can be played (e.g., Real). Progressive streaming is not a good solution for long movies or material where the user may want random access. Progressive streaming is also not suited for "live" on-demand multimedia. The client has no interaction with the progressively streamed multimedia and cannot fast-forward or rewind to portions of the stream. Progressive streaming does not adapt to fluctuations in the clients' bandwidth.

In real-time streaming, the multimedia file is transmitted and consumed by the client in real-time or near real-time. When the client connects to the server, the stream will begin to play out on the clients' machine automatically or after a short delay of 1 or 2 s. Real-time streaming is suited for longer videos, for example, live broadcasts, presentations, training videos, and lectures. However, real-time streaming is constrained by fluctuations in network conditions. An adaptive streaming server keeps track of the network conditions and adapts the quality of the stream to minimize interruptions and stalling. Typically, the Real-Time Transport Protocol (RTP) is used to transport encoded media data as a series of time-stamped packets whereas the server monitors the state of the network using the Real-Time Transport Control Protocol (RTCP). Typically, the client can interact with the real-time streaming server using Real-Time Streaming Protocol (RTSP). If the requested file is preencoded, the client can jump to any location in the video clip using the RTSP protocol as a network remote control for the stream. Real-time streaming can be delivered by either peer-to-peer (unicast) or broad-cast/multicast. There are two types of real-time streaming services [11,12]: on-demand or live streaming.

- *On-demand streaming*. Precompressed, streamable multimedia content is archived on a storage system such as a hard disk drive and transmitted to the client on demand by the streaming server. Client requests are asynchronous, that is, clients can watch different parts of the media at the same time.
- *Live streaming*. Multimedia stream comes from a live source such as a video camera and is compressed in real-time by the encoder. There is no notion of duration as these streams are live. Clients cannot avail fast-forward or rewind functions.

There are many factors, which affect the delivery of multimedia traffic over BE wired and wireless IP networks. Among them are heterogeneity, congestion, bandwidth fluctuations due to congestion or mobility, delays due to congestion, packet loss or retransmissions, noise, and interference (particularly evident in wireless networks). These are discussed in details in Section 11.4.1 (QoS when delivering multimedia).

11.3.2 Adaptive Multimedia Streaming: Principle

Bursty loss and variable delays have a negative effect on multimedia deliveries, severely affecting the end-users' perceived quality. Consequently, any effort that aims at reducing the variability of the delays and at lowering the loss rate helps to increase the quality of the remote multimedia presentations. This is also the main objective of the adaptive solutions for multimedia delivery.

The primary goal of adapting multimedia is to ensure graceful quality adaptation and maintain a smooth continuous play out. Multimedia servers should be able to adapt quality to bandwidth variations. If the available bandwidth becomes lower than the sending rate of the multimedia, the server should tailor multimedia stream intelligently rather than arbitrarily drop packets, for example, to match the available network resources.

Adaptation should behave fairly in a shared network. When there is excess bandwidth, the competing traffic flows share the excess bandwidth in a fair manner. A Transmission Control Protocol (TCP) rate control mechanism is not practical for real-time multimedia traffic as TCP multiplicatively decreases the window size on encountering a packet loss. The consequent sudden decrease of the window size may cause a sharp bit rate variation, which can seriously degrade the perceived quality at the receiver. Also multimedia applications are loss-tolerant and time-sensitive, thus retransmission of lost packets is not very efficient and per-packet acknowledgements impose a large overhead. Instead, multimedia flows should only behave in a TCP-friendly manner. They should allow non-TCP flows to receive the same share of bandwidth as TCP flows would obtain over the same link [13].

In the development of adaptation algorithms, there are a number of design and system issues, which must be considered [14]:

- Signaling or feedback mechanisms to convey congestion information between client and server
- Frequency of feedback
- Specific rate control mechanism used in response to feedback
- The responsiveness of the congestion control scheme in detecting and reacting to network congestion
- Capability of the adaptation algorithm to accommodate heterogeneous receivers with differing network connectivity, the amount of congestion on their delivery paths, and their need for transmission quality
- Scalability of the adaptation algorithm in a multicast session with a large number of receivers
- Fair sharing of bandwidth with competing flows, particularly TCP
- Perceived quality of received multimedia streams

11.3.3 Adaptive Multimedia Streaming: Solutions

Broadly speaking, adaptation techniques attempt to reduce network congestion by matching the rate of the multimedia stream to the available network bandwidth. Without a rate control mechanism, any data transmitted exceeding the available bandwidth would be discarded, lost, or corrupted in the network. Adaptation techniques can be classified by the place where adaptation occurs, that is, receiver- and sender-based hybrid schemes. Equally, they can be classified by the method in which the adaptation is performed [15,16]:

- Rate control: transport level
  - Sender-based, client-based, or hybrid
- Rate shaping: transport and encoding level
  - Sender-based, client-based, or hybrid
- Encoder-based adaptation: encoding level
  - Sender-based
- Transcoder: network level

Often in adaptive transmissions congestion control is used. One of the main objectives of congestion control is to prevent or minimize packet loss. However, packet loss is often unavoidable in the BE IP networks and may have significant impact on perceptual quality. Error control mechanisms are used to maximize multimedia presentation quality in the presence of packet loss. These mechanisms can be classified as:

- Forward error correction (FEC)
- Retransmission
- Error resilience
- Error concealment

In general, end-to-end schemes, either sender-based or client-based, have been and are still a popular topic of research. In particular, much of the research has focused on rate control solutions, especially sender-based rate control schemes. End-to-end schemes are easy to deploy and can be applied to most systems. Encoder-based and rate shaping solutions are constrained by the functionality of the codec being used. Transcoder-based schemes require assumptions and interactions with the network. Currently transcoder-based schemes are not very popular and there is very little new research in this area. Despite the efficacy of an adaptation algorithm, there may be situations in which there is unavoidable loss of data. Owing to this reason, often adaptation in a system is employed in conjunction with some error control techniques.

TCP retransmission and acknowledgements introduce unnecessary delays and overhead in the network that are not beneficial for real-time multimedia applications. User Datagram Protocol (UDP) is usually employed as the transport protocol for real-time multimedia streams. However, UDP is not able to provide congestion control and overcome the lack of service guarantees in the network. Therefore, it is necessary to implement control mechanisms above the UDP layer to prevent congestion and adapt accordingly.

To prevent congestion, two types of control can be employed:

- *Window-based control [17].* Probes for the available network bandwidth by slowly increasing a congestion window; when congestion is detected (indicated by the loss of one or more packets), the protocol reduces the size of the congestion window. The rapid reduction of the window size in response to congestion is essential to avoid network collapse.
- *Rate-based control [18].* Sets the sending rate based on the estimated available bandwidth in the network; if the estimation of the available network bandwidth is relatively accurate, the rate-based control could also prevent network collapse.

Rate-based control is usually employed for transporting real-time multimedia. Rate control can take place at the server, receiver, or a hybrid scheme can be used.

11.3.3.1 TCP-Friendly Rate Control and TFRC Protocol

TFRC [19] is a congestion control algorithm for unicast traffic, which explicitly adjusts its sending rate as a function of the measured loss event rate, where a loss event consists of one or more packets dropped within a round trip time (RTT). Receivers calculate the loss event rate p and RTT and send feedback to the sender at least once per RTT. The server does not reduce the sending rate in half in response to a single loss event, but reduces the sending rate in half in response to several successive loss events. If the sender has not received feedback after a number of RTTs, then the sender reduces its sending rate. Otherwise, the senders set their sending rate according to a TCP throughput equation. TFRC Protocol (TFRCP) presented in Ref. 20 is a rate-adjustment congestion control protocol whose sender works in rounds of M time units. At the beginning of each round, the sender computes a TCP-friendly transmission rate and sends packets at this rate. Each packet carries a sequence number and a sending time stamp. The receiver acknowledges each packet by sending a response ACK that carries the sequence number of the packets received correctly and receiving time stamp. From these, the sender can calculate the RTT and recovery time objective (RTO), and obtain the number of packets dropped, numdrop, and the number of acknowledged packets, numACKed, of each round. During periods of loss, the sender reduces its transmission rate to the equivalent TCP rate calculated using the TCP throughput equation, and during periods of no loss, the rate is doubled.

This scheme behaves in a TCP-friendly manner during periods of loss. However, its rate increase behavior may result in an unequal bandwidth distribution due to the possibility of increasing the transmission rate faster than competing TCP connections. TFRC smoothes the reduction of sending rate in response to packet loss and avoids the oscillation of rate when they are used for real-time multimedia applications. However, the model relies on a single TCP connection with a steady-loss ratio. But it is not as effective, if the RTT is affected by queuing delays or when the bottleneck router is shared among competing connections, especially when the dropping rate is large.

11.3.3.2 Enhanced Loss Delay Adjustment Algorithm

LDA+ [21] is an enhanced version of LDA [22,23], which adapts the transmission rate of UDP-based multimedia flows to the congestion situation in the network in a TCP-friendly manner. LDA+ controls the transmission rate of a sender based on end-to-end feedback information about loss rate, delay level, and bandwidth capacity measured by the receiver. The adaptive scheme was designed for unicast transmissions and it bases its functionality on using RTP for data delivery and RTCP for feedback. LDA+ is an additive increase multiplicative decrease (AIMD) algorithm that changes its transmission rate with values dynamically computed based on the current network situation and the share of the bandwidth a flow is already utilizing. In loss situations, the rate is decreased multiplicatively, but the final values should not be lower than the one the TCP model equation would suggest for the transmission rate in the same network conditions. If zero loss is recorded, the rate is increased additively with an increment computed as the minimum between three values. The first value is computed in inverse relation with the share of the bandwidth the current flow utilizes. The second value is meant to limit the increase to the bottleneck link bandwidth as it converges to zero when this happens. The third value is determined in such a manner that, at no time, the rate increases faster than that of a TCP connection sharing the same link.

11.3.4 Adaptive Multimedia Streaming: Discussion

In most adaptation algorithms, there is no definition of quality. The basic question is what is quality in terms of the multimedia content being streamed? Next, quality is discussed in the context of some multimedia delivery schemes' design issues.

- *Dependency on algorithm control parameters.* Several algorithms have a strong dependency on the choice of control parameters used within the algorithm. These parameters have an inexplicable origin and have been empirically optimized. Consider rate control schemes with an AIMD-like behavior whereby the transmission rate is increased by some additive increase factor, a, and reduced multiplicatively by a decrease constant, b. The main disadvantage of AIMD algorithms is that they have more abrupt changes and oscillations in sending rate as the operation of these algorithms relies heavily on the values chosen for a and b. If a is chosen to be too large, then increasing the transmission rate could push the system into causing congestion. This in turn causes the client to experience loss to which the server multiplicatively reduces its transmission rate. Thus, the system operation is heavily dependent on the value for a. If a is too small, the server is very slow to make use of the extra capacity. One solution would be to use a mechanism to use dynamic values for a and b. For example, a could have initially a small value and be increased exponentially maximizing the capacity faster. Similarly, b dramatically reduces the transmission rate on congestion, which can significantly degrade the perceived quality. b should have a value, which best reflects the estimated bottleneck bandwidth of the client. The use of dynamic values is the approach taken in LDA.

- *TCP throughput model and estimation of network state parameters such as RTT, loss, etc.* Algorithms using TCP throughput models have limited utility, as they do not define which version of TCP they are modeling. They are not effective for high loss rates over 16 percent and are constrained by the accuracy of the estimation of RTT and other parameters. Further, the TCP throughput equation is based only on a single TCP connection with a steady-loss ratio.
- *Over-reaction based on immediate feedback.* If there are short bursty losses as seen in wireless networks, the system reacts; however, the question is, should the server reduce the transmission rate in response to a short bursty loss? It would seem more sensible for the server to maintain a history of the clients' feedback and monitor both in terms of the clients' long- and short-term experience. For example, a client is receiving 64 kbps, but periodically there is a short bursty loss (e.g., caused by some object in the environment periodically sending signals which interfere with a client's reception), overall the client's receiving bit rate is 64 kbps, if the client's transmission rate is reduced, this will not eliminate the short bursty losses. This was addressed in QOAS [48], which uses a specially designed solution to avoid reactions to sudden bursty losses, and in TFRC, which uses loss events consisting of multiple consecutive losses as an indicator of congestion. However, most algorithms react only on immediate feedback.
- *Responsiveness of the algorithm.* The converse argument of reacting based on both short- and long-term feedback is that when there are sudden jumps of congestion sustained over a longer period of time,

the algorithms are slower to react and recover from the congestion. For RTCP-based feedback, the minimum interval between feedback messages is 5 s. The server generally only reacts on receipt of a new feedback message. Is this adaptation interval short or long enough to adapt quickly and effectively to changing network conditions? What happens if this RTCP packet is lost, then adaptation at a potentially crucial moment will only occur at a minimum 10-s interval. Lost feedback messages should be eliminated when the rate control algorithm requires it for adaptation, thus hybrid transport-layer protocols should be used, for example, RTP/UDP/IP and RTCP/TCP/IP.

- *Translation of rate into real video encoding parameters.* Consider a simple rate control algorithm, a server delivers video at 50 kbps; and based on feedback, the algorithm indicates that the transmission rate can be increased to 55 kbps. There are two main questions: How should the extra 5 kbps be achieved? Should the frame rate be increased, or the resolution? Even if this is known, how is it achieved, does it require interaction with the encoder or the server will process the bitstream only? Also for example, if the server can increase the frame rate, what happens if the increased frame rate determines a new transmission rate, which exceeds the one suggested by the algorithm? Would a rate as close as possible to the algorithm-computed transmission rate be enough? This is not mentioned in any of the algorithms proposed.
- *Codec constraints.* Adaptive encoding techniques are constrained by the adaptability of the encoder being used. Not all codecs support dynamic quantization parameters, spatial filtering, or temporal filtering. Adaptive encoding is only really effective for live transmissions; otherwise, a precompressed stream must be decompressed and then recompressed on the fly, so that the adaptive encoding policy can be applied. This may cause a serious processing overload and delay on the server. This method is not really suitable for unicast scenarios where there may be hundreds of users connected to the one server requesting different content. If applied to multicast scenarios, the flexibility of adaptation is traded off with the heterogeneity of the receivers.
- *User perception.* Many adaptation algorithms do not consider user perception as a factor in their decision-making process. The user is the primary entity affected by adaptation and therefore should be given priority in the adaptation decision-making process. For example, once again, if a clip is being streamed at a particular encoding configuration and the system needs to degrade the quality being delivered, how this adaptation occurs should be dictated by the users' perception. The way to degrade should be such to have the least negative impact on the users' perception.

11.4 Multimedia Streaming User-Perceived Quality

11.4.1 QoS-Related Factors Affecting Multimedia Streaming

11.4.1.1 Heterogeneity

Both network and receiver heterogeneity affects multimedia streaming. Network heterogeneity refers to the subnets in the network having different characteristics and resources (e.g., processing, bandwidth, storage, and congestion control policies). Network heterogeneity determines different packet loss and delay characteristics, which affect multimedia streaming. Receiver heterogeneity refers to clients having different delay requirements, visual quality requirement, end device (e.g., PDA, laptop, and mobile phone), or processing capability. In multicast sessions, the heterogeneity is of most significance, as all the receivers will have to get the same content at the same quality.

11.4.1.2 Congestion

Congestion occurs when the amount of data in the network exceeds the capacity of the network. As traffic increases, routers are no longer able to cope with the load and this results in lost packets. Congestion can be caused by several factors. Queues can build up due to high data rate applications. If there is not enough memory to hold all the data at the router, packets will be lost. But even if queues had an infinite length, this cannot eliminate congestion because by the time a packet is at the top of the queue, the packet has already timed out and duplicates have already been sent. All these packets are then forwarded onto the next router, increasing the load all the way to the destination. Slow processors can also cause congestion. If the routers' CPU are slow at performing its tasks of queuing buffers, updating tables, etc., queues can build up although there is excess capacity. Bottleneck bandwidth links also cause congestion.

11.4.1.3 Bandwidth Fluctuations

To achieve acceptable presentation quality, transmission of real-time multimedia typically has certain minimum bandwidth requirement. However, due to the BE nature of the Internet and wireless IP networks, there is no bandwidth reservation to meet such a requirement. In wireless networks there may be several reasons for bandwidth fluctuations:

- When a mobile terminal moves between different networks (e.g., from WLAN to wireless WAN), the available bandwidth may vary drastically (e.g., from a few megabits per second to a few kilobits per second).

- When a handover happens, a base station may not have enough unused radio resource to meet the demand of a newly joined mobile host.
- The throughput of a wireless channel may be reduced due to multipath fading, cochannel interference, and noise disturbances.
- The capacity of a wireless channel may fluctuate with the changing distance between the base station and the mobile host.

11.4.1.4 Delay

There are many sources of delay in a transmission system. In the network itself, in addition to propagation delays, further delays are incurred at each router along its path due to queuing and switching at various routers. At the end-points, delays are incurred in obtaining the data to be transmitted and in packetising it. Real-time multimedia is particularly sensitive to delay, as multimedia packets require a strict bounded end-to-end delay. That is, every multimedia packet must arrive at the client before its play out time, with enough time to decode and display the packet. If the multimedia packet does not arrive on time, the play out process will pause, or the packet is effectively lost. Congestion in a BE IP network can incur excessive delay, which exceeds the delay requirement of real-time multimedia. In a wireless network, there are additional sources of delay such as retransmissions on the radio link layer.

11.4.1.5 Jitter

A variable delay on the IP networks is known as jitter. Jitter is mainly due to queuing and contention of packets at intermediate routers, but can also happen when packets take a different path to the destination. To combat jitter, play out buffering is used at the receiver. Jitter does not have a huge impact on streaming multimedia transmission, which is in general not a highly interactive application and as the delay bounds are less stringent than those for interactive sessions such as Voice-over-IP (VoIP) [24]. The trade off is that for higher interactivity, there is less receiver buffering which means that the effects of network jitter increases. However, for applications with lower interactivity, there is greater receiver buffering and so the effects of network jitter become less. In receiver buffering, packets that arrive faster than expected are placed in a buffer for play out at a later time. Smooth quality of a received multimedia signal depends on appropriate buffering. The receiver buffer must be large enough to account for network jitter and if enough data is buffered, it also enables the receiver to sustain momentary drops in the sending rate by playing out of its buffer at a higher rate than the sender is currently sending. The buffer accumulates when the sender is transmitting faster than the receiver is playing out, by not aggressively increasing the sending bandwidth when the available bandwidth capacity is increased.

11.4.1.6 Loss

For streamed multimedia applications, loss of packets can potentially make the presentation displeasing to the users, or, in some cases, make continuous play out impossible. Multimedia applications typically impose some packet loss limits. Specifically, the packet loss ratio is required to be kept below a threshold (e.g., 1 percent) to achieve acceptable visual quality. Despite the loss requirement, the current Internet does not provide any loss guarantee. In particular, the packet loss ratio could be very high during network congestion, causing severe degradation of multimedia quality. Even if the network allows for retransmission of lost packets, as is the case for wireless IP networks, the retransmitted packets must arrive before their play out time. If packets arrive too late for their play out time, they are useless and effectively lost.

Congestion at routers in the network often results in queue overflow, which generally results in packets being dropped from the queue. Often consecutive packets in a queue are from the same source and belong to the same media stream. So, when packets are being dropped, they often belong to the same stream. Thus, this behavior of packet dropping in the network can be seen as bursty losses. Packet loss and delay can exhibit temporal dependency or burstiness [25]. This translates to burstiness in network losses and late losses, which may worsen the perceptual quality compared to random losses at the same average loss rate. Bursty losses can affect the efficacy of error resilience and error concealment schemes. It affects performance of error concealment techniques such as FEC. FEC can recover a lost packet only if other necessary packets belonging to the same block are received. In this way, the loss pattern affects the effectiveness of loss concealment. Although bursty losses are likely, packet interleaving reduces this effect [26].

11.4.1.7 Noise

Compared with wired links, wireless channels are typically much more noisy and have both small- (multipath) and large-scale (shadowing) fades [27], making the bit error rates (BER) very high. The resulting bit errors can have a devastating effect on multimedia presentation quality [28].

11.4.2 Objective Video Quality Metrics

Objective methods aim at determining the quality of a video sequence in the absence of the human viewer. They are based on very different principles such as the comparison between the original and the distorted version of the same sequence, the statistical assessment of a large set of analyzed cases, the analysis of the effect of possible interferences with the video streams, and the relationships from the video contents and subjective testing results. The researchers divide the metrics associated to these objective methods into mathematical- and model-based [29,30]. The mathematical metrics rely on mathematical formulae or functions based on intensive psychovisual experiments. The model metrics are based on complex models of the human visual system. In Ref. 31, the authors list three classes of approaches according to the requirement of the existence of the source video for quality assessment: full-reference methods (FR; also called picture comparison-based), reduced reference solutions (RR; or feature extraction-based), and no-reference methods (NR; also called single-ended).

Since 1997 the Video Quality Expert Group (VQEG) [32] has studied extensively possible assessment of video quality. One of its goals was to propose a quality metric for ITU-T standardization. Thereafter more than 26,000 subjective opinion scores were generated based on 20 different source sequences at bit rates between 768 kbps and 50 Mbps, processed by 16 different video systems and evaluated at 8 laboratories; they could not recommend any method for ITU standardization.

11.4.2.1 Mathematical Metrics

The most popular mathematical metric is the peak-signal-to-noise-ratio (PSNR), which is computed based on the mean square error between the values of the pixels located in the same position in the reference and modified video stream respectively, for each frame.

Another mathematical metric is weighted signal to noise ratio (WSNR), which takes into account some human visual system properties through weighting as, for example, the weighted noise power density as a function to the eye sensitivity. Although they are very simple, many studies have shown that PSNR and WSNR are poorly correlated to human vision; not taking into account, for example, visual masking. For example, this leads to similar decreases in scores regardless if the human subjects perceive or not the difference from the original. Another problem is that these metrics are applied on frame-by-frame bases, not considering temporal correlation between frames.

Picture appraisal rating (PAR) [33] was proposed by Snell & Wilcox as a no-reference method of estimating the picture quality of an MPEG-2 video by measuring the distortion introduced by the MPEG encoding process. Based on PAR Mosalina [34], an offline, single-ended monitoring software that automatically detects possible picture quality problems in MPEG-2 streams and MVA200, a real-time MPEG bitstream analyzer was launched. Although being a no-reference objective method, PAR is best suited for in-service applicability; for example, the algorithm was not built to directly detect artifacts that might be introduced by a decoder in response to problems in the transmitted stream. As PAR is based on PNSR, its correlation to the human visual system is limited.

11.4.2.2 Model-Based Metrics

Image Evaluation based on Segmentation (IES) model was proposed by CPqD and bases its operation on scenes' segmentation into plane, edge and texture regions, and on the assignment of a number of objective parameters to each of these components. A perceptual-based model that predicts subjective ratings based on the relationship between existing subjective test results and their objective assessment is used to obtain an estimated impairment level for each parameter. The final result is achieved through a combination of estimated impairment levels, based on their statistical reliabilities. An added scene classifier ensures scene-independent evaluation. This model is very complex and its reliability is limited, which makes its applicability difficult, especially in real-time.

Picture quality rating (PQR) is a joint Tektronix/Sarnoff metric based on Sarnoff's Human Vision Model (HVM) that simulates the responses of human spatiotemporal visual system taking into account the perceptual magnitudes of differences between source and processed sequences. From these differences, an overall metric of the discriminability of the two sequences is calculated based on their proprietary JNDmetrix (just noticeable difference). The model was designed under the constraint of high-speed operation in standard image-processing hardware and thus represents a relatively straightforward, easy-to-compute solution. More details about PQR can be found in Ref. 35.

Video quality metric (VQM) [36,37] was proposed by the Institute for Telecommunication Sciences, National Telecommunications and Information Administration (NTIA) (United States) as a full-reference metric, which uses reduced bandwidth features that are extracted from spatial–temporal regions of processed input and output video scenes. These features characterize spatial detail, motion, and color existent in the video sequences and are used for the subjective quality-rating computation.

Full-reference and no-reference moving pictures quality metric (MPQM) proposed at EPFL Switzerland [38,39] relies on a basic multichannel human visual model that takes into consideration modeling of contrast sensitivity and intrachannel masking. The NR metric estimates MPQM based on *a priori* knowledge about the encoding scheme (MPEG) and the effect of the loss on the encoded stream [40]. Its simple formula makes it easy to be used for in-service monitoring of video quality.

Other models or metrics include NHK/Mitsubishi model, Kokusai Denshin Denwa (KDD) model [41], Swiss Perceptual Distortion Metric

(PDM) [42], NASA's Digital Video Quality (DVQ) model and metric [43,44], and Perceptual Video Quality Measure (PVQM) [45,46].

11.5 Optimum Adaptation Trajectory-Based Quality-Oriented Adaptation Scheme

11.5.1 User Perception in Adaptive Video Streaming

In general, video quality adaptation and video quality evaluation are distinct activities. Most adaptive delivery mechanisms for streaming multimedia content do not explicitly consider user-perceived quality when making adaptations; video quality evaluation techniques are not designed to evaluate quality when it is changing over time. We propose that an OAT through the set of possible video encodings exists, and that it indicates how to adapt encoding quality in response to changes in network conditions to maximize user-perceived quality. The subjective and objective tests we carried out to find such trajectories for a number of different MPEG-4 video clips are described. Experimental results illustrate that using a two-dimensional adaptation strategy based on the OAT outperforms one-dimensional adaptation schemes, giving better short- and long-term user-perceived quality.

11.5.2 Quality-Oriented Adaptation Scheme Principle

As with any adaptive scheme for multimedia streaming, the QOAS [47,48] relies on the fact that random losses have a greater impact on the end-user-perceived quality than a controlled reduction in quality [49]. Therefore, the end-to-end sender-driven adaptation mechanism employed by QOAS controls the adjustment of both the quality of the streamed multimedia content and the transmission rate so that it maximizes the end-user-perceived quality in existing delivery conditions. This intrastream adaptation controls the quality, and consequently the quantity, of streamed multimedia-related data and is based on information received from the client.

The QOAS-based system architecture includes multiple instances of the end-to-end QOAS adaptive client and server applications. These exchange video data and control packets through the IP-based delivery network. The QOAS client continuously monitors some transmission parameters and estimates the end-user-perceived quality, and its quality of delivery grading scheme (QoDGS) regularly computes quality of delivery scores (QoD-scores) that reflect the multimedia streaming quality in current delivery conditions. These grades are sent as feedback to the QOAS server, whose server arbitration scheme (SAS) analyzes them and proposes adjustment decisions to be taken to increase the end-user-perceived quality in existing

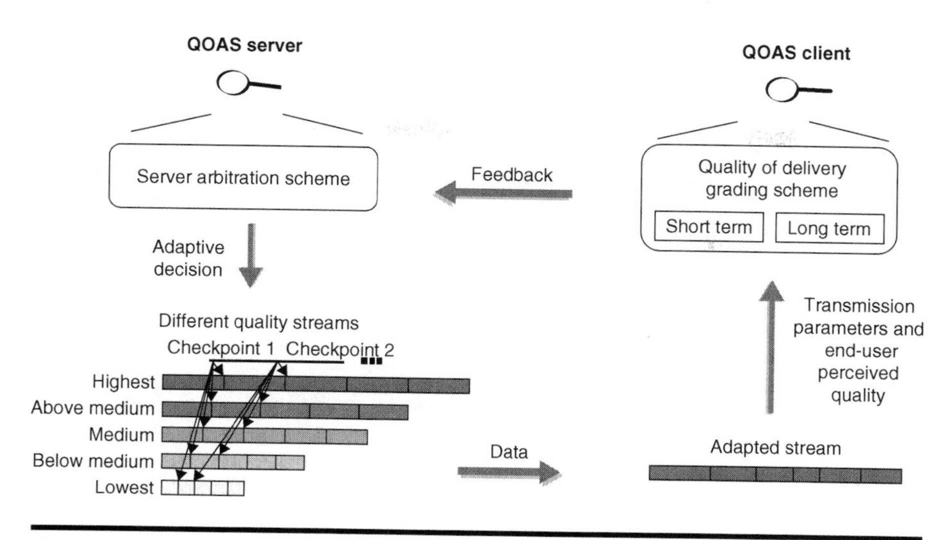

Figure 11.2 Schematic description of QOAS' adaptation principle.

conditions. The QOAS adaptation principle is schematically described in Figure 11.2. For each QOAS-based multimedia streaming process, a number of different quality states are defined at the server (e.g., the experimental tests have involved a five-state model). Each such state is then assigned to a different stream quality. The stream quality versions differ in terms of compression-related parameters (e.g., resolution, frame rate, and color depth) and therefore have different bandwidth requirements. They also differ in end-user-perceived quality. The difference between the average bit rates of these different quality streams is denoted as "adaptation step." During data transmission the client-located QoDGS computes QoDscores that are sent via feedback to the QOAS server, which dynamically varies its quality state based on suggestions made by SAS. When the delivery conditions cause excessive delays or loss, the client reports a decrease in end-user quality and the server switches to a lower quality state, reducing the bit rate of the streamed multimedia. Consequently this may reduce the delays and the loss, increasing the end-user-perceived quality. If the QOAS client reports improved streaming conditions, the server increases the quality of the delivered stream. These switches to higher- and lower quality states, respectively, are performed gradually with the granularity of the QOAS adaptation step. The smaller the adaptation step, the less noticeable to the viewer is the effect of the bit rate modification. However, the higher the adaptation step, the faster is the convergence of the algorithm to the bit rate best suited in existing network conditions. The client-located QoDGS evaluates the effect of the delivery conditions on end-user-perceived quality. It monitors both short- and long-term variations of packet loss rate, delay, and delay jitter, which have the most significant impact on the received quality [50,51] and estimates the end-user-perceived quality. Monitoring short-term variations helps to learn quickly about transient effects, such as sudden traffic changes. Long-term variations are monitored to track slow changes in the delivery environment, such as new users in the system. These short- and long-term periods are considered, respectively, an order and two orders of magnitude greater than the feedback-reporting interval. The end-user quality is estimated using the no-reference MPQM [40], which maps the joint impact of bit rate and data loss on video quality onto the ITU-T R.P.910 5-point grading scale [52]. The SAS assesses the values of a number of consecutive QoDscores received as feedback to reduce the effect of noise in the adaptive decision-taking process. Based on these scores, SAS suggests adjustment decisions. This process is asymmetric, requiring fewer QoDscores to trigger a decrease in the server's quality state than for an increase. This ensures a fast reaction during bad delivery conditions and helps to eliminate its cause. The increase is performed only when there is enough evidence that the network conditions have improved. This asymmetry helps also to maintain system stability, by reducing the frequency of quality variations. When QOAS is used to stream multimedia to multiple viewers, an interstream adaptation scheme complements the intrastream adaptation and aims for a finer adjustment in the overall adaptation process. The interstream adaptation is responsible for preventing QOAS-based adaptive processes from reacting simultaneously to variations in the delivery network. It selectively allows some of the QOAS-based sources of multimedia data to react to the received feedback, in a step-by-step process, achieving near-optimal link utilization and long-term fairness between the clients.

11.5.3 Optimum Adaptation Trajectory Principle

There should be an optimal way in which multimedia transmissions should be adapted in response to network conditions to maximize the user-perceived quality. This is based on the hypothesis that within the set of different ways to achieve a target bit rate, there exists an encoding configuration that maximizes the user-perceived quality. If a particular multimedia file has n independent encoding configurations, then there exists an adaptation space with n dimensions. When adapting the transmission from some point within that space to meet a new target bit rate, the adaptive server should select the encoding configuration that maximizes the user-perceived quality for that given bit rate. When the transmission is adjusted across its full range, the locus of these selected encoding configurations should yield an OAT within that adaptation space [53,54].

This approach is applicable to any type of multimedia content. The work presented here focuses for concreteness on the adaptation of MPEG-4 video streams within a two-dimensional adaptation space defined by frame rate and spatial resolution. These encoding variables were chosen as they

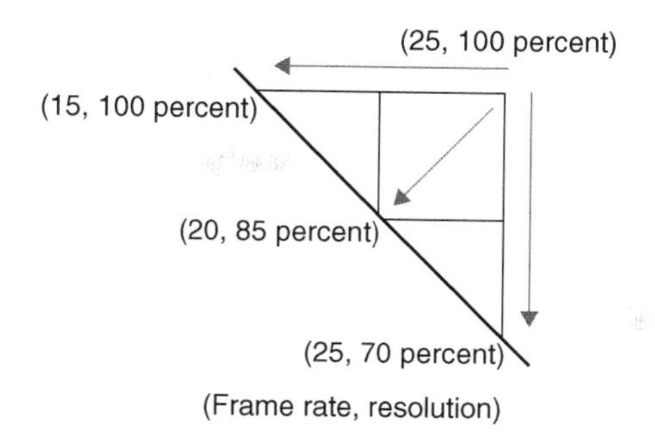

Figure 11.3 Encoding quality options.

most closely map to the spatial and temporal complexities of the video content. The example shown in Figure 11.3 indicates that, when degrading the content quality expressed as a frame rate, resolution pair from position (25, 100 percent), there are a number of possibilities such as (15, 100 percent), (20, 85 percent), or (25, 70 percent), which all lie within a zone of equal average bit rate (EABR). The clips falling within a particular zone of EABR have different, but similar bit rates. For example, consider, the bit rates corresponding to the encoding points (17, 100 percent), (25, 79 percent), and (25, 63 percent) were 85, 88, and 82 kbps, respectively. To compare clips of exactly the same bit rate would require a target bit rate to be specified, and then the encoder would use proprietary means to achieve this bit rate by compromising the quality of the encoding in an unknown manner. Using zones of EABR effectively quantizes the bit rate of different video sequences with different encoding configurations. The boundaries of these zones of EABR are represented as linear contours for simplicity, because their actual shape is irrelevant for this scheme.

The OAT indicates how the quality should be adapted (upgraded or downgraded) so as to maximize the user-perceived quality. The OAT may be dependent on the characteristics of the content. There is a content space in which all types of video content exist in terms of spatial and temporal complexity (or detail and action). Every type of video content within this space can be expanded to an adaptation space as shown in Figure 11.4.

11.5.3.1 Test Methodology

User perception of video quality may vary with the content type; for example, viewers may perceive action clips differently from slow-moving clips. Thus, there may exist a different OAT for different types of content based on their spatial and temporal characteristics. To characterize content in terms of its spatial and temporal complexity, a spatial–temporal grid was constructed, as shown in Figure 11.5. The spatial and temporal perceptual

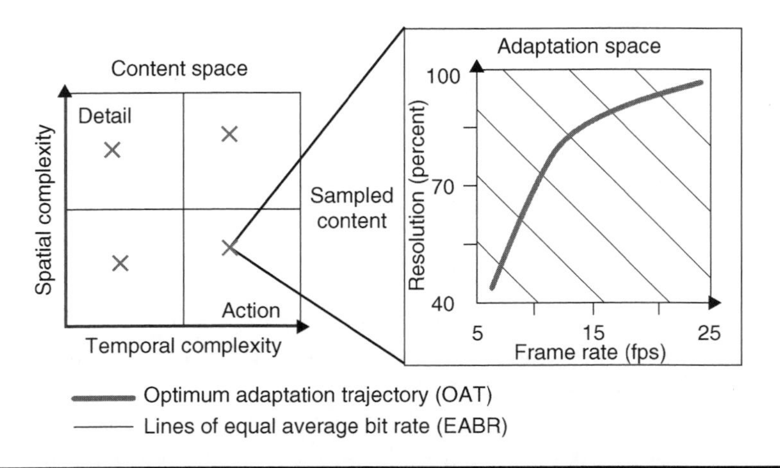

Figure 11.4 Adaptation space.

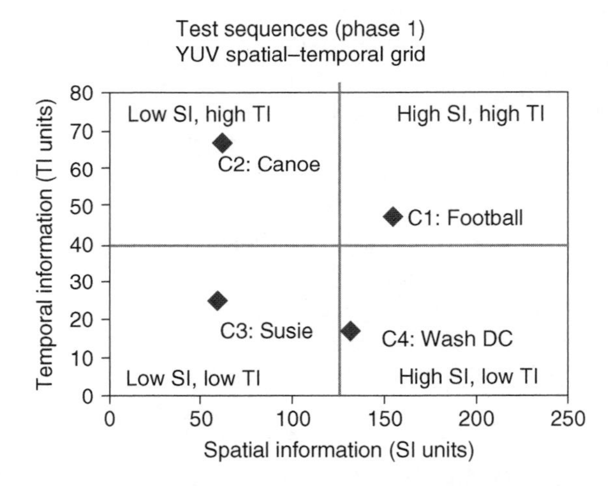

Figure 11.5 Spatial–temporal grid.

information of the content was determined using the metrics, spatial information (SI) and temporal information (TI). The SI parameter is based on the Sobel filter, and is implemented by convolving two 3×3 kernels over the video frame and taking the square root of the sum of the squares of the results of these convolutions [52]. The TI parameter is based on the motion difference feature, which is the difference between the pixel values (of the luminance plane) at the same location in space but at successive times or frames [52]. Various content types were sampled from each quadrant of the spatial–temporal grid. The sampled content taken from this grid was then encoded to form an adaptation space as shown in Figure 11.6.

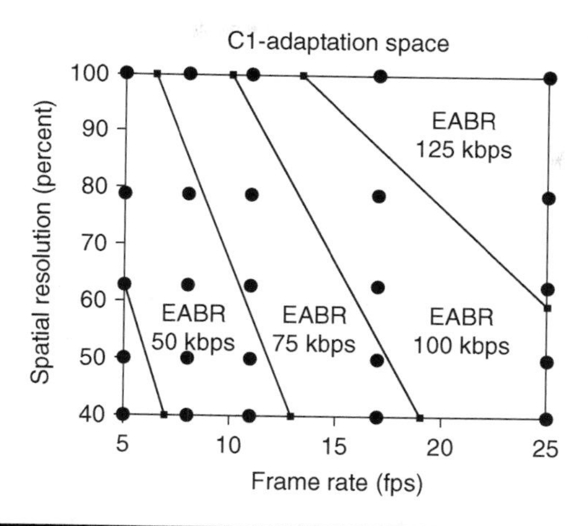

Figure 11.6 Adaptation space for a single content type.

There are a number of methodologies proposed by the ITU-T, including absolute category rating (ACR), which is a single stimulus method, and degraded category rating (DCR) and pair comparison (PC), which are double stimulus methods. DCR uses a 5-point impairment scale; ACR uses a 5-point quality grading scale or alternatively the continuous quality scale (CQS) [52]. There is another methodology being developed for subjective testing of video by the European Broadcasting Union (EBU) called multiple stimulus hidden reference and anchors (MUSHRA) [55], which has been used as a subjective test methodology for audio in MPEG-4 Audio Version 2 verification tests [56,57]. However, with all these methodologies, different subjects may interpret the associated rating scale in different ways. The forced choice methodology is often employed in cognitive science, and PC is one of its applications. In the forced choice method, the subject is presented with a number of spatial or temporal alternatives in each trial. The subject is forced to choose the location or interval in which their preferred stimulus occurred. Using the forced choice method, the bias is binary, which simplifies the rating procedure and allows for reliability, verification, and validation of the results. Our subjective test consisted of a subject watching every combination of pairs of clips from each EABR zone and making a forced choice of the better quality or preferred encoding configuration. Intrareliability and interreliability of a subject were factored into our test procedure by including repetition of the same test sequence presentation.

11.5.3.2 Test Sequence Preparation

The test sequences were acquired from the VQEG in YUV format [58]. These were then converted to a reference test sequence with MPEG-4 encoding, using the most accurate best quality compression at 25 fps with a key frame every 10 frames and QCIF (176 × 144) spatial resolution. This reference test sequence was then used to obtain samples of the adaptation space with various combinations of spatial resolution and frame rate. To achieve a spatial resolution of, for example, 90 percent, the original was spatially subsampled in the DCT domain with a fixed quantization parameter at 90 percent QCIF maintaining the aspect ratio, that is, 158 × 130, and then the image was scaled to maintain the original display size without any modification to the bit rate or encoding parameters of the test sequence; the actual display size of the sequence was not varied. This emulates the behavior of an adaptive transmitter that uses spatial resolution to reduce its output bit rate, with a client that rescales so that the actual display size is not varied during play out. To achieve a frame rate of, for example, 17 fps, the reference test sequence was reencoded with the exact same frame sequence and frame rate parameters—a key frame every 10 frames. Thus, the frame-dropping policy was the same for all test sequences. During the preparation of the test sequences for the subjective testing, the encoding method used was the "most accurate": no target bit rate was specified, and the encoder followed the supplied encoding parameters as closely as possible regardless of the resulting bit rate.

11.5.3.3 Results

The subjective testing consisted of two independent testers performing identical test procedures and test sequences on subjects taken from a student population ranging in ages from 18 to 30. Some test results were rejected due to the subject either being knowledgeable about video quality assessment or having visual impairments. The correlation between the tester results was 97 percent, which indicates that the results of the independent testers can be combined.

The forced choice methodology consists of a series of binomial experiments. Where there was a multinomial experiment depending on the number of encoding configurations falling into a zone of EABR, each possible combination of sequences within that zone of EABR was tested. The null hypothesis for binomial/multinomial experiments is that there is no preference for one encoding over another, thus each encoding should have equal probability. The chi-square test with 95 percent confidence was performed to measure the degree of disagreement between the data and the null hypothesis. In most cases the chi-squared test indicated a high degree of disagreement with the null hypothesis.

The subjective tests were conducted in two phases. Phase 1 considered four test sequences—one from each quadrant of the SI–TI grid. Adaptation space was sampled using a logarithmic scale to reflect Weber's law of just noticeable difference. The frame rates tested were 5, 7, 11, 17, and 25 fps, and the spatial resolutions were 100, 79, 63, 50, and 40 percent. During

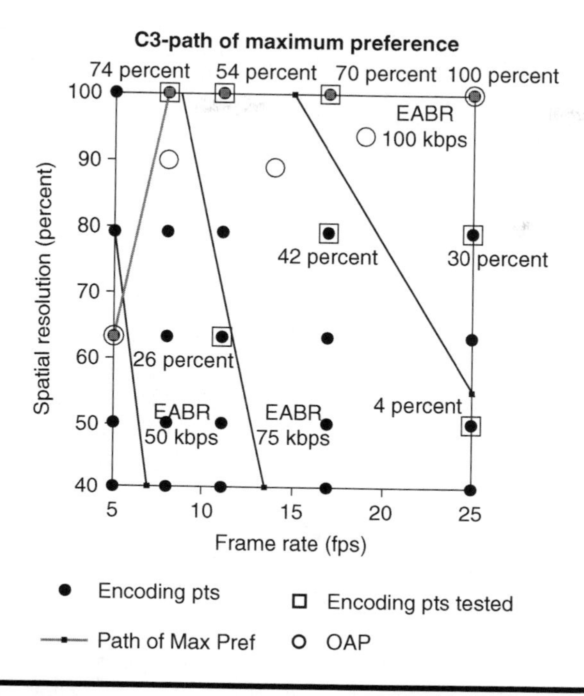

Figure 11.7 Subjective test results for C3.

Phase 1, 120 subjects were tested. Phase 2 considered four different test sequences with similar SI–TI values to those used for Phase 1. However, this time, the adaptation space was sampled using a linear scale. The frame rates tested were 5, 10, 15, 20, and 25 fps, and the spatial resolutions were 100, 85, 70, 55, and 40 percent. During Phase 2, 40 subjects were tested. The main objective of having two different test phases was to verify and validate the results from Phase 1. In addition, using different encoding scales, it could be ascertained that the OAT was similar in shape regardless of whether a linear or logarithmic scale was used, and regardless of the encoding points tested.

Figure 11.7 shows a grid of circular encoding points where the frame rate is on the x-axis and the resolution is on the y-axis. Through these encoding points are diagonal gray lines denoting the zones of EABR, ranging from 100 to 25 kbps. The encoding points marked with a percentage preference value are those points that were tested within a zone of EABR. For example, in EABR-75 kbps, there were two encoding configurations tested: (17, 100 percent) and (25, 79 percent). Seventy percent of the subjects preferred encoding configuration (17, 100 percent), whereas the remaining 30 percent preferred encoding configuration (25, 79 percent).

The clips falling within a particular zone of EABR have different (but similar) bit rates. For example, the bit rates corresponding to the encoding points (17, 100 percent), (25, 79 percent), and (25, 63 percent) were 85, 88, and 82 kbps, respectively. To compare clips of exactly the same bit rate would require a target bit rate to be specified, and then the codec would use proprietary means to achieve this bit rate by compromising the quality of the encoding in an unknown manner. By using zones of EABR, the bit rate of different test sequences with different encoding configurations is effectively quantized, which in turn dramatically reduces the number of test cases. To further reduce the number of test cases, it was decided that comparing (25, 79 percent) and (25, 63 percent) was redundant as it was considered counterintuitive that a user would prefer a clip with a lower spatial resolution when all other encoding factors are the same.

The dashed line between the encoding points is the path of maximum user preference through the zones of EABR. Weighted points were then used to obtain the optimal adaptation perception (OAP) points. The weighted points were interpolated as the sum of the product of preference with encoding configuration. For example, 70 percent of subjects preferred encoding (17, 100 percent) and 30 percent preferred encoding point (25, 79 percent). The weighted vector of these two points is (70 percent [17] + 30 percent [25], 70 percent [100 percent] + 30 percent [79 percent]), which equals OAP point (19.4, 93.7 percent). The weighted path of preference is the path joining the OAPs. There are two possible paths, which can be used to represent the OAT, the path of maximum user preference and the weighted path of preference.

It seems likely that by using the weighted path of preference, the system can satisfy more users. Using the same subjective testing methodology, the OAPs were compared against the maximum preferred encoding and the other encoding configurations in each zone of EABR. In all cases, the interpolated OAP did not have a statistically significant preference from the maximum preferred encoding indicating that this simple weighted vector approach is acceptable. It was observed that there was a higher incidence of forced choices when the maximum preferred encoding and the OAP were close together. This implies that either the path of maximum user preference or the interpolated weighted path of preference can be used as the OAT for adaptively streaming video. From Figure 11.8, it can be clearly seen from the paths of maximum user preference that when there is high action (C1 and C2), the resolution is less dominant regardless of whether the clip has high spatial characteristics or not. This implies that the user is more sensitive to continuous motion when there is high TI in the video content. Intuitively this makes sense as, when there is high action in a scene, often the scene changes are too fast for the user to be able to assimilate the scene detail. Conversely, when the scene has low temporal requirements (C3 and C4), the resolution becomes more dominant regardless of the spatial characteristics.

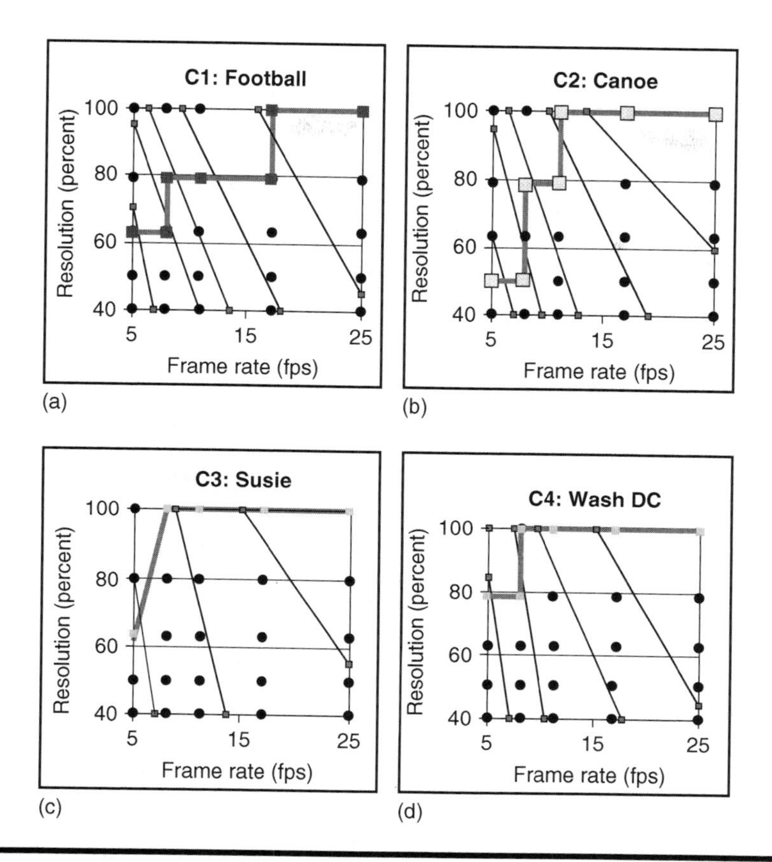

Figure 11.8 Path of maximum user preference.

From Figure 11.9, it can be seen that the best-fit curve (obtained by a least squares fit) arising from the various content types is a logarithmic curve with the general form $y = C\ln(x) + B$.

When the OATs for each of the content types were compared in Figure 11.10, the curves tended toward one another. The outlier is C4 (Wash DC), which is a slow-moving highly detailed clip. Its OAT deviates strongly toward resolution, indicating that when degrading the quality of the video being delivered, the frame rate should be sacrificed for resolution for this particular content type. Conversely for high action clips, frame rate should be given higher precedence to ensure smooth, continuous motion.

The usefulness of the OAT is dependent on the contents' spatial and temporal characteristics known *a priori* or calculated in real-time. Often this is not possible as the SI and TI metrics are highly computationally expensive. For real-time streaming, it may be sufficient to use a globally averaged OAT.

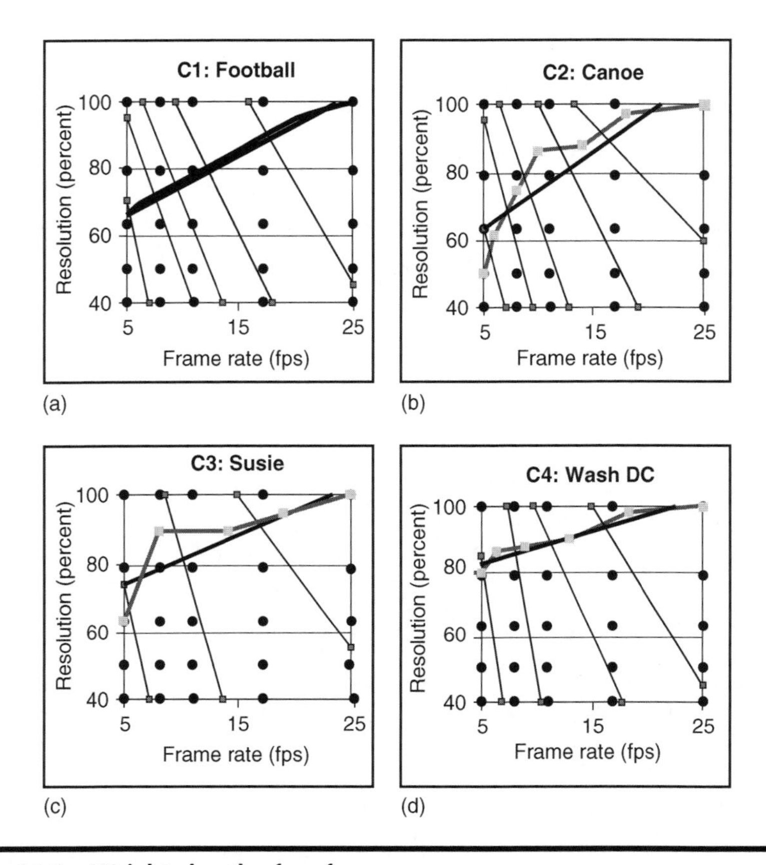

Figure 11.9 Weighted path of preference.

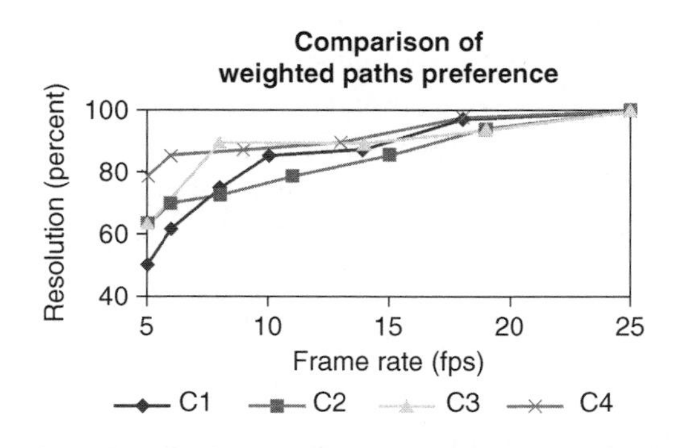

Figure 11.10 Comparison of the weighted path of preference.

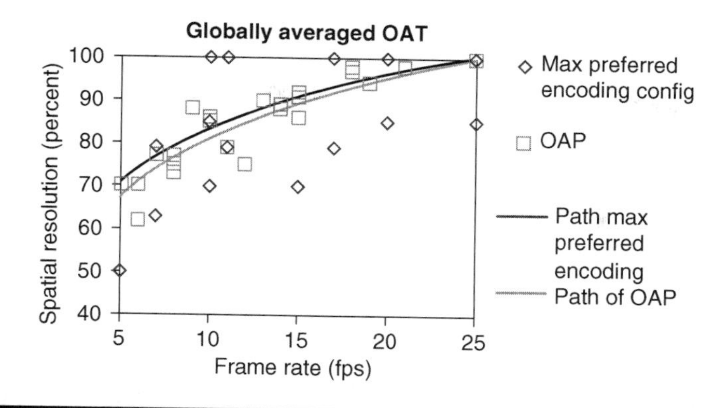

Figure 11.11 Globally averaged OAT.

To determine the globally averaged OAT (Figure 11.11), both the maximum preferred encoding configurations and interpolated OAPs for all content types are analyzed together and plotted on a scattergram. A least squares fit curve is plotted for both the maximum preferred encoding configurations and the interpolated OAPs. The correlation between the predicted max preference encoding least squares equation and OAP encoding least squares equation is above 99 percent.

11.5.3.4 Discussion

It was shown that when adapting the quality of a stream in response to network conditions, there is an OAT that maximizes the user-perceived quality. More specifically, within the set of different ways to achieve a target bit rate, there exists an encoding that maximizes the user-perceived quality. We have described one way of finding OATs through subjective testing, and applied it to finding OATs for various MPEG-4 video clips. Further work is being conducted to analyze the relationship between content characteristics and the corresponding OAT to determine the sensitivity of an OAT to the particular video being transmitted. Further subjective testing will be carried out to gauge the usefulness of our adaptation scheme in a practical adaptive system.

11.6 Quality-Oriented Multimedia Streaming Evaluation

To verify and validate the performance of the quality-oriented multimedia streaming employing OAT and QOAS, objective testing was performed, involving Network Simulator 2 (NS-2) and NS-2-built simulation models. These simulations aim at showing that the OAT-QOAS-based solution achieves significant performance in different delivery conditions including when cross traffic of various types, sizes, and variation patterns is encountered. Next the network topology, multimedia clips, simulation models, and performance assessment results are presented.

11.6.1 Simulation Setup

Simulations use NS-2 and No Ad Hoc (NOAH) wireless routing agent that support only direct communication between base stations and mobile nodes. Table 11.3 presents the MAC settings used. Figure 11.12 shows the simulation topology based on an IEEE 802.11b WLAN. It involves a smart access point (SAP) streaming multimedia content to a number of N clients (deployed at nodes Ci, i = 1, N). SAP includes N senders deployed at nodes Si, i = 1, N. Si-B1 (bandwidth = 100 Mbps, propagation delay = 5 ms) and B1-B2 (200 Mbps, 5 ms) links are overprovisioned so that only packet drops and significant delays are due to WLAN delivery. Buffering at intermediate nodes uses drop-tail queues. Clients' buffers are set in such a manner that they accommodate all received data and no loss occurs due to their limited sizes.

Table 11.3 MAC Settings Used during Simulations

MAC Settings			
Bit rate	11 Mbps	SIFS	10 μs
CWmin	21	Preamble length	144 bits
CWmax	1023	Short retry limit	7
Slot time	20 μs	Long retry limit	4

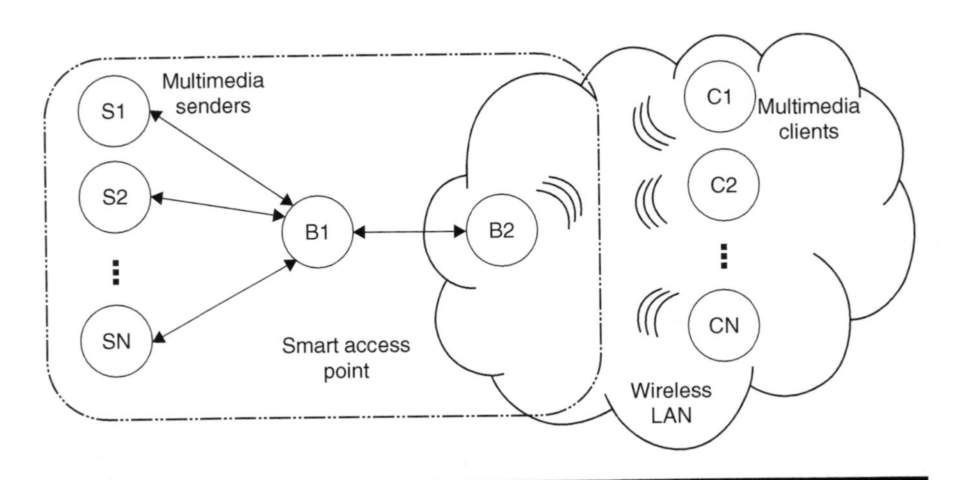

Figure 11.12 Simulation setup.

Table 11.4 Peak/Mean Bit Rate Ratios for All MPEG-4-Encoded Quality Versions of the Clips Used during Simulations

	MPEG-4: Average Rate (kbps)				
Clip	*64*	*128*	*256*	*384*	*512*
DH	3.92	3.85	4.46	4.56	4.46
RE	6.86	4.50	4.32	4.31	4.31
DW	4.18	3.91	3.90	3.90	3.90
JP	4.63	3.26	3.20	3.19	3.19
FM	4.75	3.79	3.78	3.78	3.78

11.6.2 Multimedia Clips

Five five-minute long multimedia clips with different degrees of motion content were considered: DH—high, RE—average/high, JP—average, DW—average/low, and FM—low. They were encoded at five different rates using the MPEG-4 scheme. The MPEG-4 clips have their average bit rates evenly distributed between 64 and 512 kbps, respectively. Among these different bit rate versions, the content is selected during adaptive streaming. All the clips have a frame rate of 25 fps, IBBP frame pattern, and 9 frames per group of pictures. Peak/mean bit rate ratios for all multimedia sequences used during testing are presented in Table 11.4.

11.6.3 Simulation Models

Testing was performed using NS-2-built QOAS, LDA+, TFRC, and NoAd models that follow the descriptions made in Ref. 59. To increase feedback accuracy, QOAS employs very high interfeedback intervals (100 ms) and makes use of small feedback report packets (40 B). This balances the need for the most up-to-date information with the requirement of low overhead. TFRC implementation had a 5 s update interval as suggested in Ref. 19 for delays greater than 100 ms, as in our setup. LDA+ implementation used an RTCP feedback interval of 5 s as suggested in Ref. 21.

11.6.4 Testing Scenarios

QOAS, LDA+, TFRC, and NoAd approaches were used in turn for streaming MPEG-4-encoded clips to a number of N clients. In successive tests, N was increased from 1 to 15 in a step-wise fashion increasingly loading the delivery network. This number of clients ensures that the average-user Quality of Experience (QoE) when using the best streaming solution—QOAS—is still above the "good" perceptual level on the ITU-T R. P.910 5-point subjective quality scale. This level of end-user perceptual quality is considered in this chapter as the minimum level of interest. However, the solution can support even a higher number of simultaneous streaming sessions, but their quality is expected to be lower. The clients randomly select both the movie clip and the starting point from within the chosen clip. This ensures that all movie types are used during testing. It also provides independence of the simulation results from the natural bit rate variation in time within each of the streamed movies. Consecutive streaming sessions were started at intervals of 25 s and the transitory periods are not considered when processing the results reported in this chapter. During the stable periods of 600 s, the loss rate, one-way delay, delay jitter, end-user QoE, and total throughput were measured and analyzed. The end-user-perceived quality was estimated using the non-reference moving picture quality metric and expressed on the ITU-T R. P.910 5-point quality scale.

11.6.5 Testing Results

Table 11.5 presents the complete set of testing results when QOAS, LDA+, TFRC, and NoAd streaming solutions were used in turn for delivering multimedia clips over the IEEE 802.11b WLAN. The tests were performed with increasing number of simultaneous clients and for each case the end-user-perceived quality, loss rate, delay, and the total throughput are monitored and averages are computed and indicated in the table.

Figure 11.13 plots the average viewers' perceived quality as a function of the increase in the number of simultaneously served clients. The graph clearly shows how QOAS outperforms all the other streaming approaches, including LDA+ that has the best results among the other approaches. For example, when streaming multimedia to one client, the perceived quality is 4.50 when using QOAS, 4.34 when using LDA+, 4.46 when TFRC is employed for streaming, and 3.99 when nonadaptively delivering the multimedia content. These values are recorded on the 1–5 ITU-T 5-point scale. As the number of simultaneous clients increases, the scores begin to diverge more significantly. For example, when the multimedia content is streamed simultaneously to ten clients over the IEEE 802.11b WLAN. When using QOAS the end-user-perceived quality reaches 4.20 in comparison to only 4.03 when using LDA+, 3.84 when TFRC-based streaming, and the lowest value of 1.00 when NoAd approach was used.

The most significant results are obtained in highly loaded delivery conditions. For 15 clients, when using QOAS the end-user-perceived quality is still at the "good" quality level, whereas for all the other approaches it drops much below this level: 3.85—LDA+, 3.68—TFRC, and 1.00—NoAd.

Figure 11.14 shows how the average loss rate increases with the increase in the number of concurrent sessions when different streaming approaches are employed for multimedia delivery. It can be clearly seen that QOAS and LDA+ outperform the other solutions when streaming MPEG-4-encoded content. They successfully maintain an average loss rate below or in the

Table 11.5 Performance Results When Streaming MPEG-4 Clips over IEEE 802.11b WLAN Using QOAS, NoAd, LDA+, and TFRC

Number of Clients		*1*	*3*	*5*	*7*	*9*	*11*	*13*	*15*
Quality (1–5)	QOAS	4.50	4.42	4.38	4.29	4.22	4.15	4.07	4.00
	NoAd	3.99	1.44	1.00	1.00	1.00	1.00	1.00	1.00
	LDA+	4.34	4.37	4.24	4.21	4.09	4.04	3.96	3.85
	TFRC	4.46	4.26	4.09	3.96	3.88	3.80	3.71	3.68
Loss rate (percent)	QOAS	0.02	0.17	0.19	0.35	0.48	0.74	1.05	1.51
	NoAd	3.11	22.03	38.58	44.36	47.76	>50.00	>50.00	>50.00
	LDA+	0.00	0.01	0.16	0.06	0.29	0.52	0.70	1.47
	TFRC	0.27	1.39	2.36	3.32	3.93	4.58	5.43	5.60
Delay (ms)	QOAS	12.64	16.31	22.52	28.30	33.32	41.58	53.03	51.75
	NoAd	26.40	72.00	106.40	117.80	124.20	>127.00	>127.00	>127.00
	LDA+	12.15	14.16	17.86	20.73	28.55	41.15	43.68	62.18
	TFRC	16.10	26.53	42.62	54.26	63.74	71.32	77.38	84.39
Jitter (ms)	QOAS	0.51	4.23	7.05	10.18	11.83	14.11	17.42	16.56
	NoAd	1.14	3.55	3.02	2.73	2.73	>2.97	>2.97	>2.97
	LDA+	0.25	2.75	4.53	6.33	7.56	9.49	10.12	10.40
	TFRC	1.86	8.17	13.03	14.41	14.72	14.95	14.45	14.29
Throughput (Mbps)	QOAS	0.48	1.26	1.95	2.34	2.66	3.00	3.17	3.43
	NoAd	0.50	1.20	1.57	1.99	2.41	<2.59	<2.59	<2.59
	LDA+	0.34	1.08	1.45	1.88	2.04	2.35	2.51	2.66
	TFRC	0.49	1.35	1.89	2.42	2.76	2.96	3.21	3.42

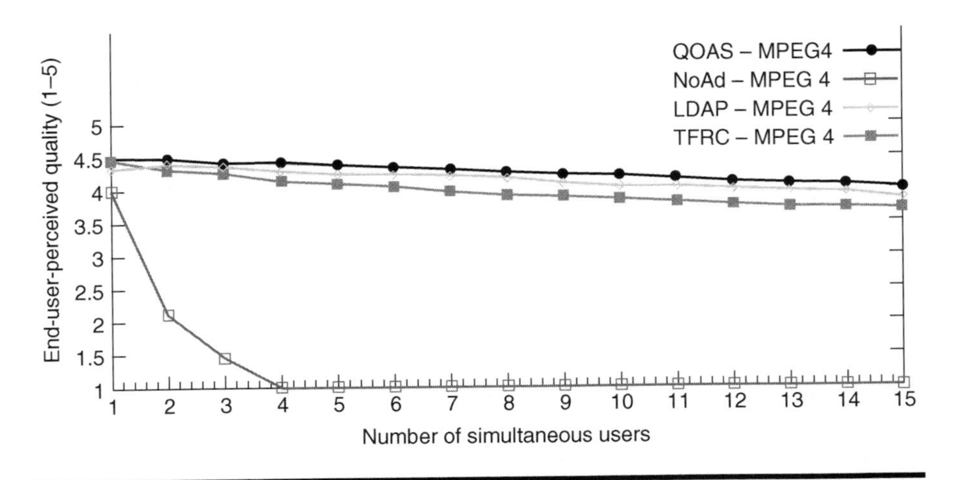

Figure 11.13 Comparative quality.

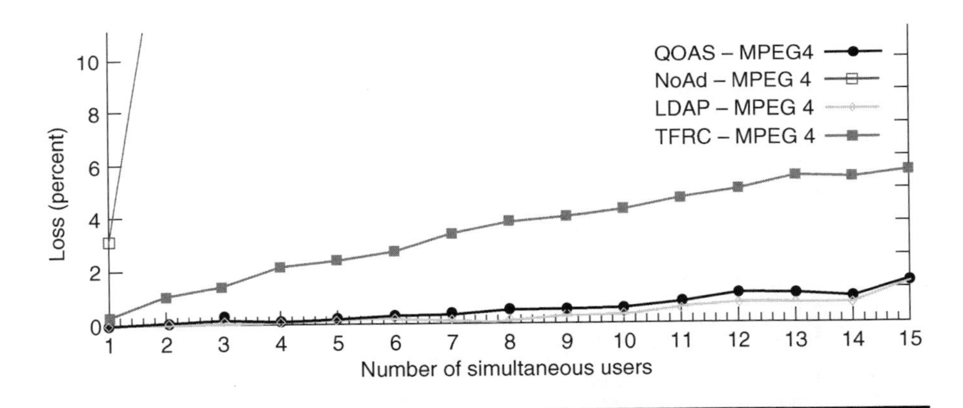

Figure 11.14 Comparative loss rate.

region of 1 percent much lower than those recorded when TFRC and NoAd approaches are used. For example, when streaming multimedia content to 10 clients the average loss rates are 0.56 percent when using QOAS, 0.32 percent for LDA+, 4.19 percent when using TFRC, and 49.38 percent for the NoAd approach.

Figure 11.15 presents the total throughput recorded when using different streaming approaches for delivering multimedia to an increasing number of simultaneous viewers. The figure shows how QOAS and TFRC achieve much better total throughput than the other approaches, regardless of the number of multimedia clients. For example, when streaming clips to 10 simultaneous clients using QOAS and TFRC approaches the total throughput is 2.90 and 2.85 Mbps, respectively, in comparison with

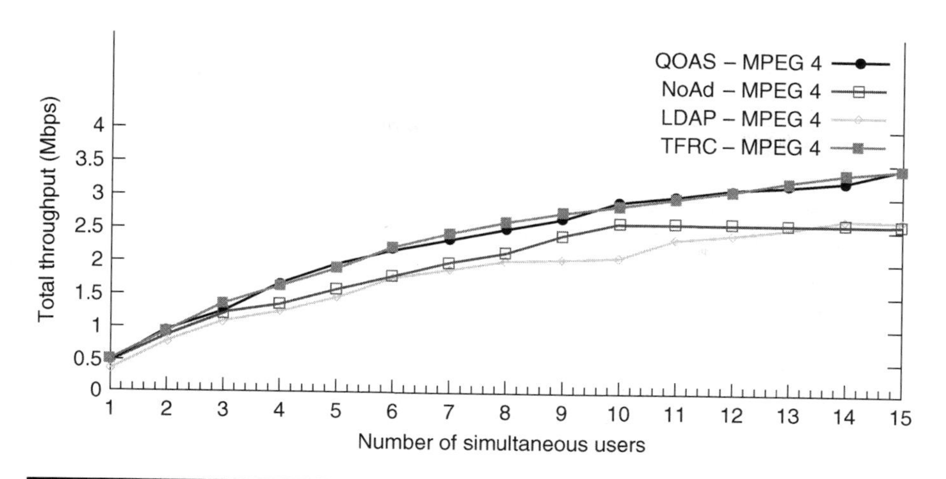

Figure 11.15 Comparative total throughput rate.

that measured when using LDA+ and NoAd which is 2.08 and 2.59 Mbps, respectively. Both QOAS and TFRC achieve similar high total throughput when delivering multimedia content to 15 clients. In this situation the total throughput is 3.43 and 3.42 Mbps, respectively, compared to only 2.66 Mbps when using LDA+.

Looking at other performance parameter values presented in details in Table 11.5, it can be seen that when using QOAS, the delay remains at very low levels in spite of the high increase in the number of simultaneous multimedia sessions. In contrast, streaming using the other approaches such as TFRC and NoAd incurs a significant increase in the delay that may eventually affect end-user QoE. The slight increase in jitter when using QOAS with the high increase in overall traffic over the IEEE 802.11b WLAN in comparison with when the other streaming solutions are employed can be coped with using client buffering and is not expected to affect the viewers.

These results show significant performance gains when using QOAS for streaming multimedia over IEEE 802.11b WLAN mainly in terms of viewers' perceived quality, loss, and total throughput in comparison to when other schemes are employed. Of particular importance is the achieved increase in WLAN's overall delivery capacity and efficiency when using QOAS, which is a highly desirable attribute as a higher number of simultaneous viewers who experience the same "good" perceived quality can be served from an existing infrastructure.

11.7 Conclusions

This chapter provides an overview of WLAN solutions and describes the main challenges for multimedia streaming over WLAN. The limitations of existing adaptive wireless streaming solutions are discussed. The chapter shows how knowledge of video quality perception can be integrated and used to optimally adapt the video quality so as to maximize end-user-perceived quality.

In this context, we demonstrate that significant improvements in terms of both end-user-perceived quality and bandwidth efficiency for wireless video streaming could be achieved by using adaptive multimedia streaming solutions. Therefore adaptive multimedia systems are becoming increasingly important. The fundamental question addressed in this chapter is how to adapt video quality in terms of video encoding parameters and user-perceived quality for streamed video over BE wireless IP networks. By gaining a better understanding of user perception, this knowledge can be integrated into video quality adaptation techniques such as the proposed OAT-QOAS.

The novel contribution of this chapter is complementing the user quality-oriented rate adaptation solution—QOAS that suggests target transmission bit rate adjustments of the video stream, whereas the OAT indicates how this target bit rate can be achieved in terms of encoding parameters to optimize user-perceived quality in given delivery conditions. The rate adaptation-based optimization is performed in a two-dimensional space defined by frame rate and resolution.

Through extensive subjective testing it has been found that an OAT through the set of possible video encoding configurations exists. The OAT indicates how the video quality should be adapted in terms of the encoding configuration parameters to maximize the end-user-perceived quality. The OAT varies according to the time-varying characteristics of the video content. For example, when the bit rate of the video stream must be reduced to achieve a target transmission bit rate, for high-action video clips the resolution should be reduced and the frame rate preserved because users cannot assimilate all the spatial detail but are more sensitive to continuity of motion within the scene. In contrast, for low action clips the frame rate can be reduced because there is low temporal activity within the scene and continuity of motion can be preserved in spite of a reduced frame rate; however, users are more sensitive to the loss of SI for this content type.

Along other QoS-related parameters, the QOAS considers end-user-perceived quality estimates in the adaptive multimedia streaming process. Based on client feedback, QOAS performs bit rate adaptation of the video stream and achieves increased streaming quality in highly variable and bandwidth-constrained network delivery conditions. As part of the OAT-QOAS solution, QOAS uses the OAT to determine how the video quality should be adapted to achieve the target transmission bit rate. Experimental results illustrate that using a two-dimensional adaptation strategy based on the OAT outperforms one-dimensional adaptation schemes that affect either multimedia stream's resolution or its frame rate, providing better user-perceived quality. Simulation-based test results are presented showing how by using QOAS for video streaming loss rate decreases, average throughput increases, and the user-perceived quality improves. When the OAT is used in conjunction with QOAS, end-user-perceived quality is maximized, improving the overall delivery of the video streams over wireless networks.

References

1. Halsall, F., *Computer Networking and the Internet*, 5th edition, Addison Wesley, Reading, MA, 2005.
2. Canet, M. J., F. Vicedo, V. Almenar, J. Valls, and E. R. De Lima, A Common FPGA Based Synchronizer Architecture for Hiperlan/2 and IEEE 802.11a WLAN Systems, *15th IEEE International Symposium on Personal, Indoor and Mobile Radio Communications*, 1, 531–535, September 5–8, 2004.
3. IEEE 802.11 Wireless Local Area Networks, http://grouper.ieee.org/groups/802/11/.
4. Geier, J., 802.11 Alphabet Soup, WiFi Planet, August 2002, http://www.wi-fiplanet.com/tutorials/article.php/1439551.
5. Ergen, M., *IEEE 802.11 Tutorial*, Department of Electrical Engineering and Computer Science, University of California Berkeley. Available at http://wow.eecs.berkeley.edu/ergen/docs/ieee.pdf, June 2002.
6. WhatIs, The Leading IT Encyclopedia and Learning Center, http://whatis.techtarget.com/definitionsAlpha/0,289930,sid9,00.html.
7. Ni, Q., L. Romdhani, and T. Turletti, A survey of QoS enhancements for IEEE 802.11 wireless LAN, *Journal of Wireless Communications and Mobile Computing*, 4(5), 547–566, 2004.
8. Intel Corp, Understanding Wi-Fi and WiMAX as Metro-Access Solutions, http://www.intel.com/netcomms/technologies/wimax/304471.pdf, 2004.
9. Ni, Q., Performance analysis and enhancements for IEEE 802.11e wireless networks, *IEEE Network*, 19(4), 21–27, 2005.
10. IEEE Std 802.11e, 2005 Edition, IEEE Standards for Local and Metropolitan Area Networks: Specific requirements Part 11: Wireless LAN Medium Access Control (MAC) and Physical Layer (PHY) specifications Amendment 8: Medium Access Control (MAC) Quality of Service Enhancements, 2005.
11. Li, M. M. Claypool, R. Kinicki, and J. Nichols, Characteristics of Streaming Media Stored on the Internet, WPI-CS-TR-03-18, Computer Science Technical Report Series, Worcester Polytechnical Institute, MA, May 2003.
12. Liu, J., Signal Processing for Internet Video Streaming: A Review, *Proceedings of SPIE Image and Video Communications and Processing*, San Jose, CA, January 2000.
13. Gupta, A., Rate Based Flow and Congestion Control Schemes, Technical Report, Department of Computer Science, University of Arizona, Tucson, AZ, 2002.
14. Wang, X. and H. Schulzrinne, Comparison of adaptive internet applications, *Proceedings of IEICE Transactions on Communications*, E82-B(6), 806–818, 1999.

15. Wu, D., Y. T. Hou, W. Zhu, H.-J. Lee, T. Chiang, Y.-Q. Zhang, and H. J. Chao, On end-to-end architecture for transporting MPEG-4 video over the Internet, *IEEE Transactions on Circuits and Systems for Video Technology*, 10(6), 923–941, 2000.
16. Wu, D., T. Hou, W. Zhu, H.-J. Lee, T. Chiang, Y.-Q. Zhang, and H. J. Chao, MPEG-4 video transport over the internet: A summary, *IEEE Circuits and Systems Magazine*, 2(1), 43–46, 2002.
17. Jacobson, V., Congestion Avoidance and Control, *Proceedings of ACM SIGCOMM 1988*, ACM, Stanford, CA, August 1988, pp. 314–329.
18. Turletti, T. and C. Huitema, Videoconferencing on the Internet, *IEEE/ACM Transactions on Networking*, 4(3), 340–351, 1996.
19. IETF RFC 3448, TCP Friendly Rate Control (TFRC): Protocol Specification, January 2003, ftp://ftp.rfc-editor.org/in-notes/rfc3448.txt.
20. Padhye, J., J. Kurose, D. Towsley, and R. Koodli, A Model Based TCP-friendly Rate Control Protocol, *Proceedings of NOSSDAV'99*, Basking Ridge, June 1999.
21. Sisalem, D. and A. Wolisz, LDA+: A TCP-Friendly Adaptation Scheme for Multimedia Communication, *Proceedings of IEEE International Conference on Multimedia and Expo III*, New York, July–August 2000.
22. Sisalem, D. and H. Schulzrinne, The Loss-Delay Based Adjustment Algorithm: A TCP-friendly Adaptation Scheme, *Proceedings of NOSSDAV '98*, Cambridge, UK, July 1998.
23. Sisalem, D., F. Emanuel, and H. Schulzrinne, The Direct Adjustment Algorithm: A TCP-friendly Adaptation Scheme, Technical Report GMD-FOKUS, August 1997.
24. ITU-T, Study-Group 12, COM 12-D98-E, Analysis, Measurement and Modeling of Jitter, Internal ITU-T Report.
25. Moon, S., J. Kurose, P. Skelly, and D. Towsley, Correlation of Packet Delay and Loss in the Internet. Technical Report, University of Massachusetts, USA, January 1998.
26. Jiang, W. and H. Schulzrinne, Modelling of Packet Loss and Delay and Their Effect on Real-time Multimedia Service Quality, *Proceedings of NOSSDAV*, Chapel Hill, NC, 2000.
27. Sklar, B., Rayleigh fading channels in mobile digital communication systems part I: Characterization, *IEEE Communications Magazine*, 35(7), 90–100, 1997.
28. Villasenor, J., Y.-Q. Zhang, and J. Wen, Robust video coding algorithms and systems, *Proceedings of the IEEE*, 87(10), 1724–1733, 1999.
29. Van den Branden, Lambrecht, C. J., Perceptual Models and Architectures for Video Coding Applications, Ph.D. Thesis, L'Ecole Polytechnique Federale de Lausanne (EPFL), Lausanne, Switzerland, 1996.
30. Frossard, P., Robust and Multiresolution Video Delivery: From H.26x to Matching Pursuit Based Technologies, Ph.D. Thesis, L'Ecole Polytechnique Federale de Lausanne (EPFL), Lausanne, Switzerland, 2001.
31. Video Quality Metrix—Frequently Asked Questions, Sarnoff Web Site, http://www.sarnoff.com.
32. The Video Quality Experts Group (VQEG), Final Report, April 2000, ftp://ftp.crc.ca/crc/vqeg/phase1-docs/final_report_april00.pdf.
33. Knee, M., *The Picture Appraisal Rating (PAR)—A Single-ended Picture Quality Measure for MPEG-2*, Snell & Wilcox, IBC, Amsterdam, The Netherlands, 2000.

34. Technical Guide for MPEG Quality Monitoring, Snell & Wilcox, http://www.snellwilcox.com/community/knowledge_center/engineering_guides/mpeg_qm.pdf.
35. Tektronix, A Guide to Maintaining Video Quality of Service for Digital Television Programs, White Paper, http://www.tek.com/Measurement/applications/video/mpeg2.html, 2000.
36. Wolf, S. and M. H. Pinson, In-Service Video Quality Measurement System Utilizing an Arbitrary Bandwidth Ancillary Data Channel, U.S. Patent No. 6,496,221, December 2002.
37. Webster, A. A., C. T. Jones, M. H. Pinson, S. D. Voran, and S. Wolf, Objective video quality assessment system based on human perception, *SPIE Human Vision, Visual Processing, and Digital Display IV*, 1913(February), 15–26, 1993.
38. Van den Branden, Lambrecht, C. J. and O. Verscheure, Perceptual quality measure using a spatio-temporal model of the human visual system, *Proceedings of the SPIE*, 2668(February), 450–461, 1996.
39. Van den Branden, Lambrecht, C. J., Color moving pictures quality metric, *Proceedings of International Conference on Image Processing ICIP'96*, 1885–888, 1996.
40. Verscheure, O., P. Frossard, and M. Hamdi, User-oriented QoS analysis in MPEG-2 video delivery, *Journal of Real-Time Imaging*, 5(5), 305–314, 1999.
41. Pixelmetrix and KDD Media to Jointly Market VP Series Picture Quality Analyzer, Pixelmetrix Press Release, http://www.pixelmetrix.com/rel/press%20release/1kdd.pdf, 2000.
42. Winkler, S., A Perceptual Distortion Metric for Digital Color Video, *Proceedings of Human Vision and Electronic Imaging SPIE*, vol. 3644, San Jose, CA, January 1999, pp. 175–184.
43. Watson, A. B., J. Hu, and J. F. III. McGowan, Digital video quality metric based on human vision, *Journal of Electronic Imaging*, 10(1), 20–29, 2001.
44. Watson, A. B., Method and Apparatus for Evaluating the Visual Quality of Processed Digital Video Sequences, U.S. Patent No. 6,493,023, December 2002.
45. Hekstra, A. P., J. G. Beerends, D. Ledermann, F. E. de Caluwe, S. Kohler, R. H. Koenen, S. Rihs, M. Ehrsam, and D. Schlauss, PVQM—A perceptual video quality measure, *Journal of Signal Processing: Image Communication*, 17(10), 781–798, 2002.
46. ITU-T R. P.861, Objective Quality Measurement of Telephone-Band (300–3400 Hz) Speech Codecs, February 1996.
47. Muntean, G.-M., P. Perry, and L. Murphy, A new adaptive multimedia streaming system for all-IP multi-service networks, *IEEE Transactions on Broadcasting*, 50(1), 1–10, 2004.
48. Muntean, G.-M., Efficient delivery of multimedia streams over broadband networks using QOAS, *IEEE Transactions on Broadcasting*, 52(2), 230–235, 2006.
49. Ghinea, G. and J. P. Thomas, QoS Impact on User Perception and Understanding of Multimedia Video Clips, *Proceedings of ACM Multimedia*, ACM, Bristol, United Kingdom, 1998, pp. 49–54.
50. Zhang, L., L. Zheng, and K. S. Ngee, Effect of delay and delay jitter on voice/video over IP, *Computer Communications*, 25(9), 863–873, 2002.
51. Bolot, J.-C. and T. Turletti, Experience with control mechanisms for packet video in the internet, *ACM Computer Communication Review*, 28(1), 4–15, 1998.

52. ITU-T Recommendation P.910, Subjective Video Quality Assessment Methods for Multimedia Applications, September, 1999.
53. Cranley, N., L. Murphy, and P. Perry, User perception of adapting video quality, *International Journal of Human-Computer Studies (IJHCS)*, 64(8), 637–647, 2006.
54. Cranley, N., L. Murphy, and P. Perry, Dynamic content based adaptation of streamed multimedia over best-effort IP networks, *Journal of Network and Computer Applications, Intelligence Based Adaptation for Ubiquitous Multimedia Communications* (Elsevier), 30(3), 983–1006, 2007.
55. Mason, A. J., The MUSHRA Audio Subjective Test Method, BBC R & D White Paper WHP 038, September 2002.
56. ISO/IEC JTC1/SC29/WG11 N3075, Report on the MPEG-4 Audio Version 2 Verification Test, 1999.
57. Pereira, F. and T. Ebrahimi, *The MPEG-4 Book*, Prentice Hall PTR, USA, 2002.
58. VQEG Test Sequences, ftp://ftp.crc.ca/crc/vqeg/TestSequences/.
59. Muntean, G.-M. and N. Cranley, Resource efficient quality-oriented wireless broadcasting of adaptive multimedia content, special issue on mobile multimedia broadcasting, *IEEE Transactions on Broadcasting*, 53(1), 362–368, 2007.

QUALITY OF SERVICE AND ENABLING TECHNOLOGIES

Chapter 12

End-to-End QoS Support for Video Delivery over Wireless Internet

Qian Zhang, Wenwu Zhu, and Ya-Qin Zhang

Contents

12.1 Introduction....410
12.2 Network-Centric Cross-Layer End-to-End QoS Support....414
12.2.1 Network QoS Provisioning for Wireless Internet....415
12.2.2 Cross-Layer QoS Support for Video Delivery over Wireless Internet....416
12.2.2.1 Wireless Network Modeling....416
12.2.2.2 Prioritized Transmission Control....418
12.2.2.3 QoS Mapping and QoS Adaptation....418
12.3 End System–Centric QoS Support....419
12.3.1 Network Adaptive Congestion Control....420
12.3.1.1 End-to-End Packet Loss Differentiation and Estimation....422
12.3.1.2 Available Bandwidth Estimation....422
12.3.2 Adaptive Error Control....423
12.3.3 Joint Power Control and Error Control....424
12.3.4 Rate Distortion–Based Bit Allocation....426

12.4 Conclusion and Further Remarks 427
References 428

Providing end-to-end quality-of-service (QoS) support is essential for video delivery over the next-generation wireless Internet. This chapter addresses several key elements in end-to-end QoS support for video delivery including scalable video representation, network-aware end system, and network QoS provisioning. There are generally two approaches in QoS support, that is, the network- and the end system–centric solutions. The fundamental problem in a network-centric solution is how to map QoS criterion at different layers, respectively, and optimize total quality across these layers. This chapter first presents a general framework of cross-layer network-centric solution, and then describes the recent advances in network modeling, QoS mapping, and QoS adaptation, which are the key components for network-centric solution. The key targets in end system–centric approach are network adaptation and media adaptation. This chapter first presents a general framework of end system–centric solution and investigates the recent developments. Specifically, for network adaptation, we review the available bandwidth estimation and efficient video transport protocol; for media adaptation, we describe the advances in error control, power control, and corresponding bit allocation. Finally, we highlight several advanced research directions.

12.1 Introduction

With the rapid growth of wireless networks and great success of Internet video, wireless video services are expected to be widely deployed in the near future. As different types of wireless networks are converging into all Internet Protocol (IP) networks, that is, the Internet, it is important to study video delivery over the wireless Internet. The current trends in the development of real-time Internet applications and the rapid growth of mobile systems indicate that the future Internet architecture will need to support various applications with different QoS [1]* requirements [2]. QoS support is a multidisciplinary topic involving several areas, ranging from applications, terminals, networking architectures to network management, business models, and finally the main target—end users.

Enabling end-to-end QoS support in Internet is difficult, and becomes more challenging when introducing QoS in an environment involving

* Note that the definition of QoS may be somewhat confusing and has different implications. We adopt the definition "the ability to ensure the quality of the end user experience" [1] in this chapter.

mobile hosts under different wireless access technologies, because available resources (e.g., bandwidth and battery life) in wireless networks are scarce and change dynamically over time. For wireless networks, because the capacity of a wireless channel varies randomly with time, providing deterministic QoS (i.e., zero QoS violation probability) will likely result in extremely conservative guarantees and waste of resources, which is hardly useful. Thus, in this chapter, we only consider statistical QoS [3]. To support end-to-end QoS for video delivery over wireless Internet, there are several fundamental challenges:

- *QoS support encompasses a wide range of technological aspects.* To be specific, many technologies, including video coding, high-performance physical and link layers support, efficient packet delivery, congestion control, error control, and power control affect the overall QoS.
- *Different applications have very diverse QoS requirements in terms of data rate, delay bound, and packet loss probability.* For example, unlike non-real-time data packets, video services are very sensitive to packet delivery delay but can tolerate some frame losses and transmission errors.
- *Different types of networks inherently have different characteristics.* This is also referred to as network heterogeneity. It is well known that Internet is based on IP, which only offers the best effort (BE) services. Specifically, network conditions, such as bandwidth, packet loss ratio, delay, and delay jitter, vary from time to time. An important characteristic of wireless networks in the future is that there shall be mixtures of heterogeneous wireless access technologies that co-exist, such as wireless local area network (WLAN) access, 2.5G/3G cellular access, Bluetooth, WiMax, and ultrawideband (UWB) networks. It is well known that bit error rate (BER) in a wireless network is much higher than that in a wireline network. Moreover, link-layer error control scheme, such as automatic repeat request (ARQ) and forward error correction (FEC), is widely used to overcome the varying wireless channel errors. This will further increase the dramatic variation of bandwidth and delay in wireless networks. To make things even more complicated, the end-to-end packet loss in wireless Internet can be caused by either congestion loss occurred due to buffer overflow in wireline network or the erroneous loss occurred in the wireless link due to channel error.
- *There is a dramatic heterogeneity among end users.* End users have different requirements in terms of latency, video visual quality, processing capabilities, power, and bandwidth. It is thus a challenge to design a video delivery mechanism that not only achieves high

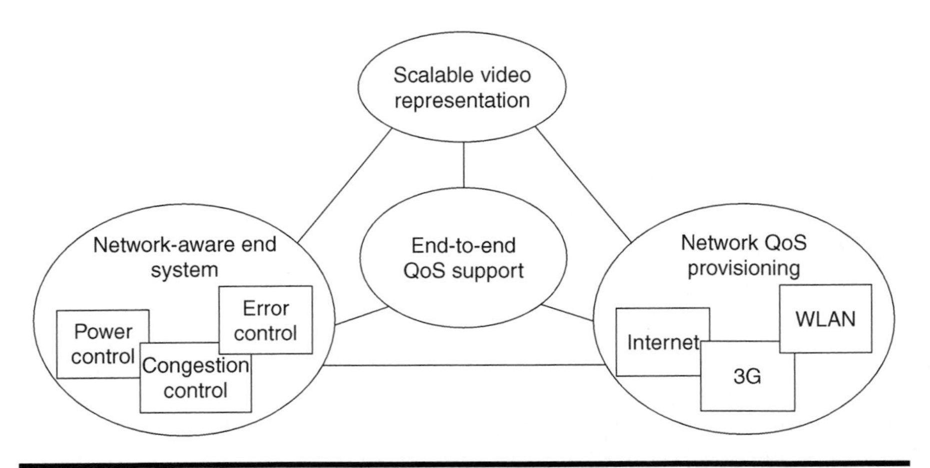

Figure 12.1 Fundamental components for end-to-end QoS support.

efficiency in network bandwidth but also meets the heterogeneous requirements of the multiple end users.

To address the mentioned challenges, one should support the QoS requirement in all components of the video delivery system from end to end, which include QoS provisioning from networks, scalable video presentation from applications, and network adaptive congestion/error/power control in end systems. Figure 12.1 illustrates key components for end-to-end QoS support.

- *QoS provisioning from networks.* The BE nature of Internet has promoted the Internet Engineering Task Force (IETF) community to seek for efficient QoS support through network-layer mechanisms. The most well-known mechanisms are Integrated Services (IntServ) [4] and Differentiated Services (DiffServ) [5]. The approaches in providing QoS in wireless networks are quite different from their Internet counterparts. General Packet Radio Service (GPRS)/Universal Mobile Telecommunications System (UMTS), IEEE 802.11, and IEEE 802.16 have totally different mechanisms for QoS support.
- *Multilayered scalable video coding from applications.* In scalable coding, the signal is separated into multiple layers of different visual importance. The base layer can be independently decoded and it provides basic video quality. The enhancement layers can only be decoded together with the base layer and they further refine the video quality. Enhancements on layered scalable coding have proposed to provide further fine granularity scalability (FGS) [6–8]. Scalable video representation provides fast adaptation to bandwidth variations as

well as inherent error resilience and complexity scalability properties that are essential for efficient transmission over error-prone wireless networks.

- *Network adaptive congestion/error/power control in end systems.* When network condition changes, the end systems can employ adaptive control mechanisms to minimize the impact on user-perceived quality. Power-, congestion-, and error control are the three main mechanisms to support QoS for robust video delivery over wireless Internet. Power control is performed collectively from the group point of view by controlling transmission power and spreading gain for a group of users so as to reduce interference [9]. Congestion- and error control are conducted from the individual user's point of view to effectively combat the congestions and errors occurred during transmission by adjusting the transmission rate and allocating bits between source and channel codings [10,11].

There are two types of approaches in providing end-to-end QoS support: the first one is network-centric QoS provisioning, in which routers/switches or base stations/access points in the networks provide prioritized QoS support to satisfy data rate, delay bound, and packet loss requirements by different applications. In the prioritized transmission, QoS is expressed in terms of the probability of buffer overflow or delay violation at the link layer. However, at the video application layer, QoS is measured by the mean squared error (MSE) or peak-signal-to-noise ratio (PSNR). Thus, one of the key issues for end-to-end QoS provisioning using network-centric solution is the effective QoS mapping across different layers. More specifically, one needs to consider how to model the varying network and coordinate effective adaptation of QoS parameters at video application layer and prioritized transmission system at link layer. Section 12.2 describes a general framework of a cross-layer architecture of a network-centric end-to-end QoS support solution and reviews recent developments in individual components including network QoS support, channel modeling, QoS adaptation, and QoS mapping.

The second type of approach in providing end-to-end QoS support is solely end system–centric. In particular, the end systems employ various control techniques, which include congestion-, error-, and power control, to maximize the application-layer video quality without any QoS support from the underlying network. The advantage of end-system control is that there are minimum changes required in the core network. The main challenge, however, is how to design efficient power-, congestion- error control mechanisms. Section 12.3 presents a framework that targets at minimizing the end-to-end distortion or power consumption, and reviews the recent studies on various mechanisms.

12.2 Network-Centric Cross-Layer End-to-End QoS Support

As stated earlier, different layers (e.g., application- and link-/network layer) have different metrics to measure QoS, which bring challenge for end-to-end QoS provisioning. Figure 12.2 shows the general block diagram of end-to-end QoS support for video delivery in the network-centric cross-layer solution. This solution considers an end-to-end delivery system for a video source from the sender to the receiver, which includes source video encoding, cross-layer QoS mapping and adaptation, prioritized transmission control, adaptive network modeling, and video decoder/output modules. To support end-to-end QoS with network-centric approach, a dynamic QoS management system is needed for video applications to interact with underlying prioritized transmission system to handle service degradation and resource constraint in time-varying wireless Internet. Specifically, to offer a good compromise between video quality and available transmission resource, the key is how to provide an effective cross-layer QoS mapping and an efficient adaptation mechanism.

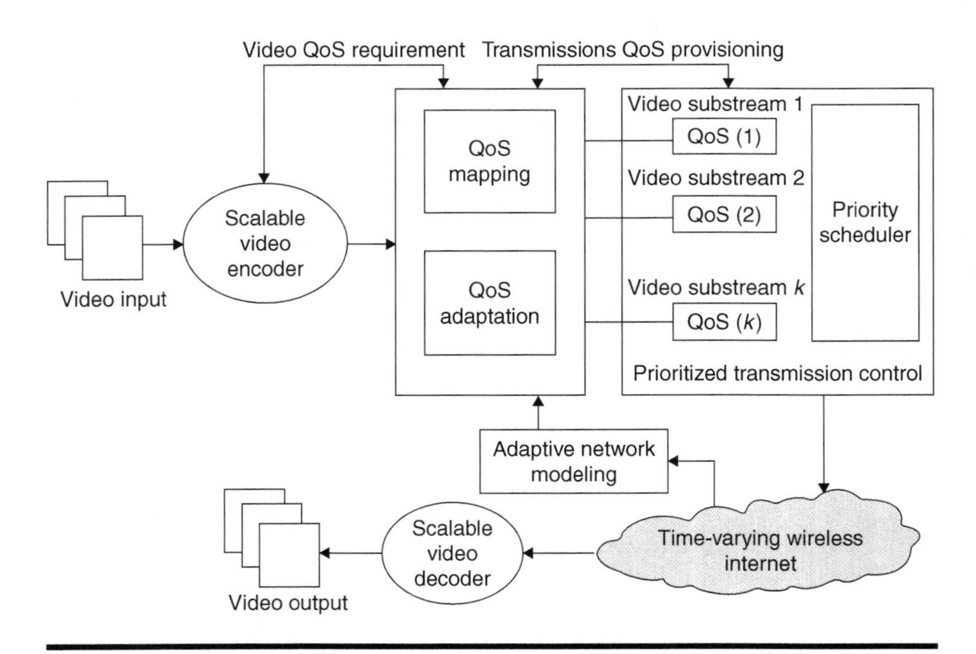

Figure 12.2 General framework of end-to-end QoS support for video over wireless Internet with network-centric solution.

12.2.1 Network QoS Provisioning for Wireless Internet

QoS provisioning for the Internet has been a very active area of research for many years. Two different approaches have been introduced in IETF, which are IntServ [4] and DiffServ [5], respectively. IntServ was introduced in IP networks to provide guaranteed and controlled services in addition to the existing BE service. IntServ and reservation protocols, such as Resource Reservation Protocol (RSVP), have failed to become a practical end-to-end QoS solution because of lack of scalability and the difficulty in having all elements in the network RSVP-enabled. DiffServ was proposed to provide a scalable and manageable network with service-differentiation capability. In contrast to the per-flow-based QoS guarantee in the Intserv, Diffserv networks provide QoS assurance on a per-aggregate basis.

The Internet research community has been proposing and investigating different approaches to achieve differentiated services. In particular, significant efforts have been made to achieve service differentiation in terms of queuing delay and packet loss [12,13], both of which are of primary concern for multimedia applications. Many QoS control mechanisms, especially in the areas of packet scheduling [14,15] and queue management algorithms [16,17], have been proposed in recent years. Elegant theories such as network calculus [18] and effective bandwidths [19] have also been developed. Firoiu et al. [20] provided a comprehensive survey on a number of recent advances in Internet QoS provisioning.

There have also been many studies related to QoS provision in wireless networks. Third Generation Partnership Project (3GPP; www.3GPP.org) is the main standard body that defines and standardizes a common QoS framework for data services, particularly IP-based services. It has defined a comprehensive framework for end-to-end QoS covering all subsystems, from radio access network (RAN) through core network to gateway node (to the external packet data network) within a UMTS network [21]. It has also defined four different UMTS QoS classes according to delay sensitivity: conversational, streaming, interactive, and background classes.

In WLANs, the original IEEE 802.11 communication modes, namely, distributed coordination function (DCF) and point coordination function (PCF), do not differentiate traffic types. IEEE is proposing enhancements in 802.11e to both coordination modes to facilitate QoS support [22]. In enhanced distribution coordination function (EDCF), the concept of traffic categories is introduced. EDCF establishes a probabilistic priority mechanism to allocate bandwidth based on traffic categories. Aiming to extend the polling mechanism of PCF, hybrid coordination function (HCF) is proposed. A hybrid controller polls stations during a contention-free period. The polling grants each station a specific start time and a maximum transmit duration. In the mean time, a group of vendors have proposed wireless multimedia enhancements (WMEs) to provide an interim QoS solution for 802.11 networks [22]. WME uses four priority levels in negotiating communication between wireless access points and client devices.

The demand for QoS in wireless has been addressed by the IEEE 802.16 standard [23]. What specifically sets this standard apart is a polling-based Media Access Control (MAC) layer that is more deterministic than the contention-based MAC used by 802.11. The 802.16's MAC layer enables classification of QoS- and non-QoS-dependant application flows and maps them to connections with distinct scheduling services, enabling both guaranteed handling and traffic enforcement. Each connection is associated with a single scheduling data service, and each data service is associated with a set of QoS parameters that quantify aspects of its behavior. There are four types of scheduling services supported by 802.16, which are unsolicited grant service (UGS), real-time polling service (rtPS), non-real-time polling service (nrtPS), and BE, respectively. The key QoS metrics associated with each of these four scheduling services are the maximum sustained rate (MSR), minimum reserved rate (MRR), maximum latency, and maximum jitter and priority. The MRR is associated with different scheduling services, acting as the "guarantee," although the MSR serves to limit the rate of a connection or "enforcement" to a maximum sustained rate.

12.2.2 Cross-Layer QoS Support for Video Delivery over Wireless Internet

An efficient QoS mapping scheme that addresses cross-layer QoS provision issues for video delivery over wireless Internet includes the following important building blocks: (1) wireless network modeling that can effectively model time-varying and nonstationary behavior of the wireless networks, (2) prioritized transmission control scheme that can derive and adjust the rate constraint of a prioritized transmission system, and (3) QoS mapping and adaptation mechanism that can optimally map video application classes to statistical QoS guarantees of a prioritized transmission system to provide the best trade-off between the video application quality and the transmission capability under time-varying wireless networks.

12.2.2.1 Wireless Network Modeling

One can model a communication channel at different layers, that is, physical- and link layer (see Figure 12.3). Physical-layer channel can be further classified into radio-, modem-, and codec-layer channel.

Among them, radio-layer channel models can be classified into large-scale path loss and small-scale fading [24]. Large-scale path loss models

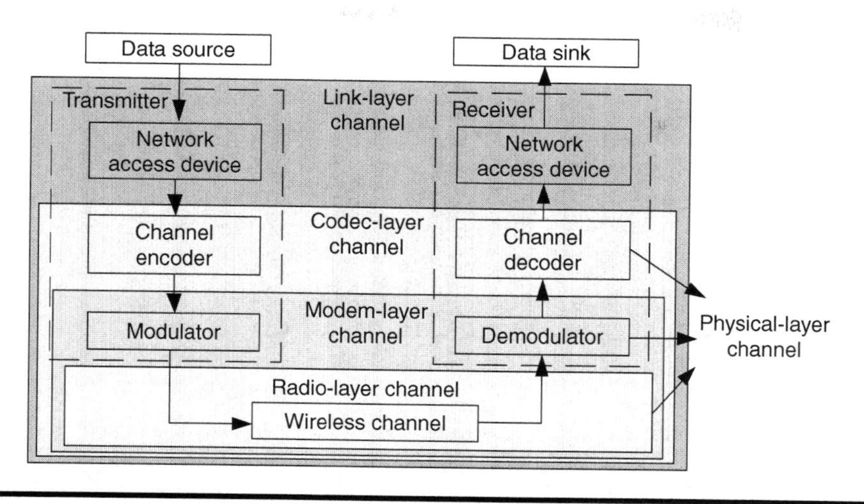

Figure 12.3 Different channel models.

characterize the underlying physical mechanisms (i.e., reflection, diffraction, and scattering) for specific paths. Small-scale fading models describe the characteristics of generic radio paths in a statistical fashion. Modem-layer channel can be modeled by a finite-state Markov chain [25], whose states are characterized by different bit error rates. A codec-layer channel can also be modeled by a finite-state Markov chain, whose states can be characterized by different data rates, or a symbol being error-free/in-error, or a channel being good/bad [26]. Zorzi et al. [26] demonstrated that Markov model is an approximation on block transmission over a slowly fading wireless channel.

In general, on the basis of existing physical-layer channel models, it is very complex to characterize the relationship between the control parameters and the calculated QoS measures. This is because the physical-layer channel models do not explicitly characterize the wireless channel in terms of the link-level QoS metrics such as data rate, delay, and delay violation probability.

Recognizing the limitation of physical-layer channel models in QoS support, that is, the difficulty in analyzing link-level performances, attempts have been made to move the channel model up in the protocol stack, from physical- to link layer [27,28]. In Ref. 27, an effective capacity (EC) channel model was proposed. The model captures the effect of channel fading for the link-queueing behavior using a computationally simple yet accurate model; thus can be a critical tool for designing efficient QoS provisioning mechanisms.

12.2.2.2 Prioritized Transmission Control

To achieve differentiated services, a class-based buffering and scheduling mechanism is needed in the prioritized transmission control module. In particular, K QoS priority classes are maintained with each class of traffic being maintained in separate buffers. A priority scheduling policy is employed to serve packets of these classes. Under this class-based buffering and priority scheduling mechanism, each QoS priority class can obtain certain level of statistical QoS guarantees in terms of probability of packet loss and packet delay. The next step is to translate the statistical QoS guarantees of multiple priority classes into rate constraints based on the effective capacity theory [27]. The calculated rate constraints in turn specify the maximum data rate that can be transmitted reliably with statistical QoS guarantee over the time-varying wireless channel. Consequently, video substreams can be classified into classes and bandwidth can be allocated accordingly for each class.

The rate constraint of multiple priority classes under a time-varying service rate channel can be derived according to the guaranteed packet loss probabilities and different buffer sizes of each priority class [28]. The statistical QoS guarantee of each priority class is provided in terms of packet loss probability based on the effective service capacity theory. Kumwilaisak et al. [28] derived the rate constraint of substreams under a simplest strict (non-preemptive) priority scheduling policy.

12.2.2.3 QoS Mapping and QoS Adaptation

QoS mapping and QoS adaptation are the key components to achieve cross-layer QoS support in the video delivery architecture. Unlike the adaptive channel modeling module and prioritized transmission control module, the QoS mapping and QoS adaptation are application-specific. Because the QoS measure at the video application layer (e.g., distortion and uninterrupted video service perceived by end users) is not directly related to QoS metrics at the link layer (e.g., packet loss/delay probability), a mapping and adaptation mechanism must be in place to more precisely match the QoS criterion across different layers. Specifically, at the video application layer, each video packet is characterized based on its loss and delay properties, which contributes to the end-to-end video quality and service. Then, these video packets are classified and optimally mapped to the classes of link transmission module under the rate constraint. The video application layer QoS and link-layer QoS are allowed to interact with one another and adapt to the wireless channel condition, whose objective is to find the QoS trade-off, which simultaneously provides a desired video service of the end users with available transmission resources.

There have been many studies on the cross-layer design for efficient multimedia delivery with QoS assurance over wired and wireless networks in recent years [13,28–31]. The focus has been on the utilization of the differentiated service architecture to convey multimedia data. The common approach is to partition multimedia data into smaller units and then map these units to different classes for prioritized transmission. The partitioned multimedia units are prioritized based on its contribution to the expected quality at the end user, although the prioritized transmission system provides different QoS guarantees depending on its corresponding service priority. Servetto et al. [32] proposed an optimization framework to segment a variable bit rate source to several substreams that are transmitted in multiple priority classes. The objective is to minimize the expected distortion of the variable bit rate source. Shin et al. [13] proposed to prioritize each video packet based on its error propagation effect, if it is lost. Video packets were mapped differently to transmission priority classes with the objective of maximizing the end-to-end video quality under the cost or price constraint. Tan and Zhakor [30] examined the same problem as that formulated in Ref. 13 with different approaches for video prioritization. Other types of multimedia delivery over DiffServ network, such as prioritized speech and audio packets, were considered by Martin [33] and Sehgal and Chou [29].

The stochastic behavior of wireless networks [34,35] introduced a cross-layer design with adaptive QoS assurance for multimedia transmission where absolute QoS was considered. Xiao et al. [34] studied the rate-delay trade-off curve derived from the lower-layer protocol to the applications. The application layer selected the operating point from this curve as a guaranteed QoS parameter for transmission. These curves are allowed to change with the change in the wireless network environment. In Ref. 35, the dynamic QoS framework to adaptively adjust QoS parameters of the wireless network to match with time-varying wireless channel condition is investigated, where the application was given the flexibility to adapt to the level of QoS provided by the network. Targeting at scalable video codec and considering the interaction between layers to obtain the operating QoS trade-off points, in Ref. 28, the QoS mapping and adaptation for wireless network was addressed in the following two steps. First, find the optimal mapping policy from one group of picture (GOP) to K priority classes such that the distortion of this GOP is minimized. Second, find a set of QoS parameters for the priority network, such that the expected video distortion is minimized.

12.3 End System–Centric QoS Support

To provide end-to-end QoS with end-system solution, the video applications should be aware of and adaptive to the variation of network condition in wireless Internet. This adaptation consists of network- and media adaptation. The network adaptation refers to how many network resources

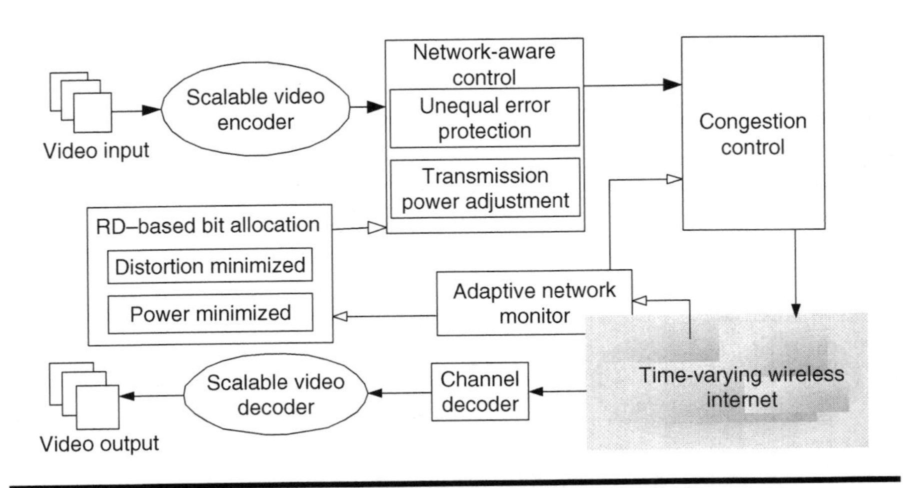

Figure 12.4 General framework for end-to-end QoS provisioning for video over wireless Internet with end system–centric solution.

(e.g., bandwidth and battery power) a video application should utilize for its video content, that is, to design an adaptive media transport protocol for video delivery. The media adaptation controls the bit rate of the video stream based on the estimated available bandwidth and adjusts error and power control behaviors according to the varying wireless Internet conditions.

The general diagram for end system–centric QoS provisioning is illustrated in Figure 12.4. To address network adaptation, an end-to-end video transport protocol is needed to handle congestion control in wireless Internet. More specifically, the adaptive network monitor deals with probing and estimating the dynamic network conditions. The congestion control module adjusts sending rate based on the feedback information.

For media adaptation, considering that different parts of compressed scalable video bitstream have different importance level, network-aware unequal error protection (UEP) module protects different layers of scalable video against congestive packet losses and erroneous losses according to their importance and network status. Network-aware transmission power adjustment module adjusts the transmission power of the end system to affect the wireless channel conditions. Rate distortion (RD)–based bit allocation module performs media adaptation control with two different targets, that is, distortion minimization and power consumption minimization.

12.3.1 Network Adaptive Congestion Control

Bursty loss and excessive delay have a devastating effect on perceived video quality, and are usually caused by network congestion. Thus, congestion-control mechanism at end systems is necessary to reduce packet loss and delay. Typically, for conferencing and streaming video, congestion control takes the form of rate control. Rate control attempts to minimize the possibility of network congestion by matching the rate of the video stream to the available network bandwidth.

To deliver media content, several protocols are involved in which some are proprietary solutions. These protocols include the Real-Time Transport Protocol (RTP) and Real-Time Control Protocol (RTCP) [36], Session Description Protocol (SDP) [37], Real-Time Streaming Protocol (RTSP) [38], Stream Control Transmission Protocol (SCTP) [39], Session Initiation Protocol (SIP) [40], and Hypertext Transport Protocol (HTTP).

Because a dominant portion of today's Internet traffic is TCP-based, it is very important for multimedia streams to be "TCP-friendly," which means that a media flow generates similar throughput as a typical TCP flow along the same path under the same condition with lower latency. There are two existing types of TCP-friendly flow-control protocols for multimedia delivery applications: sender-based rate adjustment and model-based flow control. Sender-based rate adjustment [10,41] performs additive increase and multiplicative decrease (AIMD) rate control in the sender as in TCP. The transmission rate is increased in a steplike fashion in the absence of packet loss and reduced multiplicatively when congestion is detected. This approach usually requires the receiver to send frequent feedback to detect congestion indications, which may potentially degrade the overall performance. Model-based flow control [42,43], however, uses a stochastic TCP model [44] that represents the throughput of a TCP sender as a function of packet loss ratio and round trip time (RTT). One issue to be considered for this type of approach is that the estimated packet loss ratio is not for the next time interval so as to affect the accuracy of the throughput calculation.

Although TCP friendliness is a useful fairness criterion in today's Internet, it is possible that future network architectures (in which TCP is either no longer the predominant transport protocol or has a very bad performance) will allow or require different definitions of fairness. For example, fairness definition for wireless networks is still subject to research because TCP performance in wireless networks still needs to be improved.

When designing a transport protocol for video transmission over wireless Internet, several issues related to network condition estimation should be considered. The most important one is the estimation of congestion loss ratio. In wireless Internet, the end-to-end packet loss can be caused by either congestion loss due to buffer overflow or the erroneous loss occurred in the wireless link. Traditional TCP and TCP-friendly media transport protocols [45,46] treat any lost packet as a signal of network congestion and adjust its transmission rate accordingly. However, this rate reduction is unnecessary if the packet loss is due to the error occurred in wireless network, which in turn causes bad performance for end-to-end delivery quality. The second issue is the RTT estimation. There is large variation in end-to-end delay in wireless Internet [47]. Sending only a single acknowledgement to measure the RTT during a predefined period of time may be inaccurate and fluctuates greatly. The third issue is the available bandwidth estimation. There are many studies on available bandwidth estimation in Internet; how to apply those schemes for transport protocol design in wireless networks is now attracting many attentions [46,48].

12.3.1.1 End-to-End Packet Loss Differentiation and Estimation

As stated earlier, the key issue of designing an efficient media transport protocol is to correctly detect whether the network is in congestion or not. Generally there are two types of methods to distinguish the network status [49], which are split connection and end-to-end methods. In the split connection method, it requires an agent at the edge of wired and wireless network to measure the conditions of two types of networks separately [50,51]. Specifically, an agent is needed at every base station in the entire wireless communication system, which adds excessive complexity in the actual deployment. The end-to-end method focuses on differentiating the congestive loss from the erroneous packet loss by adopting some heuristic methods such as interarrival time or packet pair [52–54]. This type of solution expects a packet to exhibit a certain behavior under wireless Internet. It is known that a specific behavior of a packet in the network reflects the joint effect of several factors. Considering that the traffic pattern in the Internet is a complicated research topic, how to find out a good pattern to predict the behavior of packets in wireless Internet still needs some fundamental research.

Yang et al. [55] proposed a different mechanism to use the combined link layer and sequence number information to differentiate the wireless erroneous loss and congestive loss. The arrival time of the erroneous packets is used to derive the distribution of lost packets among the erroneous packets between two back-to-back correctly received packets.

12.3.1.2 Available Bandwidth Estimation

There are two types of approaches for available bandwidth estimation in media transport protocols.

The first type of approach calculates the available bandwidth based on the estimated RTT and packet loss ratio. Padhye et al. [44] proposed a formula to calculate the network throughput that has been widely adopted [42,47].

The second type of approach calculates the available bandwidth using the receiver-based packet pair (RBPP) method [56]. RBPP requires the use of two consecutively sent packets to determine a bandwidth share sample.

The most recognized scheme in this category is TCP-Westwood [48], which maintains two estimators, along with a method to identify the predominant cause of packet loss. Depending on the cause of the loss, the appropriate estimator is "adaptively" selected. One estimator, called bandwidth estimator, considers each ACK pair separately to obtain a bandwidth sample, filters the samples, and returns to the (short-term) bandwidth share that the TCP sender is getting from the network. The other estimator, called rate estimator (RE), measures the amount of data acknowledged during the latest interval *T*. RE tends to estimate the (relatively longer-term) rate that the connection has recently experienced. Several media transport protocols such as SMCC [57] and VTP [58], proposed recently, follow the idea of TCP-Westwood.

12.3.2 Adaptive Error Control

There are two basic error correction mechanisms, namely, ARQ and FEC. ARQ has been shown to be more effective than FEC. However, FEC has been commonly suggested for real-time applications due to their strict delay requirements. Hybrid ARQ scheme proposed in Ref. 59 can achieve both delay bound and rate effectiveness by limiting the number of retransmissions. Other hybrid FEC and delay-constrained ARQ schemes are discussed in Refs 60–62.

Girod and Farber [63] reviewed the existing solutions for combating wireless transmission errors. Although their focus was on cellular networks, most presented protection strategies can also be applied to the transmission of video over the other types of wireless networks. Shan and Zakhor [64] presented an integrated application-layer packetization, scheduling, and protection strategies for wireless transmission of nonscalable coded video. Cote et al. [65] presented a survey of the different video-optimized error resilience techniques that are necessary to accommodate the compressed video bitstreams. Various channel/network errors can result in a considerable damage to or loss of compressed video information during transmission; effective error concealment strategies become vital for ensuring a high quality of the video sequences in the presence of errors/losses. A review of the existing error concealment mechanisms was given by Wang and Zhu [66]. Majumdar et al. [67] addressed the problem of resilient real-time video streaming over IEEE 802.11b WLANs for both unicast and multicast transmissions. For the unicast scenario, a hybrid ARQ algorithm that efficiently combines FEC and ARQ was proposed. For multicast, progressive video coding based on MPEG-4 FGS was combined with FEC.

Scalable video has received lots of attention in recent years due to its fast adaptation characteristic. For scalable video, one way to efficiently combat channel errors is to employ UEP for information of different importance.

More specifically, strong channel-coding protection is applied to the base-layer data stream whereas weaker channel-coding protection is applied to the enhancement layer parts. Studying how to add FEC to a scalable video coding is of great interest, recently. Joint work on scalable video coding with UEP for wired network [68,69] and wireless communication [70–72] has been proposed. In Ref. 72, a network adaptive application-level error control scheme using hybrid UEP and delay-constrained ARQ was proposed for scalable video delivery. Current and estimated RTT is used at the sender side to determine the maximum number of retransmission based on delay constraint. Van der Schaar and Radha [73] discussed the combination of MPEG-4 FGS with scalable FEC for unicast and multicast applications, and a new UEP strategy referred as fine grained loss protection (FGLP) was introduced.

It has been shown that, under general wireless environments, different protection strategies exist at the various layers of the protocol stack, and hence a joint cross-layer consideration is desirable to provide an optimal overall performance for the transmission of video. In Ref. 74, avertical system integration, referred as "cross-layer protection," was introduced, which enabled the joint optimization of the various protection strategies existing in the protocol stack. Xu et al. [75] developed a cross-layer protection strategy for maximizing the received video quality by dynamically selecting the optimal combination of application-layer FEC and MAC retransmissions based on the channel conditions.

12.3.3 Joint Power Control and Error Control

In general, there exists trade-off between maintaining good quality of video application and reducing average power consumption, including processing- and transmission power at end systems. From network point of view, multipath fading and multiple access interference (MAI) in wireless network necessitate the use of high transmission power. From video coding point of view, to decrease transmission power and maintain a desired video quality, more complex compression algorithms and more powerful channel coding schemes can be applied to source- and channel coding, respectively.

The motivation of jointly considering power- and error control for video communication comes from the following observations on the relationship among rate, distortion, and power consumption.

- *Case 1.* According to the rate-distortion theory (Figure 12.5; 1), the lower the source coding rate R, the larger the distortion D. More generally, it can be represented as $D = D(R)$.
- *Case 2.* When video compression is performed with a given power constraint P, the power-constrained distortion includes both the distortion caused by the source rate control and the power constraint

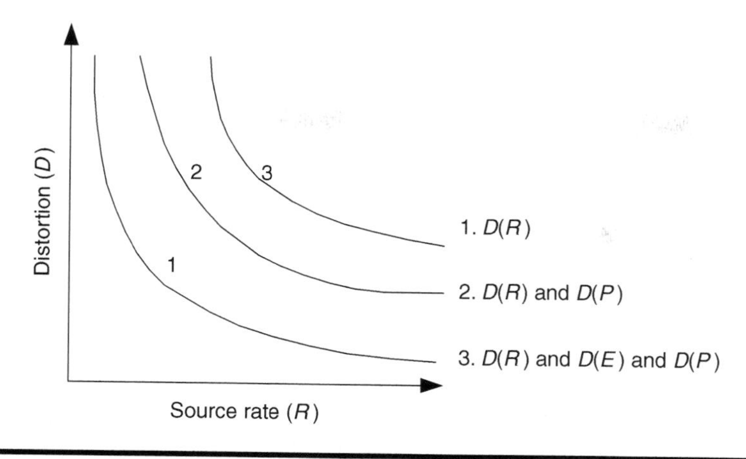

Figure 12.5 An illustration of RD with/without considering power constraint and transmission error.

(Figure 12.5; 2). More generally, it can be denoted as $D = D(R)$ and $D(P)$.

- *Case 3*. Considering a more specific scenario, a video bitstream is transmitted over wireless links with a given bit error rate E and a limited power constraint P, the end-to-end distortion is composed of the distortion caused by the source rate control, channel errors, and power constraint (Figure 12.5; 3). More generally, it can be denoted as $D = D(R)$ and $D(E)$ and $D(P)$.

From the individual user point of view, some studies on allocating available bits for source and channel coders are aiming at minimizing the total processing power consumption under a given bandwidth constraint. Specifically, a low-power communication system for image transmission was investigated in Ref. 76. A power-optimized joint source-channel coding (JSCC) approach for video communication over wireless channel was proposed in Ref. 77.

From the group user point of view, power control adjusts a group of users' transmission powers to maintain their video quality requirements. Recently, the focus has been on adjusting transmission powers to maintain a required signal-to-interference ratio (SIR) for each network link using the least possible power. It is also referred to as resource management based on the power control technique discussed in Refs 9, 78, and 79, where it is formulated as a constrained optimization problem to minimize the total transmission power or maximize the total rate subject to the SIR and bandwidth requirements. The key observation Eisenberg et al. [80] and Zhang et al. [81] made independently is that when the transmission power of one user is changed to achieve its minimal power consumption, its interference to other users varies accordingly. This interference variation will alter other users' receiving SIRs and may result in their video quality requirements not being achieved, and then in turn deviate from the optimal state of their power consumptions. Therefore, due to the multiple access interference, the global minimization of power consumption must be investigated from the group point of view.

12.3.4 Rate Distortion–Based Bit Allocation

For video delivery over wired or wireless network, the most common metrics used to evaluate video quality are the expected end-to-end distortion D_T and expected end-to-end power consumption P_T. Here D_T consists of source distortion D_s and channel distortion D_c. The source distortion is caused by source coding such as quantization and rate control. The channel distortion occurs when the packet loss due to network congestion or the errors in wireless channel is caused during the transmission. P_T consists of processing power on the source coding P_s, processing power on the channel coding P_c, and the transmission power for data delivery P_t.

It is well known that channel bandwidth capacity is highly limited in wireless Internet. Thus, it is very important to adequately allocate the bits between the source- and channel coding for error protection, under a given fixed bandwidth capacity so as to achieve the minimal expected end-to-end distortion or end-to-end power consumption [69,80]. More specifically, the resource allocation problem can be formulated as follows.

$$\min D_T(D_s, D_c) \quad \text{s.t.} \quad R_T \leq R_0$$

where R_T is the total bandwidth assigned to source coding and channel protection, whereas R_0 is the total bandwidth budget. Or

$$\min P_T(P_s, P_c, P_t) \quad \text{s.t.} \quad R_T = R_0 \quad \text{and} \quad D_T = D_0$$

where D_0 is the end-to-end distortion budget.

In all the schemes aforementioned, the erroneous and congestive losses are treated the same; and only one type of packet loss is considered. As discussed earlier, in wireless Internet, the packet losses consist of both congestive and erroneous losses, which in turn have different loss patterns in wireless and wired network parts. Considering that different loss patterns lead to different perceived QoS at application level [82], Yang et al. [55] presented a loss differentiated RD–based bit allocation scheme [51], in which the channel distortion is caused by two parts: one is caused during the transmission over wired-line part of the connection, $D_{c,\text{wired}}$, and the other during the transmission over the wireless channel, $D_{c,\text{wireless}}$.

12.4 Conclusion and Further Remarks

This chapter reviews recent advances in providing end-to-end QoS support for video delivery over wireless Internet from both network-centric and end system–centric perspectives. In the network-centric solution, we present the general cross-layer QoS support architecture for video delivery over wireless Internet. This architecture enables to perform QoS mapping between statistical QoS guarantees at the network level to a corresponding priority class with different video quality requirements. In the end system–centric approach, we describe the framework that includes network- and media adaptation and reviewed several key components in this framework. More specifically, recent developments in congestion-, error-, and power control, and RD–based bit allocation schemes were addressed.

Cross-layer design of heterogeneous wireless Internet video systems is a relatively new and active field of research, in which many issues need further examination. Optimally allocating resources in this heterogeneous setting presents many challenges and opportunities. To solve the cross-layer optimization problems for video transmission, several components, such as (1) adaptive modulation and channel coding, (2) adaptive retransmission, and (3) adaptive source rate control, need to be jointly optimized to achieve better performance. Moreover, this chapter is primarily focused on QoS support in a unicast scenario. Efficient end-to-end QoS support for multicast video transmission systems [83–86] is an area that still requires considerable work.

It has been recognized that the Internet interdomain routing algorithm, Border Gateway Protocol (BGP), is not always able to provide good quality routes between domains. More recently, there have been proposals to establish application-level overlay networks for multimedia applications. Examples of overlay networks include application-layer multicast [87–90], Web content distribution networks, and resilient overlay networks (RONs) [91]. Recently, there has been investigation on providing QoS support mechanism in overlay networks similar to the one in the Internet. OverQoS [92] aimed to provide architecture to offer QoS using overlay network. Service overlay networks [93] purchase bandwidth with certain QoS guarantees from individual network domains via bilateral service level agreement (SLA) to build a logical end-to-end service delivery infrastructure on top of existing data transport networks. Unlike the work on network-based QoS, research for QoS provisioning in application-layer overlay has been pursued in an *ad hoc* manner. Thus, there is a considerable room for improvement, especially in considering the video delivery requirement.

Enabling video transport over *ad hoc* multihop networks is another challenging task. The wireless links in an *ad hoc* network are highly error prone and can go down frequently because of node mobility, interference, channel fading, and the lack of infrastructure. In the initial stage, *ad hoc* multihop protocol design is largely based on a layered approach, where each layer in the protocol stack is designed and operated independently. Mao et al. [94] proposed to combine multistream coding with multipath transport to show that path diversity provides an effective way to combat transmission error in *ad hoc* networks. QoS routing [95] and QoS-aware MAC [96] are two types of approaches to provide QoS for *ad hoc* networks from networking point of view. However, the inflexibility and suboptimality of this layered paradigm result in poor performance for *ad hoc* multihop networks in general, especially when the application has high bandwidth needs or stringent delay constraints. To meet these QoS requirements, recent study on *ad hoc* multihop networks has demonstrated that cross-layer design can significantly improve the system performance [97–100]. Extending the cross-layer framework to exploit the video delivery over *ad hoc* networks is quite an interesting direction for research.

References

1. Hébert, P.-Y., End-to-end QoS in the user's point of view. *ITU Workshop on End-to-End Quality of Service. What Is It? How Do We Get It?*, Geneva, October 1–3, 2003.
2. Wu, D., Y. T. Hou, and Y.-Q. Zhang, Transporting real-time video over the Internet: Challenges and approaches. *Proc. IEEE*, 88(December), 1855–1877, 2000.
3. Liebeherr, J., A framework for analyzing networks with deterministic and statistical QoS. *Comet Group Seminar*, Columbia University, 2000.
4. Wroclawski, J., The use of RSVP with IETF integrated services. *RFC 2210*, September 1997.
5. Grossman, D., New terminology and clarifications for Diffserv. *RFC 3260*, April 2002.
6. Li, W., Overview of fine granularity scalability in MPEG-4 video standard. *IEEE Trans. CSVT*, 11(3), 301–317, 2001.
7. Wu, F., S. Li, and Y.-Q. Zhang, A framework for efficient progressive fine granularity scalable video coding. *IEEE Trans. CSVT*, 11(March), 332–344, 2001.
8. Van der Schaar, M. and H. Radha, Adaptive motion compensation fine-granular-scalability (AMC-FGS) for wireless video. *IEEE Trans. CSVT*, 12(6), 360–371, 2002.
9. Sampath, A., P. S. Kumar, and J. M. Holtzman, Power control and resource management for a multimedia CDMA wireless system. *IEEE PIMRC'95*, 1, 21–25, 1995.
10. Rejaie, R., M. Handley, and D. Estrin, Quality adaptation for congestion controlled video playback over the Internet. *ACM SIGCOMM 99*, 29(4), 189–200, 1999.
11. Qian, L., D. L. Jones, K. Ramchandran, and S. Appadwedula, A general joint source-channel matching method for wireless video transmission. *Proceedings of the IEEE DCC'99*, pp. 414–423, 1999.

12. Dovrolis, C., D. Stiliadis, and P. Ramanathan, Proportional differentiated services: Delay differentiation and packet scheduling. *IEEE/ACM Trans. Network.*, 10, 12–26, 2002.
13. Shin, J., J. Kim, and C.-C. Jay Kuo, Quality-of-service mapping mechanism for packet video in differentiated services network. *IEEE Trans. Multimedia*, 3(2), 219–231, 2001.
14. Zhang, H., Service disciplines for guaranteed performance service in packet switching networks. *Proc. IEEE*, 83(October), 1374–1396, 1995.
15. Zhang, Z.-L., Z. Duan, and Y. T. Hou, Virtual time reference system: A unifying scheduling framework for scalable support of guaranteed services. *IEEE J. Select. Area. Commun.*, 18(December), 2684–2695, 2000.
16. Misra, V., W. Gong, and D. Towsley, Fluid-based analysis of a network of AQM routers supporting TCP flows with an application to RED. *ACM SIGCOMM*, pp. 151–160, 2000.
17. Kunniyur, S. and R. Srikant, Analysis and design of an adaptive virtual queue algorithm for active queue management. *ACM SIGCOMM 2001*, San Diego, CA, 2001.
18. Le Boudec, J.-Y., Application of network calculus to guaranteed service networks. *IEEE Trans. Inform. Theory*, 44(May), 1087–1096, 1998.
19. Chang, C.-S. and J. A. Thomas, Effective bandwidth in high-speed digital networks. *IEEE J. Select. Area. Commun.*, 13(6), 1091–1100, 1995.
20. Firoiu, V., J.-Y. Boudec, D. Towsley, and Z.-L. Zhang, Theories and models for internet quality of service. *Proc. IEEE*, 90(9), 1565–1591, 2002.
21. Quality of Service (QoS) concept and architecture, 3GPP TS 23.107, September 2003.
22. Kitchin, D., The 802.11 MAC protocol and quality of service. Lecture, Intel Corp., 2003.
23. Hayat, B., R. Mansoor, and A. Nasir, 802.16 2001 MAC Layer QoS. *Ubiquity*, 7(17), 2006.
24. Rappaport, T. S., *Wireless Communications: Principles and Practice*, Prentice Hall, New York, 1996.
25. Zhang, Q. and S. A. Kassam, Finite-state markov model for Rayleigh fading channels. *IEEE Trans. Commun.*, 47(11), 1688–1692, 1999.
26. Zorzi, M., R. R. Rao, and L. B. Milstein, Error statistics in data transmission over fading channels. *IEEE Trans. Commun.*, 46(11), 1468–1477, 1998.
27. Wu, D. and R. Negi, Effective capacity: A wireless link model for support of quality of service. *IEEE Trans. Wireless Commun*, 2(4), 630–643, 2003.
28. Kumwilaisak, W., Y. Hou, Q. Zhang, W. Zhu, C.-C. Kuo, and Y.-Q. Zhang, A cross-layer quality of service mapping architecture for video delivery in wireless networks. *IEEE JSAC*, December 2003.
29. Sehgal, A. and P. A. Chou, Cost-distortion optimized streaming media over DiffServ networks. *IEEE ICME*, Lausanne, pp. 857–860, August 2002.
30. Tan, W. and A. Zhakor, Packet classification schemes for streaming MPEG video over delay and loss differentiated networks. *Proceedings of the IEEE Packet Video workshop 2001*, Kyongju, Korea, April 2001.
31. Liu, J., B. Li, H.-R. Shao, W. Zhu, and Y.-Q. Zhang, A proxy-assisted adaptation framework for object video multicasting. *IEEE Trans. CSVT.*, 15(3), 402–411, 2005.

32. Servetto, S. D., K. Ramchandran, K. Nahrstedt, and A. Ortega, Optimal segmentation of a VBR source for its parallel transmission over multiple ATM connections. *Proceedings of the IEEE International Conference on Image Processing*, Santa Barbara, CA, pp. 5–8, October 1997.
33. De Martin, J. C., Source-driven packet marking for speech transmission over differentiated-service networks. *Proceedings of the IEEE International Conference on Acoustics, Speech, and Signal Processing*, pp. 753–756, May 2001.
34. Xiao, L., M. Johansson, H. Hindi, S. Boyd, and A. Goldsmith, Joint optimization of communication rates and linear systems. *IEEE Trans. Automatic Control*, 48(1), 148–153, 2003.
35. Mirhakkak, M., N. Schult, and D. Thomson, Dynamic bandwidth management and adaptive applications for a variable bandwidth wireless environment. *IEEE J. Select. Area. Commun.*, 19(10), 1984–1997, 2001.
36. Schulzrinne, H., S. Casner, R. Frederick, and V. Jacobson, RTP: A transport protocol for real-time applications. *RFC 3550*, July, 2003.
37. Handley, M. and V. Jacobson, SDP: Session description protocol. *RFC 2327*, April 1998.
38. Schulzrinne, H., A. Rao, and R. Lanphier, Real time streaming protocol (RTSP). *RFC 2326*, April 1998.
39. Fu, S. and M. Atiguzzaman, SCTP: State of the art in research, products, and technical challenges. *IEEE Commun. Mag.*, 42(4), 64–76, 2004.
40. Handley, M., H. Schulzrinne, E. Schooler, and J. Rosenberg, SIP: Session initiation protocol. *RFC 2543*, 1999.
41. Jacobs, S. and A. Eleftheriadis, Streaming video using TCP flow control and dynamic rate shaping. *J. Vis. Commun. Image Represen.*, 9(3), 211–222, 1998.
42. Tan, W. and A. Zakhor, Real-time Internet video using error resilient scalable compression and TCP-friendly transport protocol. *IEEE Trans. Multimedia*, 1(June), 172–186, 1999.
43. Disalem, D. and H. Schulzrinne, The loss-delay based adjustment algorithm: A TCP-friendly adaptation scheme. *Workshop on NOSSDAV*, July 1998.
44. Padhye, J., V. Firoiu, D. Towsley, and J. Kurose, Modeling TCP throughput: A simple model and its empirical validation. *ACM SIGCOMM'98*, pp. 303–314, August 1998.
45. Floyd, S., M. Handley, J. Padhye, and J. Widmer, Equation based congestion control for unicast applications. Available at http://www.aciri.org/tfrc.
46. Widmer, J., R. Denda, and M. Mauve, A survey on TCP-Friendly congestion control. *IEEE Mag. Netw.*, 15(May/June), 28–37, 2001.
47. Zhang, Q., W. Zhu, and Y.-Q. Zhang, Resource allocation for multimedia streaming over the Internet. *IEEE Trans. Multimedia*, September 2001.
48. Gerla, M., B. K. F. Ng, M. Y. Sanadidi, M. Valla, and R. Wang, TCP Westwood with adaptive bandwidth estimation to improve efficiency/friendliness tradeoffs. *Comput. Commun. J.*, 27(1), 41–58, 2004.
49. Montenegro, G., S. Dawkins, M. Kojo, V. Magret, and N. Vaidya, Long thin networks. *RFC 2757*, January 2000.
50. Bakre, A. and B. Badrinath, I-TCP: Indirect TCP for mobile hosts. *Proceedings of the 15th International Conference on Distributed Computing Systems (ICDCS)*, Vancouver, pp. 136–143, May 30–June 2, 1995.

51. Cheung, G. and T. Yoshimura, Streaming agent: A network proxy for media streaming in 3G wireless networks. *IEEE Packet Video Workshop*, Pittsburgh, PA, April 2002.
52. Cen, S., P. Cosman, and G. Voelker, End-to-end differentiation of congestion and wireless loss. *Proceedings of the ACM Multimedia Computing and Networking*, San Jose, CA, pp. 1–15, January 2002.
53. Barman, D. and I. Matta, Effectiveness of loss labeling in improving TCP performance in wired/wireless network. *10th IEEE ICNP*, Paris, pp. 2–11, November 2002.
54. Biaz, S. and N. Vaidya, Discriminating congestion losses from wireless losses using inter-arrival times at the receiver, *Proceedings of the IEEE Symposium on Application-Specific Systems and Software Engineering and Technology*, Richardson, TX, pp. 10–17, March 1999.
55. Yang, F., Q. Zhang, W. Zhu, and Y.-Q. Zhang, End-to-end TCP-friendly streaming protocol and bit allocation for scalable video over wireless Internet. *IEEE JSAC*, 2004.
56. Lai, K. and M. Baker, Nettimer: A tool for measuring bottleneck link bandwidth. *Proceedings of the USENIX Symposium on Internet Technologies and Systems*, San Francisco, March 2001.
57. Aboobaker, N., D. Chanady, M. Gerla, and M. Y. Sansadidi, Streaming media congestion control using bandwidth estimation. *Proceedings of the IFIP/IEEE International Conference on Management of Multimedia Networks and Services*, Santa Barbara, pp. 89–100, 2002.
58. Wang, R., M. Valla, M. Y. Sanadidi, and M. Gerla, Using adaptive rate estimation to provide enhanced and robust transport over heterogeneous networks. *Proceedings of the 10th IEEE ICNP 2002*, Paris, November 12–15, 2002.
59. Zhang, Q. and S. A. Kassam, Hybrid ARQ with selective combining for fading channels. *IEEE J. Select. Area. Commun.*, 17(5), 867–880, 1999.
60. Puri, R., K. Ramchandran, and A. Ortega, Joint source channel coding with hybrid ARQ/FEC for robust video transmission. *IEEE Multimedia Signal Processing Workshop*, CA, December 1998.
61. Wu, D., Y. T. Hou, and Y.-Q. Zhang, Scalable video coding and transport over broad-band wireless networks. *Proc. IEEE*, 89(1), 6–20, 2001.
62. Zhang, Q., W. Zhu, and Y.-Q. Zhang, Channel-adaptive resource allocation for scalable video transmission over 3g wireless network. *IEEE Trans. CSVT*, 2004.
63. Girod, B. and N. Farber, Wireless video. in *Compressed Video over Networks*, Marcel Dekker, New York, 2001.
64. Shan, Y. and A. Zakhor, Cross layer techniques for adaptive video streaming over wireless networks, *IEEE ICME*, pp. 277–280, August 2002.
65. Cote, G., F. Kossentini, and S. Wenger, Error resilience coding. in *Compressed Video over Networks*, Marcel Dekker, New York, 2001.
66. Wang, Y. and Q.-F. Zhu, Error control and concealment for video communications: A review. *Proceedings of the IEEE*, 86(5), 974–997, 1998.
67. Majumdar, A., D. Sachs, I. Kozintsev, K. Ramchandran, and M. Yeung, Multicast and unicast real-time video streaming over wireless LANs. *IEEE Trans. CSVT*, 12(June), 524–534, 2002.

68. Zhang, T. and Y. Xu, Unequal packet loss protection for layered video transmission. *IEEE Trans. Broadcast.*, 45(June), 243–252, 1999.
69. Cheung, G. and A. Zakhor, Bit allocation for joint-source channel coding of scalable video. *IEEE Trans. Image Proc.*, 9(March), 340–356, 2000.
70. Horn, U., B. Girod, and B. Belzer, Scalable video coding for multimedia applications and robust transmission over wireless channels. *7th Workshop on Packet Video*, Brisbane, pp. 43–48, 1996.
71. Hagenauer, J., T. Stockhammer, C. Weiss, and A. Donner, Progressive source coding combined with regressive channel coding for varying channels. *Proceedings of the 3rd ITG Conference on Source and Channel Coding*, Munich, pp. 123–130, January 2000.
72. Wang, G., Q. Zhang, W. Zhu, and Y.-Q. Zhang, Channel-adaptive unequal error protection for scalable video transmission over wireless channel. *Proceedings of the SPIE VCIP*, San Jose, CA, pp. 648–655, January 2001.
73. Van der Schaar, M. and H. Radha, Unequal packet loss resilience for fine-granular-scalability video. *IEEE Trans. Multimedia*, December 2001.
74. Krishnamachari1, S., M. Schaar, S. Choi, and X. Xu, Video streaming over wireless LANs: A cross-layer approach. *Packet Video Workshop 2003*, 2003.
75. Xu, X., M. Schaar, S. Krishnamachari, S. Choi, and Y. Wang, Adaptive error control for fine-granular-scalability video coding over IEEE 802.11 wireless LANS. *IEEE ICME 03*, pp. 669–672, 2003.
76. Goel, M., S. Appadwedula, N. R. Shanbhag, K. Ramchandran, and D. L. Jones, A low-power multimedia communication system for indoor wireless applications. *IEEE Workshop on SiPS'99*, pp. 473–482, 1999.
77. Zhang, Q., W. Zhu, Z. Ji, and Y.-Q. Zhang, A power-optimized joint source and channel coding for scalable video streaming over wireless channels. *IEEE ISCAS'01*, 5(May), 137–140, 2001.
78. Soleimanipour, M., W. Zhuang, and G. H. Freeman, Modeling and resource allocation in wireless multimedia CDMA systems. *IEEE VTC'98*, 2, 1279–1283, 1998.
79. Kim, S. L., Z. Rosberg, and J. Zander, Combined power control and transmission rate selection in cellular networks. *IEEE VTC'99*, 3, 1653–1657, 1999.
80. Eisenberg, Y., C. E. Luna, T. N. Pappas, R. Berry, and A. K. Katsaggelos, Joint source coding and transmission power management for energy efficient wireless video communications. *IEEE Trans. CSVT*, 12(6), 411–424, 2002.
81. Zhang, Q., Z. Ji, W. Zhu, and Y.-Q. Zhang, Power-minimized bit allocation for video communication over wireless channels. *IEEE Trans. CSVT*, 2002.
82. Jiang, W. and H. Schulzrinne, Modeling of packet loss and delay and their effect on real-time multimedia service quality. *Proceedings of the 10th International Workshop on NOSSDAV*, Chapel Hill, North Carolina, June 2000.
83. Bolot, J., T. Turleeti, and I. Wakeman, Scalable feedback control for multicast video distribution in the Internet. *ACM SIGCOMM*, pp. 58–67, 1994.
84. Jacobson, V., S. Mccanne, and M. Vetterli, Receiver-driven layered multicast. *ACM SIGCOMM'96*, Stanford, CA, pp. 117–130, August 1996.
85. Lee, T.-W. A., S.-H. G. Chan, Q. Zhang, W. Zhu, and Y.-Q. Zhang, Allocation of layer bandwidths and FECs for video multicast over wired and wireless networks. *IEEE Trans. CSVT*, 12(12), 1059–1070, 2002.

86. Liu, J.-C., B. Li, and Y.-Q. Zhang, An end-to-end adaptation protocol for layered video multicast using optimal rate allocation. *IEEE Trans. Multimedia*, 7(6), 87–102, 2004.
87. Ratnasamy, S., M. Handley, R. Karp, and S. Shenker, Application-level multicast using content-addressable networks. *3rd International Workshop on Networked Group Communication (NGC '01)*, London, 2001.
88. Pendarakis, D., S. Shi, D. Verma, and M. Waldvogel, Almi: An application level multicast infrastructure. *3rd USENIX Symposium on Internet Technologies and Systems (USITS)*, pp. 49–60, 2001.
89. Castro, M., M. B. Jones, A.-M. Kermarrec, A. Rowstron, M. Theimer, H. Wang, and A. Wolman, An evaluation of scalable application-level multicast built using peer-to-peer overlays. *IEEE Infocom 2003*, San Francisco, CA, April 2003.
90. Padmanabhan, V. N., H. J. Wang, and P. A. Chou, Supporting heterogeneity and congestion control in peer-to-peer multicast streaming. *3rd Workshop for IPTPS*, San Diego, CA, 2003.
91. Andersen, D., H. Balakrishnan, M. Kaashoek, and R. Morris. Resilient overlay networks. *ACM SOSP*, Chateau Lake Louise, Banff, CA, 2001.
92. Subramanian, L., I. Stoica, H. Balakrishnan, and R. H. Katz, OverQoS: Offering QoS using overlays. *1st Workshop on Hop Topics in Networks (HotNets-I)*, Princeton, New Jersey, 2002.
93. Duan, Z., Z.-L. Zhang, and Y. T. Hou, Service overlay networks: SLA, QoS and bandwidth provisioning. *Proceedings of the ICNP'02*, Paris, France, 2002.
94. Mao, S., S. Lin, S. S. Panwar, Y. Wang, and E. Celebi, Video transport over ad hoc networks: Multistream coding with multipath transport. *IEEE J. Select. Area. Commun.*, December 2003.
95. Lin, C. and J. Liu, QoS routing in ad hoc wireless networks. *IEEE J. Select. Area. Commun.*, 17(8), 1999.
96. Kumar, S., V. S. Raghavan, and J. Deng, QoS-aware MAC protocols for ad-hoc wireless networks: A survey. *Elsevier Ad-Hoc Network J.*, 4, 326–358, 2006.
97. Goldsmith, A. and S. B. Wicker, Design challenges for energy-constrained ad hoc wireless networks. *IEEE Wireless Commun. Mag.*, 9(4), 8–27, 2002.
98. Qu, Q., Y. Pei, J. W. Modestino, X. Tian, and B. Wang, Cross-layer QoS control for video communications over wireless ad hoc networks. *EURASIP J. Wireless Commun. Network.*, 5, 743–756, 2005.
99. Setton, E., T. Yoo, X. Zhu, A. Goldsmith, and B. Girod, Cross-layer design of Ad Hoc networks for real-time video streaming. *IEEE Wireless Commun. Mag.*, 12(4), 59–65, 2005.
100. Zhang, Q. and Y.-Q. Zhang, Cross-layer design for QoS support in multi-hop wireless networks. *Proc. IEEE (invited)*, 96(1), 64–76, 2008.

Chapter 13

Handoff Management of Wireless Multimedia Services: A Middleware Approach

P. Bellavista, A. Corradi, and L. Foschini

Contents

13.1 Introduction....437
13.2 Mobile Multimedia Handoff Management: Challenges, Requirements, and Design Guidelines....439
13.2.1 Full Context Awareness for Handoff Management....439
13.2.1.1 Handoff Awareness....441
13.2.1.2 Location Awareness....442
13.2.1.3 Quality of Service Awareness....443
13.2.2 Handoff Management Requirements....445
13.2.3 Handoff Middleware for Service Continuity: Design Guidelines....446
13.3 An Overview of State-of-the-Art Handoff Management Solutions....450
13.3.1 Data-Link/Network/Transport Layers....451

13.3.2 Application Layer 452
13.3.2.1 Media-Independent Preauthentication....................452
13.3.2.2 Interactive Mobile Application Session Handoff..453
13.3.2.3 Seamless Media Streaming over Mobile Internet Protocol–Enabled Wireless LAN...............454
13.3.2.4 Mobiware..455
13.3.2.5 Vertical Handover in Future Wireless Networks 456
13.3.2.6 Session Initiation Protocol–Based Handoff Solutions..457
13.3.2.7 Mobile Agent–Based Ubiquitous Multimedia Middleware 458
13.3.3 Discussion..459
13.4 Handoff Management: Next Steps 462
13.4.1 Ongoing Industry Efforts 462
13.4.1.1 IEEE 802.21 Media-Independent Handover...........462
13.4.1.2 Unlicensed Mobile Access 463
13.4.1.3 Internet Protocol Multimedia Subsystem 463
13.4.2 Open Research Issues 464
13.4.2.1 Handoff Management Support for Mobile Networks 464
13.4.2.2 Multimedia Content Provisioning for Wireless Mesh Networks..465
13.4.2.3 Open Handoff Management for Internet Protocol Multimedia Subsystem 466
13.5 Conclusion 466
References 467

Advances in wireless networking and content delivery systems are enabling new challenging scenarios. A growing number of users require continuous access to their multimedia services, for example, audio and video streaming, while moving between different points of attachment to the Internet, possibly with different connectivity technologies (Wi-Fi, Bluetooth [BT], and third generation [3G] cellular). To grant service continuity, it is crucial to dynamically personalize service provisioning to the characteristics of wireless provisioning environment and to smooth possible discontinuities in wireless resource availability during handoff. Among the several solutions proposed in the field, middleware approaches working at the application layer are increasingly demonstrating their effectiveness and applicability. However, a set of common and standardized design guidelines for the development of such middleware infrastructures is still missing. This chapter proposes a general framework to address main handoff-related problems and requirements, identifies main handoff management design guidelines, presents an extensive survey of state-of-the-art application-layer middlewares for handoff management, and concludes by sketching current industry handoff standardization directions and open research issues.

13.1 Introduction

Today the wireless Internet (WI) is a more common deployment scenario: WI extends the traditional wired Internet and its services with wireless connectivity supported by access points (APs) working as bridges between fixed hosts and wireless devices [1,2]. Although device and network capabilities are increasing, the development of WI applications still remains a very challenging task, particularly for "mobile multimedia services," that is, applications that distribute time-continuous multimedia flows with stringent quality-of-service (QoS) requirements (data arrival time, jitter, data losses, etc.) toward WI mobile users. QoS constraints for the whole duration of service delivery make provisioning also a complex task in the traditional fixed Internet [3,4]. The WI advent and the high heterogeneity of personal portable devices complicate the scenario even further by introducing additional complexities such as bandwidth fluctuations, limited display size or resolution, small client-side memory, and especially temporary loss of connectivity, which may occur when client devices dynamically change their wireless APs due to user movements [5]. The final event is usually called "handoff."

This chapter addresses one of the most challenging issues for current mobile multimedia provisioning—service continuity maintenance during handoffs. We define "service continuity" as the capability to continue service provisioning and avoid flow interruptions during handoffs by minimizing or eliminating handoff delays and packet losses. Indeed, guaranteeing service continuity is a complex task as it cuts across different layers of the network protocol stack and requires handling various provisioning aspects (client mobility, QoS management, multimedia data transmission, etc.). Complexity mainly stems from the high heterogeneity of employed wireless technologies, spanning from IEEE 802.11 (Wi-Fi) and BT to 3G cellular, which exhibit very different handoff behavior due to different data-link-layer approaches and from the high number of competing mobility protocols at network and upper layers, for example, Mobile Internet Protocol (MIP) and Session Initiation Protocol (SIP). The management of all these technical issues complicates WI application development and slows down the implementation, deployment, and diffusion of mass-market mobile multimedia services. Therefore, a large number of research proposals and practical solutions have emerged recently to tackle service continuity—each one with specific goals, advantages, and limitations.

Although different handoff-related research efforts in the literature typically share similar functional requirements and adopt similar mechanisms, most of them, especially the efforts at lower protocol layers (data-link, network, and transport layers), are usually statically optimized for specific WI deployments and service requirements. Some recent research work, particularly at the application layer, has started to explore opportunities and problems connected to the creation of more general middleware solutions—"handoff middlewares"—so as to ease the design and implementation of mobile multimedia services by providing flexible and dynamic solutions for service continuity and to relieve WI applications from the burden of handoff management [6–12]. However, a set of common and standardized design guidelines for the development of such middleware supports is still missing.

This chapter aims to bridge this gap by pointing out continuity management challenges and the main design requirements for novel handoff middlewares. The proposed design solution is based on the core idea that only the full awareness of both WI environment and handoff processes enables effective and efficient service continuity. Generally, context awareness means full visibility of all the characteristics describing service execution environments to enable management operations that adapt service provisioning to current system conditions. By focusing on specific context awareness requirements for service continuity, the visibility of handoff-related context is essential to operate effective handoff management operations. It should include handoff types and characteristics, changes in local provisioning environment due to client mobility between WI access localities, and transient or definitive QoS degradation of provided multimedia flows associated with handoff occurrence. In particular, we claim that handoff middlewares should include three enabling awareness properties: (i) "handoff awareness" to enable effective management actions by full visibility of employed handoff procedures and parameters; (ii) "location awareness" to enable runtime decisions based on client mobility, network topology, and current resource position; and (iii) "QoS awareness" to actively participate in the management of multimedia flows, for example, by adopting buffering and retransmission techniques, according to service requirements and QoS degradations monitored during handoffs.

The full visibility of relevant context information enables the execution of appropriate handoff management countermeasures necessary to guarantee service continuity. In particular, after an in-depth survey of state-of-the-art research efforts in the field, we have recognized four main handoff management activities: "interoperable handoff initiation management" to monitor technology-dependent handoff management processes at the data-link layer, for example, to possibly predict a handoff situation due to user roaming; "handoff decision management" to select an appropriate access network and a handoff management strategy to comply with specific mobile multimedia application requirements; "data flow continuity management" to massage multimedia flow provisioning so as to mask handoff impairment effects, for example, to smooth handoff-related packet losses by employing buffer-and-retransmit techniques; and "mobility management" to redirect ongoing multimedia flows through new points of attachment, that is, to rebind communication endpoints as a client changes its access network and IP address.

The remainder of this chapter is organized as follows. Section 13.2 introduces handoff management challenges and motivates the need for handoff, location, and QoS awareness; reports main handoff requirements; and finally suggests primary handoff middleware design guidelines. Section 13.3 is devoted to an overview of the most relevant and recent activities about mobile multimedia handoff management: first, it briefly presents data-link, network, and transport-layer proposals; then, it focuses on application-layer solutions such as media-independent preauthentication (MPA), interactive mobile application session handoff (iMASH), Mobiware, and mobile agent–based ubiquitous multimedia middleware (MUM) [6,7,11,12]. Open research issues and ongoing industry standardization efforts are presented in Section 13.4. Finally, Section 13.5 gives conclusive remarks and ends the chapter.

13.2 Mobile Multimedia Handoff Management: Challenges, Requirements, and Design Guidelines

In the following, a taxonomy aimed to highlight main handoff challenges is proposed and the whole handoff management process is described; thereafter, handoff management requirements are introduced; and finally, the design guidelines for the development of novel handoff middlewares for service continuity are defined.

13.2.1 Full Context Awareness for Handoff Management

We define "handoff management" as the execution of all the actions needed to maintain service continuity when a mobile device changes its AP of attachment to the infrastructure network. Figure 13.1 reports the first example of handoff in a WI network to introduce main handoff steps: a mobile device, equipped with Wi-Fi and BT, is initially attached to Wi-Fi AP1; in response to its movement, a handoff procedure is triggered (step A); there are two possible target APs (Wi-Fi AP2 and BT AP3), and the handoff procedure identifies AP3 as the target AP (step B); handoff terminates with the

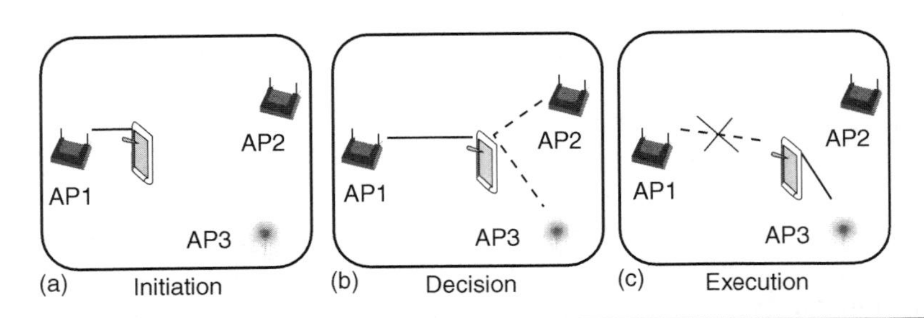

Figure 13.1 Handoff procedure steps.

reattachment of the mobile device to AP3 (step C). Technically, it can be assumed that each handoff procedure consists of three consecutive steps:

- *Initiation (see Figure 13.1a).* This phase monitors the network status to trigger the migration. Monitoring can either be performed at the client side (client-initiated handoff) or at the network side (network-initiated handoff).
- *Decision (see Figure 13.1b).* This phase takes care of the selection of a new AP among the available ones. The decision can either be managed by the client device (client-controlled handoff) or by the network (network-controlled handoff). Generally, this phase also includes the discovery of possible target APs.
- *Execution (see Figure 13.1c).* This phase is in charge of dropping the connection to the old AP and establishing the new connection with the selected target AP. Execution may also include all those QoS adaptation operations needed to guarantee service continuity, for example, the execution of multimedia downscaling to adapt deliver data flows to fit the characteristics of the target wireless access network.

Although different handoff solutions in the literature typically tackle all the preceding handoff steps by sharing similar functional requirements and adopting similar mechanisms (possibly at different layers), there is no classification framework to analyze and classify all different handoff requirements and approaches. Moreover, the lack of an appropriate framework makes it difficult to compare different systems and contributes to motivate why a set of common and standardized design guidelines for the development of novel infrastructures for WI handoff is still missing. Therefore, it is useful to sketch a taxonomy for clarifying main handoff challenges and handoff characteristics. In particular, we propose a novel taxonomy that adopts handoff, location, and QoS awareness as the three main classification criteria.

13.2.1.1 Handoff Awareness

Let us consider Maria who moves with her BT- and Wi-Fi-enabled personal digital assistant (PDA), getting out from her BT-covered office, and entering a large hall served with Wi-Fi while making a phone call. Handoff awareness is the ability to have full visibility of supported handoff types and handoff management strategies. Handoff awareness is crucial to decide effective handoff management actions depending on local WI environment and enable the automatic execution of management operations necessary to grant service continuity.

By focusing on wireless infrastructures and the "direction" of handoff, it is possible to distinguish horizontal and vertical handoffs [13]. "Horizontal handoff" occurs within one homogeneous wireless infrastructure, for example, when a Wi-Fi mobile device moves between two Wi-Fi cells and changes its AP. "Vertical handoff" occurs when a device with different network interfaces operates in an area served by various heterogeneous APs and decides to make a handoff from one wireless infrastructure to another, for example, to switch from a Wi-Fi network to a BT one. Vertical handoffs can be further distinguished into "upward" and "downward" handoffs: upward/downward vertical handoff occurs when a device performs a handoff to a destination network with wider/narrower coverage. For instance, handoffs from BT to Wi-Fi or from Wi-Fi to the cellular network are examples of vertical upward handoffs.

Depending on the initiation approach, handoff schemes can be classified as "reactive" and "proactive" [14]. Reactive approaches, usually based on broken link recognition, initiate handoffs as soon as the current AP becomes unavailable. Conversely, proactive initiation tries to predict and start handoff operations before disconnection (when the origin AP is still available). Proactive methods can trigger a handoff initiation by monitoring data-link quality indicators, for example, receiver signal strength indicator (RSSI).

Finally, one further distinction is between "hard" and "soft" handoffs. If mobile devices are allowed to have two or more simultaneous connections to different APs, then the handoff is said to be soft; otherwise it is defined as hard [15]. This distinction applies to the decision and execution steps: decision/execution can be performed either while the device is already disconnected from the old AP (hard handoff) or when the device is still connected to the old AP (soft handoff). This definition usually applies to the data-link layer; however, we can also use it for upper layers, and in particular for the middleware or application layer. When compared to a strict physical or data-link-layer solution, upper-layer solutions can adaptively take advantage of both handoff schemes. Specifically, an application can either use multiple network interfaces (soft handoff) or adopt only one

Table 13.1 Handoff Evaluation Criteria

Evaluation Criteria	*Related Issues and Characteristics*
Handoff awareness	Direction: horizontal/vertical (H/V) Initiation: proactive/reactive (P/R) Intercell procedure: soft/hard (S/Ha)
Location awareness	Geographical scope: micro/macro/global (Mi/Ma/G) Service rebind Context transfer
QoS awareness	Latency Data loss Content adaptation

network interface at a time (hard handoff) depending on its requirements and runtime conditions.

When Maria gets out of her office, several multiple wireless network are available to permit horizontal/vertical handoff; her PDA, equipped with several network interfaces, can perform both soft and hard application-layer handoff management; hence, given the strict voice call requirements (especially in terms of delay), the handoff management infrastructure can decide to adopt soft management along with proactive handoff initiation to grant service continuity. The first part of Table 13.1 reports the handoff parameters cited earlier and introduces the acronyms used in the rest of the chapter to differentiate handoff solutions.

13.2.1.2 Location Awareness

Let us consider Maria who, when accessing a news service, moves from university to home; during her roaming, she changes her WI provider switching from her department Wi-Fi network to the public cellular network. Handoff management infrastructures should maintain service provisioning by passing context information from the old to the new provider and by configuring local resources available in the new provider domain. Location awareness is the visibility of client location, WI access domain, and local resources, that is, which domain or network configuration changes will occur due to handoff, which resources are locally available to support service continuity, and how roaming clients can exploit them. Location awareness also permits to transfer context data in response to client mobility.

By adopting a classification typically used to accommodate MIP extensions, it is possible to distinguish three "geographical scopes" for handoffs: micro, macro, and global [8,15]. "Microhandoff" (intrasubnet handoff) includes only data-link-layer handoff and relates to clients who roam between two different APs without changing their IP addresses. "Macro-handoff" (intradomain handoff) refers to clients, which move between two wireless APs attached to different IP subnets and includes network-layer handoff with changes in client IP address. "Global handoff" (interdomain handoff) relates to mobile clients who roam between two APs attached to different Internet domains and require not only address change but also transfer of user authentication, authorization, and accounting (AAA) data needed when entering a new access domain.

"Service rebind" is the ability to dynamically reconnect clients to resources and service components at the WI access localities where the clients are moving into. In fact, client mobility, especially during macro and global handoffs, may undermine established network and transport connections. Hence, service rebind refers to the possibility to transparently discover, locate, and dynamically connect/reconnect roaming clients to local and remote resources. For instance, specific local mediators such as enhanced wireless APs or application proxies could be available within WI localities to assist roaming clients during handoffs with buffering, data retransmission, and content adaptation. Handoff supports should be able to autonomously bind clients to those local mediators [16].

"Context transfer" is the ability to move context information between fixed infrastructure and mobile nodes and from origin to target localities. Context transfer is crucial to complete reconfiguration operations (spanning from client node readdressing and AAA operations to mediator activation) before handoff execution. Although context transfer is considered an enabling factor toward WI location-aware distributed infrastructures by academia and industry [17,18], context definitions in the literature are often rather poor and incomplete. They usually include only AAA information, but we claim that there is also the crucial need to include (and update) all context data describing handoff status, that is, handoff, location, and QoS information.

When Maria moves from work at the university to home, a global handoff occurs, and the university WI domain transfers Maria's context data to the target public WI domain; in addition, the handoff infrastructure automatically readdresses her PDA and rebinds ongoing communications. The second part of Table 13.1 sums up location parameters.

13.2.1.3 Quality of Service Awareness

Let us consider Maria who roams from one source Wi-Fi cell to one BT target; management infrastructures should grant service continuity by eliminating QoS degradations during vertical handoff, that is, disconnections and packet losses, and by adapting contents to smaller BT network bandwidth. In general, QoS management requires performing network and system management operations relating to different layers [4]. However, in this chapter we focus only on QoS impairments introduced by handoff process execution. Hence, we define QoS awareness as the full visibility of temporary degradation and permanent variations of QoS during handoffs. QoS awareness permits to react to handoff by performing countermeasures that span from data buffering and retransmission techniques to content downscaling operations.

"Handoff latency" is the main handoff-related QoS impairment. We call handoff latency the time needed for the completion of decision and execution steps. In particular, handoff latency consists of the following main factors (see Ref. 19 for an exhaustive discussion of related issues): "handoff discovery time," that is, the time for probing available wireless infrastructures to discover possible target APs; "handoff decision time," that is, the time for selecting the best wireless target technology and the target AP where the device will attach, and for deciding appropriate handoff management countermeasures (including also application-layer management such as multimedia flow downscaling during vertical handoff); and "handoff execution time," that is, the time for reestablishing ongoing multimedia session with the target AP.

The other main impairment introduced by handoff is "packet loss" defined as the quantity of transmitted data packets or frames lost during handoff transitory. Packet loss depends on both handoff latency and adopted buffering techniques. For instance, several intermediate network buffers could be interposed between server and client to smooth packet losses. Hard handoff procedures provoke temporary disconnections and are, thus, more prone to packet losses. Handoff latency and packet loss largely depend on technological constraints (e.g., as discussed in Refs 20 and 21, Wi-Fi handoff discovery and decision phase are highly dependent on wireless card vendor and model) and employed handoff management protocols and procedures at all different layers. During the handoff latency period, both packet losses (especially for hard handoffs) and excessive packet delivery delays (for hard and soft handoffs) may be experienced. Vertical handoffs exacerbate the issue because they require either macro or global handoff management and include a wider set of management actions, thus, increasing latency.

Finally, "content adaptation" is the ability of dynamically tailoring contents in response to drastic bandwidth changes; typically due to vertical handoffs. For instance, a vertical downward handoff from Wi-Fi to BT provokes a sudden bandwidth drop: it requires a quick QoS reduction of all multimedia flows exceeding BT network capacity to avoid congestion by reducing multimedia flow bandwidth occupation through dynamic content downscaling.

When Maria is leaving the Wi-Fi cell, the handoff infrastructure exploits intermediate infrastructure-side buffers to smooth handoff latencies and packet losses due to (possibly long) BT handoff discovery times; in addition, it dynamically tailors delivered multimedia flows, especially bandwidth consuming video flows, to fit limited BT capabilities.

Based on the preceding evaluation of the different aspects of handoff, location, and QoS awareness, we propose the taxonomy in Table 13.1. The remainder of this chapter uses this taxonomy to draw the design guidelines for the development of context-aware handoff management solutions and to clearly position surveyed research work as well as our original contribution.

13.2.2 Handoff Management Requirements

We claim that the full visibility of handoff, location, and QoS information is essential to dynamically tune service continuity depending on current executing conditions (handoff type, latency, geographical scope, etc.). The three core requirements for novel context-aware handoff solutions are proactivity, adaptiveness, and reconfigurability.

First, handoff solutions should be "proactive"; in other words, should foresee handoff occurrence. Because handoff management operations can be long, service continuity management should prepare actions and countermeasures as early as possible. Proactivity strongly calls for full handoff awareness; in particular, handoff solutions should be able to predict horizontal and vertical handoff events to trigger handoff management actions in advance. For instance, some handoff management solutions have started to explore the possibility to exploit predictions to proactively move multimedia data chunks received at the old cell toward new destinations; the purpose is to minimize QoS degradations due to handoff by quickly restarting multimedia provisioning as client devices reconnect after handoff [6,16,22].

Handoff solutions should be "reconfigurable," that is, able to autonomously adapt their deployment and settings when client devices change their access networks or WI providers. Reconfiguration deals with account mobility management and lower-level configuration operations. In particular, geographical scope awareness, that is, visibility of micro/macro/global handoffs, and context transfer are crucial to locate WI target networks, to move context data there, and to complete configuration operations required by client roaming (readdressing, reauthentication, etc.). In addition, service rebind is crucial to reestablish ongoing sessions by reconnecting client components to locally available resources, possibly through local mediators (APs, MIP Foreign Agents [FA], application-level proxies, etc.).

Finally, handoff solutions should be "adaptive" by exploiting full QoS awareness. Adaptation focuses on data flow and content management. In particular, mobile multimedia provisioning requires a deep change in the traditional QoS management perspective: traditional (fixed) multimedia systems adopt a reservation-based approach and treat QoS adaptation as an infrequent management event that occurs only during overload or congestion situations [23]. We claim that novel mobile multimedia systems should instead consider QoS management by adaptation as the rule [24]. We use the term adaptation to indicate the ability to dynamically (and possibly proactively) adjust and massage multimedia flow provisioning to obtain service continuity. Handoff latency awareness is crucial, coupled with visibility of service requirements, to select and dimension handoff management procedures, for example, to choose hard- or soft handoff, with or without multimedia flow buffering and retransmission. Moreover, adaptive handoff solutions should perform multimedia flow content adaptation (format, frame rate, etc.) by considering the current capabilities of target wireless networks.

13.2.3 Handoff Middleware for Service Continuity: Design Guidelines

Notwithstanding context awareness, service continuity management remains a complex task that includes several non-trivial operations and requires a deep understanding of many technological details at different layers, which intrinsically depend on underlying executing platforms, used protocols, and employed wireless technologies. The effort required to learn and manage these technological aspects resulted in a slowdown of the deployment of continuous services and their exploitation in novel WI applications.

This chapter, therefore, claims that WI applications should delegate service continuity management to specialized support infrastructures—what we call handoff middlewares. Handoff middlewares should relieve WI applications from the handoff management burden by transparently taking over service continuity responsibility, and by effectively handling WI handoffs, irrespective of underlying wireless technologies. For this purpose, for the sake of flexibility, handoff middlewares should be implemented at the application level. The application level is recognized as the only suitable to provide flexible solutions to crucial mobility issues with enough generality by allowing application-specific caching and filtering, QoS management, and interoperable session control [2,25]. In other words, only application-layer handoff middlewares present the necessary flexibility and expressiveness to enable effective service coordination, tailored to specific application domains. Let us note that lower-layer handoff management solutions could not take service-dependent decisions; instead, the adoption of an application-level approach allows handoff middlewares to perform some handoff management operations selectively, for example, only for WI multimedia flows, and enables diverse handoff treatments depending on service-specific requirements.

However, to develop effective handoff middleware solutions and cut off handoff middleware development and administration costs, it is crucial to identify main handoff management aspects by clearly determining the basic building blocks that separately collaborate toward the common goal of service continuity. In the following, we introduce four main management aspects, which from the analysis of state-of-the-art literature and our experience, we identified as the core handoff middleware facilities. In particular, we identify (i) "interoperable handoff initiation" to enable open, portable, and proactive handoff initiation, and to complete all needed data-link handoff management actions; (ii) "handoff decision" to collect context information data as well as service requirements and user preferences, and to decide target wireless networks and application-layer handoff management strategies (e.g., hard or soft); (iii) "data flow continuity" to schedule and coordinate multimedia flow transmissions and data buffering operations to enable hard or soft handoff management execution; and (iv) "mobility management" to realize handoff protocols and session control functions to enable fast client reconfiguration and service component rebind, which are necessary any time a client enters a new WI network or access domain. Let us preliminarily note that the presentation in the following primarily focuses on the handoff management aspects relevant at the application level and closely related to service continuity; for the sake of briefness, we will not consider other important system-related handoff management aspects such as radio resource allocation and call admission control. For an extensive survey of these issues, see Ref. 26.

"Interoperable handoff initiation" management is the process of monitoring, on a per-mobile-client basis, the behavior of all available wireless interfaces to trigger handoff. In particular, handoff initiation for service continuity should be proactive and should be realized at the application layer to include both horizontal and vertical handoff predictions. Proactive initiation can be triggered by several events, such as monitored variations of data-link quality indicators, for example, RSSI values of all WI APs in client visibility [14,16] or changes in a set of defined quality parameters, for example, network bandwidth or delay [27]. However, it is widely recognized in the literature that initiation techniques based on data-link quality parameters are the most proactive ones [12,22,27,28]. By focusing on interoperability, each wireless technology, and sometimes even each wireless card model and driver, implements its own set of parameters, mechanisms, and handoff procedures (hard or soft, proactive or reactive, etc.) and adopts different QoS scales and values. For instance, Wi-Fi implements only hard handoff, universal mobile telecommunications system (UMTS) supports soft and hard handoffs, and BT lets application implement hard and soft schemes [29–32]. Most state-of-the-art research proposals enable interoperability and portability by introducing specific controllers and adaptors that mask the application layer from all data-link handoff management aspects. Moreover, they include specific application-layer methods, for example, fuzzy controllers, filtering, and prediction methods, to smooth gathered raw data-link quality data and implement specific comparison functions, for example, threshold- or hysteresis-based, to evaluate handoff predictions [14,16]. Recent standardization efforts, such as the IEEE 802.21 Media Independent Handover (MIH), are trying to determine a minimum set of functions to enable data-link handoff interoperability over heterogeneous network types [33,34].

"Handoff decision" management has the main goal of selecting the most appropriate middleware countermeasures for guaranteeing service continuity. Several research efforts in the literature have tackled handoff decision by focusing on parameters such as network quality, user preferences, and service requirements (such as mean and peak bandwidth values) to be always best connected (ABC) to the wired Internet [35,36]. However, this is not enough for service continuity. In fact, effective handoff management systems, especially for heterogeneous WI vertical handoffs, should be able to autonomously choose not only the best wireless network, but also the best handoff management strategy and dynamically tune all necessary distributed system resources, that is, flow data buffers, to meet service-level requirements such as maximum packet losses and delays. As discussed in detail in the following, intermediate buffers along the server-to-client path effectively decouple service endpoints and enable novel multimedia management possibilities such as stream adaptation and delay compensation. The dimensioning of those buffers is a crucial part of handoff decision management: only correct buffer sizes permit to grant service continuity and save precious system resources; however, only a few research proposals have started facing this open challenge [12,37,38].

"Data flow continuity" management represents the main handoff execution action to guarantee continuous seamless delivery of multimedia data at clients. In particular, due to their isochronous nature, multimedia services, such as video surveillance and video on demand (VoD), are more affected by handoff impairments than other service classes. In fact, multimedia services usually adopt connectionless protocols, such as Real-Time Protocol over UDP (RTP-over-UDP); hence, handoff middlewares should be able to manage data retransmissions directly at the application level to react to high jitter and packet losses. Several solutions in the literature have explored the use of buffering infrastructures [10,22,37,39–41], and among them the introduction of multiple levels of buffering was found to be interesting [39,40]. However, each additional buffering level increases end-to-end packet delay and requires additional memory [42]. To obtain maximum advantage of

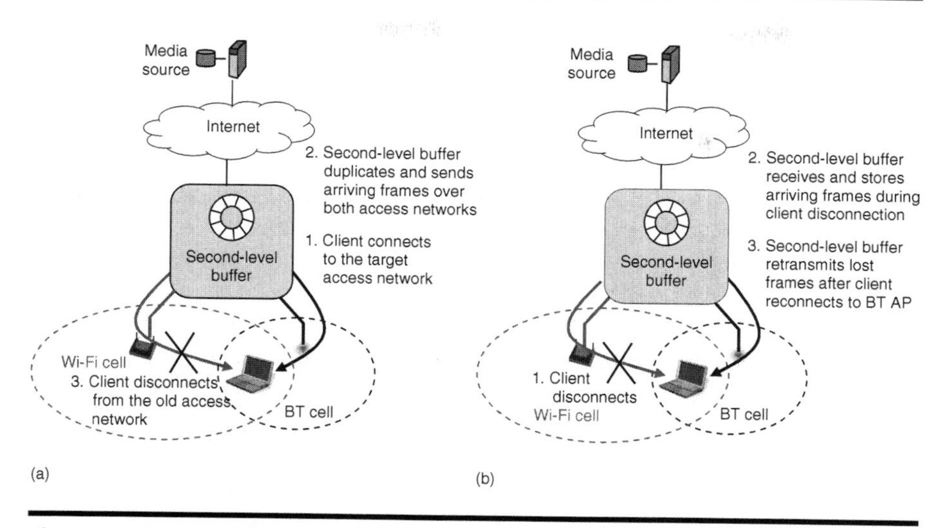

Figure 13.2 Second-level buffer for soft/hard handoff management. (a) Soft handoff management, (b) hard handoff management.

buffer interposition and limit their introduced delay, recent handoff proposals adopt two-level buffering schemes. Indeed, the introduction of a second-level buffer starts to be widely recognized as an effective mechanism for advanced handoff management operations within client wireless access localities [10,22,37,41]. In particular, two main handoff strategies for data flow continuity management at the application layer are proposed in the literature, as shown in Figure 13.2, which reports a vertical handoff from a Wi-Fi cell to a BT one (for soft and hard handoff). Soft handoff management locally supports the duplication (and simultaneous transmission) of multimedia flows over multiple wireless interfaces in the last wired–wireless hop (see Figure 13.2a). Hard handoff management, instead, uses second-level buffers to receive and store all the incoming packets that would be lost during handoff disconnection (step 2 in Figure 13.2b), and when the client connects at the target AP, enables the local retransmission of all those packets (step 3).

"Mobility management" includes the dynamic reconfiguration and update of all session information needed to enable automatic renegotiation of ongoing sessions at handoff occurrences. In particular, in the actual WI scenario several different mobility management solutions coexist, spanning from network-layer ones—for example, Dynamic Host Control Protocol (DHCP), DHCP-Relay, MIP, and MIP evolutions—to application-layer ones, SIP being the most notable example. (See Refs 15, 43, and 44 for an exhaustive survey of available mobility management solutions in the field.) As interoperable handoff initiation management hides data-link handoff peculiarities, mobility management should mask mobility implementation details to enable interoperable and open mobility management. Mobility management represents another very active handoff-related research area: current research efforts mainly focus on the exploitation of SIP, which is gaining more interest as the main session control protocol for next-generation all-IP IP multimedia subsystem (IMS) networks [45].

Finally, let us consider that the development of novel handoff middlewares asks for evolving fixed-Internet distributed infrastructures based on the traditional client/server paradigm. First, classical end-to-end monitoring mechanisms for flow or congestion control are not effective over wireless links; in particular, they cannot identify possible handoff situations that occur in the last wired–wireless hop. Second, the round-trip time between client and server limits both the applicability and the frequency of data flow management actions that require prompt intervention in the case of local wireless link changes, such as for data retransmissions during hard handoffs. Finally, because most handoff solutions adopt two-level buffering architectures, there is the need to enable buffering management functions in the last wired–wireless hop. Therefore, the architecture of handoff management middlewares should be "proxy-based": middleware proxies should be deployed at wireless network edges and should actively participate in handoff management by interacting with mobile clients. Let us note that the traditional Internet is already crowded by many kinds of proxies (e.g., for caching, authentication, and redirecting duties). The introduction of proxies to split direct client-to-server connections in wireless client proximity is an effective solution to reduce signaling traffic on service paths and to personalize service delivery [9,16,22,27,46].

13.3 An Overview of State-of-the-Art Handoff Management Solutions

This section aims to classify the most relevant recent work on handoff management. We first present some research efforts at lower layers of the Open Systems Interconnection (OSI) protocol stack (data-link, network, and transport); then, we introduce an extensive overview of application-layer solutions, among which MUM, our handoff middleware for context-aware service continuity, is rapidly described. We partition handoff management solutions according to OSI layering (for the sake of presentation, we have grouped session- and application-layer approaches under the term application layer). Because solutions may cut across different OSI layers, our classification positions each research activity at the layer that plays the key role in handoff management. Moreover, we will put more emphasis on application-layer solutions because, as detailed in the rest of section, we believe they are more suitable to grant service continuity in the highly heterogeneous and complex open WI environment; in addition, they represent an emerging novel trend in the research field and are not yet adequately surveyed in the literature. To describe the research proposals, we will employ the context awareness taxonomy criteria and the design guidelines introduced in Section 13.2.

13.3.1 Data-Link/Network/Transport Layers

Let us note preliminarily that research efforts at the data-link-, network-, and transport layer aim to improve handoff latency and packet loss through the static optimization of low-level parameters (timeouts, number of handoff detection probes, buffer size, etc.). These proposals are forced to statically decide and dimension handoff countermeasures because they lack application-layer visibility of handoff-related context. At the data-link layer, handoff management has been widely investigated in cellular systems [47]. Recent studies have focused on Wi-Fi and BT wireless technologies. Refs 20 and 21 propose two improvements to decrease the time needed for Wi-Fi handoff detection and decision, whereas Ref. 48 reduces packet losses by extending Wi-Fi Inter-Access Point Protocol (IAPP) to support buffering and retransmission of fixed-length data chunks stored at Wi-Fi APs. Ref. 31 proposes the first BT-based infrastructure called BLUEtooth Public ACcess (BLUEPAC); Ref. 32 improves BLUEPAC handoff latency, whereas Ref. 49 employs statically dimensioned soft handoff techniques to eliminate packet losses.

Network-layer proposals generally aim at improving MIP performance [50]. Refs 13 and 51 represent two pioneer research activities—developed in the context of the Bay Area Research Wireless Access Network (BARWAN) project [52]—to reduce handoff latency by introducing local monitoring of beacon messages emitted by target APs to proactively detect horizontal and vertical handoffs and to alleviate packet losses through multicasting. However, similar to the network-layer solutions below, it only supports hard handoff management, that is, it cannot exploit multiple network interfaces at the same time. Ref. 53 proposes a reactive macro handoff procedure for Wi-Fi to reduce MIP handoff detection time, but without addressing packet loss. Refs 15 and 43 survey a number of micro, macro, and global mobility management enhancements to MIP. Those research efforts resulted in two IETF protocol proposals: Hierarchical MIP (HMIP) and Fast Handover for MIP (FMIP) [28,54]. HMIP imposes a hierarchical network management infrastructure to reduce network signaling due to client relocation. FMIP exploits some form of handoff awareness, that is, data-link layer triggers (when available), to proactively initiate network routes and client readdressing, and to forward the packets potentially lost during disconnection/reconnection from/to origin/target APs. However, FMIP is aware of neither specific QoS characteristics nor service requirements; therefore, it can only dimension retransmission buffers of target APs in a completely static way. Seamless MIP (S-MIP) applies FMIP to Wi-Fi macro handoffs: S-MIP uses RSSI monitoring for handoff detection and proposes simultaneous multicast (simulcast) to alleviate packet loss during handoff transitory, but suffers the same problems as FMIP [22].

At the transport layer, early proposals are mainly focused on the maintenance of Transmission Control Protocol (TCP) end-to-end semantics [55,56], whereas only the recent approaches tackle the wider goal of service continuity. Ref. 57 adopts different handoff-aware methods for Wi-Fi and for general packet radio service (GPRS) to proactively detect vertical handoffs; the main problem of the proposed handoff-aware decision model is that it is bound to one specific simulation scenario and unable to dynamically adapt to different executing conditions. Video Transfer Protocol (VTP), presented in Ref. 27, supports vertical handoff and content adaptation through proxy interposition and UDP tunneling. Ref. 58 completes Ref. 27 with a smart handoff decision model based on connectivity costs, link capacity, and power consumption, but the adopted end-to-end transparent approach has no visibility of effective QoS handoff degradations, does not tackle packet loss, and focuses on macro handoffs only.

13.3.2 Application Layer

Several recent research efforts have achieved relevant results by addressing service continuity at the application layer. In the following, we survey several state-of-the-art systems with the overall goal to draw all primary evolutionary trends in mobile multimedia handoff management. For each solution, its main design and architectural guidelines are presented by pointing out primary pros and cons of the adopted choices.

13.3.2.1 Media-Independent Preauthentication

MPA is a highly interoperable framework for secure and proactive horizontal and vertical hard handoff [6]. In particular, MPA focuses on handoff mobility management and exploits proactivity, that is, client-side handoff initiation, to reduce handoff latency due to macro and global handoff reconfiguration. In particular, MPA enables fast client node readdressing and reauthentication when a client moves to a new target network or access domain. In addition, it proposes a portable reauthentication method, which can be deployed on any all-IP wireless network [59]. To support these functions, MPA proposes a proxy-based architecture consisting of a set of interworking agents deployed at target wireless access networks to decentralize mobility management load and to promptly assist roaming clients.

Delving into finer details, MPA enables dynamic discovery of destination networks and deploys at each wireless network three main agents, namely, the authentication agent (AA), configuration agent (CA), and access router (AR), to enable proactive preauthentication, preconfiguration, and rebinding of all connections established by the client with multimedia servers (proactive handover preswitching). From an implementation point of view, MPA adopts a cross-layer design approach and operates at different OSI layers. For instance, MPA uses data-link layer monitoring to trigger proactive handoff. It exploits network-layer techniques for fast readdressing, that is, DHCP-Relaying, and for proactive rebinding, that is, after handoff prediction MPA tunnels all incoming/outgoing data from/to multimedia servers through the target network. In addition, it employs application-layer signaling to portably (and proactively) move AAA client information between old and new AAs.

MPA confirms the effectiveness and the increasing diffusion of both proactive approaches and proxy-based architectures for handoff management. However, the proposed solution also presents some open issues. First, MPA does not specify how to obtain interoperable handoff initiation. Second, similar to network-layer approaches, MPA only focuses on the goal of reducing handoff latency rather than on the wider challenge of data flow continuity. Hence, even in the case where MPA completes all reconfiguration operations before actual data-link handoff execution, it only implements hard handoff and does not provide buffering techniques to guarantee lossless handoffs during possible switching of wireless technologies. Finally, MPA neither tackles advanced handoff decision management nor includes content adaptation facilities. Some of these issues are part of ongoing research efforts by the same group; for instance, a recent draft of MPA has proposed buffering at endpoints to avoid packet loss during data-link handoff [60].

13.3.2.2 Interactive Mobile Application Session Handoff

IMASH is one of the first attempts to provide application-layer support for horizontal and vertical handoff [7]. In particular, this system addresses handoff decision and content adaptation management during handoff by employing a proxy-based infrastructure. At handoff occurrence, iMASH proxies adapt session content to target execution environment by interacting with lightweight client-stubs deployed at mobile devices. Let us stress that client-stub deployment at client nodes is a promising design choice that paves the way to several management possibilities, such as client-initiated handoff management and coordinated data-flow continuity management (and has been widely adopted in our MUM original solution).

IMASH was primarily designed for nonmultimedia applications with looser interactivity requirements and for supporting not only service continuity during handoff (called middleware-only application session handoff

[MASH]), but also service session mobility between different client devices (called client ASH [CASH]). Therefore, for the sake of generality, iMASH adopts a rather static data management approach that focuses more on preserving session data consistency before and after handoff (savepointing) rather than respecting real-time delivery constraints. Before the occurrence of handoff, iMASH freezes and saves the state of ongoing service sessions at proxies; after handoff completion, iMASH resumes saved session state, by possibly adapting it if necessary, for example, if the user has changed the device or after a vertical handoff from Wi-Fi to GPRS. In the case of content adaptation, iMASH proposes a general and simple architecture: it describes content data types and client device/user/network profiles through XML-based descriptors; it employs a simple heuristic to determine which adaptations to apply; it autonomously adapts delivered content at proxies; finally, it exploits middleware client stubs to receive and present adapted content at the client side. Some preliminary research efforts by the same group have started to extend the basic iMASH architecture to support multimedia streaming: they followed the same general architecture for content adaptation and proposed a reactive protocol to eliminate handoff packet losses at the client side for MASH and CASH [61].

IMASH proposes a comprehensive handoff middleware architecture and recognizes the opportunity to support several different types of handoff, including session mobility between different devices. In particular, iMASH-related research verified the opportunity of using application-layer middleware proxies and client stubs to facilitate WI application development by taking over handoff and content adaptation management. However, iMASH proxies cannot grant service continuity because they neither have full visibility of data-link handoff and QoS parameters (handoff direction, handoff latency, etc.) nor realize advanced data flow continuity management such as advanced buffering and prefetching solutions.

13.3.2.3 Seamless Media Streaming over Mobile Internet Protocol–Enabled Wireless LAN

Ref. 37 presents a handoff solution to overcome packet losses of multimedia flows due to Wi-Fi hard handoff for MIP. The proposed system includes client-side handoff initiation functions to proactively trigger handoff management. In addition, it supports data flow continuity through a two-level buffering technique that exploits MIP FAs as local proxies, which can coordinate with a data flow continuity manager (prebuffering manager) deployed at the client side.

Ref. 37 preserves service continuity by proactively extending client buffers before handoff disconnection to prefetch a number of multimedia data chunks to sustain the multimedia playback during handoff latency. When the client-side handoff initiation monitor foresees a possible handoff, it emits a prediction toward the multimedia server, which replies by increasing the payback rate to push toward its client a number of frames sufficient to guarantee service continuity for the whole Wi-Fi data-link handoff process. In addition, to avoid retransmissions of frames lost during disconnection, MIP FAs are enriched with application-layer buffering capabilities: as a client moves to a new subnet, prebuffering manager rebinds the multimedia flow from the multimedia server toward the new FA, and the system forwards the packets to the new FA through the old FA.

Ref. 37 represents a notable example of systems that exploit handoff initiation indications to proactively activate data flow continuity countermeasures for service continuity in hard handoff situations. In particular, client-side proactive prebuffering, coupled with buffering-and-retransmission at proxies, is the most suitable data flow management technique because it can guarantee lossless handoffs by eliminating server-to-proxy retransmissions (thus decreasing network overhead). The main open problem is that the proposed solution lacks the visibility of context necessary to correctly and dynamically dimension/redimension the exploited buffers: this produces either waste of resources or discontinuities of multimedia flow delivery.

13.3.2.4 Mobiware

Mobiware is a programmable mobile middleware platform that supports soft and hard handoff [11]. Mobiware further extends the concept of proxy-based infrastructure to include all the entities along the service distribution path from server to client. In fact, it assumes to have complete control of all the distributed entities that participate to mobile multimedia delivery (including mobile clients, wireless APs, and switches/routers), and utilizes Common Object Request Broker Architecture (CORBA) distributed objects that provide a set of programmable interfaces to control handoff management aspects at different OSI layers.

To be more specific, Mobiware permits to specify service-level requirements through the definition of utility functions that identify both bandwidth ranges permitting service usability and adaptation policies. Given these specifications, Mobiware autonomously decides necessary handoff management actions and uses its distributed CORBA objects to enforce handoff management actions. In particular, it exploits application-layer QoS adaptation proxies to dynamically downscale and massage ongoing multimedia flows. It also enforces QoS adaptation at the lower layers through the so-called active transport objects. Finally, it controls the whole data-link handoff process through CORBA objects, thus reducing handoff latency.

Mobiware confirms the need for highly distributed handoff middleware solutions with application-layer visibility of both handoff and QoS.

In particular, it demonstrates how it is possible to realize effective and simple handoff management by requiring only the service level to declare its needs and delegating the whole burden of content adaptation and mobility management to the middleware level. Nonetheless, Mobiware follows a proprietary approach that does not fit well with the characteristics of the open WI environment. In fact, Mobiware, assumes that the underlying wireless infrastructure is Mobiware-enabled, that is, each involved node should run an implementation of all Mobiware-specific CORBA APIs, and wireless interfaces should be programmable at the physical and data-link layers. In other words, in current wireless systems Mobiware cannot support interoperable handoff initiation, which should be able to adapt to the characteristics of destination wireless technology. Finally, similar to MPA, Mobiware concentrates more on the reduction of handoff latency than on the wider goal of continuity provisioning.

13.3.2.5 Vertical Handover in Future Wireless Networks

Ref. 9 is a proactive QoS-aware framework for vertical handoff, which proposes an advanced context-aware handoff decision model. This system covers main handoff management aspects: it adopts a proxy-based infrastructure consisting of three kinds of entities deployed at each wireless access network. The three middleware entities are context repository, which is responsible for context management, monitoring, and storage; adaptation manager for handoff initiation and decision; and proxy for data flow continuity.

In the case of handoff initiation and decision, Ref. 9 defines a rich context model that includes service requirements and several additional parameters such as user and network profiles. Differing from the proposals considered previously, Ref. 9 employs network-side handoff initiation triggered by the adaptation manager: Adaptation manager is notified by context repository of relevant handoff/location/QoS context changes and predicts possible handoff events. Handoff decision decides the best target network during vertical handoff; the adopted decision method first evaluates and predicts user location changes, then adopts a multilevel hierarchical structure with several handoff objective criteria spanning from user preferences (per-byte transmission cost) to service and network requirements (maximizing available bandwidth and minimizing network delay and jitter). In addition, proxies support content adaptation and minimize packet loss through a soft handoff strategy exploiting multiple wireless interfaces during vertical handoff.

This solution represents a good example of the current trends in context-aware handoff decision management, which should consider articulated and flexible decision models based not only on the WI provisioning environment but also on service/user requirements to tackle the complex WI scenario. However, Ref. 9 focuses more on context management itself rather than context-aware service continuity. In addition, let us observe that this is the only proposal exploiting network-side handoff initiation. In fact, network-side handoff initiation is usually more prone to errors because it assumes that wireless quality parameters monitored at AP coincide with the values monitored at clients and that discrepancy could relevantly affect handoff initiation effectiveness.

13.3.2.6 Session Initiation Protocol–Based Handoff Solutions

Before presenting our original MUM handoff middleware, we conclude this application-layer solution survey with a brief overview of main SIP-based proposals for handoff management. These solutions integrate with and exploit the SIP framework. By following the SIP framework guidelines, they adopt distributed proxy-based infrastructures and are mainly focused on session signaling for mobility management.

SIP-based application-layer mobility management was first proposed in Ref. 25, which suggests to use the SIP re-INVITE message to rebind ongoing sessions. Thereafter, several SIP-based research proposals have worked on similar solutions, which mainly focus on fast rebinding [15,62]. Nonetheless, these solutions usually do not guarantee zero packet losses and incur in rather long handoff delays. Therefore, other SIP-related solutions specifically work on reducing or eliminating packet losses. Ref. 63 considers and evaluates the delays introduced by SIP-based vertical handoffs between 3G and Wi-Fi. Ref. 64 describes a reactive SIP-based macro handoff solution for Wi-Fi that exploits aggressive router selection to reduce network handoff latency. It adopts a reactive approach, insufficient to strongly reduce handoff delays and does not eliminate packet losses. Ref. 65 proposes a proxy-based infrastructure that obtains optimized SIP fast-handoff by employing standard Linux tools, for example, IPCHAINS and IPTABLES, and local multicast techniques. By focusing on SIP-related solutions with data redundancy, Ref. 10 proposes to adopt soft vertical handoff to eliminate packet loss through proxy-interposition. Experimental results show that Ref. 10 spends long time for data path activation, up to 20 s, which can easily disrupt service continuity.

SIP-based solutions probably represent the strongest direction of evolution in mobility management solutions. In particular, SIP approaches have pioneered the trend of open and interoperable mobility management infrastructures on an all-IP network infrastructure. This is the main reason why SIP has been chosen as the session control protocol for next-generation IMS infrastructures [45].

13.3.2.7 Mobile Agent–Based Ubiquitous Multimedia Middleware

MUM is a context-aware handoff middleware for service continuity. The MUM application-layer approach enables modular, flexible, and highly configurable handoff management, and permits to tackle all handoff-related technical challenges by following the full context-aware approach and the design guidelines proposed in Section 13.2. MUM provides application-layer visibility of handoff implementation details, potential handoff-related quality degradations, and client origin and target access localities (handoff/location/QoS awareness); thanks to this awareness, it enables the transparent execution of proper handoff management countermeasures at the middleware level. Middleware proxies—MUM session proxies—deployed in the last hop of server-to-client path collaborate with lightweight middleware components deployed at mobile clients, that is, MUM client stubs, to achieve service continuity. In particular, MUM session proxies guarantee data flow continuity by adapting requested multimedia flows depending on both static elements such as mobile device characteristics and especially dynamically monitored WI conditions (horizontal and vertical handoff latencies).

By concentrating on MUM design, MUM provides the middleware components necessary to complete all the four main handoff management aspects discussed earlier: interoperable handoff initiation, handoff decision, data flow continuity, and mobility management. Following are the finer details:

- MUM provides an "interoperable WI handoff predictor" that dynamically monitors the signal quality for all the wireless network interfaces available at client, and predicts possible horizontal and vertical handoff occurrences using our RSSI-Grey Model prediction technique [16]. In particular, MUM exploits full awareness of handoff behavior of wireless cards and drivers to properly interpret monitored RSSI data, and employs an original prediction technique to obtain effective horizontal and vertical handoff predictions for heterogeneous wireless technologies and vendors.
- MUM implements advanced "handoff decision" methods that autonomously choose the data continuity strategy (soft or hard) able to grant service continuity by also reducing memory resource consumption due to second-level buffering. In addition, it simplifies the specification of service-level requirements through the introduction of easy-to-use service-level specifications (delay and data-loss requirements) and supports service differentiation [12].
- MUM supports the execution of soft and hard handoff management, thus enabling data flow continuity for a wide range of streaming

services, from video surveillance to VoD. In particular, it proposes an original adaptive second-level buffer dimensioning technique for hard handoff, which improves system efficiency by enlarging second-level buffers (to store incoming packets from the server) only during handoffs. In addition, it supports service continuity for both powerful clients, for example, full-fledged laptops, and memory-constrained clients with limited client buffers, for example, PDAs and smart phones [12,66].

- MUM proposes original SIP-based extensions to enable open and advanced session control functions for mobility management. MUM integration with standard mechanisms and protocols, for example, DHCP-Relay and SIP, is expected to leverage MUM adoption and simplify its integration with currently available multimedia platforms [16,67].

13.3.3 Discussion

In Table 13.2, we compare all the presented proposals to provide a summarized overview of current state-of-the-art and to better delineate emerging solution trends. Table 13.2 sums up all handoff management solutions, analyzed by means of the evaluation criteria (and acronyms) of Table 13.1. Solutions at the data-link layer try to optimize and boost existing protocols primarily to reduce horizontal handoff latency. Network-layer proposals exploit some data-link trigger, that is, data-link handoff awareness, to reduce handoff latency, and use buffering and multicast techniques to alleviate data losses. However, approaches at data-link and network layers lack the QoS awareness and have difficulties in complying with application-specific requirements: their naming spaces uniquely map one host to one IP address; this excludes soft handoff management and flexible per-service rebind. Finally, their deployment usually requires protocol stack changes at all hosts. Transport-layer solutions enrich data-link handoff awareness with per-connection end-to-end visibility—for example, bandwidth monitoring or probing—and possibly enable content adaptation, but this tightly couples transport and application layers [27]. Application-layer solutions are more flexible and are able to comply with user and service requirements of next-generation WI infrastructures.

The first important observation is about the adopted architectural models. All surveyed application-level handoff management systems adopt a proxy-based architecture for handoff decision [7,9,11,12], data flow continuity and content adaptation [7,9–12,37], or mobility [6,10–12,25,64]. In addition, some solutions also recognize the importance to deploy lightweight middleware stubs at client devices to support handoff initiation [6,10–12,25,64] and to coordinate with their proxy counterparts for data flow continuity [9,10,12,37].

Table 13.2 Comparison of Handoff Solutions at Different Layers

		Handoff Awareness			*Location Awareness*			*QoS Awareness*		
	Solution	*Horizontal/ Vertical (H/V)*	*Soft/Hard (S/Ha)*	*Proactive/ Reactive (P/R)*	*Micro/Macro/ Global (Mi/Ma/G)*	*Service Rebind*	*Context Transfer*	*Content Adaptation*	*Data Loss Static/ Dynamic*	*Latency*
Data-link	Fast Wi-Fi [21]	H: Wi-Fi	Ha	R	Mi	—	—	—	—	✓
	Wi-Fi anal. [20]	H: Wi-Fi	Ha	P	Mi	—	✓ Neighbor Graph	—	—	✓
	IAPP ext. [48]	H: Wi-Fi	Ha	R	Mi/Ma	—	✓ IAPP	—	✓ Static	—
	Fast BT [32]	H: BT	Ha	P	Mi	—	✓ Synchronous address	—	—	✓
	Effic. BT [49]	H: BT	Ha	P	Mi/Ma	—	✓ AAA	—	✓	✓
Network	Daedalus H. [51]	H: WaveL.	Ha	P	Mi	✓ MIP	✓	—	✓ Static	✓
	Daedalus V. [13]	V	Ha	P	Ma	✓ MIP	—	—	✓ Static	✓
	MIP ext. [53]	H: Wi-Fi	Ha	R	Ma	✓ MIP	—	✓	—	✓
	FMIP [28]	H/V	Ha	P	Mi/Ma/G	✓ MIP	✓ Address + data	—	—	✓
	S-MIP [22]	H: Wi-Fi	Ha	P	Mi	✓ MIP	✓ Address + data	—	✓ Static	✓
Transport	MSOCKS [55]	H/V	S	R	Mi	✓ Propr.	—	—	—	—
	Migrate [56]	H/V	Ha	R	Mi/Ma/G	✓ Propr.	—	—	—	—
	Seam. TCP [57]	V	S	P	Mi/Ma/G	✓ Propr.	—	—	✓ Static	✓
	VTP [27], [58]	V	Ha	P	Mi/Ma	✓ Propr.	—	✓	—	—
Application	MPA [6]	H/V	Ha	P	Ma/G	—	✓ AAA	✓	—	—
	iMASH [7]	H/V	Ha	—	Mi/Ma/G	✓ Propr.	✓ Propr.	—	—	✓
	Seamless [37]	H/V	Ha	P	Ma	✓ Propr.	✓ Propr.	—	✓ Static	✓
	Mobiware [11]	H/V	Ha	P	Mi/Ma/G	✓ Propr.	✓ Propr.	✓	—	✓
	Vertical [9]	V	Ha	P	Mi/Ma/G	✓ Propr.	—	—	✓ Static	✓
	SIP [25]	H/V	Ha	P	Mi/Ma/G	✓ SIP	—	✓	—	—
	SIP lat. [64]	H: Wi-Fi	Ha	R	Ma	✓ SIP	—	✓	—	—
	SIP p. loss [10]	V	S	R	Ma	✓ SIP	—	—	✓ Static	—
	MUM [12]	H/V	S/Ha	P	Mi/Ma/G	✓ SIP	✓ SIP	✓	✓ Dynamic	✓

Notes: Anal., analysis; ext., extension; seam., seamless; lat., latency; p., packet; WaveL., WaveLAN; propr., proprietary.

By focusing on handoff management aspects, interoperable handoff initiation still represents an open research issue. Almost all solutions adopt a proactive handoff management approach. However, most of them address handoff prediction on one specific usage scenario and for only one wireless technology—only horizontal handoff predictions [11,37,64], or claim to support proactive handoff initiation but without specifying handoff initiation algorithms [6,7,9,10]. The application-layer handoff initiation approach adopted by MUM, instead, is one of the first approaches capable of evaluating both horizontal and vertical handoff predictions by also being interoperable in terms of wireless interfaces (Wi-Fi, BT, etc.) and client devices (Windows, Linux, etc.) [38].

In the case of handoff decision management, current approaches are mainly based on either vertical handoff or proper adaptation decisions to counteract QoS degradations. In particular, Refs 7, 9, and 11 realize three different and complementary approaches: Ref. 11 focuses on lower-level aspects such as dynamic enforcement and adaptation of network-layer policies depending on application-layer needs and possible QoS degradations; Ref. 9 concentrates on adapting delivered multimedia to actual WI context conditions; and finally, Ref. 7 includes the possibility to adapt the service level itself (not only delivered contents) to fit handoff conditions. However, we claim that more attention should be devoted to improve handoff decision techniques for the dynamic choice of the best handoff management strategy (soft or hard) depending on actual context conditions and service requirements. Moreover, handoff decisions should also include the dynamic resizing of client and second-level buffers, aimed to guarantee service continuity. Ref. 37 aims to dynamically adapt client buffer size depending on actual handoff latency. In the MUM project, we have already developed advanced handoff decision techniques for service continuity [12]; in addition, we are actually working to improve handoff management decisions depending on service-level requirements.

By focusing on data flow continuity, the most important consideration is that most approaches in the literature cannot completely exploit context visibility. In fact, they usually mimic the behavior of low-level solutions without taking advantage of possible full awareness of handoff, location, and QoS. Most of them statically support only one type of handoff management (soft or hard). Moreover, hard handoff management solutions do not include or statically dimension second-level buffers, thus facing either packet losses (when the buffers are underdimensioned) or memory resource waste (when buffers are overdimensioned) [6,7,11,37,64]. MUM adopts a different and original adaptive approach for data flow continuity, which differentiates continuity management depending on the applicable context and exploits all WI possibilities through different handoff solutions (soft and hard, horizontal and vertical) and dynamic buffer dimensioning [12,38].

Finally, mobility management is probably the most mature handoff management area. Within the last decade, the networking community has explored all the several different possibilities to improve mobility management, especially through proactive activation of reconfiguration protocols, for example, DHCP-Relay usage [6,11,12]. Ongoing research efforts are toward standardization and the main goal today is interoperability. Recent solutions are addressing mobility management at the application layer; thus, SIP-based mobility management [6,10,12,25,64] is becoming the primary solution and, as shown in Section 13.4, SIP is expected to spread more in the near future. Similarly, secure mobility management is required at the application layer: the security solution adopted by MPA has been recently proposed as an Internet Engineering Task Force (IETF) Internet draft. Let us conclude by stressing that application-layer handoff management solutions (such as SIP for session control management) are leading toward open context-aware handoff solutions for the next-generation WI and facilitating handoff middleware adoption in open distributed systems.

13.4 Handoff Management: Next Steps

Handoff management in WI mobile multimedia represents a lively research area as confirmed by several recent efforts by both industry and academia to design and deploy first handoff middleware infrastructures. Nonetheless, there are still several open issues and technical challenges to solve to grant service continuity in the open and highly heterogeneous WI scenario. In the following sections, all main ongoing industry and standardization efforts are reported briefly and the core open issues of the area are presented.

13.4.1 Ongoing Industry Efforts

The same approach adopted in Section 13.3 is followed here: we start from proposals based on lower protocol stack layers to conclude with the IMS that completely works at the application layer.

13.4.1.1 IEEE 802.21 Media-Independent Handover

MIH is an ongoing standardization effort by the IEEE 802.21 working group to facilitate smooth interaction and media-independent handoff between IEEE 802 and other access technologies [34]. MIH is mainly focused on handoff initiation and recognizes the importance to obtain interoperability with diverse wireless technology through the adoption of an open approach that aims to integrate MIH with existing technologies by purposely leaving handoff decision and execution out of the standard.

With a closer view at the architecture, MIH defines the MIH function (MIHF), a cross-layer entity (logically located between data-link and network layers) that provides mobility support through well-defined service APs with three main types of services: an "event service" to notify MIH users of relevant data-link layer events from both local and remote interfaces, for example, Link_available; a "command service" to provide commands for controlling MIH handoff, for example, MIH_Link_Switch; and an "information service" that maintains all context information to make more effective handoff decisions, for example, information on the list of available networks or neighbor APs. MIHF is deployed at both client and network sides and closely recalls a client-stub and proxy-based architecture. Finally, at the network side the MIH network entity enables the communication between various MIHFs (available for each wireless access network) over a transport that supports MIH services to facilitate the centralized gathering of relevant context information. Service continuity is still widely unaddressed by the MIH proposal; some preliminary research efforts have recently proposed possible solutions to provide seamless MIH-based handoff [68,69].

13.4.1.2 Unlicensed Mobile Access

Unlicensed mobile access (UMA) is a joint standardization effort of main telecom companies started in 2004 and concluded in 2005, with the UMA inclusion in the 3rd Generation Partnership Project (3GPP) specification for generic access [70]. UMA focuses mainly on data flow continuity and data flow rebind, and enables vertical handoffs of 3G cellular mobile services over unlicensed spectrum technologies, that is, BT and Wi-Fi, by using mobile devices equipped with multiple wireless technologies.

The UMA approach is highly controlled and network-centric. In particular, the UMA architecture defines the UMA network controller (UNC), which acts as a gateway providing standard interfaces to enable secure access to 3G mobile services through the unlicensed wireless network. UNC controls vertical handoff management (network-controlled): UNC establishes a secure IP tunnel with the mobile device—opened over wireless (unlicensed spectrum) network interface—for session signaling and to deliver ongoing RTP-over-UDP multimedia data flows by switching them from 3G cellular network.

13.4.1.3 Internet Protocol Multimedia Subsystem

IMS has been defined by a large group of standardization entities that range from 3GPP (and 3GPP2) to IETF and open mobile alliance (OMA) [45]. IMS mainly focuses on session control and data flow rebind: it provides an overlay architecture for all-IP next-generation networks and enables secure and efficient provision of an open set of potentially high integrated multimedia services (instant messaging, VoD, Voice-over-IP [VoIP], etc.) over highly heterogeneous WI networks. To obtain this, IMS adopts a proxy-based architecture that decentralizes session and handoff management processes within wireless access networks and provides the needed functions for AAA and session control.

IMS obtains openness and interoperability by adopting application-layer solutions, mainly by using SIP for session control. In addition, it defines a wide set of gateways that enable intercommunication between IMS core infrastructure and other external infrastructures as well as specific IMS entities called application servers (ASs) to introduce novel services and extensions. At its current stage, however, IMS does not include an effective support for service continuity during handoff (especially for vertical handoff) [71]; various ongoing efforts are aiming to solve this problem within 3GPP. For instance, the voice call continuity (VCC) standard has been proposed to enable session-data handoff between circuit switch–based and IMS-based networks [72]. In addition, a recent draft by the Service Architecture Evolution (SAE) technical specification group (part of the ongoing research work on the 3GPP long-term evolution [LTE] initiative [73]) is exploring the broader goal of multimedia session continuity [74]. Moreover, as shown in Section 13.4.2, many ongoing academic research efforts are actually tackling the same issues.

13.4.2 Open Research Issues

The encouraging results obtained in service continuity research field are stimulating further investigation. In particular, we foresee three main challenging research directions: handoff management support for mobile networks as a new application area; multimedia content provisioning within mesh networks; and open handoff management for service continuity for IMS systems intended as next-generation networks.

13.4.2.1 Handoff Management Support for Mobile Networks

Mobile networks consist of a set of hosts that move together as it happens for networks made available on ships, aircrafts, and trains. Mobile networks further complicate service continuity provisioning and handoff management by introducing multiple levels of mobility: the whole mobile network mobility together with the mobility of nodes roaming within the mobile network (nested mobile networks). As a possible application scenario, consider a train (mobile network) connected through a satellite network crossing a tunnel served by 3G cellular relaying APs; handoff middlewares should be able to seamlessly support vertical handoffs that occur when the train enters and exits the tunnel.

Earlier research efforts in this area have concentrated at the network layer producing the definition of the IETF NEtwork MObility (NEMO) basic support protocol that transparently manages mobile network mobility with an approach similar to MIP. However, most application-layer issues are still open. Much still remains to be investigated: often mobile network mobility patterns are well known; hence, handoff management can be highly optimized by taking into account these patterns. For instance, in the train example, application-layer visibility of train route and tunnel position enables location-aware prebuffering techniques to smooth possible packet losses during vertical handoffs. Moreover, when client devices within the mobile network move themselves, multiple buffering levels could grant data flow continuity not only in the service path between WI fixed infrastructure and the mobile network, but also in the last mobile network-mobile client hop. In this case, coordinated buffering dimensioning should be developed to improve overall resource usage.

13.4.2.2 Multimedia Content Provisioning for Wireless Mesh Networks

Recent technological advances have motivated the development of the so-called wireless mesh networks, that is, heterogeneous wireless networks where nodes may play the differentiated role of mesh routers and mesh clients: mesh routers wirelessly communicate with one another and constitute the backbone of the network; mesh clients roam seamlessly by dynamically changing their mesh routers. Typical application scenarios include emergency and disaster-recovery situations; in addition, with the close integration of IEEE 802.16 (WiMax) and Wi-Fi networks, wireless mesh networks are expected to provide last-hop wireless connectivity broadband services out of metropolitan areas, especially to rural communities. Handoff management represents one of the crucial mesh network support facilities, especially when dealing with multimedia service provisioning for disaster-recovery application scenarios.

The multihop nature of wireless mesh networks service paths complicates handoff management. Wireless mesh APs support mesh client mobility by dynamically coordinating one another for the delivery of the required multimedia contents; however, they have to employ the same (limited) wireless link resources used for data delivery. One research challenge is the design of optimized and local coordination algorithms with visibility of service requirements and handoff/location/QoS information to reduce network overhead and boost handoff mobility management. Moreover, multihop path reconfiguration introduces long handoff latencies; hence, novel application-layer solutions should have visibility of low-level management actions to proactively activate application-level data flow continuity management. Finally, heterogeneous mesh networks with wireless mesh routers supporting multiple wireless technologies complicate the mentioned scenario further. Only application-layer approaches may enable dynamic reconfiguration of multiple (and possibly overlapping) multihop service paths established over different wireless meshes and may be able to plan the adaptation operations to grant service continuity.

13.4.2.3 Open Handoff Management for Internet Protocol Multimedia Subsystem

IMS supports client mobility between different operator networks and access technologies, but it cannot guarantee interoperable session handoff of ongoing multimedia services. To overcome this, application scenarios should include either vertical handoff of ongoing voice calls as users roam from 3G cellular infrastructures to their home wireless networks or global handoff of a VoD service when the WI provider changes.

In particular, to fully enable the application scenarios described earlier, IMS should solve two main issues: the support of interdomain end-to-end QoS management and data flow continuity. While focusing on the first issue, the current IMS architecture does not include resource reservation functions suitable to modify resource bookings along established service paths by dynamically transferring current session context from old to target WI domains. Hence, there are the open issues of interaccess network QoS conversion and dynamic end-to-end QoS provisioning. In particular, application-layer service path reconfiguration protocols should exploit local context awareness to leash signaling overhead to scale WI dimension. In the second issue, it is necessary to design and integrate special ASs with advanced context-aware handoff management capabilities, which act as handoff middleware proxies within the core IMS infrastructure. With regard to the second aspect, we are currently extending the MUM prototype to support IMS and grant IMS-based continuous service delivery. To do this, we have implemented a special IMS-enabled MUM proxy (acting as IMS AS) and a MUM client stub that implement the novel VCC standard [72] and combine different application and communication infrastructures by interacting through standard IMS session signaling. In particular, our implementation is based on the open IMS core, an open IMS infrastructure, and the IMS communicator, an open source IMS client that we modified to guarantee seamless data flow continuity during IMS handoff [75,76]. We have deployed our prototype in a real testbed and, are currently testing it in several different handoff scenarios, including vertical handoff of VoIP flows generated by 3G infrastructures as RTP-over-UDP flows over Wi-Fi.

13.5 Conclusion

This chapter presents the handoff management efforts needed in WI to achieve service continuity of mobile multimedia services. In addition to having reestablished that the complex task of continuity requires strong coordination to handle all differently related aspects, we have proposed a full context-aware approach to clarify handoff challenges and requirements. This model is based on full handoff/QoS/location awareness and suggests the design guidelines for the development of novel WI handoff middlewares. We have also used the same framework to survey several handoff management solutions with special focus on the promising approaches at the application layer. Future handoff management research directions need to lead the evolution of current handoff middleware to answer the service continuity challenge raised by the ever-increasing diffusion of novel wireless solutions and network architectures. The main impulse to new research directions is expected to stem from standardization efforts and novel issues driven by emerging and peculiar application areas.

References

1. Corson, M.S., J.P. Macker, V.D. Park, Mobile and wireless internet services: Putting the pieces together, *IEEE Communications Magazine*, 39(6), 148–155, 2001.
2. Banerjee, N., W. Wei, S.K. Das, Mobility support in wireless internet, *IEEE Wireless Communications*, 10(5), 54–61, 2003.
3. Aurrecoechea, C., A.T. Campbell, L. Hauw, A survey of QoS architectures, *ACM/Springer Multimedia Systems Journal*, 6(3), 138–151, 1998.
4. Jin, J., K. Nahrstedt, QoS specification languages for distributed multimedia applications: A survey and taxonomy, *IEEE Multimedia Magazine*, 11(3), 74–87, 2004.
5. Ramanathan, P., K.M. Sivalingam, P. Agrawal, S. Kishore, Dynamic resource allocation schemes during handoff for mobile multimedia wireless networks, *IEEE Journal on Selected Areas in Communications (JSAC)*, 17(7), 1270–1283, 1999.
6. Dutta, A., T. Zhang, Y. Ohba, K. Taniuchi, H. Schulzrinne, MPA Assisted Optimized Proactive Handoff Scheme, *IEEE International Conference on Mobile and Ubiquitous Systems (MobiQuitous)*, San Diego, CA, 2005.
7. Bragodia, R. et al., iMASH: Interactive Mobile Application Session Handoff, *ACM International Conference on Mobile Systems, Applications, and Services (MobiSys)*, San Francisco, CA, 2003.
8. Campbell, A.T., M.E. Kounavis, R.R.-F. Liao, Programmable mobile networks, *Computer Networks and ISDN Systems*, 31(7), 741–765, 1999.
9. Balasubramaniam, S., J. Indulska, Vertical handover supporting pervasive computing in future wireless networks, *Computer Communication*, 27(8), 708–719, 2004.
10. Banerjee, N., S.K. Das, A. Acharya, Seamless SIP-based mobility for multimedia applications, *IEEE Network*, 20(2), 6–13, 2006.
11. Kounavis, M.E., A.T. Campbell, G. Ito, G. Bianchi, Design, implementation, and evaluation of programmable handoff in mobile networks, *Mobile Networks and Applications*, 6(5), 443–461, 2001.

12. Bellavista, P., A. Corradi, L. Foschini, Proactive Management of Distributed Buffers for Streaming Continuity in Wired-Wireless Integrated Networks, *IEEE/IFIP Network Operations and Management Symposium (NOMS)*, Vancouver, Canada, 2006.
13. Stemm, M., R.H. Katz, Vertical handoff in wireless overlay networks, *Mobile Networks and Applications*, 3(4), 335–350, 1998.
14. McNair, J., Z. Fang, Vertical handoffs in fourth-generation multinetwork environments, *IEEE Wireless Communications*, 11(3), 8–15, 2004.
15. Saha, D. et al., Mobility support in IP: A survey of related protocols, *IEEE Network*, 18(6), 34–40, 2004.
16. Bellavista, P., A. Corradi, L. Foschini, Application-Level Middleware to Proactively Manage Handoff in Wireless Internet Multimedia, *IEEE/IFIP International Conference on Management of Multimedia Networks and Services (MMNS)*, Spinger, Berlin, 2005.
17. Bartolini, N., P. Campegiani, E. Casalicchio, S. Tucci, A Performance Study of Context Transfer Protocol for QoS Support, *IEEE/IFIP International Symposium on Computer and Information Sciences (ISCIS)*, Springer, Berlin, 2004.
18. Loughney, J., M. Nakhjiri, C. Perkins, R. Koodli, Context Transfer Protocol (CXTP), RFC 4067, IETF, 2005.
19. Pahlavan, K., P. Krishnamurthy, A. Hatami, M. Ylianttila, J.P. Makela, R. Pichna, J. Vallstron, Handoff in hybrid mobile data networks, *IEEE Personal Communications*, 7(2), 34–47, 2000.
20. Mishra, A., M. Shin, W. Arbaugh, An empirical analysis of the IEEE 802.11 MAC layer handoff process, *ACM Computer Communication Review*, 33(2), 93–102, 2003.
21. Velayos, H., G. Karlsson, Techniques to Reduce IEEE 802.11b Handoff Time, *IEEE International Conference on Communications (ICC)*, Paris, France, 2004.
22. Hsieh, R., Z.G. Zhou, A. Seneviratne, S-MIP: A Seamless Handoff Architecture for Mobile IP, *IEEE Annual Joint Conference of the IEEE Computer and Communications Societies (INFOCOM)*, San Francisco, CA, 2003.
23. Steinmetz, R., K. Nahrstedt, *Multimedia Systems*, Spinger, Berlin, 2004.
24. Kounavis, M.E., A.T. Campbell, Seamless connectivity in infrastructure-based networks, *The Handbook of Mobile Middleware* (eds. P. Bellavista and A. Corradi), Chapman & Hall, London, 547–568, 2005.
25. Schulzrinne, H., E. Wedlund, Application-layer mobility using SIP, *ACM Mobile Computing and Communications Review*, 4(3), 47–57, 2000.
26. Ghaderi, M., R. Boutaba, Impact of mobility on resource management in wireless networks, *The Handbook of Mobile Middleware* (eds. P. Bellavista and A. Corradi), Chapman & Hall, London, 639–662, 2005.
27. Chen, L., G. Yang, T. Sun, M.Y. Sanadidi, M. Gerla, Enhancing QoS Support for Vertical Handoffs Using Implicit/Explicit Handoff Notifications, *IEEE International Conference on Quality of Service in Heterogeneous Wired/Wireless Networks (QShine)*, Orlando, FL, 2005.
28. Koodli, R., Fast Handovers for Mobile IPv6, RFC 4068, IETF, 2005.
29. Wireless LAN Medium Access Control (MAC) and Physical Layer (PHY), Standard 802.11, IEEE, 1999.
30. Ekemark, S., Radio Resource Control (RRC)—Protocol specification, Technical Specification (TS) 25.331, 3GPP, 2006.

31. Baatz, S. et al., Handoff Support for Mobility with IP over Bluetooth, *IEEE International Conference on Local Computer Networks (LCN)*, Tampa, FL, 2000.
32. Chung, S., H. Yoon, J. Cho, A Fast Handoff Scheme for IP over Bluetooth, *IEEE International Conference on Parallel Processing Workshops (ICPPW)*, Vancouver, Canada, 2002.
33. Bellavista, P., M. Cinque, D. Cotroneo, L. Foschini, Integrated Support for Handoff Management and Context Awareness in Heterogeneous Wireless Networks, *The International Workshop on Middleware for Pervasive Ad-Hoc Computing (MPAC)*, Grenoble, France, ACM Press, New York, NY, 2005.
34. Media Independent Handover Services, IEEE 802.21, http://www.ieee802.org/21/.
35. Gazis, V., N. Alonistioti, L. Merakos, Toward a generic "always best connected" capability in integrated WLAN/UMTS cellular mobile networks (and beyond), *IEEE Wireless Communications*, 12(3), 20–29, 2005.
36. Gustafsson, E., A. Jonsson, Always best connected, *IEEE Wireless Communications*, 10(1), 49–55, 2003.
37. Lee, D., C. Lee, J.W. Kim, Seamless Media Streaming Over Mobile IP-Enabled Wireless LAN, *IEEE Consumer Communications and Networking Conference (CCNC)*, Las Vegas, NV, 2005.
38. Bellavista, P., A. Corradi, L. Foschini, Context-aware handoff middleware for transparent service continuity in wireless networks, *Pervasive and Mobile Computing Journal*, 3(4), 439–466, 2007.
39. Jeon, W.J., K. Nahrstedt, QoS-aware middleware support for collaborative multimedia streaming and caching service, *Microprocessors and Microsystems Journal*, 27(2), 65–72, 2003.
40. Bruneo, D., M. Villari, A. Zaia, A. Puliafito, VOD Services for Mobile Wireless Devices, *IEEE International Symposium on Computers and Communications (ISCC)*, Antalya, Turkey, 2003.
41. Dutta, A., E. Van den Berg, D. Famolari, V. Fajardo, Y. Ohba, K. Taniuchi, T. Kodama, H. Schulzrinne, Dynamic Buffering Control Scheme for Mobile Handoff, *IEEE International Symposium on Personal, Indoor and Mobile Radio Communications (PIMRC)*, Helsinki, Finland, 2006.
42. Mir, N.F., *Computer and Communication Networks*, Prentice Hall, New York, 2006.
43. Campbell, A.T. et al., Comparison of IP micromobility protocols, *IEEE Wireless Communications*, 9(1), 72–82, 2002.
44. Dutta, A., H. Schulzrinne, K.D. Wong, Supporting continuous services to roaming clients, *The Handbook of Mobile Middleware* (eds. P. Bellavista and A. Corradi), Chapman & Hall, London, 599–638, 2005.
45. Camarillo, G. et al., *The 3G IP Multimedia Subsystem (IMS)*, Wiley, New York, 2005.
46. Banerjee, N., S. Das, A. Acharya, SIP-based Mobility Architecture for Next Generation Wireless Networks, *IEEE International Conference on Pervasive Computing and Communications (PerCom)*, Kauai, Hawaii, 2005.
47. Tripathi, N., J.H. Reed, H.F. Van Landingham, Handoff in cellular systems, *IEEE Personal Communications*, 5(6), 26–37, 1998.
48. Chou, C., K.G. Shin, An enhanced inter-access point protocol for uniform intra and intersubnet handoffs, *IEEE Transactions on Mobile Computing*, 4(4), 321–334, 2005.

49. Chen, M., J. Chen, P. Yao, Efficient Handoff Algorithm for Bluetooth Networks, *IEEE International Conference on Systems, Man and Cybernetics*, Waikoloa, Hawaii, 2005.
50. Johnson, D., C. Perkins, J. Arkko, Mobility Support in IPv6, RFC 3775, IETF, 2004.
51. Seshan, S., H. Balakrishnan, R.H. Katz, Handoffs in cellular wireless networks: The daedalus implementation and experience, *Wireless Personal Communications*, 4(2), 141–162, 1997.
52. Katz, R., E. Brewer, The UC Berkeley BARWAN Research Project (1998), http://daedalus.cs.berkeley.edu/.
53. Sharma, S., N. Zhu, T. Chiueh, Low-latency mobile IP handoff for infrastructure-mode wireless LANs, *IEEE Journal on Selected Areas in Communications (JSAC)*, 22(4), 643–652, 2004.
54. Soliman, H., C. Castelluccia, K. El Malki, L. Bellier, Hierarchical Mobile IPv6 Mobility Management (HMIPv6), RFC 4140, IETF, 2005.
55. Maltz, D.A., P. Bhagwat, MSOCKS: An Architecture for Transport Layer Mobility, *IEEE Annual Joint Conference of the IEEE Computer and Communications Societies (INFOCOM)*, San Francisco, CA, 1998.
56. Snoeren, A.C., H. Balakrishnan, An End-to-End Approach to Host Mobility, *ACM International Conference On Mobile Computing and Networking (MobiCom)*, Boston, MA, 2000.
57. Guo, C., Z. Guo, Q. Zhang, W. Zhu, A seamless and proactive end-to-end mobility solution for roaming across heterogeneous wireless networks, *IEEE Journal on Selected Areas in Communications (JSAC)*, 22(5), 834–847, 2004.
58. Chen, L., T. Sun, B. Chen, V. Rajendran, M. Gerla, A Smart Decision Model for Vertical Handoff, *International Workshop on Wireless Internet and Reconfigurability (ANWIRE)*, Athens, Greece, 2004.
59. Dutta, A., V. Fajardo, Y. Ohba, K. Taniuchi, H. Schulzrinne, A Framework of Media-Independent Pre-Authentication (MPA) (draft-ohba-mobopts-mpa-framework-04.txt), Internet Drafts, IETF, 2007.
60. Dutta, A., Y. Ohba, K. Taniuchi, V. Fajardo, H. Schulzrinne, Media-Independent Pre-Authentication (MPA) Implementation Results, (draft-ohba-mobopts-mpa-implementation-04.txt), Internet Drafts, IETF, 2007.
61. Lin, J., G. Glazer, R. Guy, R. Bagrodia, Fast Asynchronous Streaming Handoff, *The Joint International Workshop on Interactive Distributed Multimedia Systems/Protocols for Multimedia Systems (IDMS/PROMS)*, Coimbra, Portugal, 2002.
62. Vali, D., S. Paskalis, A. Kaloxylos, L. Merakos, An efficient micro-mobility solution for SIP networks, *IEEE GLOBECOM*, San Francisco, CA, 2003.
63. Wu, W., N. Banerjee, K. Basu, S.K. Das, SIP-based vertical handoff between WWANs and WLANs, *IEEE Wireless Communications*, 12(3), 66–72, 2005.
64. Nakajima, N., A. Dutta, S. Das, H. Schulzrinne, Handoff Delay Analysis and Measurement for SIP Based Mobility in IPv6, *IEEE International Conference on Communications (ICC)*, Seattle, WA, 2003.
65. Dutta, A., S. Madhani, W. Chen, O. Altintas, H. Schulzrinne, Fast-Handoff Schemes for Application Layer Mobility Management, *IEEE International Symposium on Personal Indoor and Mobile Radio Communications (PIMRC)*, Barcelona, Spain, 2004.

66. Bellavista, P., A. Corradi, L. Foschini, Java-Based Proactive Buffering for Multimedia Streaming Continuity in the Wireless Internet, Poster Paper Proceeding of IEEE International Symposium on a World of Wireless, *Mobile and Multimedia Networks (WoWMoM)*, Giardini Naxos, Italy, 2005.
67. Bellavista, P., A. Corradi, L. Foschini, SIP-Based Proactive Handoff Management for Session Continuity in the Wireless Internet, *IEEE International Workshop on Services and Infrastructure for the Ubiquitous and Mobile Internet (SIUMI), Held in Conjunction with the International Conference on Distributed Computer Systems (ICDCS)*, Lisbon, Portugal, 2006.
68. Dutta, A., S. Das, D. Famolari, Y. Ohba, K. Taniuchi, V. Fajardo, R. Marin Lopez, T. Kodama, H. Schulzrinne, Seamless proactive handover across heterogeneous access networks, *Wireless Personal Communications: An International Journal*, Kluwer Academic Publisher, Hingham, MA, 43(3), 837–855, 2007.
69. Mussabbir, Q.B., W. Yao, Z. Niu, X. Fu, Optimized FMIPv6 Using IEEE802.21 MIH services in vehicular networks, *IEEE Transactions on Vehicular Technology*, 56(6), 3397–3407, 2007.
70. Vikberg, J., Generic Access (GA) to the A/Gb Interface; Mobile GA Interface Layer 3 Specification, Technical Specification (TS) 44.318, 3GPP, 2007.
71. Gouveia, F., J. Floroiu, T. Magedanz, Seamless Service Access for Mobile SIP-based Applications and IMS Services on Top of Different Access Networks, *IEEE International Workshop on Applications and Services in Wireless Networks (ASWN)*, Berlin, Germany, 2006.
72. Bennett, A., Voice Call Continuity (VCC) Between Circuit Switched (CS) and IP Multimedia Subsystem (IMS), Technical Specification (TS) 23.206, 3GPP, 2006.
73. Ekstrom, H., A. Furuskär, J. Karlsson, M. Meyer, S. Parkvall, J. Torsner, M. Wahlqvist, Technical solutions for the 3G long-term evolution, *IEEE Communications Magazine*, 44(3), 38–45, 2006.
74. Song, J.S., Feasibility Study on Multimedia Session Continuity, Technical Specification (TS) 23.893, 3GPP, 2007.
75. Open IMS Core, Fokus Berlin, http://www.openimscore.org/.
76. IMS Communicator, PT Inovao, http://imscommunicator.berlios.des.

Chapter 14

Packet Scheduling in Broadband Wireless Multimedia Networks

Rong Yu, Yan Zhang, Zhi Sun, and Shunliang Mei

Contents

14.1 Introduction....474
14.2 Models in Wireless Scheduling....475
14.2.1 Network Model....475
14.2.2 Channel Model....476
14.2.3 Traffic Model....477
14.3 Major Issues in Wireless Scheduling....479
14.3.1 Challenges....479
14.3.2 Objectives....481
14.4 Existing Scheduling Algorithms....482
14.4.1 Wireline Networks....482
14.4.2 Algorithms under Gilbert–Elliot Channel Model....485
14.4.3 Algorithms under Multistate Channel Model....487
14.4.4 Summary of Existing Algorithms....488
14.5 Learning-Based Packet Scheduling....490
14.5.1 Semi-Markov Decision Process Formulation....490
14.5.2 Construction of Cost Function....492

14.5.3 Neuro-Dynamic Programming Solution 494
14.5.3.1 Function Approximation Architecture 495
14.5.3.2 Online Parameter Tuning 496
14.5.4 Performance Evaluation 497
14.6 Future Directions 500
14.6.1 Inaccuracy of System Information 500
14.6.2 Collaboration with Admission Control and Congestion Control 501
14.6.3 Compatibility with Standards 501
References 501

The design of broadband wireless multimedia networks introduces a set of challenging technical issues. This chapter focuses on packet-scheduling algorithms. The state of the art for packet scheduling in wireless multimedia networks is presented here, including the system models (e.g., network-, wireless channel–, and traffic model), challenges (e.g., diverse traffic properties and high link variability), and objectives (e.g., quality-of-service [QoS] differentiation and guarantee, maximizing bandwidth utilization, and fairness provisioning). In the overview of existing packet-scheduling algorithms, scheduling algorithms applied in wireline and wireless environment are briefly introduced. On the basis of summary of existing scheduling algorithms, a new scheduling framework, which is learning based, is proposed. The central idea of the proposed scheduling mechanism is to employ machine learning method in the scheduling process and make the scheduler adaptable to the complicated scheduling environment in wireless multimedia networks.

14.1 Introduction

With an increasing demand for multimedia communication and the rapid growth of popularity for wireless personal services, the design and deployment of broadband wireless multimedia networks have become emergent tasks. It is expected that broadband wireless multimedia networks should be able to support high-speed and high-quality, real-time and non-real-time applications, such as video, image, voice, text, and data. Therefore, providing QoS for heterogeneous traffic with diverse properties is the basic requirement in emerging broadband wireless multimedia networks. Being an indispensable component of the entire entity of communication network protocol, packet scheduling plays a key role in QoS provisioning and resource sharing. Because packet scheduling has such a dominant impact on the ability of the networks to deliver QoS guaranteed services, the research on packet scheduling has attracted much attention in recent years.

Usually, packet-scheduling algorithms have the responsibility of managing the bandwidth allocation and multiplexing at the packet level. They operate across different sessions (i.e., connections or traffic flows) to guarantee that all sessions receive services according to the contract established when sessions are set up. Being different from traditional wireline scheduling problems, packet scheduling in wireless multimedia networks will encounter two major difficulties—the high variability of wireless channel conditions (due to signal attenuation, interference, fading, user mobility, etc.) and the unknown and complicated model of packet arrival processes (due to the diverse properties of different classes of traffic). Besides the difficulties, packet-scheduling algorithms for wireless multimedia networks should concurrently achieve three performance objectives—QoS differentiation and guarantee, maximizing wireless bandwidth utilization, and fairness provisioning. These special difficulties and objectives make the scheduling problem in wireless multimedia networks considerably challenging.

The work in this chapter consists of two main parts. The first part presents an overview of the wireless scheduling problems, including the system models, major issues, and exiting algorithms. In the second part, a new learning-based scheduling algorithm for wireless multimedia networks is presented, where the methodology of neuro-dynamic programming (NDP) is involved to address the packet-scheduling problem.

14.2 Models in Wireless Scheduling

14.2.1 Network Model

We consider a general shared-channel wireless cellular network model in which the whole service area is divided into many cells. A base station is built in each cell and serves all the mobile hosts in the cell. Different base stations are connected by wireline networks. The wireless links between mobile hosts and the base station are independent of one another. The communication between a mobile host and the base station may contain more than one session (or traffic flow). The scheduler within the base station is in charge of the packet scheduling in both downlinks (from base station to mobile hosts) and uplinks (from mobile hosts to base station). For downlink transmissions, all the users' packets will directly queue in the base station. Thus, the base station is supposed to have full knowledge about the status of downlink packets. For the uplink transmissions, when a mobile host has packets to transmit, it needs to send a request message to the base station, reporting the queuing status in its inbuilt buffer. The base station collects these request messages and establishes virtual queues in its own buffer. On the basis of these virtual queues, the base station is also supposed to have the knowledge about the status of uplink packets.

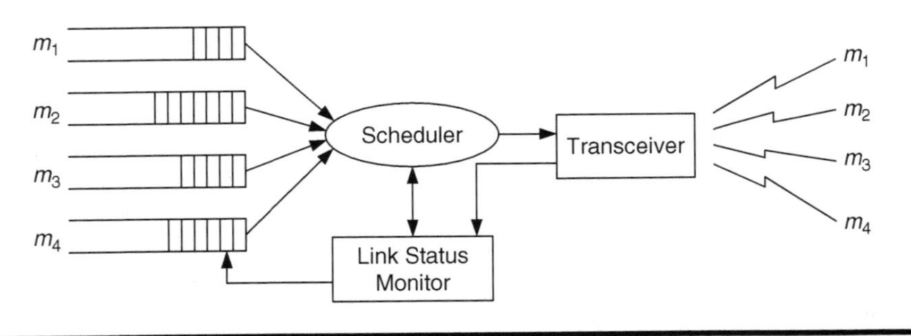

Figure 14.1 A general scheduler model.

A general scheduler model is shown in Figure 14.1, where the multiple-packet queues are served by a single scheduler. Each queue represents one session between a certain mobile host and the base station. Packet status such as queuing delay and size are recorded and made available for the scheduler. The link status monitor (LSM) [1] is responsible for monitoring the state of channels between the base station and mobile hosts. The method to implement LSM in high data rate (HDR) systems is introduced in Ref. 2, which is a good reference for generalizing LSM in other real systems. The usage of LSM enables the scheduler to acquire wireless link conditions of all sessions. With the information of all the session's packet queuing status and wireless link conditions, the scheduler is able to perform centralized packet scheduling.

14.2.2 Channel Model

One of the main differences between wireless networks and wireline networks is the high error rate of wireless channels. The wireless link error is also one of the reasons that wireless scheduling algorithms should differ from wireline scheduling algorithms. Setting up an accurate wireless channel model is an important precondition for the design of wireless scheduling algorithms. In previous works, the following two main types of wireless channel models are discussed.

- *GE model.* The classical two-state Gilbert–Elliot (GE) model [3,4] was initially used to model wireless link variation. The channel is simply described to be either in good (error free) or in bad (error prone) state in this model. The channel condition moves between the two states according to a certain transition probability matrix. A packet is successfully received if and only if the link stays in the good state throughout the packet transmission time.
- *Multistate Markov channel model.* It is also called finite state Markov channel (FSMC) in Refs 5 and 6. In this channel model, the channel

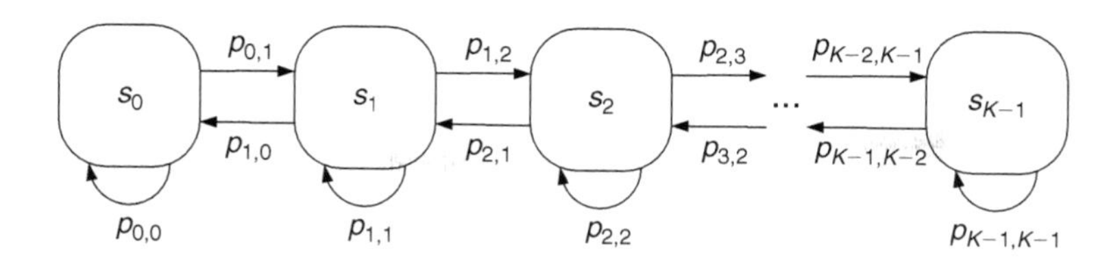

Figure 14.2 Multistate Markov channel model.

condition is categorized into multiple (more than two) states. At any time, the channel condition will stay in one of these states, but the condition may change with time. As shown in Figure 14.2, the state transition only takes place between two adjacent states. Because FSMC is derived by partitioning the received signal-to-noise ratio (SNR) into multiple intervals, there is a maximum transmission rate in each state with respect to the actual requirement of average bit-error rate (BER) in the real system.

The aforementioned two channel models are different in two ways. First, the number of states are different in these two models. The GE model fixes the state number by two, but the FSMC model allows the state number to be an arbitrary finite number. Obviously, by relaxing the limit on state number, the FSMC model has the ability to represent the actual channel condition with higher accuracy. The two-state GE model is usually too coarse to describe the wireless link variability, especially in the situations that channel quality varies dramatically. Second, the channel capacity are different in these two models. With the enlarged state number, the FSMC has the potential of providing higher channel capacity than the GE model. The fact is reported and analyzed in Ref. 5. Following this point, packet-scheduling algorithms based on the FSMC model have the underlying advantage over those based on the GE model.

14.2.3 Traffic Model

As an input of the scheduler, the traffic flows have a large impact on the performance of the scheduling system. The study of traffic model is then helpful for the design of packet-scheduling algorithms. First, a suitable description of traffic in the actual network is necessary for the traffic simulation in the computer experiments. Nowadays, computer simulation has become a strong tool for the protocol design and performance analysis of wireless networks. To make the simulation experiments more close to the actual situation and more persuasive, the mathematical model that is employed in the simulation should be correct. So, an accurate traffic model is indispensable for a good traffic simulation. Second, many existing packet-scheduling algorithms are put forward for their specific application fields. These algorithms are supposed to work under their specific traffic models. It is no guarantee for them to provide good work under other traffic models. In other words, scheduling algorithms are usually customized. It is hard (if not impossible) to find out a special scheduling algorithm that always has satisfying performance under different traffic models. Therefore, given an application background, the study on the traffic model plays a key role in designing an effective scheduling algorithm. The following are some popular traffic models in previous works.

- *Constant-rate model.* Constant-rate model is a model where packets are assumed to arrive with a constant rate. As known, the audio traffic is usually described by the constant-rate model. The merit of constant-rate model lies in its simplicity, which brings great convenience in the theoretical analysis and simulation experiments. However, the constant-rate model is still too weak to capture the characteristics of many other types of traffic.
- *Poisson model.* Poisson model is a well-known model. It is extensively used in many practical fields. There are two reasons for it having drawn so much attention. First, Poisson model is a suitable model to describe the traffic arrival process in many practical problems. For instance, in telecommunication networks, the number of telephone call occurring at the switchboard follows the Poisson distribution. Second, Poisson model has some special properties (e.g., the interval of two adjacent arrivals is exponentially distributed and therefore memoryless), which makes the theoretical analysis simplified.
- *Interrupted Poisson process (IPP) model and its variants.* IPP model is an extension of Poisson model. In IPP model, the source that generates traffic is supposed to have two states—ON and OFF. In state ON, the packets are produced by the original Poisson model; but in state OFF, no packet is produced. The source's dwelling time in state ON and OFF follows exponential distribution. IPP model has some variants called interrupted deterministic process (IDP) and interrupted renewal process (IRP). Detailed description on them could be found in Ref. 7.
- *Greedy model.* Greedy model generates packets "greedily" by its source. It always keeps all the queues overflowed. Specifically, when one or several packets are transmitted and some of the queues are not totally filled, the source will immediately generate some new packets to fill the queuing buffer. One typical application of greedy model is the File Transfer Protocol (FTP) services. In FTP services, users usually submit a transmission request of a large amount of data, which is

delay-tolerant, to the server. In this case, it is suitable to use greedy model as the traffic model.

- *General model.* General model does not make any assumption of the packet arrival processes. For this reason, it can be used to depict any type of traffic. But due to the fact that the general model does not provide any feature information about the traffic, the analysis and discussions on this model are very difficult. Generally, the general model is considered only when the designers have no knowledge about the traffic or they hope to obtain results that are applicable in general cases.

14.3 Major Issues in Wireless Scheduling

Owing to the unique characteristics of wireless multimedia applications, packet scheduling in such networks involves a series of underlying challenges and performance objectives.

14.3.1 Challenges

- *Heterogeneous traffic properties.* Being different from traditional telecommunication networks, wireless multimedia networks are envisioned to support the transmission of heterogeneous traffic with diverse QoS requirements. Typical traffic flows include the following service classes:
  1. *Constant bit rate (CBR).* The CBR service class is intended for real-time applications, that is, those requiring tightly constrained delay and delay variation, as would be appropriate for voice and video applications. The consistent availability of a fixed quantity of bandwidth is considered appropriate for CBR service.
  2. *Non-real-time variable bit rate (nrt-VBR).* The nrt-VBR service class is intended for non-real-time applications, which have "bursty" traffic characteristics. It requires better than best effort (BE) service, for example, bandwidth intensive file transfer and digital video. These applications are time-insensitive and require minimum bandwidth allocation.
  3. *Real-time VBR (rt-VBR).* The rt-VBR service class is intended for real-time applications that are sensitive to delay and delay variation, such as interactive compressed voice and video applications. Sources are expected to transmit at a rate that varies with time, that is, bursty traffic. Applications such as videoconference belong to this class. Late packets that miss the deadline will be useless.
  4. *Available bit rate (ABR).* The ABR service class is employed to describe those types of traffic, which have the ability to reduce

their information transmitting rate if the network requires them to do so. Likewise, they can increase their information transmitting rate if there is extra bandwidth available within the network. There may not be deterministic parameters because the users are willing to continue with unreserved bandwidth.

5. *Unspecified bit rate (UBR).* The UBR service class is intended for delay-tolerant or non-real-time applications, that is, those that do not require tightly constrained delay and delay variation. Traditional computer communications applications, such as FTP, Telnet, and e-mail, are supported by UBR. UBR does not specify traffic-related service guarantees.

   For these five classes of traffic, very different requirements of QoS parameters are defined for every class [8]. For instance, the typical bandwidth requirements are 32 or 64 kbps for voice traffic (CBR) and 1–10 Mbps for videoconference applications (nrt-VBR). The loss rate requirement of voice traffic is 10^{-2}, but the FTP services (UBR) should at least guarantee loss rate at the level of 10^{-8}. Apparently, for all these classes of traffic, different scheduling rules should be constituted. The coordination of different scheduling rules is the key point for designing an efficient scheduling algorithm.

- *Wireless link variability.* Owing to the high quality of the transmission media, packet transmissions on wireline networks enjoy very low error rate. However, channel conditions of mobile wireless communications are more error prone and suffer from interference, path loss, shadowing, and multipath fading. Actually, the most significant difference between wireless and wireline links is the fact that the capacity of wireless links may experience considerably high variation. During some severe bursty error period, a wireless link can be so bad that no successful packet transmissions can be achieved. Besides this "time-dependent" problem, wireless link capacity is also related to the users' locations. At a particular time instance, a base station can communicate with more than one mobile host simultaneously. Because of the different physical locations, some mobile hosts may enjoy error-free communication with the base station, whereas others may not be able to communicate at all. This is the so-called "location-dependent" error. Moreover, mobility of the hosts also increases the variability of the transmission links. It is notable that the links between the base station and different users are independent from one another. As a result, the channel conditions for different users fluctuate randomly and asynchronously. For a system with a large number of users, it is with high probability that at least one user has error-free channel. Therefore, how to take advantage of the link independence of different users and integrate some dynamic mechanisms to deal with the time- and location-dependent burst errors is the key concern in wireless scheduling problems.

14.3.2 Objectives

- *QoS*. Wireless multimedia networks are expected to provide services for heterogeneous classes of traffic with diverse QoS requirements. As a result, QoS differentiation and guarantee must be supported. To achieve this target, some additional mechanisms for QoS support should be integrated into the scheduling algorithms. QoS support in wireless scheduling problem is dictated by the service model. For DiffServ [9] type of services, at least priority-based scheduling service for aggregated traffic with QoS differentiation should be implemented in scheduling algorithms; whereas for IntServ [10] type of services, support for per-flow-based guaranteed QoS performance, such as delay or jitter bound, should be offered by the scheduling algorithms. When a wireless link suffers from drastic channel degradation, it is very difficult to provide QoS guarantee for the traffic flows using this link. But for other links with slight channel degradation, QoS guarantee of traffic flows could still be achieved on them, if the channel conditions are better than some certain levels.
- *Bandwidth utilization*. The limited bandwidth is usually the bottleneck in wireless networks to provide satisfying broadband and smooth multimedia services. It is no doubt that bandwidth is the most precious resource in wireless networks. Therefore, to provide as much as possible services, an efficient wireless scheduling algorithm should aim to minimize unproductive transmissions on error links and increase the effective service delivered, so as to maximize the utilization of the wireless channels.
- *Fairness*. Scheduling fairness in wireline networks is usually achieved by dedicating different service rates to different flows. The scheduling algorithm should prevent different flows from interfering with one another. Because wireline environment could be considered error free, the service rate allocated to a particular flow is actually the amount of service share that will be received by the flow. However, the fairness issue in wireless environment is much more complicated. It may happen that a certain flow is assigned to transmit a packet according to the scheduling rule to get its fair share of the bandwidth, but the flow's link is actually in an error period. If the packet is transmitted, it will be corrupted and the transmission will waste bandwidth resources. In this case, postponing the transmission of the flow until its link recovers from the error period is clearly a reasonable choice. As a result, this error-prone flow has to temporarily lose its share of the wireless bandwidth. To guarantee fairness, the flow should be

compensated later. But, developing a rule to compensate the loss of services is not an easy task. The definition and objectives of fairness become more ambiguous in the wireless networks. Some existing definitions of fairness in wireless environment include proportional fairness (PF) [11], maximum–minimum fairness [12], and long- and short-term [13] fairness. The appropriate interpretation of fairness for wireless scheduling should depend on the service model, traffic type, and channel characteristics.

14.4 Existing Scheduling Algorithms

Existing packet-scheduling algorithms could be broadly categorized into two classes—algorithms applied in wireline and wireless environment. Moreover, wireless scheduling algorithms could be further classified into two types—those under the GE wireless channel and multistate wireless channel model. This session presents a brief introduction of existing packet-scheduling algorithms. See Refs 8, 14, and 15 for more detailed discussion of existing scheduling algorithms.

14.4.1 Wireline Networks

Wireline scheduling algorithms have been developed and analyzed in the research community for some time. Scheduling in a wireline environment is relatively easy because the link capacity is always considered to be invariable. Therefore, the target of the algorithms is to allocate this fixed amount of bandwidth fairly among all the traffic flows, sometimes taking into account the different QoS requirements agreed when the connection is established. We mention some popular wireline scheduling algorithms as follows:

- *First come first served (FCFS)*. The FCFS or first in first out (FIFO) is the simplest scheduling scheme. All packets that arrive are enqueued in a single queue and served according to the order of their arrival time. If the queue is filled completely, all further packets will be dropped until there are some buffer spaces available again. This property is called "tail dropping." Whenever the network device is ready to transmit the next packet, the first packet in the queue, which is called "head-of-line" (HOL) packet, is dequeued and transmitted.

  This simple scheduling scheme is not suitable for providing any QoS guarantees because flows are not isolated and the scheduler does not take any of the specific requirements of a flow (e.g., delay, throughput) into account. Moreover, the property of tail dropping is inadequate for delay-sensitive flows where packets become useless once their queuing delay exceeds the required deadline.

- *Generalized processor sharing* (*GPS*). GPS [16] is an ideal scheduling scheme, which cannot be implemented in packet-based networks. Many other scheduling algorithms try to approximate GPS as far as possible.

  Assume that there are N flows to be scheduled by the algorithm. In a GPS system, every flow is assigned a weight ϕ_i, which indicates its share of the total bandwidth, satisfying the condition $\sum_{i=1}^{N} \phi_i = 1$. Let $s_i(\tau_1, \tau_2)$ denote the amount of service that a flow i has received in the time interval (τ_1, τ_2), then for every flow i continuously backlogged within the interval (τ_1, τ_2), a GPS server ensures that

$$\frac{s_i(\tau_1, \tau_2)}{s_j(\tau_1, \tau_2)} \geq \frac{\phi_i}{\phi_j}, \quad i, j \in [1, N] \tag{14.1}$$

  The server with rate r serves flow i with a guaranteed service share of $g_i = \phi_i r$. In addition, if there are some nonbacklogged flows, their abundant capacity will be distributed to those backlogged ones, proportional to the weights ϕ_i. A GPS server has the following properties, which make it very suitable for traffic scheduling:
  1. It is work conserving, which means that the server is busy whenever there is a packet stored in one of the queues. No capacity is lost because of the scheduling.
  2. It is very flexible. For example, by assigning a small weight ϕ_i to an unimportant background flow, most of the channel capacity can be used by flows of higher priorities (e.g., $\phi_j \gg \phi_i$) if they are backlogged, whereas the remaining bandwidth can be used by the background traffic.
  3. The maximum queuing delay that a flow may experience is bounded.
- *Weighted round robin* (*WRR*). WRR [16,17] is the simplest approximation of GPS in packet-based networks. It is based on the assumption that the average packet size is known by the scheduler. Every flow is given an integer weight w_i corresponding to the service share it is supposed to get. On the basis of these weights, a server with rate r precomputes a service scheduling frame, which serves session i at a rate of $\frac{w_i}{\sum_j w_j} r$ assuming the average packet size. The server then polls the queues in the order of the scheduling frame, all empty queues will be skipped. Although this scheme can be implemented very easily and requires only $\mathbf{O}(1)$ work (i.e., constant time complexity) to process a packet, it has several drawbacks. When an arriving packet misses its transmission opportunity in the frame, it has to delay the transmission significantly in a heavy loaded system, because the waiting time for the next opportunity in the

next frame could be fairly long. The algorithm is also not suitable when the average packet size is unknown or highly varying, because in the worst case, a flow can consume up to L_{max}/L_{min} times the rate that it was supposed to receive, where L_{max} and L_{min} are, respectively, the maximum and minimum packet length of the flow.

- *Weighted fair queuing (WFQ)*. WFQ or packet-by-packet generalized processor sharing (PGPS) [16] is another popular scheduling scheme that approximates GPS on a packet level. The algorithm makes the following assumptions. First, packets should be transmitted as entities. Second, the next packet to depart under GPS may not have arrived yet when the server becomes free. Hence, the server cannot process the packets in increasing order of their GPS departure time.

In WFQ, the scheduler will select and transmit the packet, which would leave an ideal GPS server next if no other packets were received. Therefore, the difference of maximum packet queuing delay between PGPS and GPS is bounded by the ratio of maximum packet size L_{max} over the server's rate r, that is, L_{max}/r. The amount of service in PGPS also does not fall behind the one in the corresponding GPS system by more than one packet of maximum packet size L_{max}. Although a packet in WFQ will not finish more than L_{max}/r time units later than in GPS, it can finish much earlier than in GPS. This specific issue is addressed by worst-case fair weighted fair queuing (WF^2Q), which is introduced later.

- *Start-time fair queuing (SFQ)*. SFQ [18] takes a slightly different approach to process packets in the order of their start time instead of scheduling packets in the sequence of finish tags. As a result, SFQ is better suited than WFQ for integrated service networks because it is able to provide a smaller average and maximum delay for low-throughput flows. In addition, SFQ also has better fairness and smaller average delay than deficit round robin (DRR) and simplifies the computational complexity of the virtual time $v(t)$.

  When a packet arrives, it is assigned a start- and finish tag based on the arrival time of the packet and the finish tag of the previous packet of the same flow, respectively. The virtual time $v(t)$ of the server is then defined as the start tag of the packet currently in service. In an idle period, $v(t)$ is set to be equal to the maximum finish tag of all packets processed before. All packets are arranged to transmit in the increasing order of their start tags.
- *WF^2Q*. WF^2Q is an extension of WFQ to address the problem that the service provided by WFQ can be significantly ahead of the GPS

service it approximates [19]. It can keep the amount of service neither ahead nor behind by more than one packet with the maximum length. By avoiding oscillation between high- and low-service states for a flow, WF^{2}Q is also much more suitable for feedback-based congestion control algorithms.

In a WF^{2}Q system, the server chooses the next packets for transmission only among those packets, which should have started their service in the corresponding GPS system. In this group of packets, the one with minimum GPS departure time will be selected. Being extended from WFQ, WF^{2}Q has the same performance aspects concerning delay and work completed for a session as WFQ. In addition, because a packet of flow i that departs from the WF^{2}Q system should start its service in GPS system, the amount of service for flow i in WF^{2}Q will not fall behind the one in the corresponding GPS system by $(1 - r_i/r)L_{i,\max}$, where r_i and $L_{i,\max}$, respectively, are the rate and maximum packet length of flow i.

14.4.2 Algorithms under Gilbert–Elliot Channel Model

- *Channel state dependent packet scheduling (CSDPS)*. CSDPS [1] is a wireless scheduling framework, which allows the use of different service disciplines such as round robin (RR), longest queue first (LQF), or earliest time stamp first (ETF). The key idea behind CSDPS is to avoid bursty errors at the link layer instead of relying on the transport or application layer for error recovery. The channel conditions of all sessions are monitored by the LSM. If the channel condition for a specific session is too bad for transmission, the transmission of its packet will be postponed. CSDPS does not include the concept of lead and lag, and hence does not provide any compensation for lagging sessions.
- *Idealized wireless fair queuing (IWFQ)*. IWFQ [20] is a realization of PGPS with a compensation mechanism for error-prone sessions. Besides PGPS, WF^{2}Q can also be used as the error-free reference system. Each session is provided with a service tag that is set equal to the virtual finish time of its HOL packet. If the session is not backlogged, the service tag is set to be ∞. Every incoming packet is stamped a start tag and a finish tag as in SFQ and then queued. The algorithm services sessions with good channels in the increasing order of their service tags.

  Lagging sessions have the lowest service tag values; thus, whenever their links recover from bad states, they have higher priorities to access the channel than leading or synchronous sessions. During this period, nonlagging (leading or synchronous) sessions are unable to transmit. To control this "blackout" period, IWFQ sets bounds for the lead and lag of each session. One drawback of IWFQ is that it does not provide graceful service degradation for leading sessions, because the channel will be occupied by a lagging session until it has all its lag compensated.

- *Channel-condition independent fair queuing* (*CIF-Q*). CIF-Q [13] uses SFQ as its error-free reference system because the authors interpret that with location-dependent channel errors, it is easier to base scheduling on the start tag rather than the finish tag. CIF-Q is able to achieve (i) short-time fairness and throughput/delay guarantees for error-free sessions, (ii) long-time fairness for sessions with bounded channel error, and (iii) graceful degradation of leading flows. In CIF-Q, every session is attached a lead and lag counter that keeps track of the number of packets by which the session is leading or lagging its error-free SFQ reference system. CIF-Q introduces a parameter $\alpha \in [0, 1]$ to control the rate at which a leading session gives up its lead to a lagging session. In particular, a leading session i retains a fraction α of its service and relinquishes the balance of its service to a lagging session. The bound of a lead or lag counter is not needed because the parameter α can be carefully set to provide graceful service degradation for leading sessions.
- *Server-based fairness approach* (*SBFA*). SBFA [21] provides a generalized framework for adapting arbitrary wireline packet fair queuing algorithms to the wireless environment. SBFA introduces the concept of a long-term fairness server (LTFS), which is a hypothetical session and shares a portion of the outgoing bandwidth as other normal sessions. The purpose of setting up LTFS is to compensate lagging sessions and maintain the long-term fairness guarantee. SBFA also borrows the concepts of separate slots and packet queues as in wireless fair service (WFS), which is presented next. Once a session is scheduled, it is allowed to transmit if it has a good channel; otherwise, a slot for this session will be generated and put into the queue of the LTFS session, and a new session with good channel condition is chosen for transmission. Once the HOL slot from the LTFS session is selected for transmission, the corresponding original session is allowed to transmit if its channel is in a good state; otherwise the next slot will be selected. No lead or lag counter is required in SBFA, because the compensation rates dedicated to lagging sessions are determined by the reserved bandwidth in LTFS session.
- *WFS*. WFS is a scheduling algorithm proposed in Ref. 22. WFS is designed to decouple the bandwidth and delay requirements of flows so as to treat error- and delay-sensitive flows differently. Besides, WFS can also avoid disturbing these synchronous flows because of

service compensation, if possible. The error-free reference system, which WFS runs, is an extension of WFQ.

In WFS, each flow is assigned a weight for rate r_i and delay ϕ_i. Each incoming packet is labeled by a start- and finish tag. The service tag of each session is set equal to the virtual finish tag of its HOL packet. Of all the packets whose start tags are not larger than the current virtual time $v(t)$ by more than the "look-ahead" parameter ρ, the one with the least finish tag is selected for transmission. The look-ahead parameter ρ here is an important parameter that determines the scheduling interval. For instance, for session i, if $\rho = \infty$ and $r_i = \phi_i$, the error-free reference system is WFQ; if $\rho = 0$ and $r_i = \phi_i$, the error-free reference system is WF^2Q. Counters for tracking the lead and lag of flows are used to achieve fairness as well as graceful service degradation for leading sessions.

14.4.3 Algorithms under Multistate Channel Model

Compared with those under the GE channel model, scheduling algorithms under multistate channel model are usually much more complex to design. One common characteristic of most of this class of algorithms is that, among all the sessions that can transmit, the scheduler chooses the one with the maximum value of constructed key metric for transmission [23]. Different definitions of the key metric lead to different scheduling disciplines. Some representatives are mentioned as follows.

Let $k^*(t)$ be the session selected for transmission at time t, $\tau_k^1(t)$ the HOL packet delay of session k at time t, $r_k(t)$ the average throughput of session k until time t, r_k the long-term average throughput of session k, and $R_k(t)$ the measured maximum possible instant channel rate for session k at time t (which is an independent and stationary random sequence), and each session can feedback this channel information to the BS in an error-free manner. Then, existing scheduling algorithms under multistate channel model can be characterized as follows:

- Algorithm proposed in Refs 24 and 25 directly lets the maximum instant channel rate $R_k(t)$ to be the key metric, and chooses the session for the next transmission $k^*(t)$ to be the one that has the largest value of instant channel rate at each scheduling epoch, that is, $k^*(t) = \arg\max_k R_k(t)$. As a result, the scheduler maximizes the network throughput at each scheduling epoch, as well as the total throughput $\sum_k r_k$. However, this scheduling discipline does not take into account any QoS issue or fairness criteria.
- Algorithm proposed in Ref. 26 looks on the ratio of the instant channel rate $R_k(t)$ over the received service rate $r_k(t)$ as the key metric,

and selects the session for the next transmission to be the one with the largest value of ratio at each scheduling epoch, that is, $k^*(t) = \arg\max_k R_k(t)/r_k(t)$. It turns out that, the algorithm will try to maximize the product of the long-term average throughput $\prod_k r_k$, which means that it achieves PF.

- Algorithm proposed in Ref. 27 defines the sum of the instant channel rate $R_k(t)$ and an offset v_k as the key metric, and picks the session with the largest sum for the next transmission at each scheduling epoch, that is, $k^*(t) = \arg\max_k(R_k(t) + v_k)$, where the offset v_k is determined to satisfy a given time fraction assignment requirement. Work in Ref. 27 states that, when this requirement is met, this algorithm will maximize the average total throughput in the given time fraction.
- Algorithm proposed in Ref. 2 treats the weighted instant channel rate $a_k R_k(t)$ as the key metric, and tries to find a session with the largest weighted channel rate for the next transmission at each scheduling epoch, that is, $k^*(t) = \arg\max_k a_k R_k(t)$, where the weight a_k is regulated to make each normalized long-term average throughput identical, for example, $r_1/b_1 = r_2/b_2 = \cdots = r_K/b_K$. Work in Ref. 2 proves that, when this requirement is met, the algorithm will maximize the average total throughput and hence, achieve the "Pareto-optimal" throughput.
- Algorithm called modified largest weighted delay first (M-LWDF) in Ref. 28 defines the product of instant channel rate $R_k(t)$, HOL packet delay $\tau_k^1(t)$, and positive constants γ_k as the key metric, and chooses the session with the largest product for the next transmission at each scheduling epoch, that is, $k^*(t) = \arg\max_k \gamma_k \tau_k^1(t) R_k(t)$. The choice of parameters γ_k allows one to control packet delay distributions for different sessions. Work in Ref. 28 analytically proves that, by using this constructed key metric, M-LWDF could achieve throughput optimal.

14.4.4 Summary of Existing Algorithms

In the foregoing text, we have introduced three classes of scheduling algorithms—those applied in wireline environment, under GE wireless channel model, and under multistate wireless channel model. Now we make a summary of existing algorithms. Our concentration will focus on the basic ideas and the designing methodology of these algorithms.

For wireline scheduling algorithms, GPS is generally treated as the optimal scheduling discipline in the fluid mode. Therefore, many wireline packet-scheduling algorithms try to approximate it at the packet level. The way to approximate GPS can be various, and thus, different algorithms are devised, such as WRR, PGPS/WFQ, and WF^2Q. Under the GE channel model, the condition of the wireless channel is supposed to be either good or bad. If channel condition always stays in the good state, the scheduling problem will degrade into a wireline one. With this in mind, the basic idea of most GE-model-based scheduling algorithms is to simulate an error-free reference system and try to schedule packets under error-prone channels in the same order of the reference systems. With different selections of reference system and different lagging-flow compensation mechanisms, a broad class of algorithms is proposed, including IWFQ, CIF-Q, SBFA, and WFS. As multistate channel model is involved, the conditions of wireless channel are discretized into multiple levels. Possible channel rates are not only zero and full rate, but also several intermedial rates. In this case, error-free reference systems cannot help any more. A general solution is to create a self-defined key metric for each session and schedule the session with maximum key metric for transmission. The key metric is often constructed by combining some important network parameters, for example, the channel rate, session throughput, packet delay, and also some constants. Different definitions of key metric result in different scheduling algorithms, such as in Refs 2, 24, and 26–28.

As discussed earlier, we come to the conclusion that traditional solutions for packet-scheduling problem are mostly based on heuristic methods. Either introducing the virtual/ideal service models as references or defining some key metrics as criteria, many existing algorithms are built on the observation, judgement, intuition, and experience of the designers. For instance, by involving error-free reference systems, the designers hope to obtain the scheduling guideline via observing the running of the reference systems; whereas the selection of key metrics is usually done according to the intuition and experience of designers. Although heuristic algorithms have the merit of simplicity, they have two underlying drawbacks. First, heuristic algorithms are generally proposed with respect to some given special outside conditions, which means that they are of less adaptivity. When the practical conditions do not match the provided ones, it is hard for heuristic algorithms to maintain their scheduling performance. For example, the GE-model-based scheduling algorithms can work very well under two-state wireless channels. But there is no guarantee for this class of algorithms to have the same satisfying performance under the multistate wireless channels. Second, designing a heuristic scheduling algorithm is easy when the problem is relatively simple, but could be rather tough when the scheduling environments and objectives become complicated. For scheduling problem in wireless multimedia networks, the scheduler is required to achieve three main objectives simultaneously, and at the same time, combat highly variable channels and sophisticated traffic models. Such a complex scheduling problem is far out of the ability of a heuristic algorithm to handle.

To overcome the disadvantages of heuristic algorithms, this chapter also introduces a new scheduling framework, which is learning-based, for the scheduling problem in wireless multimedia networks. It is well known that machine learning is an effective method in the research field of artificial intelligence (AI). Having the inbuilt capacity of studying from outside environment and adapting to the practical environment, machine learning method is outstanding in its adaptivity. Furthermore, the well-developed framework of machine learning allows the designer to cast their attentions to the objectives and performance of the scheduler itself. Designers just need to connect the practical problem with the ingredients of machine learning, for example, translating scheduling objectives to the system's reward/cost. Other details like how to train the scheduler is not the concern of the designers. This feature makes machine learning very suitable for the complex scheduling problem in wireless multimedia networks. Actually, work in Refs 2 and 27 have realized that fixed and unadaptable scheduling schemes are incapable in complicated scheduling problem. Therefore, tunable constants are integrated into the key metrics so that the scheduler can adjust its strategy to some extent. Nonetheless, the learning abilities of these algorithms are very limited, and they should still be looked on as heuristic algorithms.

14.5 Learning-Based Packet Scheduling

This section discusses learning-based solution for the scheduling problem in wireless multimedia networks. We cast the practical scheduling problem into the framework of semi-Markov decision process (SMDP) and then employ the methodology of NDP, also known as reinforcement learning (RL), to solve the corresponding SMDP problem. Feature-based linear approximation architecture and temporal-difference (TD) learning technique are introduced here. Simulation experiments are also carried out to evaluate the proposed algorithm.

14.5.1 Semi-Markov Decision Process Formulation

Consider a general shared-channel wireless cellular network model and use a scheduler model as shown in Figure 14.1. The scheduler is assumed to have full knowledge of the scheduling system state, including the queuing delay of all packets, channel conditions of all sessions, and practical service rates of all sessions. Suppose there are M sessions in the system, and each session has a queuing buffer that can store K packets at the most.* The queuing delay of the kth packet in the mth session is denoted by τ_m^k. For the mth session, let R_m denote the maximum available channel rate, $\hat{r}_m$,

* Without loss of generalization, we assume that all queues can store same number of packets. A more practical assumption may be that the buffer length of the mth session is K_m in packets.

the nominal service rate; and r_m, the practical average service rate. We now connect the packet-scheduling problem with main ingredients of the SMDP as follows:

- *Stage*. The scheduling process is naturally divided into a series of stages by events $(\omega_0, \omega_1, \omega_2, \ldots)$, which represent packet arrivals and departures (we say that a packet "departs" if its service completes). Stages are indexed by integer $i = 0, 1, 2, \ldots$.
- *State*. A state x consists of the queuing status of the whole system, including the current queuing delay of all packets, wireless link conditions of all sessions, and actual service rates of all sessions. In particular, the state at stage i is given by

$$x_i = (\tau_m^k, R_m, r_m | m \in \{1, 2, \ldots, M\}, k \in \{1, 2, \ldots, K\}) \tag{14.2}$$

  where the actual service rates may be calculated by the approach in Ref. 26. The set of all possible states is called "state space," denoted by X, which is a $(K+2)M$ dimension space.
- *Decision*. At stage i, with the occurrence of event ω_i, the scheduler should make a decision u_i according to the state x_i. The decision is defined by the session number of the next transmission, that is, $u_i = m,\ m \in \{1, 2, \ldots, M\}$. The decision space is denoted by $U = \{1, 2, \ldots, M\}$. At stage i, if event ω_i is packet departure, because the wireless channel is idle, the decision subset $U(x_i)$ contains all sessions that can transmit.* If event ω_i is packet arrival, when the channel is busy (being used for transmission), $U(x_i)$ is a singleton; when the channel is idle (as no session can transmit, e.g., an empty system), the decision subset $U(x_i)$ contains all sessions that can transmit.
- *Policy*. Policy $\pi = (\mu_0, \mu_1, \mu_2, \ldots)$ is a sequence of decision functions, where decision function $\mu_i : X \mapsto U$ is a mapping from state space X to the decision space U. A policy is said to be "stationary" if $\forall i,\ \mu_i \equiv \mu$, which means the decision function does not change with stages. The space of stationary policy is denoted by $\prod_s$. We only consider stationary policies as candidate policies in this chapter, and without confusion, we simply use μ as the denotation of stationary policy. Note that, under policy μ, the decision at stage i is given by $u_i = \mu(x_i)$.
- *Cost function*. The definition of cost in packet-scheduling problem should take into account the three objectives: (i) QoS differentiation

* A session that "can transmit" is a session with nonempty queue and nonzero channel rate.

and guarantee, (ii) maximizing wireless bandwidth utilization, and (iii) fairness provisioning. Because cost function has major influence on the scheduler's performance, we place the construction of cost function in a single subsection, which is presented in Section 14.5.2. In the ith stage, the cost with state x_i and decision u_i is represented by $g(x_i, u_i)$. Under policy μ, the average cost of the system is found to be

$$v(\mu) = \lim_{N\to\infty} \frac{1}{t_N} \sum_{i=1}^{N} g(x_i, \mu(x_i)) \tag{14.3}$$

where N is the total number of stages and t_i is the time when event ω_i happens. The target of the SMDP problem is to find the optimal policy, which leads to the minimum average cost. A policy μ^* is said to be optimal if

$$v(\mu^*) \leq v(\mu) \tag{14.4}$$

for every other policy μ. Usually, the average cost under the optimal policy is denoted by v^*.

14.5.2 Construction of Cost Function

The definition of cost function directly relates to the performance of the scheduling algorithm. The cost function of interest should have the properties as follows:

- *Simplicity*. A simple form of cost function can significantly reduce the computation complexity of the whole scheduling algorithm.
- *Effectiveness*. Because the wireless packet scheduler is expected to concurrently address issues of QoS, bandwidth utility, and fairness, the definition of cost function should take into account all these three aspects.
- *Scalability*. In different practical applications, the requirements on every single objective may be quite different. Therefore, the structure of the cost function may have some tunable coefficients, which are determined by the network supervisor according to actual situations.

To start with, we consider a "draining problem," where given a number of packets in the system and assuming no more packet arrivals, the scheduler tries to arrange the order of packet transmissions so that the total cost is minimized. Draining problem is indeed a simplified version of the original scheduling problem. By simplification, it is more analyzable than the original one. Because the cost function should simultaneously consider the three objectives, it is natural to have the thought of dealing with only one objective at one time. Specifically, consider a draining problem with the following setup: there are M sessions in the system; the queue length of session m is K_m in packets, $m \in \{1, 2, \ldots, M\}$; at the starting moment, all the packets' queuing delay are supposed to be zero.

- Consider the following definition of cost function. At stage i, let

$$g(x_i, u_i) = \sum_{m=1}^{M} \tau_m^1 \mathbf{1}_{\{u_i=m\}} \tag{14.5}$$

  where τ_m^1 is the HOL delay of session m; $\mathbf{1}_A$ is the indicator function, satisfying that $\mathbf{1}_A = 1$ when A is true, $\mathbf{1}_A = 0$ when A is false. By this definition, the scheduler will always assign the session with the highest channel rate to transmit, so as to minimize the sum of all packets's delay. Obviously, this scheduling policy will result in maximizing bandwidth utilization.
- Consider the second definition of cost function. At stage i, let

$$g(x_i, u_i) = \sum_{m=1}^{M} W_m \tau_m^1 \mathbf{1}_{\{u_i=m\}} \tag{14.6}$$

  where W_m is the weight with session m. The existence of weights makes the scheduling rule different from the one defined in Equation 14.5. Consider two sessions m_1 and m_2 with weights W_{m_1} and W_{m_2}, respectively, and $W_{m_1} > W_{m_2}$. Suppose at a certain stage, $R_{m_1} = R_{m_2}$ and that both are the highest channel rates of all sessions. The scheduler will choose to assign m_1 for transmission first, because for the same period of waiting time, cost of m_1 is larger than m_2. Clearly, different weights give the session different priority level. Following the definition of cost in Equation 14.6, the scheduler can achieve QoS differentiation as well as maximizing channel utilization.
- Consider the third definition of cost function. At stage i, let

$$g(x_i, u_i) = \sum_{m=1}^{M} \frac{\hat{r}_m}{r_m} \mathbf{1}_{\{u_i=m\}} \tag{14.7}$$

  where $\hat{r}_m$ and r_m are, respectively, the nominal and actual average service rate of session m. In this definition, the scheduler is insensitive to the bandwidth utilization. It is whether every traffic gets its ordered service that concerns the scheduler. If all sessions have the same channel rate R, under the "fluid limit" [29], the proportion of service that session m receives from the scheduler is given by*

$$\varepsilon_m = \frac{\hat{r}_m}{\sum_{n=1}^{M} \hat{r}_n} \tag{14.8}$$

where the sum of service proportion $\sum_{m=1}^{M} \varepsilon_m = 1$. The result in Equation 14.8 shows that each traffic will get its fair share of service being proportional to the square root of its nominal service rate.

Although none of the preceding three cost functions can simultaneously address all the three performances, they give us helpful guideline to construct an appropriate one. Actually, by simply combining the three features, we obtain the structure of the cost function, which is found as

$$g(x_i, u_i) = \sum_{m=1}^{M} W_m F_m(r_m) \tau_m^1 \mathbf{1}_{\{u_{t_i}=m\}} \tag{14.9}$$

where W_m's are constants called the priority factors and F_m's are called the fairness factors, defined by

$$F_m(r_m) = \left(\frac{r_m}{\hat{r}_m}\right)^{-n_m} \qquad m \in \{1, 2, \ldots, M\} \tag{14.10}$$

where n_m is a constant with session m related to the fairness issue. The specific values of W_ms and n_ms are left to be determined according to the practical situations by the network supervisor. It is easy to see that the form of cost function defined in Equation 14.9 has the characteristics mentioned at the beginning of this subsection—simplicity, effectiveness, and scalability.

14.5.3 Neuro-Dynamic Programming Solution

Generally, the optimal policy μ^* of the SMDP problem can be obtained by solving the "Bellman optimality equation." In the average cost problem, Bellman optimality equation takes the following form [30]:

$$v^*\mathbf{E}\{\Delta t|x\} + h^*(x) = \mathbf{E}\{\min_{u\in U(x)} [g(x, u) + h^*(y)]\} \tag{14.11}$$

* A detailed discussion is skipped here; the key idea is to use Lagrangian multiplier method to solve the related constrained problem.

$$h^*(\hat{x}) = 0 \tag{14.12}$$

Here, $\mathbf{E}\{\cdot\}$ stands for expectation, Δt the duration of current stage, $U(x)$ the decision set at current time, and y the state at next stage. The value v^* is the optimal average cost, and $h^*(\cdot)$ is called the optimal differential cost, which has the following interpretation: when we operate the system under an optimal policy, then $h^*(x) - h^*(x')$ equals to the expectation of the difference of the total cost (over the infinite horizon) for a system with initial state x, compared with a system with initial state x'. State $\hat{x}$ is a recurrent state with zero differential cost (e.g., the state in which all queues are empty).

Once the optimal differential cost function $h^*(\cdot)$ is available, the optimal scheduling policy can be obtained by

$$\mu^*(x) = \arg\max_{u \in U(x)} [g(x, u) + h^*(y)] \tag{14.13}$$

As shown in Equation 14.13, the optimal policy is constituted by a series of optimal decisions at all stages, which maximizes the value of the right-hand side of Equation 14.13.

The desire for the exact solution of $h^*(\cdot)$ in Equation 14.11 is unrealistic because of the huge size of state space and the overwhelming computational requirement. Hence, we employ the methodology of NDP [31,32] to produce a near optimal solution. The central idea is to approach $h^*(\cdot)$ by a parameterized approximating function $\tilde{h}(\cdot, \theta)$. Online learning is carried out to continuously regulate the parameter vector θ and make $\tilde{h}(\cdot, \theta)$ more close to $h^*(\cdot)$.

14.5.3.1 Function Approximation Architecture

A common approximation architecture of NDP is shown in Figure 14.3. The architecture consists of two components—the feature extractor and function approximator. The feature extractor is used to produce a feature vector $f(x)$, which is capable of capturing the most important aspects of state x. Features are usually handcraft, based on human intelligence or experience. Our choice of feature vector will be presented shortly in this subsection.

The function approximation architectures could be broadly categorized into two classes—the nonlinear and linear structures. A typical nonlinear

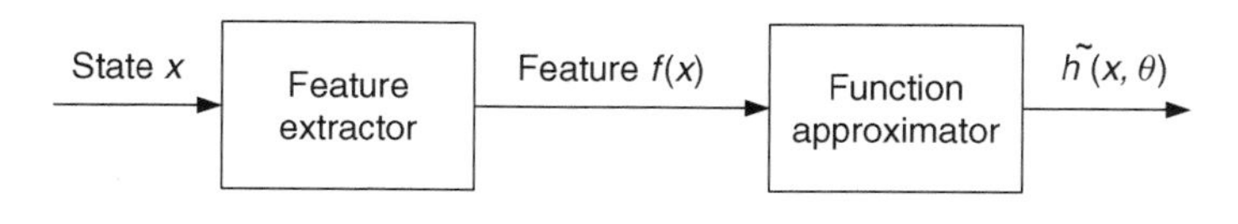

Figure 14.3 Architecture of NDP system.

architecture is the "multilayer perceptron" or "feedforward neural network" with a single hidden layer (see Ref. 31). In this architecture, the output approximating function $\hat{h}(\cdot, \theta)$ is the nonlinear combination of input feature vector $f(x)$ with internal tunable weight vector θ. This architecture is powerful because it can approximate arbitrary functions of $h^*(\cdot)$. The drawback, however, is that the dependence on θ is nonlinear and tuning can be time-consuming and unreliable.

Alternatively, a linear feature–based approximation architecture is much simpler and easier to implement, where the output approximating function $\hat{h}(\cdot, \theta)$ is the linear combination of the input feature $f(x)$, given by

$$\tilde{h}(x, \theta) = \theta^T f(x) \tag{14.14}$$

where the superscript T represents transpose. In this chapter, we choose linear approximation architecture in the NDP framework.

Because the linear combination of feature vector should approach the optimal differential average cost, the definition of feature vector explicitly relates to the definition of cost function. As Equation 14.9 shows, the cost at each stage is described by the weighted queuing delay of the packet being assigned to transmit. Considering this point, we can define one feature component for each packet in the system. In particular, for the kth packet in session m, its feature component is defined by

$$\xi_m^k = W_m F_m(r_m)\left(\tau_m^k + \frac{kl_m}{r_m}\right) \tag{14.15}$$

where l_m is the packet size of session m and kl_m/r_m has the interpretation of estimation of the remaining waiting time for the kth packet in session m. Note that by extending Equation 14.14, we get $\tilde{h}(x, \theta) = \sum_{k,m} \theta_m^k \xi_m^k$, where θ_m^k is the parameter variable associated with feature ξ_m^k. The total number of tunable parameters in $\hat{h}(\cdot, \theta)$ is equal to MK.

14.5.3.2 Online Parameter Tuning

After establishing the approximating architecture, we employ the technique of TD learning [32] to perform online parameter tuning. The algorithm starts with an arbitrary initial parameter vector θ_0 and improves the approximation as more and more state transitions are observed. At each stage, a greedy decision is made based on current x and θ. Specifically, the decision should be chosen by

$$u^* = \arg \min_{u \in U(x)} g(x, u) + \tilde{h}(\tilde{y}, \theta) \tag{14.16}$$

where $\tilde{y}$ is the estimation of state at next stage. The computation of $\tilde{y}$ is trivial if we simply assume that no packet arrives during the current stage. The update of parameter vector θ and scalar $\tilde{v}$ is carried out stage by stage according to

$$\theta_i = \theta_{i-1} + \gamma_i \, d_i \nabla_\theta \tilde{h}(x_{i-1}, \theta_{i-1}) \tag{14.17}$$

$$\tilde{v}_i = \tilde{v}_{i-1} + \eta_i [g(x_{i-1}, u_{i-1}) - \tilde{v}_{i-1} \Delta t_i] \tag{14.18}$$

where γ_i and η_i are small step size parameters, and the temporal difference d_i is found as

$$d_i = g(x_{i-1}, u_{i-1}) - \tilde{v}_{i-1} \Delta t_i + \tilde{h}(x_i, \theta_{i-1}) - \tilde{h}(x_{i-1}, \theta_{i-1}) \tag{14.19}$$

Note that we use the linear approximation structure as Equation 14.14, which leads to $\nabla_\theta \tilde{h}(x, \theta) = f(x)$ in Equation 14.17. Clearly, the complexity of gradient computation is significantly reduced in the linear feature–based approximation architecture.

14.5.4 Performance Evaluation

Because the methodology of NDP is the central idea of the proposed scheduling algorithm, we name the algorithm as NDP scheduling (NDPS). In this section, we evaluate the performance of NDPS by comparing it with two classical algorithms—CSDPS-WRR [1] and CIF-Q [13]. As listed in Table 14.1, there are five sessions in the simulation. All traffic flows are generated based on the models presented in Ref. 7. Link variations of data-2 and data-3 sessions are characterized by a six-state Markov channel model [5,6]. Transmission rates of link states s_1–s_6 are 100, 80, 60, 40, 20, 0 percent of the full channel rate (2 Mbps), respectively. The steady state probability vector for error pattern-1 is $\pi_1 = (0.52, 0.10, 0.10, 0.10, 0.10, 0.08)$, whereas the one for pattern-2 is $\pi_2 = (0.17, 0.17, 0.17, 0.17, 0.17, 0.15)$. During

Table 14.1 Properties of the Five Sessions in the Simulation

Session	*Audio*	*Video*	*Data-1*	*Data-2*	*Data-3*
Source model	IDP	2IRP	4IPP	4IPP	4IPP
Packet size	66B	188B	192B	192B	192B
Mean rate	22.4 kbps	0.19 Mbps	0.8 Mbps	0.8 Mbps	0.8 Mbps
Peak rate	64 kbps	0.4 Mbps	1.8 Mbps	1.8 Mbps	1.8 Mbps
Link variation	None	None	None	Pattern-1	Pattern-2

the 1200 s periods of our simulation, channel errors occur only in the first 400 s. The remaining error-free time is to demonstrate the long-term fairness property of NDPS. We empirically set $n_{\{1,2\}} = 1$, $n_{\{3,4,5\}} = 3$, and $W_{\{1,2\}} = 1$, $W_{\{3,4,5\}} = 10^{-2}$ in the simulation.

The simulation results are reported in Figures 14.4 through 14.6. The convergence process of the learning algorithm is plotted in Figure 14.4, from which we can see that the convergence rate of the learning algorithm is rapid in the simulation. For instance, it takes about 400 s for the scalar $\tilde{v}$ to reach a suitable near optimal value. Furthermore, when the conditions of scheduling environment change (i.e., channel conditions change at time 400 s), the learning algorithm could adaptively regulate the tunable parameters according to the actual environment.

We know from Figure 14.5a and 14.5b that both the average and maximum delay of audio and video sessions in NDPS are clearly smaller than in either of the other two algorithms. This demonstrates that NDPS is able to provide better QoS guarantee for real-time traffic flows. Furthermore, it is stressed that the satisfying QoS guarantee for audio and video sessions in NDPS is not obtained by sacrificing the throughput of the other sessions. Actually, as Figure 14.5c depicts, the three data sessions receive more service in NDPS. In particular, over the first 400 s, the total throughput of data sessions in NDPS (1.62 Mbps) is about 37 percent higher than in CSDP-WRR

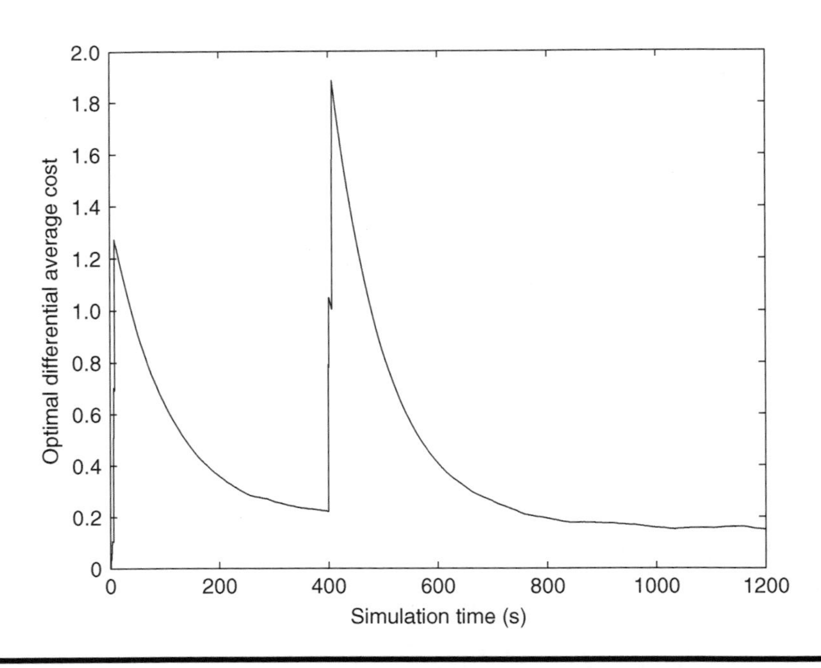

Figure 14.4 Convergence process of the learning-based scheduler.

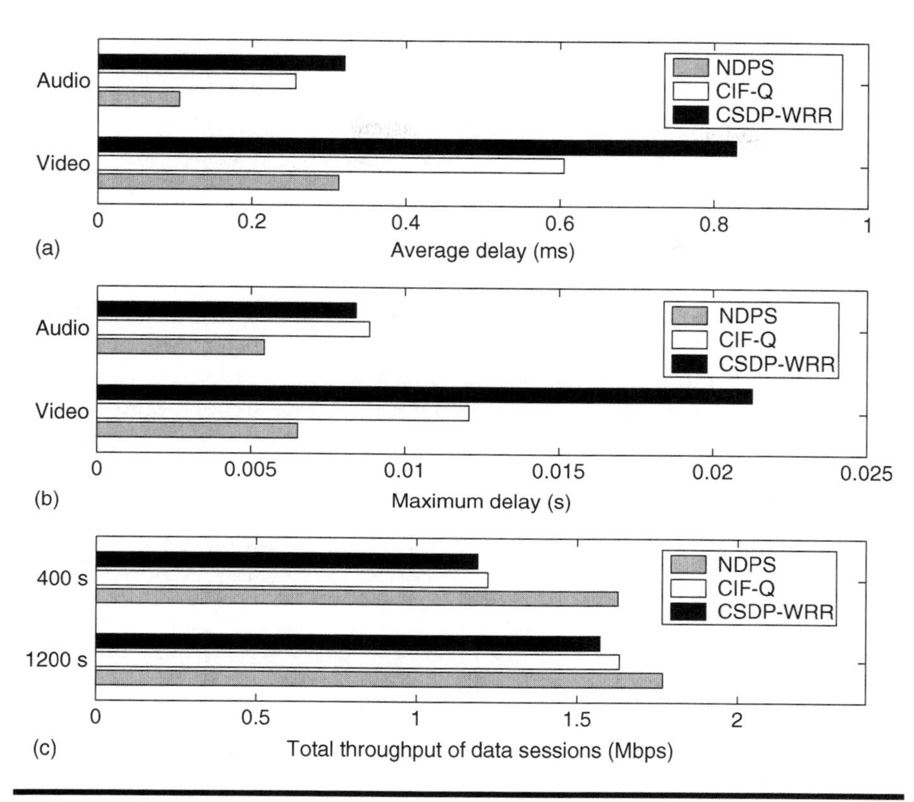

Figure 14.5 Performance comparison among CSDP-WRR, CIF-Q, and NDPS.

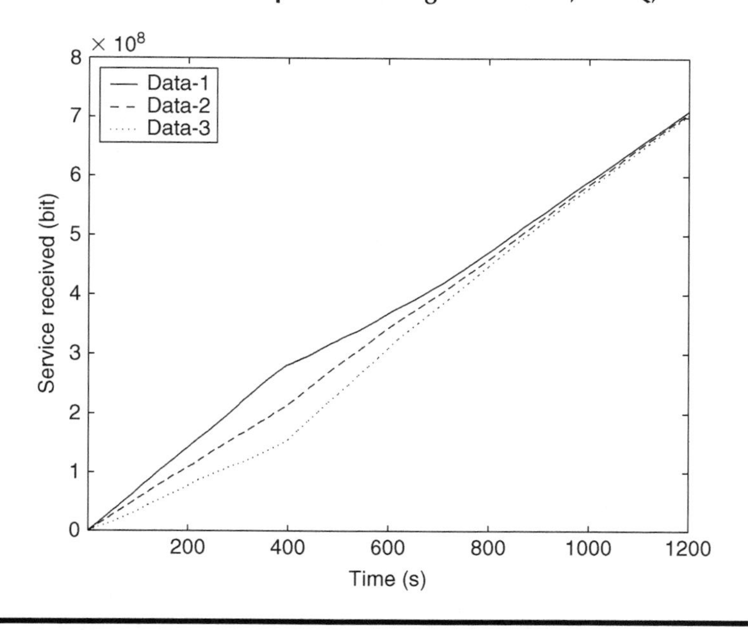

Figure 14.6 Service received by data sessions in NDPS.

(1.19 Mbps), and about 33 percent higher than in CIF-Q (1.22 Mbps). These figures indicate that NDPS has a considerable improvement in channel utilization. If we calculate the throughput over the whole simulation time (including a long error-free term), NDPS (1.77 Mbps) still has an advantage of about 13 and 9 percent over CSDP-WRR (1.57 Mbps) and CIF-Q (1.63 Mbps), respectively.

To demonstrate the short- and long-term fairness properties of NDPS, we present the service received by the data sessions over the whole simulation time in Figure 14.6. It is observed that the curves diverge moderately in the error-prone period and then converge gradually in the error-free period. We give the interpretation as follows. To provide short-term fairness, NDPS does not force the leading sessions to give up all their leads, but just makes them degrade gracefully. Thus, data-1 session receives relatively more service in the first 400 s. However, when the system becomes error free, the lagging sessions will gradually get back their lags, and finally all data sessions obtain the same throughput. This observation validates that NDPS offers both short- and long-term fairness guarantee in the sense that it provides short- and long-term fairness for error-free sessions and error-prone sessions, respectively. This exactly coincides with the spirit of CIF properties proposed in Ref. 13.

14.6 Future Directions

Although packet-scheduling problem has been analyzed and discussed for a long period of time and many feasible solutions have been put forward, there are still some issues that have not been addressed and are potential future research topics.

14.6.1 Inaccuracy of System Information

In most existing work on packet scheduling, it is assumed that the scheduler has full knowledge of the system status, including the instant channel rate and downlink and uplink packet queuing delay. However, a more realistic assumption is that these types of system status information could be imprecise or not immediately available for the scheduler. For example, in the uplink scheduling, if there are packets to transmit, a mobile host should first send a request message to the base station to apply for the bandwidth. On receiving this message, the scheduler will add a virtual packet to the virtual queue of the corresponding session. As a result, there is a difference of queuing delay between the virtual queue in the base station and the real queue in the mobile host. In addition, for the information of instant channel rate, there are always deviation and latency between the real channel conditions and the ones reported to the scheduler. Therefore, one of the possible future works on wireless scheduling is to take into account this imperfect system information and develop scheduling mechanisms that are robust to them.

14.6.2 Collaboration with Admission Control and Congestion Control

Because the wireless channel is bandwidth constrained and highly variable, to provide guaranteed QoS to mobile hosts, packet-scheduling algorithms should be supported by appropriate admission control and congestion control schemes. The mechanisms of admission control, congestion control, and packet scheduling are all important components that constitute the architecture of wireless resource management. The performance of a scheduling algorithm is tightly dependent on the other two components. Topics on how a packet-scheduling algorithm is affected by the admission control and congestion control policies in a wireless environment, and how to integrate admission control, scheduling, and congestion control in an optimal way require and deserve further study.

14.6.3 Compatibility with Standards

In recent years, the development of broadband wireless networks is very rapid, especially the broadband wireless access networks (BWANs) such as Worldwide Interoperability for Microwave Access (WiMAX) and Wireless Fidelity (WiFi) systems. As an important platform to transmit multimedia flows, BWANs can be viewed as a specific form of multimedia network. Some industry standards of BWANs, for example, IEEE 802.16 standards, left the QoS-based packet-scheduling algorithms (which determine the uplink and downlink bandwidth allocation) undefined. Considering this point, developing specific scheduling algorithms, which are compatible with existing standards, has great practical meaning. Some related works have been done in Ref. 33.

References

1. Bhagwat, P., A. Krishna, and S. Tripathi, Enhancing throughput over wireless LANs using channel state dependent packet scheduling, *Proceeding of IEEE INFOCOM96*, pp. 1133–1140, San Francisco, March 1996.
2. Borst, S. and P. Whiting, Dynamic rate control algorithms for HDR throughput optimization, *Proceeding of IEEE INFOCOM*, Anchorage, AK, 2001.
3. Gilbert, E. N., Capacity of a burst-noise channel, *Bell Syst. Tech. J.*, 39(9), 1253–1265, 1960.
4. Elliott, E. O., Estimates of error rates for codes on burst-noise channels, *Bell Syst. Tech. J.*, 42(9), 1977–1997, 1963.

5. Wang, H. S. and N. Moayeri, Finite-state Markov channel: A useful model for radio communication channels, *IEEE Trans. Veh. Tech.*, 44, 163–171, 1995.
6. Zhang, Q. and S. A. Kassam, Finite-state Markov model for Rayleigh fading channels, *IEEE Trans. Commun.*, 47, 1688–1692, 1999.
7. Baugh, C. R. and J. Huang, Traffic Model for 802.16 TG3 MAC/PHY Simulations, IEEE 802.16 working group document, 2001-03-02. http://wirelessman.org/tg3/contrib/802163c-01_30r1.pdf.
8. Fattah, H. and C. Leung, An overview of scheduling algorithms in wireless multimedia networks, *IEEE Wireless Commun.*, 9(5), 76–83, 2002.
9. Blake, S., D. Black, M. Carlson, E. Davies, Z. Wang, and W. Weiss, An architecture for differentiated services, RFC2475, August 1998.
10. Wroclawski, J., The use of RSVP with IETF integrated services, RFC2210, September 1997.
11. Kelly, F., Charging and rate control for elastic traffic, *Eur. Trans. Telecommun.*, 8, 33–37, 1997.
12. Bertsekas, D. and R. Gallager. *Data Networks*, 2nd edition. Upper Saddle River, NJ, Prentice-Hall, 1993.
13. Eugene Ng, T. S., I. Stoica, and H. Zhang, Packet fair queuing algorithms for wireless networks with location-dependent errors, *Proceeding of IEEE INFOCOM98*, pp. 1103–1111, San Francisco, March 1998.
14. Cao, Y. and V. O. K. Li, Scheduling algorithm in broad-band wireless networks, *Proc. IEEE*, 89(1), 76–87, 200l.
15. Wischhof, L. and J. W. Lockwood, Packet Scheduling for Link-Sharing and Quality of Service Support in Wireless Local Area, technical report, WUCS-01–35, November 2001.
16. Parekh, A. K. and R. G. Gallage, A generalized processor sharing approach to flow control in integrated services networks: the single-node case, *IEEE/ACM Trans. Networking*, 1(3), 344–357, 1993.
17. Shreedhar, M. and G. Varghese, Efficient fair queuing using deficit round robin, *IEEE/ACM Trans. Networking*, 4(3), 375–385, 1995.
18. Goyal, P., H. M. Vin, and H. Cheng, Start-Time Fair Queuing: A Scheduling Algorithm for Integrated Services Packet Switching Networks, Technical Report CS-TR-96-02, Department of Computer Science, The University of Texas, Austin, TX, vol. 1, 1996.
19. Bennett, J. C. R. and H. Zhang, WF^2Q: worst-case fair weighted fair queuing, *Proceedings of IEEE INFOCOM*, San Francisco, CA, pp. 120–128, 1996.
20. Lu, S., V. Bharghavan, and R. Sirkant, Fair scheduling in wireless packet networks, *IEEE/ACM Trans. Networking*, 7(4), 473–489, 1999.
21. Ramanathan, P. and P. Agrawal, Adapting packet fair queuing algorithms to wireless networks, *Proceedings of ACM MobiCom*, Dallas, TX, pp. 1–9, 1998.
22. Lu, S., T. Nandagopal, and V. Bharghavan, Design and analysis of an algorithm for fair service in error-prone wireless channels, *Wireless Networks*, 6(4), 323–343, 2000.
23. Park, D., H. Seo, H. Kwon, and B. G. Lee, Wireless packet scheduling based on the cumulative distribution function of user transmission rates, *IEEE Trans. Commun.*, 53(11), 1919–1929, 2005.
24. Knopp, R. and P. A. Humblet, Information capacity and power control in single-cell multiuser communications, *Proc. IEEE Int. Conf. Commun.*, 1, C331–C335, 1995.

25. Tsybakov, B. S., File transmission over wireless fast fading downlink, *IEEE Trans. Inf. Theory*, 48(8), 2323–2337, 2002.
26. Jalali, A., R. Padovani, and R. Pankaj, Data throughput of CDMA-HDR: A high efficiency-high data rate personal communication wireless system, *Proc. IEEE Veh. Technol. Conf.*, 3, 1854–1858, 2000.
27. Liu, X., E. K. P. Chong, and N. B. Shroff, Opportunistic transmission scheduling with resource-sharing constraints in wireless networks, *IEEE J. Select Areas Commun.*, 19(10), 2053–2064, 2001.
28. Andrews, M., K. Kumaran, K. Ramanan, S. Stolyar, R. Vijayakumar, and P. Whiting, Providing quality of service over a shared wireless link, *IEEE Comm. Mag.*, 39, 150–154, 2001.
29. Liu, P., R. Berry, and M. L. Honig, A fluid analysis of utility-based wireless scheduling policies, *Proceedings of IEEE Conference on Decision and Control (CDC)*, pp. 3283–3288, Orlando, FL, 2004.
30. Bertsekas, D. P., *Dynamic Programming and Optimal Control*, vol. 1 and 2. Belmont, MA: Athena Scientific, 1995.
31. Bertsekas, D. P. and J. N. Tsitsiklis, *Neuro-Dynamic Programming*. Belmont, MA: Athena Scientific, 1996.
32. Sutton, R. S., Learning to predict by the methods of temporal differences, *Mach. Learning*, 3, 9–44, 1988.
33. Wongthavarawat, K. and A. Ganz, Packet scheduling for QoS support in IEEE 802.16 broadband wireless access systems, *Int. J. Commun. Syst.*, 16(1), 81–96, 2003.

Chapter 15

The Peak-to-Average Power Ratio in Orthogonal Frequency Division Multiplexing Wireless Communication Systems

Tao Jiang, Laurence T. Yang, and Yan Zhang

Contents

15.1 Introduction 506
15.2 Characteristics of OFDM Signals 508
15.3 PAPR of the OFDM Signals 509
15.4 Distribution of the PAPR in OFDM Systems 510
15.5 PAPR Reduction Techniques in OFDM Systems 513
15.5.1 Clipping and Filtering 514
15.5.2 Coding Schemes 515
15.5.3 PTS and SLM 517

15.5.4 Nonlinear Companding Transforms 519
15.5.5 TR and TI 523
15.6 Criteria of the PAPR Reduction in OFDM Systems 524
15.7 PAPR Reduction for WiMAX and MIMO-OFDM Systems 526
15.7.1 PAPR Reduction for WiMAX 526
15.7.2 PAPR Reduction for MIMO-OFDM Systems 527
15.8 Conclusion 528
References 528

Orthogonal frequency division multiplexing (OFDM) has been widely used in many wireless communication systems due to its attractive technical features. However, some challenging issues still remain unresolved in OFDM wireless communication systems, and one of them is its sensitivity to the peak-to-average power ratio (PAPR) of transmitted signals. This chapter emphasizes on summarizing most of the existing schemes of PAPR reduction, which is based on a complete investigation of the characteristics of OFDM signals including the distribution of the PAPR in OFDM wireless communication systems. It provides an effective criteria for PAPR reduction. Moreover, it also discusses the PAPR reduction for Worldwide Interoperability for Microwave Access (WiMAX) and multiple-input multiple-output (MIMO)-OFDM broadband wireless communication systems.

15.1 Introduction

As an attractive technology for wireless communications, OFDM, which is one of the multicarrier modulation (MCM) techniques, offers a considerable high bandwidth and power efficiency. In particular, OFDM splits up a wideband into many narrow and equally spaced subchannels, and thus a wideband frequency-selective fading channel is converted into narrowband flat fading subchannels. To achieve high spectral efficiency, OFDM allows the subchannels to overlap in the frequency domain, and the subchannels are made orthogonal to one another. Moreover, OFDM introduces the guard interval and cyclic prefix to remove the intercarrier and intersymbol interference. With the development of digital signal processing and very large-scale integrated circuits, OFDM can be realized by inverse fast Fourier transform (IFFT) and fast Fourier transforms easily with low complexity. Currently, OFDM has been used in many wireless communication standards such as IEEE 802.11—a standard for wireless local area networks.

However, some challenging issues still remain unresolved in the design of OFDM wireless communication systems. One of the major problems is its sensitivity to the PAPR of transmitted signals. As a result, with a high PAPR, an OFDM receiver's detection efficiency becomes very sensitive to the nonlinear devices used in its signal processing loop, such as digital-to-analog converters (DACs), mixers, and high-power amplifiers (HPAs), which may severely impair system performance due to induced spectral regrowth and detection efficiency degradation. For example, most radio systems employ the HPA in the transmitter to obtain sufficient power transmission and it is usually operated at or near the saturation region to achieve the maximum output power efficiency, and thus the memoryless nonlinear distortion due to the high PAPR of the input signals will be introduced into the communication channels. Moreover, the nonlinear characteristic of the HPA is very sensitive to the variation in signal amplitudes. If the HPA is not operated in the linear region with large power back off, it is impossible to keep the out-of-band power below the specified limits. This situation leads to inefficient amplification and very expensive transmitters.

Therefore, it is of great importance to reduce the PAPR in OFDM wireless communication systems to utilize the technical features of the OFDM. Recently, various schemes have been proposed to deal with the PAPR problem for OFDM wireless communication systems. These approaches include clipping [1], coding schemes [2,3], nonlinear companding transforms [4–6], tone reservation (TR) and tone injection (TI) [7], partial transmission sequence (PTS) [8], and selective mapping (SLM) [9]. Among these methods, clipping is the simplest and most widely used. However, clipping causes additional clipping noise that degrades the system performance. The inherent error control capability and simplicity of implementation make the coding method more promising for practical OFDM systems' design. However, the main disadvantage of this method is the good performance of the PAPR reduction at the cost of coding rate loss. One of the most attractive schemes is nonlinear companding transform due to its good system performance including PAPR reduction and bit error rate (BER), low implementation complexity, and no bandwidth expansion. An effective PAPR reduction technique should be given the best trade-off between the capacity of PAPR reduction and transmission power, data rate loss, implementation complexity, and BER performance.

This chapter is organized as follows. First, the distribution of PAPR based on the characteristics of the OFDM signals, which is often expressed in terms of complementary cumulative distribution function (CCDF), is investigated. Then, five typical techniques of PAPR reduction is analyzed, and the criteria for PAPR reduction in OFDM wireless communication systems is proposed. Finally, the issue of PAPR in some broadband wireless systems correlative with OFDM technology including WiMAX and MIMO-OFDM systems is discussed briefly.

15.2 Characteristics of OFDM Signals

Figure 15.1 depicts a typical transmitter in OFDM wireless communication systems. First, an encoder is chosen to encode the binary input data prior to a baseband modulator such as rectangular quadrature amplitude modulation (QAM). Then, the pilot symbols for channel estimation should be inserted before the symbols are converted into parallel. Subsequently, the IFFT process converts the frequency symbols into the time domain, and the symbols are once again converted into the serial format. The resulting sequence is converted into an analog signal using a DAC, and then passed on to the radio frequency (RF) modulation stage. Finally, the resulting RF-modulated signal is transmitted to the receiver using a transmit antenna. In this chapter, we set the size of the guard interval to zero for brevity.

Let a block of N symbols $\mathbf{X} = \{X_k, k = 0, 1, \ldots, N-1\}$ be formed with each symbol modulating one of a set of subcarriers $\{f_k, k = 0, 1, \ldots, N-1\}$, where N is the number of subcarriers. The N subcarriers are chosen to be orthogonal, that is, $f_k = k\,\Delta f$, where $\Delta f = 1/(NT)$ and T is the original symbol period. Therefore, the complex envelope of the transmitted OFDM signals can be written as

$$x(t) = \frac{1}{\sqrt{N}} \sum_{k=0}^{N-1} X_k e^{j2\pi f_k t} \quad 0 \leq t \leq NT \tag{15.1}$$

where $j = \sqrt{-1}$.

Suppose that the input data stream is statistically independent and identically distributed (i.i.d.), that is, the real part $\Re e\{x(t)\}$ and the imaginary part $\Im m\{x(t)\}$ are uncorrelated and orthogonal. Therefore, based on the central limit theorem, when N is considerably large, the distribution of both $\Re e\{x(t)\}$ and $\Im m\{x(t)\}$ approaches Gaussian distribution with zero mean and variance $\sigma^2 = E[|\Re e\{x(t)\}|^2 + |\Im m\{x(t)\}|^2]/2$, where $E[x]$ is the expected value of x [5].

Figure 15.2 describes the distributions of OFDM signals and its amplitude, respectively. Note that the measurements of OFDM signals shown in

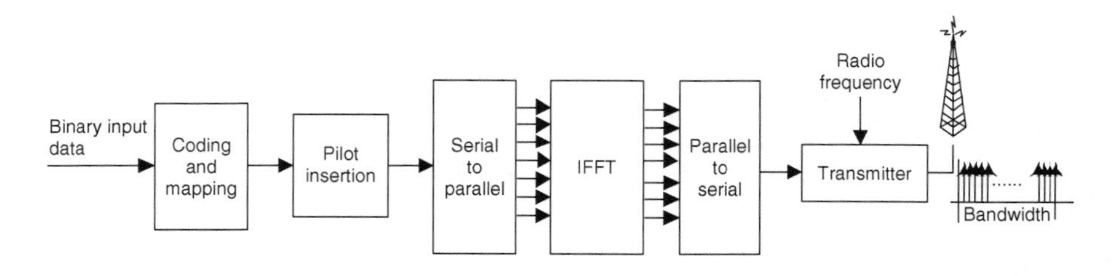

Figure 15.1 Typical transmitter in OFDM systems.

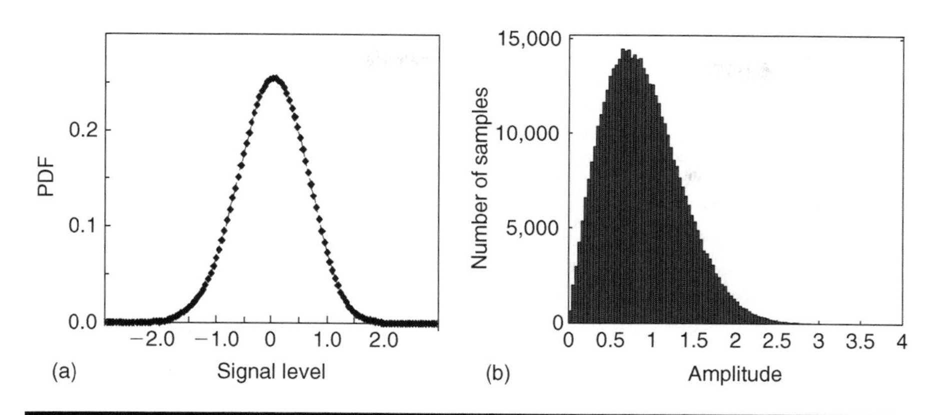

Figure 15.2 Distribution of OFDM signals modulated with 16-QAM. (a) PDF of the OFDM signals, (b) histogram for the amplitude of the OFDM signal.

Figure 15.2a fit quite well to the Gaussian normal distribution. According to the central limit theorem, OFDM signals with large N become Gaussian distributed with probability density function (PDF) as [4]

$$\Pr\{x(t)\} = \frac{1}{\sqrt{2\pi}\sigma} \exp\left\{-\frac{[x(t)]^2}{2\sigma^2}\right\} \tag{15.2}$$

where σ is the variance of $x(t)$.

Moreover, the Rayleigh nature of OFDM original signals' amplitude can be easily observed from Figure 15.2b and the corresponding PDF has been given as [7]

$$\Pr(r) = 2r \cdot \exp(-r^2) \tag{15.3}$$

where r is the amplitude of OFDM signals.

15.3 PAPR of the OFDM Signals

The PAPR of OFDM signals $s(t)$ is usually defined as the ratio between the maximum instantaneous power and its average power, namely,

$$\mathrm{PAPR}[x(t)] = \frac{\max\limits_{0\leq t\leq NT} [|x(t)|^2]}{P_{\mathrm{av}}} \tag{15.4}$$

where P_{av} is the average power of $x(t)$, and it can be computed in the frequency domain because IFFT is a (scaled) unitary transformation.

Note that, if N is large, an OFDM system usually does not employ pulse shaping because the power spectral density of the band-limited OFDM signal is approximately rectangular. Thus, the amplitude of OFDM RF signals can be expressed as $|\Re e\{x(t) \cdot e^{j2\pi f_c t}\}|$, where f_c is the carrier frequency and $f_c \gg \Delta f$. Therefore, the peak of RF signals is equivalent to that of the complex baseband signals. This chapter considers only the PAPR of the baseband OFDM signals.

In principle, we are often more concerned with the PAPR of the continuous-time OFDM signals because the cost and power dissipation of the analog components often dominates. To better approximate the PAPR of continuous-time OFDM signals, the OFDM signals samples are obtained by oversampling L times. L-times oversampled time-domain samples are LN-point IFFT of the data block with $(L-1)N$ zero padding. Therefore, the oversampled IFFT output can be expressed as

$$x_n \triangleq x\left(\frac{nT}{L}\right) = \frac{1}{\sqrt{N}} \sum_{k=0}^{N-1} X_k e^{j\frac{2\pi nk}{LN}} \quad 0 \leq n \leq LN-1 \tag{15.5}$$

It has been shown that $L \geq 4$ is sufficient to get accurate PAPR results [7]. The PAPR computed from the L-times oversampled time-domain OFDM signal samples can be defined as

$$\text{PAPR}[x(t)] = \frac{\max\limits_{0\leq n\leq NL-1}\left[|x_n|^2\right]}{E\left[|x_n|^2\right]} \tag{15.6}$$

15.4 Distribution of the PAPR in OFDM Systems

The distribution of the PAPR bears stochastic characteristics in OFDM systems, which can often be expressed in terms of CCDF. Figure 15.3 shows the distribution of the OFDM signals' PAPR with $N = 256$ and different oversampling factors $L = 1, 2, 4, 16$. As shown, the largest PAPR increase happens from $L = 1$ to $L = 2$, but does not increase significantly after $L = 4$ [7]. Therefore, all the theoretical results of the PAPR are based on $L = 4$ in this chapter.

The CCDF itself can be used to estimate the bounds for the minimum number of redundancy bits required to identify the PAPR sequences and evaluate the performance of any PAPR reduction schemes. We can also determine a proper output back-off of the HPA to minimize the total degradation according to CCDF. Moreover, we can directly apply the distribution of PAPR to calculate the BER and estimate achievable information rates. In practice, we usually adjust these design parameters jointly according to simulation results. Therefore, if we can use an analytical expression to

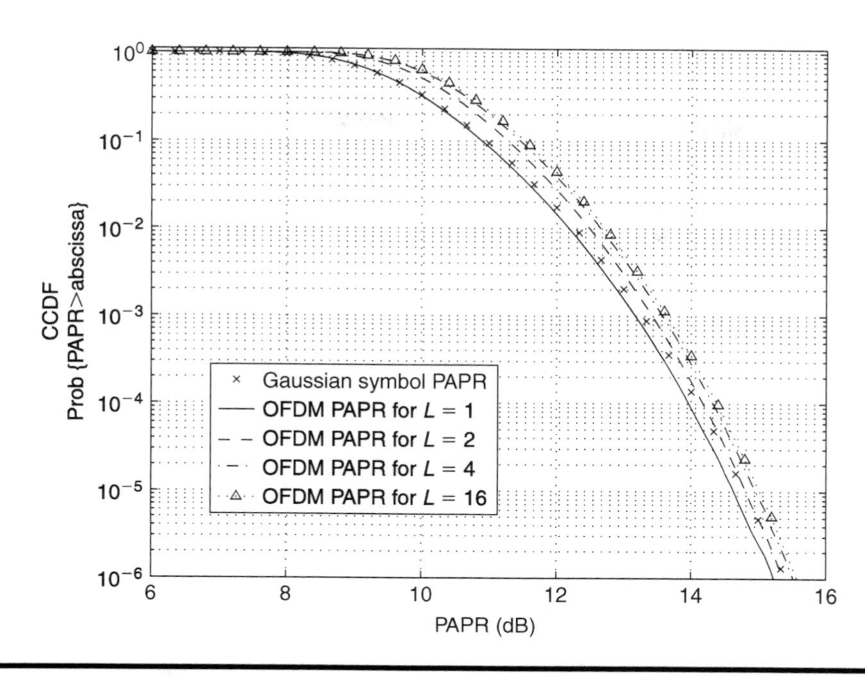

Figure 15.3 Distribution of PAPR of OFDM signal samples oversampled by different *L*. (From Tellado, J., Peak to average power ratio reduction for multicarrier modulation, PhD thesis, University of Stanford, 1999. With permission.)

accurately calculate the PAPR distribution for OFDM systems, it can greatly simplify the system design process. Therefore, it is of great importance to accurately identify the PAPR distribution in OFDM systems.

Recently, some researchers have reported on the determination of the upper and lower bounds of the PAPR distribution according to the Rayleigh distribution and Nyquist sampling rate. When the number of the subcarriers N is relatively small, the CCDF expression of the PAPR of OFDM signals can be written as [10]

$$\text{Prob}\{\text{PAPR} > \gamma\} = 1 - (1 - e^{-\gamma})^N \tag{15.7}$$

In Ref. 11, an empirical approximation expression of the CCDF of the PAPR in OFDM systems has been given as

$$\text{Prob}\{\text{PAPR} > \gamma\} = 1 - (1 - e^{-\gamma})^{2.8N} \tag{15.8}$$

For a relatively large N, the lower and upper bounds of the distribution of the PAPR have been proposed in Ref. 12, which were developed based on the earlier works in conjunction with some approximations and parameters obtained through simulations. In Ref. 13, bound analysis has also been developed for both independent and dependent subcarriers in OFDM systems. For independent subcarriers, a generic path for bounding practical constellations was used and discussed. For dependent subcarriers, some theoretical bounds of distributions of the PAPR have been obtained in terms of the Euclidian distance distributions, in which the focus was mainly on binary codes such as Bose–Chaudhuri–Hocquenghem (BCH) codes. However, the lower and upper bounds can offer little help in characterizing the distribution of the PAPR in practical OFDM systems. In fact, the accurate statistical distribution of the PAPR for generic OFDM system is what we require.

In Ref. 14, an analytical PAPR CCDF expression has been developed, which is based on the level-crossing rate approximation of the peak distribution along with the exact distribution because the envelope of an OFDM signal can always be considered as an asymptotically Gaussian process in a band-limited OFDM system. In fact, the theoretical results obtained in Ref. 14 were based on the conditional probability of the peak distribution of the OFDM signals when the reference level $\tilde{\gamma}$ is given. When the constraint provides a lower bound of $\tilde{\gamma} \geq 0.71$, the effect on the accuracy of the PAPR distribution can be numerically evaluated. Indeed, for high $\tilde{\gamma}$, the conditional probability that the peak of the OFDM signals exceeds $\tilde{\gamma}$ may be very small. In this case, the expression of the PAPR CCDF, as shown in Ref. 14, can be simplified as

$$\text{Prob}\{\text{PAPR} > \gamma\} \cong 1 - \exp\left\{-Ne^{-\gamma}\sqrt{\frac{\pi}{3}\gamma}\right\} \tag{15.9}$$

If the range of the PAPR of interest is large, the distribution can be further simplified without the loss of accuracy. In Ref. 14, it has also been shown that the statistical distribution of the PAPR of the OFDM signals is not so sensitive to the increase in the number of subcarriers.

In coded OFDM systems, it has been proven that the complex envelope of the coded OFDM signals can converge weakly to a Gaussian random process if the number of subcarriers goes to infinity [15]. In Ref. 15, a simple approximation of the CCDF of PAPR has been developed by employing the extreme value theory, and the expression can be written as

$$\text{Prob}\{\text{PAPR} > \gamma\} \cong 1 - \exp\left\{-Ne^{-\gamma}\sqrt{\frac{\pi}{3}\log N}\right\} \tag{15.10}$$

Similarly, with the help of the extreme value theory for chi-squared-2 process, a more accurate analytical expression of the CCDF of PAPR for adaptive OFDM systems with unequal power allocation to subcarriers has

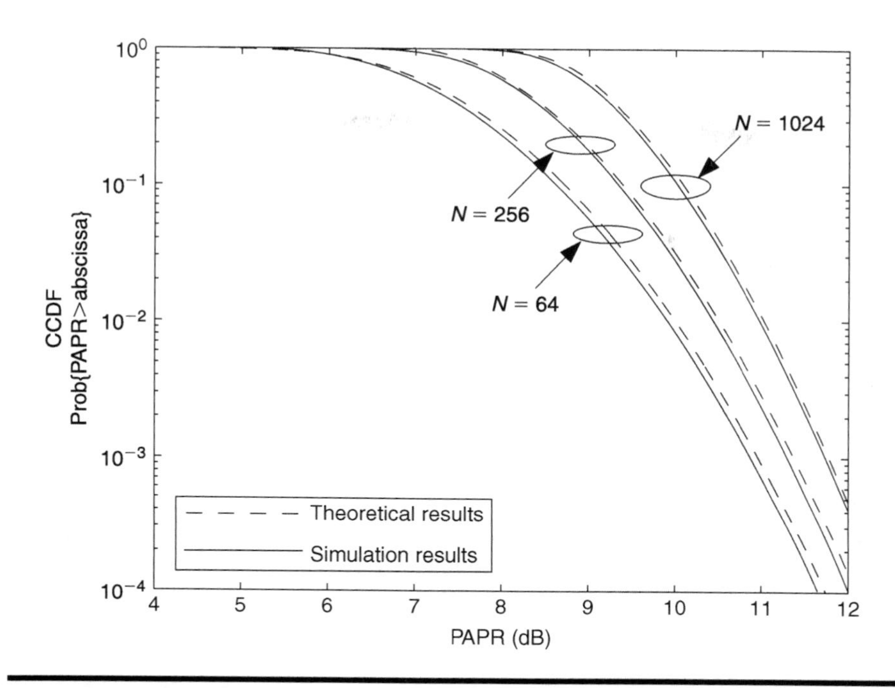

Figure 15.4 PAPR distributions with unequal power allocation to subcarriers. (From Jiang T., M. Guizani, H. Chen, W. Xiang, and Y. Wu, *IEEE Trans. Wireless Communication*, 2008. With permission.)

been derived in Ref. 28.

$$\text{Prob}\{\text{PAPR} > \gamma\} = 1 - \exp\left\{-e^{-\gamma}\sqrt{\frac{\pi\gamma \sum_{k=0}^{N-1}(2k+1-N)^2\sigma_k}{\sum_{k=0}^{N-1}\sigma_k}}\right\} \qquad (15.11)$$

where σ_k denotes the transmission power allocated to the kth subcarrier. Figure 15.4 shows the difference in the distribution of PAPR between the simulation results and the analytical results in Ref. 28, in which an incremental power based on water-filling rule is allocated to the subcarriers while normalizing the transmission power to 1.

15.5 PAPR Reduction Techniques in OFDM Systems

Figure 15.5a shows an example of an OFDM signal with $N = 128$. It is clear that the peak power occurs when independent complex variables are added with the same phase. In other words, the maximum of the PAPR in

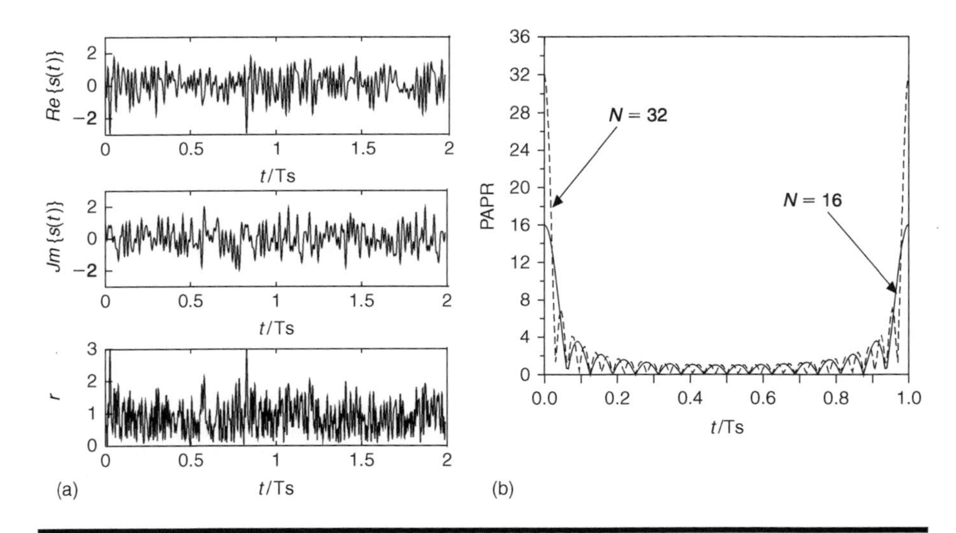

Figure 15.5 Some OFDM signals with high PAPR. (a) An OFDM signal with average power of 1, (b) PAPR of OFDM signal is equal to the number of subcarriers.

OFDM systems may be equal to the number of subcarriers N, and some examples have been shown in Figure 15.5b. Therefore, it is important to reduce the PAPR. Otherwise, when OFDM signals with large PAPR get into the nonlinear region of the HPA, signal distortion including intermodulation among the subcarriers and out-of-band radiation will be caused. This section mainly discusses five typical techniques for PAPR reduction in OFDM wireless communication systems.

15.5.1 Clipping and Filtering

The earliest and the simplest technique of PAPR reduction is to basically clip the parts of the signals that are outside the allowed region [1]. For example, using HPA with saturation level below the signal span will automatically cause the signal to be clipped. For amplitude clipping, that is,

$$C(x) = \begin{cases} x, & |x| \leq A \\ A, & |x| > A \end{cases} \tag{15.12}$$

Generally, clipping is performed at the transmitter. However, the receiver needs to estimate the clipping that has occurred and compensate the received OFDM symbol accordingly. Typically, at most one clipping occurs per OFDM symbol, and thus the receiver has to estimate two parameters: location and size of the clip. However, it is difficult to get this information. Therefore, the clipping method introduces both in-band distortion and out-of-band radiation into OFDM signals, which degrades the system performance including BER and spectral efficiency.

Although filtering can reduce out-of-band radiation after clipping, it cannot reduce the in-band distortion. However, clipping may cause some peak regrowth so that the signal after clipping and filtering will exceed the clipping level at some points. To reduce peak regrowth, a repeated clipping-and-filtering operation can be used to obtain a desirable PAPR at the cost of a computational complexity increase.

As improved clipping methods, peak windowing schemes also attempt to minimize the out-of-band radiation by using narrowband windows such as Gaussian window to attenuate peak signals.

15.5.2 Coding Schemes

When N signals are added with the same phase, they produce a peak power that is N times the average power. Of course, not all code words result in a bad PAPR. Therefore, a good PAPR reduction can be obtained when some measures are taken to reduce the occurrence probability of the same phase of the N signals, which is the key idea of the coding schemes.

A simple block coding scheme was introduced by Jones et al. [2], and its basic idea is to map 3 bit data into a 4 bit code word by adding a simple odd parity code (SOBC) at the last bit across the channels. Later, Wulich [17] applied the cyclic coding (CC) to reduce the PAPR. In 1998, Fragicomo et al. [18] proposed an efficient simple block code (SBC) to reduce the PAPR of OFDM signals. Subsequently, complement block coding (CBC) and modified complement block coding (MCBC) schemes were proposed to reduce the PAPR without the restriction of frame size [3,19].

CBC and MCBC are more attractive due to their flexibility on choosing the coding rate, frame size, and low implementation complexity. CBC and MCBC utilize the complementary bits that are added to the original information bits to reduce the probability of the peak signal's occurrence, as shown in Figure 15.6. To compare, some results of the PAPR reduction obtained with different coding schemes have been shown in Table 15.1, in which the number of subblocks is two and the coding rate $R = 3/4$ for MCBC. About 3 dB PAPR reduction can be obtained when the coding rate $R > (N-2)/N$ by using CBC with long frame size. It is also shown that the PAPR reductions obtained with CBC when the coding rate $R = (N-1)/N$ are almost the same as that when $R < (N-1)/N$. In addition, when the coding rate is 3/4, more than 3 dB, more PAPR reduction can be obtained using MCBC than the other schemes with any frame size. The flexibility in coding rate choice and low complexity make the proposed CBC and MCBC schemes attractive for OFDM wireless communication systems with large frame sizes and high coding rates.

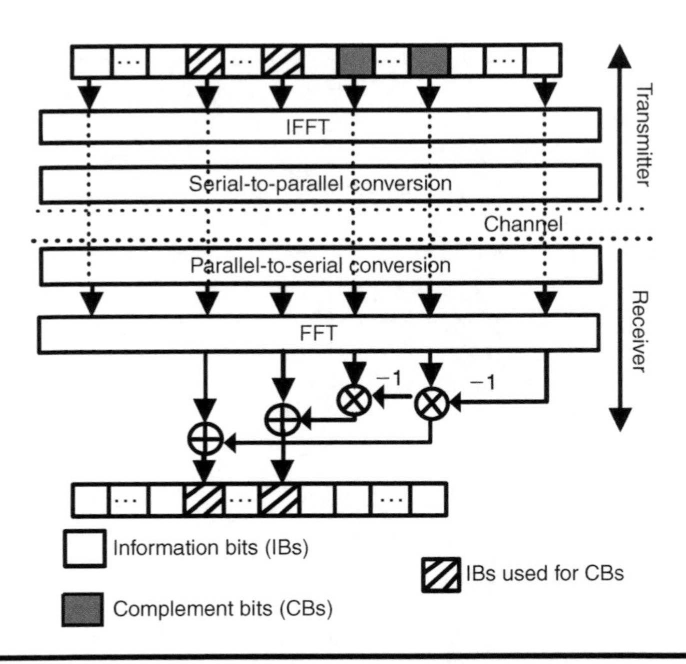

Figure 15.6 The key idea of the CBC and MCBC to reduce PAPR. (From Jiang, T. and G. X. Zhu, *IEEE Commun. Mag.* 43(9), S17–S22, 2005. With permission.)

Table 15.1 PAPR Reduction Comparison with Different Coding Schemes

			PAPR Reduction (dB)				
N	*n*	*R*	*CBC*	*SBC*	*MCBC*	*SOPC*	*CC*
4	1	3/4	3.56	3.56	–	3.56	3.56
8	1	7/8	2.59	2.52	–	2.52	3.66
	2	3/4	2.67	3.72	2.81	($R=7/8$)	($R=3/4$)
16	1	15/16	2.74	1.16	–	1.18	3.74
	2	7/8	2.74	2.52	–	($R=15/16$)	($R=3/4$)
	3	13/16	2.74	–	–		
	4	3/4	2.74	2.98	3.46		
32	1	31/32	1.16	0.55	–	0.58	–
	2	15/16	1.16	1.16	–	($R=31/32$)	
	3	29/32	2.75	–	–		
	4	7/8	2.50	2.51	–		
	5	27/32	2.75	–	–		
	8	3/4	2.75	3.00	3.45		

Notes: N, total size of the code word; n, number of complemental bits; $R=(N-n)/N$, coding rate.

15.5.3 PTS and SLM

In a typical OFDM system with the PTS approach to reduce the PAPR, the input data block in $\mathbf{X}$ is partitioned into M disjoint subblocks, which are represented by the vectors $\{\mathbf{X}^{(m)}, m = 0, 1, \ldots, M-1\}$ [8] as shown in Figure 15.7. Therefore, we can obtain

$$\mathbf{X} = \sum_{m=0}^{M-1} \mathbf{X}^{(m)} \tag{15.13}$$

where $\mathbf{X}^{(m)} = [X_0^{(m)} X_1^{(m)} \ldots X_{N-1}^{(m)}]$ with $X_k^{(m)} = X_k$ or 0 $(0 \le m \le M-1)$. In general, for the PTS scheme, the known subblock partitioning methods can be classified into three categories [8]: adjacent partition, interleaved partition, and pseudorandom partition. Then, the subblocks $\mathbf{X}^{(m)}$ are transformed into M time-domain partial transmit sequences.

$$\mathbf{x}^{(m)} = [x_0^{(m)} x_1^{(m)} \ldots x_{LN-1}^{(m)}] = \mathrm{IFFT}_{LN \times N}[\mathbf{X}^{(m)}] \tag{15.14}$$

These partial sequences are independently rotated by phase factors $\mathbf{b} = \{b_m = e^{j\theta_m}, m = 0, 1, \ldots, M-1\}$. The objective is to optimally combine the M subblocks to obtain the time-domain OFDM signals with the lowest PAPR

$$\tilde{\mathbf{x}} = \sum_{m=0}^{M-1} b_m \mathbf{x}^{(m)} \tag{15.15}$$

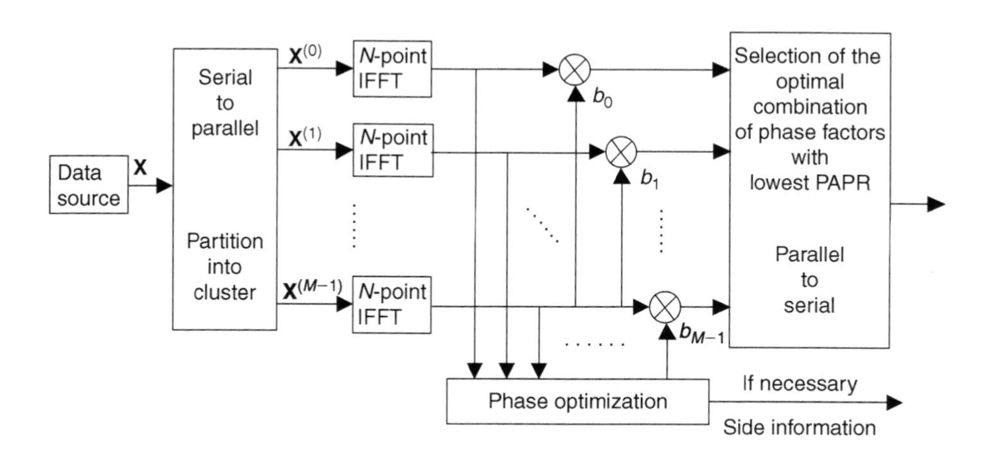

Figure 15.7 Block diagram of PTS technique. (From Muller, S. H. and J. B. Huber, *IEE Electron. Lett.*, 33(5), 36–69, 1997. With permission.)

Therefore, there are two important issues to be solved in the PTS: high computational complexity for searching the optimal phase factors and the overhead of the optimal phase factors as side information need to be transmitted to the receiver for the correct decoding of the transmitted bit sequence.

Suppose that there are W phase angles to be allowed, thus b_m can have the possibility of W different values. Therefore, there are W^M alternative representations for an OFDM symbol. To reduce the searching complexity and avoid/reduce the usage of side information, many extensions of the PTS have been developed recently [16,20]. In Ref. 16, the authors proposed a novel scheme, which is based on a nonlinear optimization approach named as simulated annealing, to search for the optimal combination b_m of phase factors with low complexity. In general, PTS needs M IFFT operations for each data block, and the number of the required side information bits is $\lfloor M \log_2^W \rfloor$, where $\lfloor x \rfloor$ denotes the smallest integer that does not exceed x.

Similarly, in SLM, the input data sequences are multiplied by each of the phase sequences to generate alternative input symbol sequences. Each of these alternative input data sequences is made for the IFFT operation, and then the one with the lowest PAPR is selected for transmission [9]. A block diagram of the SLM technique is depicted in Figure 15.8. Each data block is multiplied by V different phase factors, each of length N, $\mathbf{B}_v = [b_{v,0}, b_{v,1}, \ldots, b_{u,N-1}]^T$ $(v = 0, 1, \ldots, V-1)$, resulting in V different data blocks. Thus, the vth phase sequence after multiplied is $\mathbf{X}^v = [X_0 b_{v,0}, X_1 b_{v,1}, \ldots, X_{N-1} b_{v,N-1}]^T$ $(v = 0, 1, \ldots, V-1)$. Therefore, the OFDM signals can be written as

$$x^v(t) = \frac{1}{\sqrt{N}} \sum_{n=0}^{N-1} X_n b_{v,n} e^{j2\pi f_n t} \quad 0 \le t \le NT \quad v = 1, 2, \ldots, V-1 \quad (15.16)$$

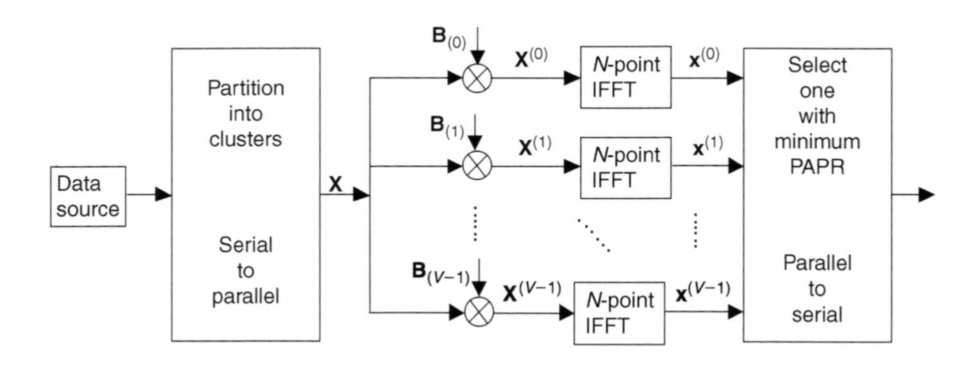

Figure 15.8 Block diagram of SLM technique. (From Bauml, R. W., R. F. H. Fisher, and J. B. Huber, *IEE Electron Lett.*, 32(22), 2056–2057, 1996. With permission.)

Among the data blocks $\mathbf{X}^v$ ($v = 0, 1, \ldots, V-1$), only the one with the lowest PAPR is selected for transmission and the corresponding selected phase factors $b_{v,n}$ also should be transmitted to the receiver as side information. For implementation of SLM OFDM systems, the SLM technique needs V IFFT operations and the number of required bits as side information is $\lfloor \log_2^V \rfloor$ for each data block. Therefore, the ability of PAPR reduction in SLM depends on the number of phase factors V and the design of the phase factors. Some extension to SLM has also been proposed to reduce the computational complexity and the number of bits for side information transmission. For example, an SLM scheme without explicit side information was proposed in Ref. 21.

Although PTS and SLM are important probabilistic schemes for PAPR reduction, it is already known that SLM can produce multiple time-domain OFDM signals that are asymptotically independent, whereas the alternative OFDM signals generated by PTS are interdependent. PTS divides the frequency vector into some subblocks before applying the phase transformation. Therefore, some of the complexity of the several full IFFT operations can be avoided in PTS, so that it is more advantageous than SLM if the amount of computational complexity is limited [8]. It is also demonstrated that the PAPR reduction in PTS performs better than that of SLM. However, the size of required bits of the side information in PTS is larger than that of SLM.

15.5.4 Nonlinear Companding Transforms

Another well-known method to reduce the PAPR of OFDM signals is the nonlinear companding transform due to its good system performance, low complexity, and no bandwidth expansion.

The first nonlinear companding transform is the μ-law companding, which is based on the speech-processing algorithm μ-law, and it has shown better performance than that of the clipping method [22]. μ-law mainly focuses on enlarging signals with small amplitude and keeping peak signals unchanged, and thus it increases the average power of the transmitted signals and possibly results in exceeding the saturation region of the HPA to make the system performance worse.

In fact, the nonlinear companding transform is also a special clipping scheme. The differences between the clipping and nonlinear companding transform can be summarized as follows: (1) Clipping method deliberately clips large signals when the amplitude of the original OFDM signals is larger than the given threshold, and thus the clipped signals cannot be recovered at the receiver. However, nonlinear companding transforms compand the original OFDM signals using the strict monotone increasing function. Therefore, the companded signals at the transmitter can be recovered correctly through the corresponding inversion of the nonlinear transform function at the receiver. (2) Nonlinear companding transforms enlarge the small signals while compressing the large signals to increase the immunity of small signals from noise, whereas the clipping method does not change the small signals. Therefore, the clipping method suffers from three major problems: in-band distortion, out-of-band radiation, and peak regrowth after digital-to-analog conversion. As a result, the system performance degradation due to clipping may not be optimistic. However, nonlinear companding transforms can operate well with good BER performance while keeping good PAPR reduction [23].

The design criteria for nonlinear companding transform has also been given in Ref. 23. Because the distribution of the original OFDM signals has been known, such as Rayleigh distribution of the OFDM amplitudes shown in Equation 15.3, we can obtain the nonlinear companding transform function through theoretical analysis and derivation according to the desirable distribution of the companded OFDM signals. For example, we transform the amplitude of the original OFDM signals into the desirable distribution with its PDF $f_{s_c}(s) = ks + b\ (k < 0,\quad b > 0)$. Therefore, the nonlinear transform function can be derived as

$$C(x) = \sqrt{6}\sigma\left[1 - \exp\left(-\frac{x^2}{2\sigma^2}\right)\right] \tag{15.17}$$

Based on this design criteria, two types of nonlinear companding transform, which are based on error function and exponential function, respectively, have been proposed in Refs 4 and 5. In Ref. 4, the error transform function has been given as

$$C(x) = \frac{2k_1}{\sqrt{\pi}}\int_0^{k_2 x} e^{-t^2}\,dt \tag{15.18}$$

The parameters k_1 and k_2 are positive numbers, which play the most important role in the transformation of the input signal. Figure 15.9 illustrates the relationship between the input and output with different k_1 and k_2. It is clear that parameter k_1 is the factor of controlling the amplitude of the companded signals, whereas k_2 is the factor of controlling the companded form in terms of nonlinearity.

Similarly, the exponential transform function has been proposed in Ref. 5.

$$C(x) = \mathrm{sgn}(x)\sqrt[d]{\alpha\left[1 - \exp\left(-\frac{x^2}{\sigma^2}\right)\right]} \tag{15.19}$$

where $\mathrm{sgn}(x)$ is the sign function and σ^2 the variance of the original OFDM signals. The positive constant α determines the average power of the output

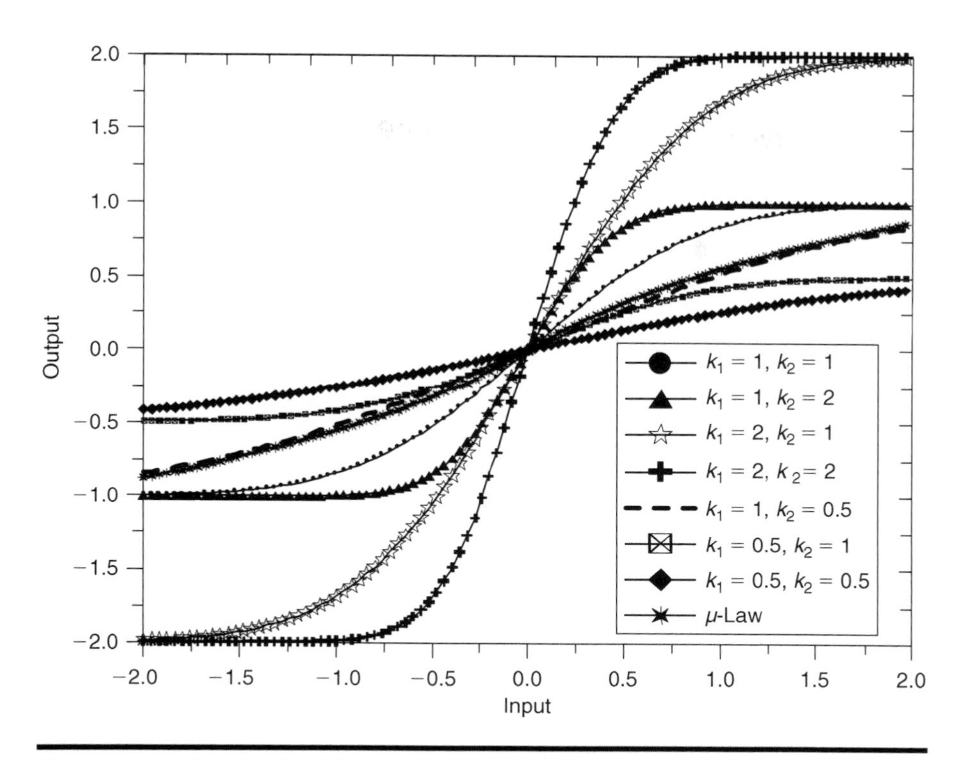

Figure 15.9 The relationship between input and output using different nonlinear transform functions. (From Jiang, T. and G. X. Zhu, *IEEE Trans. Broadcasting,* 50(3), 342–346, 2004. With permission.)

signals. Figure 15.10 shows the exponential companding function $C(x)$ with degree d as a parameter. It is also clear that the companded signals have uniformly distributed amplitudes and powers, respectively, for the cases $d=1$ and $d=2$. When $d\geq 2$, the proposed function $C(x)$ can compress large input signals and expand small signals simultaneously. As seen from Figure 15.10, the differences between exponential companding functions are ignorable when $d\geq 8$.

It is known that the original OFDM signals have a very sharp, rectangularlike power spectrum as shown in Figure 15.11. This good property will be affected by the PAPR reduction schemes, for example, slower spectrum roll-off, more spectrum side lobes, and higher adjacent channel interference. Many PAPR reduction schemes cause the generation of spectrum side lobes, but the nonlinear companding transforms cause less spectrum side lobes. As seen in Figure 15.11, error and exponential (with degree $d=2$) companding transforms have much less impact on the original power spectrum compared to the μ-law companding scheme. This is the major reason that the error and exponential companding schemes not only enlarge

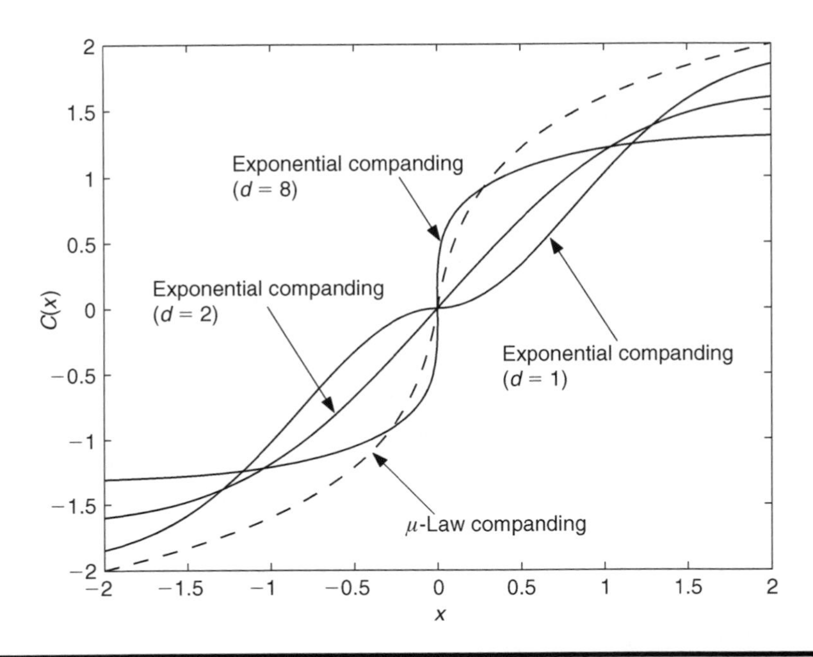

Figure 15.10 The exponential companding function $C(x)$. (From Jiang, T., Y. Yang, and Y. Song, *IEEE Trans. Broadcasting*, 51(2), 244–248, 2005. With permission.)

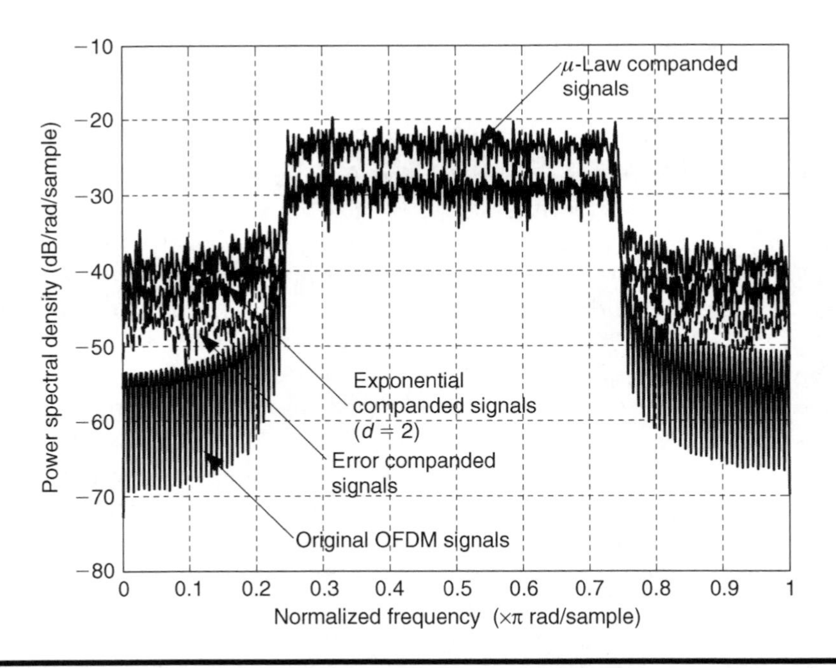

Figure 15.11 The spectrums of original OFDM signals and companded signals.

the small amplitude signals but also compress the large amplitude signals, although maintaining the average power unchanged by properly choosing parameters, which can increase the immunity of small amplitude signals from noise. However, the μ-law companding transform increases the average power level and therefore requires a larger linear operation region in the HPA.

Nonlinear companding transform is a type of nonlinear process that may lead to significant distortion and performance loss by companding noise. Companding noise can be defined as the noises that are caused by the peak regrowth after DAC to generate in-band distortion and out-of-band noise by the excessive channel noises magnified after inverse nonlinear companding transform, etc. For out-of-band noise, it needs to be filtered and oversampled. For in-band distortion and channel noises that are magnified, they need iterative estimation. Unlike additive white Gaussian noise (AWGN), companding noise is generated by a known process and that can be recreated at the receiver and subsequently be removed. In Ref. 6, the framework of an iterative receiver has been proposed to eliminate companding noise for companded and filtered OFDM systems.

15.5.5 TR and TI

TR and TI are two efficient techniques to reduce the PAPR of OFDM signals [7]. Figure 15.12 describes the block diagram of TR and TI, in which the key idea is that both the transmitter and the receiver reserve a subset of tones for generating PAPR reduction signals c. Note that these tones are not used for data transmission.

In TR, the objective is to find the time-domain signal c to be added to the original time-domain signal x to reduce the PAPR. Let $\{\mathbf{c} = c_n | n = 0, 1, \ldots, N-1\}$ denote complex symbols for TR at reserved tones. Thus, the data vector changes to $\mathbf{x} - \mathbf{c}$ after TR processing, and this results in a new modulated OFDM signal as

$$\tilde{\mathbf{X}} = \text{IFFT}(x + c) = \mathbf{X} + \mathbf{C} \tag{15.20}$$

where $\mathbf{C} = \text{IFFT}(\mathbf{c})$. Therefore, the main objective of the TR is to find out the proper $\mathbf{c}$ to make the vector $\tilde{\mathbf{X}}$ with low PAPR. To find the value of $\mathbf{c}$,

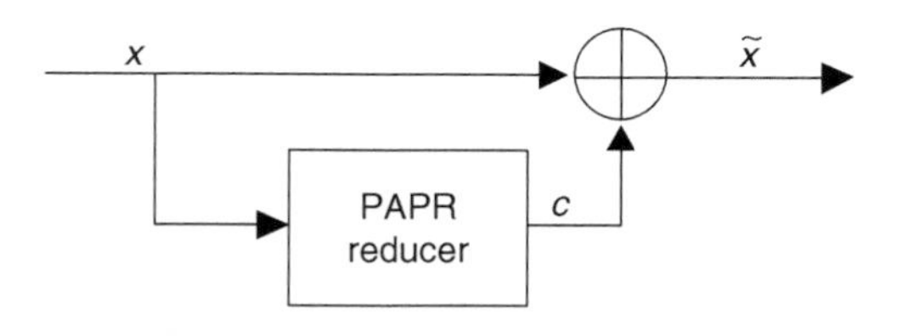

Figure 15.12 Block diagram of TR/TI approaches for PAPR reduction.

we must solve a convex optimization problem that can easily be cast as a linear programming problem.

Similarly, TI also uses an additive correction to optimize **C** in Equation 15.20. The basic idea of TI is to extend the constellation, and thus the same data point corresponds to multiple possible constellation points. One option is to replicate the original shaded constellation into several alternative ones. Therefore, **C** is a translation vector such that $\mathbf{C} = (\tilde{\mathbf{X}})\mathrm{mod}(\mathbf{X})$. Note that TI does not require the extra side information, and the receiver only needs to know how to map the redundant constellations on the original one. An alternative strategy is to move the constellation points by applying an FFT on the clipped time signals, and the same operations are repeated until all the constellation points are within the specified boundaries and the PAPR specification of the time signal is satisfied [24].

The TI technique is more problematic than the TR technique because the injected signal occupies the frequency band as the information-bearing signals. Moreover, the alternative constellation points in the TI technique have an increased energy and the implementation complexity increases for the computation of optimal translation vector.

15.6 Criteria of the PAPR Reduction in OFDM Systems

In the preceding analysis, we find that most of the existing solutions still have some drawbacks and the obvious one is the trade-off between PAPR reduction and some factors such as bandwidth. The criteria of the PAPR reduction is to find the approach that can reduce the PAPR largely and can simultaneously keep the good performance in terms of the following factors.

1. *High capability of PAPR reduction.* It is the primary factor to be considered in selecting the PAPR reduction technique with as few harmful side effects such as in-band distortion and out-of-band radiation.
2. *Low average power.* When the average power of the original OFDM signals is increased, the PAPR of the OFDM signals also can be reduced. However, it requires a larger linear operation region in the HPA and thus resulting in the degradation of BER performance.
3. *Low implementation complexity.* Generally, techniques with high complexity can exhibit better ability of PAPR reduction. However, in practice, both time and hardware requirements for the PAPR reduction should be minimal.
4. *No bandwidth expansion.* The bandwidth is a rare resource in wireless communication systems. The bandwidth expansion directly

results in the data code rate loss due to side information (such as the phase factors in PTS and complementary bits in CBC). Moreover, when the side information are received with errors, some ways of protection such as channel coding needs to be employed. As a consequence, the loss in data rate is further increased due to added information.

5. *No BER performance degradation.* The objective of PAPR reduction is to obtain better system performance including BER than that of the original OFDM system. Therefore, in practice, all the methods that have an increase in BER at the receiver should be paid more attention in practice. Moreover, if the side information is received with errors at the receiver, it may also result in whole erroneous data frame, and thus the BER performance is reduced.
6. *No additional power.* The design of a wireless system should always take into consideration the power efficiency. If an operation of the technique, which reduces the PAPR, needs additional power, it degrades the BER performance when the transmitted signals are normalized back to the original power signal.
7. *No spectral spillage.* Any PAPR reduction technique cannot destroy OFDM's attractive technical features such as immunity to the multipath fading. Therefore, the spectral spillage should be avoided in the PAPR reduction.
8. *Other factors.* More attention should be paid to the effect of the nonlinear devices used in signal processing loop in the transmitter, such as DACs, mixers, and HPAs because the PAPR reduction mainly avoids nonlinear distortion due to these memoryless devices introduced into the communication channels. At the same time, the cost of these nonlinear devices is also an important factor in designing the PAPR reduction scheme.

In Table 15.2, we summarize the five typical PAPR reduction techniques and find that nonlinear companding transform is a promising scheme due

Table 15.2 Comparison of Different PAPR Reduction Technologies

	Power Increase	*Implementation Complexity*	*Bandwidth Expansion*	*BER Degradation*
Clipping	No	Low	No	Yes
Coding	No	Low	Yes	No
PTS/SLM	No	High	Yes	No
NCR	No	Low	No	No
TR/TI	Yes	High	Yes	No

Note: NCR, nonlinear companding transform.

to large PAPR reduction, good BER performance, low complexity, and no bandwidth expansion.

15.7 PAPR Reduction for WiMAX and MIMO-OFDM Systems

15.7.1 PAPR Reduction for WiMAX

Recently, WiMAX has received more attention due to its scalability in both radio access and network architecture and high-throughput broadband connection over long distance. As standardized, WiMAX defines two different PHYs in its Point-to-Multipoint (PMP) model that can be used in conjunction with the Media Access Control (MAC) layer to provide a reliable end-to-end link: OFDM and OFDM access (OFDMA). Therefore, WiMAX also suffers from high PAPR. This section discusses only the PAPR reduction for OFDMA communication systems.

It is clear that each user modulates its data to some subcarriers in each data frame at the transmitter, and thus the PAPR can be reduced according to these schemes mentioned earlier in an OFDMA uplink. However, the PAPR reduction is more complicated in an OFDMA downlink than that in an OFDMA uplink. If the downlink PAPR reduction is achieved by some approaches that have been designed for OFDM, each user has to process the whole data frame and then demodulate the assigned subcarriers to extract their own information. Thus, it introduces additional processing for each user at the receiver. Therefore, we mainly describe some modifications of the PAPR reduction techniques for an OFDMA downlink as follows:

1. *PTS/SLM for PAPR reduction in OFDMA*. PTS and SLM techniques can easily be modified for PAPR reduction in an OFDMA downlink. For PTS, subcarriers assigned to one user are grouped into one or more subblocks, and then PTS can be applied to subblocks for all users. As side information, the selected phase factor for each subblock can also be embedded into the prereserved subcarrier in each subblock. Note that, the prereserved subcarrier does not undergo the phase rotation in each subblock. Similarly, some of the subcarriers can be used to transmit side information when the modified SLM is applied to reduce the PAPR for OFDMA. All users use the information carried by these subcarriers to obtain the phase sequence at the transmitter, and thus the data for each user can be recovered correctly.
2. *TR for PAPR reduction in OFDMA*. In the TR technique for OFDMA [25], the symbols in peak reduction subcarriers are optimized for the whole data frame in both amplitude and phase. At the same time, some peak reduction subcarriers are assigned to each user in the TR

for PAPR reduction. It has also been proven that the highest PAPR reduction gain can approach 6.02 dB and the average PAPR reduction to 1.47 dB by the TR method under the worst case in WiMAX systems.

15.7.2 PAPR Reduction for MIMO-OFDM Systems

By using multiple antennas at both the transmitter and receiver sides, MIMO technology can exploit the system's spatial dimension capability to improve the wireless link performance and system capacity. However, MIMO-OFDM systems also suffer from the inherent high PAPR because it is based on the OFDM, which entails limitations on applications.

1. *PTS/SLM for PAPR reduction in MIMO-OFDM systems.* It is easy to make some extensions of PTS/SLM for PAPR reduction in MIMO-OFDM systems. Suppose that input data symbols are converted into a number of parallel streams and the PTS/SLM scheme can be applied to each stream.
2. *STBC for PAPR reduction in MIMO-OFDM systems.* Space time block coded (STBC) is an effective approach to reduce PAPR for MIMO-OFDM systems. In Figure 15.13, 2×2 MIMO-OFDM system

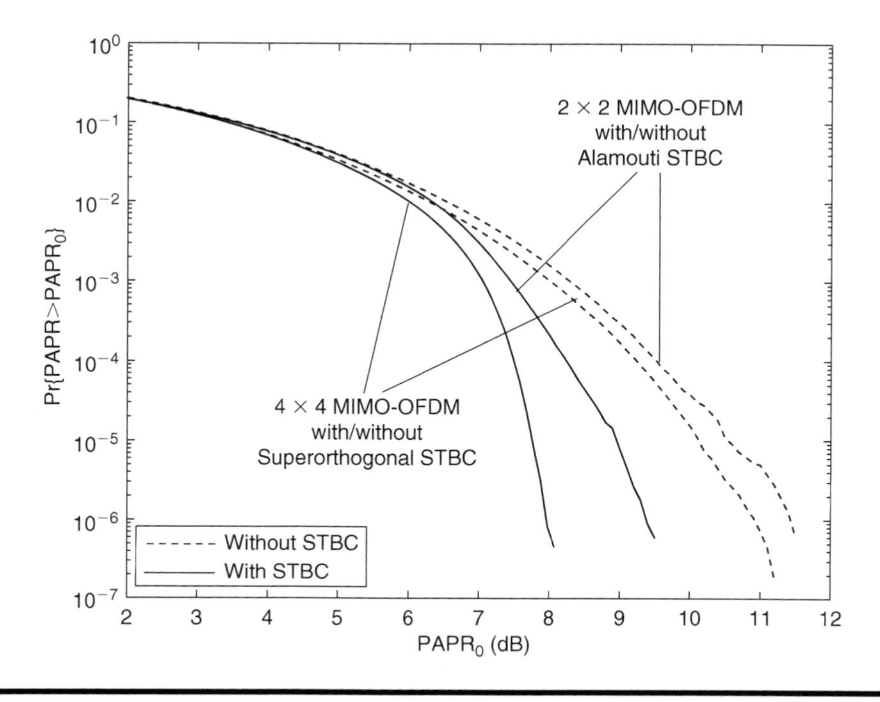

Figure 15.13 Comparisons of CCDFs in MIMO-OFDM systems with different STBCs and without STBC.

with Alamouti STBC [26] and 4×4 MIMO-OFDM system with super-orthogonal STBC [27] are considered. It is clear from each group curves (dashed and solid curves) to find that the STBC design can reduce the PAPR for MIMO-OFDM system. It can also be seen that about 2 and 3 dB of gains are observed at $\text{CCDF} = 10^{-6}$ in 2×2 and 4×4 antennas, respectively. Note that, these gains have been obtained without sacrificing spectral efficiency and computational complexity increase.

3. *TI/TR for PAPR reduction in MIMO-OFDM systems.* In TI/TR for PAPR reduction in MIMO-OFDM systems, the positions of the reserved subcarriers are the same for all transmit antennas, and they should be known to both the transmitter and the receiver. In each transmit antenna, the PAPR reduction is performed independently, and the data symbols carried in the reserved subcarriers can be discarded at the receiver.

15.8 Conclusion

OFDM is a very attractive technique for wireless communications due to its spectral efficiency and channel robustness. One of the serious drawbacks of OFDM wireless systems is that the composite transmit signal can exhibit a very high PAPR when the input sequences are highly correlated. This chapter described several important aspects, as well as provided a mathematical analysis, including the distribution of the PAPR in OFDM wireless systems. Five typical techniques to reduce the PAPR have been analyzed, all of which have the potential to provide substantial reduction in PAPR at the cost of loss in data rate, transmit signal power increase, BER performance degradation, computational complexity increase, and so on. Compared to the other methods, nonlinear companding transform is a promising and better solution for OFDM wireless communication systems. We also showed that it is possible to reduce the PAPR for WiMAX and MIMO-OFDM systems.

References

1. O'Neill, R. and L. B. Lopes, Envelope Variations and Spectral Splatter in Clipped Multicarrier Signals, *Proc. IEEE PIMRC'95*, IEEE, Toronto, Canada, pp. 71–75, Sept. 1995.
2. Jones, A. E., T. A. Wilkinson, and S. K. Barton, Block coding scheme for reduction of peak-to-average envelope power ratio of multicarrier transmission systems, *IEE Electronics Letters*, 30(8), 2098–2099, 1994.
3. Jiang, T. and G. X. Zhu, Complement block coding for reduction in peak-to-average power ratio of OFDM signals, *IEEE Communications Magazine*, 43(9), S17–S22, 2005.

4. Jiang, T. and G. X. Zhu, Nonlinear companding transform for reducing peak-to-average power ratio of OFDM signals, *IEEE Transactions on Broadcasting*, 50(3), 342–346, 2004.
5. Jiang, T., Y. Yang, and Y. Song, Exponential companding transform for PAPR reduction in OFDM systems, *IEEE Transactions on Broadcasting*, 51(2), 244–248, 2005.
6. Jiang, T., W. Yao, P. Guo, Y. Song, and D. Qu, Two novel nonlinear companding schemes with iterative receiver to reduce PAPR in multi-carrier modulation systems, *IEEE Transactions on Broadcasting*, 52(2), 268–273, 2006.
7. Tellado, J., Peak to average power ratio reduction for multicarrier modulation, PhD thesis, University of Stanford, Stanford, CA, 1999.
8. Muller, S. H. and J. B. Huber, OFDM with reduced peak-to-average power ratio by optimum combination of partial transmit sequences, *IEE Electronics Letters*, 33(5), 36–69, 1997.
9. Bauml, R. W., R. F. H. Fisher, and J. B. Huber, Reducing the peak-to-average power ratio of multicarrier modulation by selected mapping, *IEE Electronics Letters*, 32(22), 2056–2057, 1996.
10. Shepherd, S., J. Orriss, and S. Barton, Asymptotic limits in peak envelope power reduction by redundant coding in orthogonal frequency-division multiplex modulation, *IEEE Transactions on Communications*, 46(1), 5–10, 1998.
11. van Nee, R. and A. de Wild, Reducing the peak to average power ratio of OFDM, *Proceedings of the 48th IEEE Semiannual Vehicular Technology Conference*, 3, 2072–2076, 1998.
12. Dinur, N. and D. Wulich, Peak-to-average power ratio in high-order OFDM, *IEEE Transactions on Communications*, 49(6), 1063–1072, 2001.
13. Litsyn, S. and G. Wunder, Generalized bounds on the crest-factor distribution of OFDM signals with applications to code design, *IEEE Transactions on Information Theory*, 52(3), 992–1006, 2006.
14. Ochiai, H. and H. Imai, On the distribution of the peak-to-average power ratio in OFDM signals, *IEEE Transactions on Communications*, 49(2), 282–289, 2001.
15. Wei, S. Q., D. L. Goeckel, and P. E. Kelly, A modem extreme value theory approach to calculating the distribution of the peak-to-average power ratio in OFDM systems, *IEEE International Conference on Communications*, 3, 1686–1690, 2002.
16. Jiang, T., W. D. Xiang, P. C. Richardson, J. H. Guo, and G. X. Zhu, PAPR reduction of OFDM signals using partial transmit sequences with low computational complexity, *IEEE Transactions on Broadcasting*, 53(3), 719–724, 2007.
17. Wulich, D., Reduction of peak to mean ratio of multicarrier modulation using cyclic coding, *IEE Electronics Letters*, 32(29), 432–433, 1996.
18. Fragicomo, S., C. Matrakidis, and J. J. O'Reilly, Multicarrier Transmission peak-to-average power reduction using simple block code, *IEE Electronics Letters*, 34(14), 953–954, 1998.
19. Jiang, T. and G. X. Zhu, OFDM Peak-to-Average Power Ratio Reduction by Complement Block Coding Scheme and Its Modified Version, *The 60th IEEE Vehicular Technology Conference 2004 Fall*, IEEE, Los Angeles, CA, pp. 448–451, Sept. 2004.
20. Lim, D. W., S. J. Heo, J. S. No, and H. Chung, A new PTS OFDM scheme with low complexity for PAPR reduction, *IEEE Transactions on Broadcasting*, 52(1), 77–82, 2006.

21. Breiling, H., S. H. Muller-Weinfurtner, and J. B. Huber, SLM peak-power reduction without explicit side information, *IEEE Communications Letters*, 5(6), 239–241, 2001.
22. Wang, X. B., T. T. Tjhung, and C. S. Ng, Reduction of peak-to-average power ratio of OFDM system using a companding technique, *IEEE Transactions on Broadcasting*, 45(3), 303–307, 1999.
23. Jiang, T., W. D. Xiang, P. C. Richardson, D. M. Qu, and G. X. Zhu, On the nonlinear companding transform for reduction in PAPR of MCM signals, *IEEE Transactions on Wireless Communications*, 6(6), 2017–2021, 2007.
24. Jones, D., Peak Power Reduction in OFDM and DMT via Active Channel Modification, *IEEE Proceedings 33rd Asilomar Conference on Signals, Systems, and Computers*, CA, pp. 1076–1079, 1999.
25. Hu, S., G. Wu, Y. L. Guan, C. L. Law, and S. Q Li, Analysis of Tone Reservation Method for WiMAX System, *IEEE International Symposium on Communications and Information Technologies*, pp. 498–502, Oct. 2006.
26. Alamouti, S. M., A simple transmit diversity technique for wireless communications, *IEEE Journal on Selected Areas in Communications*, 16(8), 1451–1458, 1998.
27. Lee, H., M. Siti, W. Zhu, and M. P. Fitz, Super-orthogonal space-time block code using a unitary expansion, *IEEE International Conference on Vehicular Technology Conferences*, 4, 2513–2517, 2004.
28. Jiang, T., M. Guizani, H. Chen, W. Xiang, and Y. Wu, Derivation of PAPR distribution for OFDM wireless systems based on extreme value theory, *IEEE Transactions on Wireless Communications*, 2008.

Index

A

Absolute category rating (ACR), optimal adaptation trajectory, 389
Access point (AP)
  IEEE 802.11 standard, 368
    multimedia protocols, 265–267
  peer-assisted video streaming
    content discovery, 312–314
    mobility tracking, 314–315
Activity recording and storage, wireless multimedia sensor networks, 6
Adaptation architecture
  broadband multimedia services, cross-layer optimization and adaptation, 350
  end-to-end quality of service support, wireless Internet video delivery, 418–419
  mobile multimedia handoff management, 445–446
  optimal adaptation trajectory, 387–388
Adaptive Core Multicast Routing Protocol (ACMRP), *ad hoc* wireless networks, multicast system, 198–199
Adaptive Demand-Driven Multicast Routing (ADMR), *ad hoc* wireless networks, multicast system, 198–199
Adaptive error control, end-to-end quality of service, wireless Internet video delivery, 423–424
Adaptive modulation and coding (AMC) technique, wireless local area networks, PHY layer, 336
Adaptive multimedia streaming, wireless local area networks (WLAN)
  algorithm control parameters, 376–377
  algorithm responsiveness, 377–378
  codec constraints, 378
  enhanced loss delay adjustment algorithm, 376
  feedback-based over-reaction, 377
  overview, 371–372
  quality control, 372–373
  real video encoding parameters, 378
  solutions, 373–376
  TCP-friendly rate control and TFRC protocol, 375–376
  TCP throughput model, 377
  user perception, 378
Adaptive video streaming, optimal adaptation trajectory, 384

Additive Increase Multiplicative Decrease (AIMD) algorithm
  adaptive multimedia streaming, 376
  ad hoc wireless networks, rate control, 207–208
  network adaptive congestion/error/power control, 421–423
Ad hoc wireless networks
  IEEE 802.11 multimedia protocols, 265–267
    basic requirements, 365
  multicast system, 196–197
    future research, 228–229
    multiple-tree packet forwarding, 215–216
    multiple-tree video communication, 213–216
    parallel multiple nearly disjoint trees routing protocol, 218–228
    routing protocols, 198–199
    serial multiple-disjoint trees routing protocol, 216–218
    tree connectivity/similarity, 214–215
  multihop networks, end-to-end quality of service, wireless Internet video delivery, 427–428
  multiparty audioconferencing architecture, self-organized entities, 146–147
  peer-assisted video streaming, content discovery, 312–314
  testbed implementation and evaluation, 206–213
    802.11a moving nodes, 210–213
    802.11a static nodes, 208–210
    rate control scheme, 207–208
    setup, 208
    software architecture, 206–207
  unicast systems, 195–196
    concurrent packet drop probability, two node-disjoint paths, 200–201
    envisioned network model, 199–200
    future research, 228–229
    heuristic optimum multipath selection, 203–204
    packet drop probability link computation, 201–203
    routing protocols, 197–198
    simulation results, 204–206
  video communications routing, 194–195
    application-layer performance, 162–164
    bit rates and success probabilities, 164–166
    branch-and-bound approach, 175–188
      ε-optimal algorithm, 183–185
      heuristic algorithm, 182–183
      linearization, 181–182
      numerical results, 185–188
    distributed implementation, 188–189
    metaheuristic approach, 168–175
      genetic algorithms, 168–172
      performance comparisons, 172–175
    optimal double-path routing, 166–168
    overview, 158–161
Admission control
  broadband wireless local area networks
    packet scheduling, 501
    QoS algorithms, 345–347
  IEEE 802.11 standards, 273–274
Agent-based image sensor querying, wireless sensor networks, 240–241
Algorithms
  adaptive multimedia streaming
    control parameters, 376–377
    responsiveness, 377–378
  broadband wireless LANs
    Gilbert–Elliot channel model, 485–487
    multistate channel model, 487–488

packet scheduling, 482–485
caching algorithms, peer-assisted video streaming quality enhancement, 319–322
distortion minimizing smart caching algorithm, 321–322
popularity-based algorithm, 320
prefix algorithm, 320
resource-based algorithm, 319–320
selective partial caching algorithm, 321
wireless local area network quality of service, 344–347
admission control and reservation, 345–347
scheduling, 344–345
Analog television devices, digital conversion of, 88–89
Application layer
broadband wireless LAN, multimedia services, 335–336
cross-layer optimization and adaptation, 349–350
handoff management, middleware approach, 452–459
interactive mobile application session handoff, 453–454
media-independent preauthentication, 452–453
mobile agent-based ubiquitous multimedia middleware, 458–459
mobile Internet protocol, seamless media streaming, 454–455
Mobiware, 455–456
session initiation protocol-based solution, 457
vertical handover, 456–457
multiple description video performance, 162–164
wireless multimedia sensor network, 14–23
application-specific protocols, 26–27
collaborative in-network processing, 22
multimedia encoding, 16–20
open research issues, 23
pixel-domain Wyner–Ziv encoder, 18–20
software and middleware, 20–22
traffic classes, 15–16
transform-domain Wyner–Ziv encoder, 20
Arbitration interframe spacing (AIFS[AC]), IEEE 802.11 standards
performance evaluation, 276–282
prioritized QoS, 272–273
Architectures
ad hoc wireless networks, software architectures, 206–207
broadband multimedia services, adaptation architecture, 350
multiparty audioconferencing, 131–138
centralized mixing, 131–132
distributed mixing system, 136–138
distributed partial mixing, 135–136
endpoint mixing, 132–134
hierarchical mixing, 134–135
video sensor networks, 241–243
wireless multimedia sensor networks, 10–14
coverage, 13
flexibility, 9–10
internal multimedia sensor organization, 13–14
reference architecture, 10–12
single- *vs.* multitier sensor development, 12–13
Asynchronous clocks, peer-assisted video streaming, peer selection, 316
Asynchronous Layered Coding (ALC), mobile TV systems, file download protocols, 105
Asynchronous transfer mode (ATM), wireless multimedia sensor networks, evolution of, 5

Audio mixing, multiparty audioconferencing architecture, 138–139
Audio processing unit (APU), multiple audio streams, 131
Audio stream compression, multiparty audioconferencing, 124–125
Automatic repeat request (ARQ) algorithm
  adaptive error control, 423–424
  IEEE 802.11 WLANS, peer-assisted video streaming, 308–309
  video transmissions, wireless local area networks, 239
  wireless multimedia sensor networks, 34–35
Available bit rate (ABR), broadband wireless local area networks, heterogeneous traffic, 479–480

B

Bandwidth demand and utilization
  broadband wireless local area networks, packet scheduling, 481
  multimedia streaming, 379–380
  multiparty audioconferencing architecture, consumption reduction, 147–148
  network adaptive congestion, 422–423
  wireless multimedia sensor networks, 8
Beacon delay, IEEE 802.11 multimedia protocols, 267–268
Bellman optimality equation, learning-based packet scheduling, 494–497
Bernoulli process, double-description (DD) video, end-to-end success probability, 166
Bit rates, interactive mobile television, 93–94
Black box optimization, video communications routing, wireless *ad hoc* networks, 168–172
Block acknowledgment, IEEE 802.11n MAC efficiency improvement, 286–287
Block-error-rate (BLER), high speed downlink packet access
  adaptive multi rate (AMR), 54
  handover threshold, 63
  Max C/I scheduler, 55–56
  performance analysis, 51
  scheduler properties, 68–75
  transmission quality, 63–64
Border Gateway Protocol (BGP), end-to-end quality of service, wireless Internet video delivery, 427–428
Bounding step, video communications routing, branch-and-bound techniques, 185
Branch-and-bound techniques, video communications routing, wireless *ad hoc* networks, 175–188
  ε-optimal algorithm, 183–185
  heuristic algorithm, 182–183
  linearization, 181–182
  numerical results, 185–188
Broadband wireless local area networks
  multimedia services
    admission control and reservation, 345–347
    application requirements and standards, 337–339
    cellular network interworking, 354–356
    cross-layer design and adaptation, 341–342, 348–350
    DiffServe-based end-to-end QoS support, 348
    end-to-end QoS scheme, 341, 347–348
    future research issues, 356–357

IEEE 802.11 QoS support, 343–347
image transmission, 351
IntServ/DiffServe Internet QoS architectures, 347–348
IntServe-based end-to-end QoS support, 348
JPEG2000, 339
link capacity and adaptation, 340
MAC protocols, 337, 356
MAC QoS scheme, 340–341
mobility support, 342
Moving Picture Experts Group (MPEG)-4, 338–339
multihop WLAN, 357
overview, 333–339
PHY layer, 336
physical-layer enhancement, 343
research issues, 339–342
roaming support, 354
scheduling algorithms, 344–345
seamless QoS support, 354–356
technical and applications trends, 335–336
video streaming, 352–354
voice-over-IP, 352
WLAN case study, 350–354
WLAN QoS algorithms, 343–347
packet scheduling
algorithms, 482–490
bandwidth utilization, 481
channel model, 476–477
cost function construction, 492–494
fairness, 481–482
future research issues, 500–501
Gilbert–Elliot channel model, 485–487
heterogeneous traffic properties, 479–480
learning-based scheduling, 490–500
multistate channel model, 487–488
network model, 475–476
neuro-dynamic programming, 494–497
overview, 474–475
performance evaluation, 497–500
quality of service, 481
semi-Markov decision process, 490–492
traffic model, 477–479
wireless link variability, 480–481
wireless scheduling models, 475–479
wireline networks, 482–485
Broadcast distribution scheme (BDS), mobile TV systems, OMA BCAST standard, 101–103
Broadcast/multicast download and play, interactive mobile television, 93
Broadcast-Multicast Service Center (BM-SC), mobile TV systems, 3GPP MBMS standard, 96–97
Broadcast/multicast streaming, mobile television, 92–93

C

Caching, peer-assisted video streaming quality enhancement, 319–322
distortion minimizing smart caching algorithm, 321–322
popularity-based algorithm, 320
prefix algorithm, 320
resource-based algorithm, 319–320
selective partial caching algorithm, 321
Capacity, common control channel MAC performance, 293–295
Carrier Sense Multiple Access with Collision Avoidance (CSMA/CA)
broadband wireless local area networks, MAC protocols, 337
IEEE 802.11e MAC enhancement, 370–371
Cell selection, high speed downlink packet access, 57–59

Cellular networks, broadband wireless local area networks, seamless QoS support, 354–356
Centralized mixing architecture, multiparty audioconferencing, 131–132
Channel access, wireless multimedia sensor networks, MAC protocols, 29
Channel-condition independent fair queuing (CIF-Q), Gilbert–Elliot (GE) channel model, 486
Channel modeling
  broadband wireless local area networks, packet scheduling, 476–477
  end-to-end quality of service support, wireless Internet video delivery, 416–417
Channel state dependent packet scheduling (CSDPS), Gilbert–Elliot (GE) channel model, 485
Channel zapping, mobile TV systems, 110–112
Clear-to-send (CTS) message, video transmission strategy, 244–247
Clipping, peak-to-average power ratio reduction, 514–515
Clusterheads (CH)
  multiparty audioconferencing architecture, 140
  wireless multimedia sensor networks, reference architecture, 11–12
Codec constraints, adaptive multimedia streaming, 378
Coding schemes, peak-to-average power ratio reduction, 515–516
Collaborative in-network processing, wireless multimedia sensor network, 22
Common control channel (CCC) protocols
  Media Access Control performance evaluation, 292–299
  mesh networks, 291–292
Common Object Request Broker Architecture (COBA), Mobiware system, 455–456
Common pilot channel (CPICH), high speed downlink packet access mobility, 62–64
Complementary cumulative distribution function (CCDF), orthogonal frequency division multiplexing
  peak-to-average power ratio distribution, 510–513
  signal characteristics, 507
Complement block coding (CBC), peak-to-average power ratio reduction, 515–516
Conference servers (CSs), multiparty audioconferencing architecture, 141
  selection and allocation, 148
  self-organized entities, 146–147
  stream selection, 141–143
Conflict graph, *ad hoc* wireless networks, unicast systems, 196
Congestion control
  broadband wireless local area networks, packet scheduling, 501
  multimedia streaming, 379
  wireless multimedia sensor network, 26
Congestion Detection and Avoidance (CODA) protocol, wireless multimedia sensor network, 26–27
Constant bit rate (CBR)
  broadband wireless local area networks, heterogeneous traffic, 479

prefix caching algorithm, peer-assisted video streaming quality enhancement, 320
Constant-rate model, broadband wireless local area networks, packet scheduling, 478
Construction delay, parallel multiple nearly disjoint trees multicast routing protocol, 218
Container file formats, interactive mobile television, 113–114
Content adaptation, mobile multimedia handoff management, 444–445
Content discovery, peer-assisted video streaming, 311–314
Contention-based protocols, wireless multimedia sensor networks, 29–30
Contention-free protocols, wireless multimedia sensor networks
multi-channel protocols, 32–33
single-channel protocols, 30–31
Content protection
mobile TV systems, 115
peer-assisted video streaming, IEEE 802.11 WLANS, 308
Context awareness, wireless Internet continuity, 438
Context transfer, mobile multimedia handoff management, 443
CooperativeNodes, video transmission strategy, 245–247
Corruption duration, mobile TV systems, 110
Cost function construction, learning-based packet scheduling, 492–494
Coverage architecture, wireless multimedia sensor networks, 13
Critical density, tree connectivity/similarity, multicast routing protocols, 214–215
Cross-layer coupling
end-to-end quality of service, wireless Internet video delivery, network-centric techniques, 414–419
wireless multimedia sensor network functionalities, 8
Cross-layer design and adaptation, broadband wireless local area networks, 341–342
optimization and adaptation, 348–351
Cross-layer optimization
peer-assisted video streaming, 309–310
wireless multimedia sensor networks, 35–36
Crossover, genetic algorithm, video communications routing, 170–171
Cumulative distribution function (CDF)
high speed downlink packet access mobility
handover delay, 65–67
priority scheduling, 75–78
peer-assisted video streaming, data selection, 318–319
Customized mixing, wireless network interactivity, 127

D

Data flow continuity, middleware design guidelines, 448–449
Data-link layer, handoff management solutions, 451–452
Data selection, peer-assisted video streaming, 312, 317–319
DCF mechanism, IEEE 802.11 multimedia protocols, 266–267
Decision making, handoff management, full context awareness, 440

Degraded category rating (DCR), optimal adaptation trajectory, 389
Delay
  common control channel MAC system, 295–299
  multimedia streaming, 380
  multiparty audioconferencing, 129
Delay-tolerant, loss-intolerant/loss-tolerant data, wireless multimedia sensor network, 16
Delay-tolerant, loss-tolerant multimedia streams, wireless multimedia sensor network, 15
Detailed double nearly disjoint tree construction, *ad hoc* wireless networks, 220
DiffServe end-to-end quality of service (QoS) support
  broadband wireless local area networks, 347–348
  network-centric techniques, wireless Internet, 415–419
Digital television, evolution of, 88–89
Directed acyclic graph (DAG), peer-assisted video streaming, 317–319
Directional geographical routing (DGR)
  video transmission strategy, 243–247
    performance analysis, 249–254
    simulation model, 247–248
  wireless sensor networks, 237–238
Discard timer, high speed downlink packet access mobility, 67–68
Disjoint-path routing, video communications, 174–175
Dispersity routing, *ad hoc* wireless networks, unicast system, 197–198
Distance range, IEEE 802.11 standard, 368
Distortion evolution, video communications routing, 172–175
Distortion minimizing smart caching (DMSC) algorithm, peer-assisted video streaming quality enhancement, 321–322
Distributed Coordination Function (DCF)
  broadband wireless local area networks, MAC protocols, 337
  IEEE 802.11e MAC enhancement, 370–371
Distributed hash table (DHT), peer-assisted video streaming, content discovery, 312–314
Distributed implementation, video communications routing, 188–189
Distributed mixing system (DMS), multiparty audioconferencing, 136–137
Distributed partial mixing (DPM), multiparty audioconferencing, 135–136
Double-description (DD) video
  ad hoc wireless networks, 162–164
  genetic algorithm-based solutions, 168–172
  shortest-path routing, 187–188
  video bit rates, 164–166
Download and play options, interactive mobile television, broadcast/multicast download and play, 93
DVB-H, mobile TV systems, IP broadcast over, 98–99
Dynamic and interactive multimedia scenes (DIMS), interactive mobile television, 113
Dynamic Source Routing (DSR), video communications

distributed implementation, 189
Dynamic wireless channel quality, peer-assisted video streaming, 327

E

ε-optimal algorithm, video communications routing
branch-and-bound techniques, 183–186
performance analysis, 186–188
Encoding techniques, wireless multimedia sensor network, 16–20
pixel-domain Wyner–Ziv encoder, 18–20
transform-domain Wyner–Ziv encoder, 20
Endpoint mixing architecture, multiparty audioconferencing, 132–134
End system-centric techniques, end-to-end quality of service, wireless Internet video delivery, 413–414, 419–426
adaptive error control, 423–424
joint power/error control, 424–426
network adaptive congestion control, 420–423
rate distortion-based bit allocation, 426
End-to-end quality of service (QoS)
broadband wireless local area networks, 341
wireless Internet video delivery
end-system-centric support, 419–426
future research issues, 427–428
network-centric cross-layer support, 414–419
overview, 410–413
End-to-end success probability
double-description (DD) video, 165–166
video transmission performance, packet delay, 248
Energy-aware peer selection (EPS), peer-assisted video streaming, 325–326
Energy consumption per data delivery, video transmission performance, 248
Energy efficiency and awareness, peer-assisted video streaming, 322–326
Energy-efficient differentiated directed diffusion (EDDD), wireless sensor networks, quality of service time-constrained traffic, 239–240
Enhanced distributed channel access (EDCA)
broadband wireless local area networks, admission control and reservation, 345–347
common control channel MAC system, delay parameters, 295–299
IEEE 802.11e MAC enhancement, 369–370
IEEE 802.11 standards
admission control, 273–274
enhanced IEEE 802.11e capability, 268–270
performance evaluation, 274–282
prioritized QoS, 271–273
Enhanced distribution coordination function (EDCF), end-to-end quality of service support, network-centric techniques, wireless Internet, 415–419
Enhanced loss delay adjustment algorithm (LDA+), adaptive multimedia streaming, 376
Entity self-organization, multiparty audioconferencing architecture, 146–147

Environmental monitoring, wireless multimedia sensor networks, 7
Envisioned network model, *ad hoc* wireless networks, multipath selection, unicast streaming, 199
Equal average bit rate (EABR), optimal adaptation trajectory, 387–391
Error control, end-to-end quality of service, wireless Internet video delivery, 424–426
EURANE network-level simulator, high speed downlink packet access, 50–51
Evolution data optimized (EVDO)-based cellular systems, mobile TV systems, standards, 98
Execution, handoff management, full context awareness, 440
Expected transmission count (ETX), *ad hoc* wireless networks, IWM protocol, 206–207

F

Fairness
  broadband wireless LANs, packet scheduling, 481–482
  Gilbert–Elliot (GE) channel model, server-based fairness approach (SBFA), 486
Fathoming, video communications routing, branch-and-bound techniques, 185
Feasible solutions, genetic algorithms, video communications routing, 169–170
Feedback over-reaction, adaptive multimedia streaming, 377
File download protocols, mobile TV systems, 104–105
  reliability, 107–108
Filtering, peak-to-average power ratio reduction, 514–515
Fine grained loss protection (FGLP), adaptive error control, 424
Fine granularity scalability (FGS), end-to-end quality of service, wireless Internet video delivery, 412–413
Finite state Markov channel (FMSC), broadband wireless local area networks, packet scheduling, 476–477
First come first served (FCFS), broadband wireless LANs, wireline networks, 482
Fitness function, genetic algorithm, video communications routing, 170
Flexible user reachability, interactive mobile television, 90–92
Floor control
  multiparty audioconferencing architecture, 139
    loudness number, 143–145
  multiple audio streams, 129–130
  wireless network interactivity, 126
FLUTE protocol, mobile TV systems
  container file formats, 113–114
  file downloads, 105
Forward error correction (FEC)
  adaptive error control, 423–424
  ad hoc wireless networks, unicast systems, 195–196
  IEEE 802.11 WLANS, peer-assisted video streaming, 308–309
  interactive mobile television
    channel zapping, 111–112
    file download reliability, 107–108
  video sensor networks, 236–237
    architecture, 241–243
  video transmission strategy, performance analysis, 250
  wireless multimedia sensor networks, 34–35
Forwarding efficiency, *ad hoc* wireless networks, parallel multiple nearly disjoint trees protocol, 222

Frame aggregation, IEEE 802.11n MAC efficiency improvement, 283–286
Frame rate deviation, mobile TV systems, 110
Full context awareness, handoff management, 439–445
Function approximation architecture, learning-based packet scheduling, 495–496

G

Generalized processor sharing (GPS), broadband wireless LANs, wireline networks, 483
General Packet Radio Service (GPRS), broadband wireless LAN, multimedia services, 335–336
Genetic algorithms, video communications routing
  performance comparisons, 172–175
  wireless *ad hoc* networks, 168–172
Gilbert–Elliot (GE) model
  broadband wireless LANs, scheduling algorithms, 485–487
  broadband wireless local area networks, packet scheduling, 476
Gilbert two-state link model, double-description (DD) video, 164–166
Greedy model, broadband wireless local area networks, packet scheduling, 478–479
Greedy partitioning algorithm, *ad hoc* wireless networks, packet drop probability, 203
Group of blocks (GOB), video communications routing, network-centric techniques, 175

H

Handoff awareness
  defined, 438
  mobile multimedia handoff management, 441–442
Handoff decision
  middleware design guidelines, 447–448
  mobile agent-based ubiquitous multimedia (MUM) middleware, 458–459
Handoff latency, mobile multimedia handoff management, 444–445
Handoff management
  architectural modeling, 459–462
  high speed downlink packet access mobility, 62–68
    delay, 64–67
  interactive mobile television, 95
  wireless multimedia services, middleware approach
    adaptability, 445–446
    application layer, 452–459
    content provision, wireless mesh networks, 465–466
    data-link/network/transport layers, 451–452
    full context awareness, 439–445
    future development issues, 462–464
    handoff awareness, 441–442
    IEEE 802.21 media-independent handover, 462–463
    interactive mobile application session handoff, 453–454
    Internet protocol multimedia subsystem, 463–464
    Internet protocol multimedia system open handoff, 466
    location awareness, 442–443
    media-independent preauthentication, 452–453
    mobile agent-based ubiquitous middleware, 458–459
    mobile Internet protocol-enabled WLAN seamless streaming, 454–455
    mobile network support, 464–465
    Mobiware, 455–456

Handoff management (*contd.*)
overview, 436–439
proactive solutions, 445–446
quality of service awareness, 443–445
reconfigurability, 445–446
service continuity design guidelines, 446–450
session initiation protocol-based handoff solutions, 457
solution evaluations, 459–462
state-of-the-art solutions, 450–462
unlicensed mobile access, 463
vertical handover, 456–457
HCF controlled channel access (HCCA)
IEEE 802.11 standards
admission control, 274
IEEE 802.11e MAC enhancement, 269–270, 369
quality of service protocol, 270
Healthcare delivery, wireless multimedia sensor networks, 6–7
Heterogeneity
broadband wireless local area networks, traffic properties, 479–480
multimedia streaming, 379
Heuristic algorithm, video communications routing
branch-and-bound techniques, 182–183
performance analysis, 186–188
Hierarchical mixing, multiparty audioconferencing, 134–135
High data rate (HDR) systems, broadband wireless local area networks, packet scheduling, 476
High speed downlink packet access (HSDPA)
cell selection validation, 57–59
default scenario performance, 57–61
handover mobility optimization, 62–68
delay parameters, 64–67
discard timer, 67–68
threshold parameters, 62–64
Max C/I scheduler, 55–56
mixed traffic scenario, 59–61
multicell propagation model, 51–53
overview, 48–50
performance analysis, 50–54
future issues, 84–85
improvements, 82–84
proportional fair scheduler, 56–57
round-robin scheduler, 55
scheduler algorithm performance, 68–82
comparisons, 78–81
priority scheduling, 75–78
resource consumption, 81–82
specific properties, 68–75
scheduling performance, 54–57
system and user characteristics, 53–54
time-to-live scheduler, 57–58
Hop-by-hop routing, video communications distributed implementation, 188–189
H.264 standard, broadband wireless local area networks, 338
Hybrid coordination function (HCF)
end-to-end quality of service support, network-centric techniques, wireless Internet, 415–419
IEEE 802.11 standards, enhanced IEEE 802.11e capability, 268–270, 369
Hypertext transfer protocol (HTTP), interactive mobile television, user interaction, 106–107

I

Idealized wireless fair queuing (IWFQ), Gilbert–Elliot (GE) channel model, 485–486

IEEE 802.11 standard
802.11a, *ad hoc* wireless network standard
moving nodes, 210–213
static nodes, 208–210
broadband wireless local area networks, 334–335
admission control and reservation, 345–347
algorithms, 344–347
MAC protocols, 337
mechanisms for, 343–344
physical-layer enhancement, 343
quality of service support, 343–347
scheduling algorithms, 344–345
802.11e enhanced capability, multimedia quality of service support, 268–269, 283–288
end-to-end quality of service support, network-centric techniques, wireless Internet, 415–419
multimedia quality of service support
admission control, 273–274
EDCA performance analysis, 274–282
EDCA protocol, 271–273
802.11e enhanced capability, 268–269
enhanced capability in 802.11e, 268–270
future research challenges, 299
HCCA protocol, parameterized QoS, 270
inadequate legacy 802.11 support, 267–268
MAC efficiency improvement, 283–287
802.11n QoS features, 283–288
overview, 262–263
protocol requirements, 265–267
requirements, 264–265
reverse direction protocol, 287–288
scheduling algorithm design, 270–271
802.11s mesh network mechanisms, 289–299
802.11n QoS features, broadband wireless local area networks, 334–335
wireless local area networks
evolution of, 306–307
PHY layer, 336
user-perceived video streaming quality, 365–371
802.11a, 367
802.11b, 366–367
distance range/transmission rate comparisons, 368
802.11e MAC enhancement for quality of service, 368–369
enhanced distributed coordinated access, 369–371
802.11g, 367–368
IEEE 802.16 standard, end-to-end quality of service support, network-centric techniques, wireless Internet, 416
IEEE 802.21 standard, media-independent handover standard, 462–463
Image Evaluation based on Segmentation (IES) model, multimedia streaming, 383–384
Image transmission, wireless local area networks, 351
Implementation issues, multiparty audioconferencing architecture, 149
Inaccuracy systems, broadband wireless local area networks, packet scheduling, 500–501
Incentive mechanisms, peer-assisted video streaming, node cooperation, 327–328
Independent-Tree *Ad Hoc* Multicast Routing (ITAMAR) protocol, *ad hoc* wireless

Independent-Tree (*contd.*)
networks, multicast system, 198–199
Industrial process control, wireless multimedia sensor networks, 7
Infrastructure mode, IEEE 802.11 multimedia protocols, 265–267, 365
Initial buffering duration, mobile TV systems, 110
Initialization process, video communications routing, branch-and-bound techniques, 183
Initiation, handoff management, full context awareness, 440
In-network processing, wireless multimedia sensor networks, 9
collaborative methods, 22
Integrated approach, broadband multimedia services, cross-layer optimization and adaptation, 349–350
Integrated Services Digital Broadcasting Terrestrial (ISDB-T), mobile TV systems, 100–101
Interactive mobile application session handoff (IMASH), handoff management, middleware approach, 453–454
Interactive mobile television, characteristics and requirements, 90–92
Interactive pull bearer, mobile TV systems, 104
Interactivity
mobile television, 90–92
multiparty audioconferencing, 123–124
wireless network constraints, 126
Interference aWare Multipath (IWM) protocol, *ad hoc* wireless networks
moving nodes, 210–213
optimum multipath selection, 203–204
results comparisons, 204–206
software architecture, 206–207
static nodes, 209–210
unicast system, 197–198
Internal organization, wireless multimedia sensor networks, 13–14
Internet Engineering Task Force (IETF) SIP, wireless network signaling protocol, 127
Internet Protocol (IP)
multiparty audioconferencing, 120–121
wireless multimedia sensor networks, 10
Internet Protocol Multimedia System (IMS)
handoff management, 463–464
open handoff management, 466
Interoperable handoff initiation, handoff management, 447
Interrupted Poisson Process (IPP) model, broadband wireless local area networks, packet scheduling, 478
Intra coded frames (I-frames), mobile TV systems, channel zapping, 110–112
IntServe end-to-end quality of service (QoS) support
broadband wireless local area networks, 347–348
network-centric techniques, wireless Internet, 415–419
IPDC format, mobile TV systems, DVB-H broadcast, 98–99
ITU-T H.323, wireless network signaling protocol, 127

J

Jitter effects
common control channel MAC system, delay parameters, 295–299
mobile TV systems, 110
multimedia streaming, 380

multiparty audioconferencing, delay jitter, 129
wireless multimedia sensor networks, TCI/UDP protocols, 25–26
Join-query (JQ) message
ad hoc wireless networks, scenarios, 230
detailed double nearly disjoint tree construction, 220
parallel multiple nearly disjoint trees multicast routing protocol, 218–220
Joint power control, end-to-end quality of service, wireless Internet video delivery, 424–426
JPEG2000 standard, broadband wireless local area networks, 339

L

Learning-based packet scheduling, broadband wireless local area networks, 490–500
cost function construction, 492–494
neuro-dynamic programming, 494–497
performance evaluation, 497–500
semi-Markov decision process, 490–492
Lifetime performance, video transmission strategy, 248
Linear programming (LP), video communications routing, branch-and-bound techniques, 177–188
Link capacity and adaptation, broadband wireless local area networks, 340, 480–481
Link-layer error control
ad hoc wireless networks, packet drop probability, 201–203
wireless multimedia sensor networks, 34–35
Link Quality Source Routing (LQSR), *ad hoc* wireless networks
IWM protocol, 206–207
moving nodes, 210–213
static nodes, 209–210
Link status monitor (LSM), broadband wireless local area networks, packet scheduling, 476
Live streaming, adaptive multimedia streaming, 372
Local area networks (LANs), multiparty audioconferencing architecture, 139–140
Local search, video communications routing, branch-and-bound techniques, 184
Location awareness
defined, 438
mobile multimedia handoff management, 442–443
Long-term fairness server, 486
Loss of packets
multimedia streaming, 380
user-perceived quality evaluation, 399–401
Loudness number, multiparty audioconferencing architecture
bandwidth consumption reduction, 147–148
basic properties, 143–145
enhancement, 149
safety, liveness, and fairness, 145
selection algorithm, 145–146
Lower bounds, video communications routing, initialization and relaxation, 183

M

Mapping, end-to-end quality of service support, wireless Internet video delivery, 418–419
Markov process, double-description (DD) video, end-to-end success probability, 165–166
Mass distribution, interactive mobile television, 90–92

Mathematical metrics, multimedia streaming, 382–383
Max C/I scheduler, high speed downlink packet access
basic principles, 50
performance analysis, 55–56
Maximum user preference path, optimal adaptation trajectory, 393
MBMS bearer, mobile TV systems
streaming reliability, 108–110
user service announcements, 104, 106–107
Media Access Control (MAC) layer
ad hoc wireless networks, concurrent PDP, two node-disjoint paths, 201
broadband multimedia services
cross-layer optimization and adaptation, 349–350
new multimedia support, 356
protocols, 337
broadband wireless local area networks, quality of service scheme, 340–341
common control channel performance, 292–299
IEEE 802.11 standards, 263
IEEE 802.11e QoS enhancement, 368–371
IEEE 802.11n efficiency improvement, 283–287
wireless local area networks, video streaming, 353–354
wireless multimedia sensor networks, 29–35
channel access, 29
contention-based protocols, 29–30
contention-free multi-channel protocols, 32–33
contention-free single-channel protocols, 30–31
link-layer error control, 34–35
scheduling, 33–34
Media classification, interactive mobile television, 93–94
Media codecs, interactive mobile television, 112–113
Media delivery framework, mobile TV systems, 104–112
channel zapping, 110–112
file download protocols, 104–105
file download reliability, 107–108
quality of experience framework, 110
streaming protocols, 105–106
streaming reliability, 108–110
user interaction protocols, 106–107
user service announcement, 104
Media-independent handover, media-independent handover standard, 462–463
Media-independent preauthentication (MPA), handoff management, middleware approach, 452–453
Media-oriented applications, wireless local area networks, 350–354
image transmission, 351
video streaming, 352–354
voice-over-IP, 352
Media transport reliability, interactive mobile television, 94
Mesh Connectivity Layer (MCL), *ad hoc* wireless networks, IWM protocol, 206–207
Mesh deterministic access (MDA)
IEEE 802.11 multimedia standards, 263
IEEE 802.11s standards, mesh networks, 290
Mesh medium access coordination function (MCF), mesh networks, 289–290
Mesh networks
IEEE 802.11s quality of service mechanisms
CCC MAC performance, 292–299
common control channel protocol, 291–292
deterministic access, 290
overview, 289–290

multimedia content provisioning, 465–466
Mesh points (MPs), IEEE 802.11 multimedia standards, 263
Metaheuristic approach, video communications routing, wireless *ad hoc* networks, 168–175
genetic algorithms, 168–172
performance comparisons, 172–175
Microhandoff, mobile multimedia handoff management, 442–443
Middleware requirements
handoff management, wireless multimedia services
adaptability, 445–446
application layer, 452–459
content provision, wireless mesh networks, 465–466
data-link/network/transport layers, 451–452
full context awareness, 439–445
future development issues, 462–464
handoff awareness, 441–442
IEEE 802.21 media-independent handover, 462–463
interactive mobile application session handoff, 453–454
Internet protocol multimedia subsystem, 463–464
Internet protocol multimedia system open handoff, 466
location awareness, 442–443
media-independent preauthentication, 452–453
mobile agent-based ubiquitous middleware, 458–459
mobile Internet protocol-enabled WLAN seamless streaming, 454–455
mobile network support, 464–465
Mobiware, 455–456
overview, 436–439
proactive solutions, 445–446
quality of service awareness, 443–445
reconfigurability, 445–446
service continuity design guidelines, 446–450
session initiation protocol-based handoff solutions, 457
solution evaluations, 459–462
state-of-the-art solutions, 450–462
unlicensed mobile access, 463
vertical handover, 456–457
wireless multimedia sensor network, 20–22
MIMO/OFDM systems, peak-to-average power ratio reduction, 527–528
Mixed traffic scenario, high speed downlink packet access, 59–61
Mixing architectures, multiparty audioconferencing, 131–138
benefits, 137
centralized mixing, 131–132
distributed mixing system, 136–138
distributed partial mixing, 135–136
endpoint mixing, 132–134
hierarchical mixing, 134–135
limitations, 138
Mobile agent-based ubiquitous multimedia (MUM) middleware, handoff management, 458–459
Mobile Internet protocol (MIP)
data-link/network/transport layers, 451–452
seamless media streaming, 454–455
Mobile multimedia handoff management
full context awareness, 439–445
service continuity design guidelines, 446–450
system requirements, 445–446
Mobile networks, handoff management, 464–465

Mobile TV systems
broadcast/multicast download and play, 93
broadcast/multicast download-user interaction, 93
broadcast/multicast streaming, 92–93
characteristics and requirements, 90–92
container file formats, 113–114
design criteria, 93–95
digital technology and evolution of, 89–90
dynamic and interactive media scenes, 113
future research issues, 116
interactive properties, overview, 88
media codecs, 112–113
media delivery framework, 104–112
channel zapping, 110–112
file download protocols, 104–105
file download reliability, 107–108
quality of experience framework, 110
streaming protocols, 105–106
streaming reliability, 108–110
user interaction protocols, 106–107
user service announcement, 104
media transport reliability, 94
roaming and handovers, 95
security, 114–115
standards, 95–103
3GPP2 BCMCS, 96–98
3GPP MBMS, 95–96
IP datacast over DVB-H, 98–101
OMA BCAST, 101–103
supported media types and bit rates, 93–94
tune-in delay, 94
usage scenarios, 92–93
Mobility management, middleware design guidelines, 449–450
Mobility tracking
broadband wireless local area networks, 342
high speed downlink packet access, handover optimization, 62–68
multiparty audioconferencing architecture, 148–149
peer-assisted video streaming, 311, 314–315
Mobiware, handoff management middleware, 455–456
Model-based metrics, multimedia streaming, 383–384
Modified complement block coding (MCBC), peak-to-average power ratio reduction, 515–516
Monitoring applications, wireless multimedia sensor networks, 7
Moving nodes, 802.11a 802.11a, *ad hoc* wireless network standard, 208–210
Moving Picture Experts Group (MPEG)-4 standard
broadband wireless local area networks, 338
multimedia streaming user-perceived quality evaluation, 397–401
Moving pictures quality metric (MPQM), multimedia streaming, 383–384
Multicast system
ad hoc wireless networks, 196–197
future research, 228–229
multiple-tree packet forwarding, 215–216
multiple-tree video communication, 213–216
parallel multiple nearly disjoint trees routing protocol, 218–228
routing protocols, 198–199
serial multiple-disjoint trees routing protocol, 216–218

tree connectivity/similarity, 214–215
multiparty audioconferencing
availability, 124
constraints on, 128
Multicell propagation model, high speed downlink packet access, 51–53
Multihop wireless networks. *See Ad hoc* wireless networks
end-to-end quality of service, wireless Internet video delivery, 427–428
multimedia support, 357
Multilayered scalable video coding, end-to-end quality of service, wireless Internet video delivery, 412–413
Multimedia services
broadband wireless local area networks
admission control and reservation, 345–347
application requirements and standards, 337–339
cellular network interworking, 354–356
cross-layer design and adaptation, 341–342
cross-layer optimization and adaptation, 348–350
DiffServe-based end-to-end QoS support, 348
end-to-end QoS scheme, 341, 347–348
future research issues, 356–357
IEEE 802.11 QoS support, 343–347
image transmission, 351
IntServ/DiffServe Internet QoS architectures, 347–348
IntServe-based end-to-end QoS support, 348
JPEG2000, 339
link capacity and adaptation, 340
MAC protocols, 337, 356
MAC QoS scheme, 340–341
mobility support, 342
Moving Picture Experts Group (MPEG)-4, 338–339
multihop WLAN, 357
overview, 333–339
PHY layer, 336
physical-layer enhancement, 343
research issues, 339–342
roaming support, 354
scheduling algorithms, 344–345
seamless QoS support, 354–356
technical and applications trends, 335–336
video streaming, 352–354
voice-over-IP, 352
WLAN case study, 350–354
WLAN QoS algorithms, 344–347
WLAN QoS mechanism, 343–344
user-perceived quality, 379–381
evaluation process, 395–401
Multimedia surveillance, wireless multimedia sensor networks, 6
Multipacket Protocol Encapsulation (MPE) sections, mobile TV systems, 98–100
Multiparty audioconferencing
bandwidth consumption reduction, 147–148
components, 139–141
CS allocation, 148
design criteria, 138–139
entity self-organization, 146–147
implementation, 149
interactive features, 123–125
loudness number, 143–145
enhancement, 149
selection algorithm, 145–146
mixing architectures, 131–138
centralized mixing, 131–132
distributed mixing system, 136–138
distributed partial mixing, 135–136
endpoint mixing, 132–134
hierarchical mixing, 134–135
mobility, 148–149
overview, 120–123
quality improvement, 149
session mobility, 149

Multiparty (*contd.*)
streaming techniques, 129–131
stream selection, 141–143
wireless network constraints, 125–129
customized mixing, 127
interactivity, 126
multicasting availability, 128
quality controls, 129
scalability, 127–128
signaling, 127
traffic reduction, 128–129
Multipath-based real-time video communications, wireless sensor networks, 241–247
video transmission strategy, 243–247
VSN architecture, 241–243
Multipath routing
ad hoc wireless networks
software architecture, 206–207
unicast streaming, 199–200
video transmission strategy, 243–247
wireless sensor networks, 237–238
Multiple description coding (MDC), *ad hoc* wireless networks, 195
moving nodes, 210–213
static nodes, 209–210
unicast systems, 195–198
Multiple description (MD) video
ad hoc wireless networks, 158–161
application-layer performance, 162–164
double-path routing, 166–168
Multiple Disjoint Tree Multicast Routing (MDTMR) protocol, *ad hoc* wireless networks, multicast systems
serial protocols, 216–218
serial *vs.* parallel protocols, 196
Multiple radio systems, common control channel (CCC) protocol, 291–292
Multiple-tree video communication
ad hoc multicast wireless networks, 213–216
multicast packet forwarding, 215–216
Multipoint control unit (MCU), multiple audio streams, 131
Multistate Markov channel model
broadband wireless local area networks, packet scheduling, 476–477
scheduling algorithms, 487–488
Multitier sensor development, wireless multimedia sensor networks, 12–13
Mutation, genetic algorithm, video communications routing, 171–172

N

NDP scheduling algorithm, learning-based packet scheduling, performance evaluation, 497–500
Near disjointness, parallel multiple nearly disjoint trees multicast routing protocol, 218
Network adaptive congestion/error/power control, end-to-end quality of service, wireless Internet video delivery, 413, 420–423
bandwidth availability estimation, 422–423
packet loss differentiation/estimation, 422
Network-centric techniques
end-to-end quality of service, wireless Internet video delivery, 413
cross-layer support, 414–419
video communications routing, performance analysis, 174–175

Network layer
broadband wireless local area networks, applications and standards, 338–339
handoff management solutions, 451–452
wireless multimedia sensor network, 27–29
Network models, broadband wireless local area networks, packet scheduling, 475–476
Network simulator (NS)
ad hoc wireless networks, concurrent PDP, two node-disjoint paths, 201
multimedia streaming user-perceived quality evaluation, 395–401
Neuro-dynamic programming, learning-based packet scheduling, 494–497
performance evaluation, 497–500
No *Ad Hoc* (NOAH) agent, multimedia streaming user-perceived quality evaluation, 395–401
No Caching (NC) algorithm, peer-assisted video streaming quality enhancement, 322
Node cooperation, peer-assisted video streaming, incentive mechanisms, 327–328
Node-disjoint multipath (NDM) routing, *ad hoc* wireless networks, concurrent packet drop probability, unicast systems, 200–201
Node selection
802.11a 802.11a, *ad hoc* wireless network standard
moving nodes, 210–213
static nodes, 208–210
video communications routing, branch-and-bound techniques, 183
Noise, multimedia streaming, 381
Nonlinear companding transforms, peak-to-average power ratio reduction, 519–523
Non-real-time variable bit rate (nrt-VBR), broadband wireless local area networks, heterogeneous traffic, 479
Nonsimultaneous media content consumption, interactive mobile television, 90–92
Number of bad periods, *ad hoc* wireless networks
parallel multiple nearly disjoint trees protocol, 221
simulation comparisons, 205–206

O

Oline parameter tuning, learning-based packet scheduling, 496–497
OLSR protocol, video communications distributed implementation, 189
OMA BCAST standard, mobile TV systems, 101–103
security issues, 115
On-Demand Multicast Routing Protocol (ODMRP), *ad hoc* wireless networks
multicast system, 198–199
Serial Multiple Disjoint Tree Multicast Routing (MDTMR) protocol, 216–218
simulation results, 222–228
On-demand streaming, adaptive multimedia streaming, 372
Open Systems Interconnection (OSI) framework, broadband wireless local area networks, cross-layer design and adaptation, 341–342
OPNET Modeler, video transmission strategy, 247–248

Optimal adaptation trajectory (OAT)
multimedia streaming evaluation, 395–401
user-perceived video streaming quality
adaptive video streaming, 384
basic principle, 386–395
quality-oriented adaptation scheme, 384–386
results, 390–395
test methodology, 387–389
test sequence, 389–390
wireless local area networks, user-perceived video streaming quality, 362–365
Optimal multipath routing
ad hoc wireless networks
heuristic solution, 203–204
results comparisons, 204–206
unicast streaming, 199–200, 229–230
branch-and-bound techniques, 178–181
multiple description (MD) video, 166–168
genetic algorithms, 169–170
video communications routing, numerical results, 185–188
Orthogonal frequency division multiplexing (OFDM)
IEEE 802.11a standard, 367
overview, 506–507
peak-to-average power ratio
clipping and filtering, 514–515
coding schemes, 515–516
distribution, 510–513
future research issues, 528
MIMO-ODFM systems, 527–528
nonlinear companding transforms, 519–523
PTS/SLM approaches, 517–519
reduction criteria, 524–525
reduction techniques, 513–524
signal properties, 509–510
TR/TI techniques, 523–524
WiMax systems, 526–527
signal characteristics, 508–509
wireless multimedia sensor networks, 37–38
Overlay networks, end-to-end quality of service, wireless Internet video delivery, 427–428

P

Packet delay, multiparty audioconferencing, 129
Packet drop probability (PDP), *ad hoc* wireless networks
concurrent PDP, two node-disjoint paths, 200–201
link-based computation, 201–203
multipath selection, unicast streaming, 199–200
optimum multipath selection, 203–204
unicast systems, 195–196
Packet forwarding, multiple-tree multicast systems, 215–216
Packet loss percentage
end-to-end QoS, differentiation and estimation, 422
mobile multimedia handoff management, 444–445
multiparty audioconferencing, 129
Packet overhead, *ad hoc* wireless networks, parallel multiple nearly disjoint trees protocol, 221
Packet scheduling, broadband wireless local area networks
admission control collaboration, 501
algorithms, 482–490
bandwidth utilization, 481
channel model, 476–477
cost function construction, 492–494
fairness, 481–482
future research issues, 500–501
Gilbert–Elliot channel model, 485–487
heterogeneous traffic properties, 479–480
learning-based scheduling, 490–500
multistate channel model, 487–488

network model, 475–476
neuro-dynamic programming, 494–497
overview, 474–475
performance evaluation, 497–500
quality of service, 481
semi-Markov decision process, 490–492
standards compatibility, 501
system inaccuracy, 500–501
traffic model, 477–479
wireless link variability, 480–481
wireline networks, 482–485
Pair comparison, optimal adaptation trajectory, 389
Parallel multiple nearly disjoint trees multicast routing protocol, *ad hoc* wireless networks, 218–228
Partial transmission sequence (PTS), peak-to-average power ratio reduction, 507, 517–519
Partitioning
ad hoc wireless networks, packet drop probability, 202–203
video communications routing, branch-and-bound techniques, 184–185
Path pair packet drop probability (PP_PDP), *ad hoc* wireless networks, unicast systems, 195–196
PCF mechanism, IEEE 802.11 multimedia protocols, 266–267
Peak/mean bit rate ratios, multimedia streaming user-perceived quality evaluation, 397
Peak-signal-to-noise-ratio (PSNR)
802.11a 802.11a, *ad hoc* wireless network standard
moving nodes, 210–213
static nodes, 209–210
multimedia streaming, 382–383
peer-assisted video streaming, data selection, 318–319
video communications routing
network-centric techniques, 175
shortest-path performance, 187–188
video transmission performance, 249, 251–254
Peak-to-average power ratio (PAPR), orthogonal frequency division multiplexing
clipping and filtering, 514–515
coding schemes, 515–516
distribution, 510–513
future research issues, 528
MIMO-ODFM systems, 527–528
nonlinear companding transforms, 519–523
PTS/SLM approaches, 517–519
reduction criteria, 524–525
reduction techniques, 513–524
signal properties, 509–510
TR/TI techniques, 523–524
WiMax systems, 526–527
Peer-assisted video streaming, wireless local area networks
caching for quality enhancement, 319–322
content challenges, 308
content discovery, 312–314
dynamic wireless channel quality, 327
energy efficiency and awareness, 322–326
mobility tracking, 314–315
node cooperation incentives, 327–328
peer selection, 315–317
prior protection-based techniques, 308–309
quality-of-service requirements, 327
rate-distortion optimization, 317–319
resource exploitation and enhancement, 309–310
security issues, 328
streaming protocols, 311–319
Peer selection, peer-assisted video streaming, 312, 315–317
energy efficiency and awareness, 325–326

Peer update (PUPD) message, peer-assisted video streaming, content discovery, 314
Performance analysis
  ad hoc wireless networks, parallel multiple nearly disjoint trees protocol, 226–228
  learning-based packet scheduling, 497–500
  peak-to-average power ratio reduction, orthogonal frequency division multiplexing systems, 524–526
  video communications routing
    branch-and-bound algorithms, 186–188
    genetic algorithm, 172–175
  video transmission, 249–254
Performance metrics, video transmission strategy, 248–249
Person locator services, wireless multimedia sensor networks, 7
Physical layer, broadband multimedia services, wireless LAN, 336
Picture appraisal rating (PAR), multimedia streaming, 382–383
Picture quality rating (PQR), multimedia streaming, 383–384
Pixel-domain Wyner–Ziv encoder, wireless multimedia sensor network, 18–20
Point Coordination Function (PCF)
  end-to-end quality of service support, network-centric techniques, wireless Internet, 415–419
  IEEE 802.11e MAC enhancement, 370–371
Point-to-point push bearer, mobile TV systems, 104
Poisson model, broadband wireless local area networks, packet scheduling, 478
Polyhedral outer approximation, video communications routing, branch-and-bound techniques, 178–180
Popularity-based caching (PBC) algorithm, peer-assisted video streaming quality enhancement, 320
Power consumption, wireless multimedia sensor networks, 9
Predictively coded frames (P-frames), mobile TV systems, channel zapping, 110–112
Prefix caching algorithm, peer-assisted video streaming quality enhancement, 320
Prioritized transmission control, end-to-end quality of service support, wireless Internet video delivery, 418
Priority scheduling, high speed downlink packet access, 75–78
Proactive principles, mobile multimedia handoff management, 445
Probability density function (PDF), high speed downlink packet access mobility, handover delay, 65–67
PROB message, video transmission strategy, 243–247
Progressive streaming, adaptive multimedia streaming, 371–372
Proportional Fair (PF) scheduler, high speed downlink packet access
  basic principles, 50
  performance analysis, 56–57
Protection-based streaming, peer-assisted video streaming
  data selection, 318–319
  IEEE 802.11 WLANS, 308–309

Protocol data unit (PDU), IEEE
802.11 multimedia
protocols, 265–267
Protocols
IEEE 802.11 multimedia standards,
265–267
wireless multimedia sensor
networks, 23–36
application-specific and
nonstandard protocols,
26–27
channel access policies, 29
contention-based protocols,
29–30
contention-free multi-channel
protocols, 32–33
contention-free single-channel
protocols, 30–31
cross-layer protocols, 35–36
link-layer error control, 34–35
MAC layer, 29–35
network-based QoS routing,
27–28
network layer, 27–29
scheduling, 33–34
streaming routing protocols,
28–29
TCP/UCP and TCP-friendly
schemes, 24–26
traffic classes QoS routing, 28
transport-layer protocols, 24–27
Proxy-assisted systems
middleware design guidelines,
450
peer-assisted video streaming
content discovery, 313–314
mobility tracking, 314–315
Push-to-talk, multiple audio streams,
130–131

Q

Quadrature amplitude modulation
(QAM), orthogonal
frequency division
multiplexing, 508–509
Quality control
multiparty audioconferencing, 125
constraints on, 129
multiparty audioconferencing
architecture, 149
Quality of experience (QoE)
framework, mobile TV
systems, 110–111
Quality of service (QoS)
awareness of, 438
broadband wireless local area
networks
algorithms, 344–347
end-to-end scheme, 341,
347–348
MAC layer, 340–341
packet scheduling, 481
seamless support systems,
354–356
end-to-end quality of service
(QoS), wireless Internet
video delivery
end-system-centric support,
419–426
future research issues,
427–428
network-centric cross-layer
support, 414–419
overview, 410–413
IEEE 802.11 standards
admission control, 273–274
EDCA performance analysis,
274–282
EDCA protocol, 271–273
enhanced capability in 802.11e,
268–270
future research challenges, 299
HCCA protocol, parameterized
QoS, 270
IEEE 802.11e MAC
enhancement, 368–371
inadequate legacy 802.11
support, 267–268
MAC efficiency improvement,
283–287
mobile multimedia handoff
management, 443–444
mobile TV systems, 3GPP MBMS
standard, 96–97
multimedia streaming
user-perceived quality,
379–381

Quality of service (QoS) (*contd.*)
802.11n QoS features, 283–288
overview, 262–263
peer-assisted video streaming, traffic requirements, 327
protocol requirements, 265–267
requirements, 264–265
reverse direction protocol, 287–288
scheduling algorithm design, 270–271
802.11s mesh network mechanisms, 289–299
wireless Internet constraints, 437–438
wireless multimedia sensor networks
application-specific requirements, 7–8
evolution of, 4–5
network conditions, 27–28
traffic classes routing, 28
wireless sensor networks, time-constrained traffic, 239–240
Quality-oriented adaptation scheme (QOAS)
multimedia streaming evaluation, 395–401
optimal adaptation trajectory, 384–386
wireless local area networks, user-perceived video streaming quality, 362–365
Quarter common intermediate format (QCIF)
ad hoc wireless networks
simulation comparisons, 205–206
video bit rates, 164–166
optimal adaptation trajectory, 389–390
peer-assisted video streaming, data selection, 318–319
video communications routing, shortest-path performance, 187–188

R

Rate constraints
adaptive multimedia streaming, 375
ad hoc wireless networks
control algorithms, 207–208
unicast system, 197–198
Rate distortion-based bit allocation, end-to-end quality of service, wireless Internet video delivery, 426
Rate-distortion optimization (RaDiO) streaming, peer-assisted video, data selection, 317–319
Ratio of bad frames, *ad hoc* wireless networks
parallel multiple nearly disjoint trees protocol, 221
simulation comparisons, 205–206
Real-time, loss-tolerant/loss-intolerant data, wireless multimedia sensor network, 15–16
Real-time, loss-tolerant multimedia streams, wireless multimedia sensor network, 15
Real-time streaming, adaptive multimedia streaming, 372
Real-Time Transport Control Protocol (RTCP), adaptive multimedia streaming, 372
Real-Time Transport Protocol (RTP)
adaptive multimedia streaming, 372
customized mixing, 127
mobile TV systems
Quality of Experience framework, 110
streaming, 105–106
peer-assisted video streaming, 317–319
energy efficiency and awareness, 323–326
Real-time variable bit rate (rt-VBR), broadband wireless local area networks, heterogeneous traffic, 479

Real video encoding, adaptive multimedia streaming, 378
Rebuffering duration, mobile TV systems, 110
Receiver-based packet pair (RBPP), bandwidth estimation, 422–423
Reconfigurability, mobile multimedia handoff management, 445–446
Recovery time objective (RTO), TCP-friendly rate control, 375–376
Reduced interframe space (RIFS), IEEE 802.11n MAC efficiency improvement, 286–287
Reference architecture, wireless multimedia sensor networks, 10–12
Reformulation linearization technique (RLT), video communications routing
  bound-factor product constraints, 181–182
  bounding step, 185
  branch-and-bound techniques, 178–188
Reliability
  mobile TV systems
    file downloads, 107–108
    streaming sessions, 108–109
  wireless multimedia sensor networks, TCI/UDP protocols, 25–26
Request-to-send (RTS) message, video transmission strategy, 244–247
Resource-based caching (RBC) algorithms, peer-assisted video streaming quality enhancement, 319–320
Resource consumption, high speed downlink packet access, 80–82
Reverse direction protocol, IEEE 802.11n standards, 287–288
Roaming
  broadband wireless local area networks, seamless QoS support, 354
  interactive mobile television, 95
Round-robin (RR) scheduler, high speed downlink packet access
  basic principles, 49
  performance analysis, 55, 78–80
Round trip time (RTT), TCP-friendly rate control, 375–376
Routing overhead, parallel multiple nearly disjoint trees multicast routing protocol, 218

S

Scalability
  multiparty audioconferencing, 124
    architecture, 139
  wide area large-scale participation distribution, 127–128
Scalable video, adaptive error control, 423–424
Scheduler algorithms
  broadband wireless local area networks, 344–345
  high speed downlink packet access performance, 68–82
    comparisons, 78–81
    priority scheduling, 75–78
    resource consumption, 81–82
    specific properties, 68–75
  IEEE 802.11 standards, 271–272
Scheduling protocols
  high speed downlink packet access performance, 54–57
  wireless multimedia sensor networks, 33–34
Seamless QoS support, broadband wireless local area networks, 354–356
  cellular network interface, 354–356
  roaming, 354
Security issues
  mobile TV systems, 114–115
  peer-assisted video streaming, 328

Selection algorithm, multiparty audioconferencing architecture, loudness number, 145–146
Selective mapping (SLM), peak-to-average power ratio reduction, 507, 517–519
Selective partial caching algorithm, peer-assisted video streaming quality enhancement, 321
Self-organized entities, multiparty audioconferencing architecture, 146–147
Semi-Markov decision process, learning-based packet scheduling, 490–492
Sensors, wireless multimedia sensor networks, single *vs.* multitier development, 12–13
Serial Multiple Disjoint Tree Multicast Routing (MDTMR) protocol, *ad hoc* wireless networks, multicast systems, 216–218
Server-based fairness approach (SBFA), Gilbert–Elliot (GE) channel model, 486
Server-client service model, peer-assisted video streaming, 309–310
Service continuity, defined, 437–438
Service data unit (SDU), IEEE 802.11 multimedia protocols, 265–267
Service interval (SI) calculation, IEEE 802.11 standards, scheduler algorithms, 271–272
Service protection, mobile TV systems, 115
Service rebind, mobile multimedia handoff management, 443
Session initiation protocol (SIP)
  handoff management, 457
  Internet Engineering Task Force wireless network signaling, 127
  multiparty audioconferencing architecture, 139–140
  self-organized entities, 146–147
Session mobility, multiparty audioconferencing architecture, 149
Shortest path (SP) routing
  ad hoc wireless networks, multicast system, 198
  video communications, performance analysis, 174–175, 187–188
Shortest widest path (SWP) routing, *ad hoc* wireless networks, simulation comparisons, 204–206
Signaling
  multiparty audioconferencing, 124
  orthogonal frequency division multiplexing, 508–509
  wireless network interactivity, 127
Simple block coding, peak-to-average power ratio reduction, 515–516
Simple Caching Simple Fetch (SCSF), peer-assisted video streaming quality enhancement, 322
Simulation, video transmission, model properties, 247–248
Simulation results, *ad hoc* wireless networks, 204–206
  parallel multiple nearly disjoint trees protocol, 221–228
Simultaneous media content consumption, interactive mobile television, 90–92
Single description coding (SDC)
  ad hoc networks, 195
  ad hoc wireless networks
    moving nodes, 210–213
    static nodes, 209–210
Single radio systems, common control channel (CCC) protocol, 291
Single-tier sensor development, wireless multimedia sensor networks, 12–13

Sink delivery, video transmission strategy, network lifetime, 248
Software requirements
  ad hoc wireless networks, 206–207
  wireless multimedia sensor network, 20–22
Source coding, wireless multimedia sensor networks, 8–9
Source routing, video communications distributed implementation, 188–189
Spatialization, multiparty audioconferencing, 124
Spatial-temporal grid, optimal adaptation trajectory, 387–388
SPEED protocol
  wireless multimedia sensor networks, streaming regulation, 28–29
  wireless sensor networks, quality of service time-constrained traffic, 239–240
Standards. *See also* IEEE 802.11 standard
  ad hoc wireless networks
    moving nodes, 210–213
    static nodes, 208–210
  broadband wireless local area networks, 338–339
    packet scheduling compatibility, 501
  mobile TV systems, 95–103
    3GPP2 BCMCS, 96–98
    3GPP MBMS, 95–96
    IP datacast over DVB-H, 98–101
    OMA BCAST, 101–103
Start-time fair queuing (SFQ), broadband wireless LANs, wireline networks, 484
Static nodes, 802.11a 802.11a, *ad hoc* wireless network standard, 208–210
Streaming protocols
  adaptive multimedia streaming
    algorithm control parameters, 376–377
    algorithm responsiveness, 377–378
    codec constraints, 378
    enhanced loss delay adjustment algorithm, 376
    feedback-based over-reaction, 377
    overview, 371–372
    quality control, 372–373
    real video encoding parameters, 378
    solutions, 373–376
    TCP-friendly rate control and TFRC protocol, 375–376
    TCP throughput model, 377
    user perception, 378
  interactive mobile television
    broadcast/multicast streaming, 92–93
    media delivery framework, 105–106
    reliability, 108–110
  mobile Internet protocol-enabled WLAN, seamless streaming, 454–455
  multiparty audioconferencing
    multiple audio streams, 129–131
    selection criteria, 141–143
  peer-assisted video streaming quality enhancement
    caching enhancement, 319–322
    resource enhancement, 309–310
  wireless local area networks
    peer-assisted video streaming, content challenges, 308
    video streaming, 352–354
  wireless multimedia sensor networks
    routing protocols, 28–29
    TCI/UDP protocols, 25–26
Streaming session ID (SSID), peer-assisted video streaming, content discovery, 313–314
Stream mixing, multiparty audioconferencing, 125

T

Target beacon transition time (TBTT), IEEE 802.11 multimedia protocols, 266–268
TCI/UDP protocols, wireless multimedia sensor networks, 24–26
TCP-friendly rate control (TFRC)
  adaptive multimedia streaming, 375–376
  network adaptive congestion/error/power control, 421–423
  peer-assisted video streaming, peer selection, 316
TCP-Westwood, bandwidth estimation, 422–423
Terrestrial-Digital Multimedia Broadcast (T-DMB) standard, mobile TV systems, 100–101
Testbed implementation and evaluation, *ad hoc* wireless networks, 206–213
  802.11a moving nodes, 210–213
  802.11a static nodes, 208–210
  rate control scheme, 207–208
  setup, 208
  software architecture, 206–207
TFRC Protocol (TFRCP), TCP-friendly rate control, 375–376
Third Generation Partnership Project (3GPP) standard
  end-to-end quality of service support, network-centric techniques, wireless Internet, 415–419
  mobile TV systems, 95–98
Time division multiple access (TDMA), wireless multimedia sensor networks, contention-free single-channel protocols, 30–31
Time-hopping ultra wideband (UWB) communications, wireless multimedia sensor networks, 37–38
Time slicing, mobile TV systems, 98–100
Time-To-Live (TTL) scheduler
  high speed downlink packet access
    basic principles, 50
    performance analysis, 57–58
    scheduler properties, 68–75
  peer-assisted video streaming quality enhancement, distortion minimizing smart caching algorithm, 321–322
Tone reservation/tone injection (TR/TI), peak-to-average power ratio reduction, 507, 523–524
Tournament selection, genetic algorithm, video communications routing, 170
Traffic classes
  broadband wireless local area networks, packet scheduling, 477–478
  peer-assisted video streaming, quality-of-service requirements, 327
  wireless multimedia sensor network, 15–17
  wireless multimedia sensor networks, quality of service routing, 28
Traffic reduction
  multiparty audioconferencing, 124–125
    constraints on, 128–129
    mixing architectures, 137–138
  wireless multimedia sensor networks, 6
Trajectory methods, video communications routing, 172–174
Transform-domain Wyner–Ziv encoders, wireless multimedia sensor network, 20

Transmission Control Protocol (TCP)
  end-to-end semantics, data-link/network/transport layers, 452
  throughput model, adaptive multimedia streaming, 377
Transmission opportunity (TXOP)
  common control channel MAC performance, 295–299
  IEEE 802.11 standards
    enhanced IEEE 802.11e capability, 269–270
    IEEE 802.11e MAC enhancement, 371
    scheduler algorithms, 271
  mesh deterministic access, 290
Transmission quality, high speed downlink packet access mobility, handover threshold, 63–64
Transmission rate
  IEEE 802.11 standard, 368
    multimedia protocols, 268
  peer-assisted video streaming, energy efficiency and awareness, 323–326
Transmission time interval (TTI), high speed downlink packet access, 49
Transport layer, handoff management solutions, 451–452
Transport-layer protocols, wireless multimedia sensor networks, 24–27
Tree connectivity/similarity
  multicast routing protocols, 214–215
  parallel multiple nearly disjoint trees multicast routing protocol, 218
Tune-in delay, interactive mobile television, 94

U

Ultra wideband (UWB) communications
  ranging capabilities, 38–40
  wireless multimedia sensor networks, 37–40
Unicast systems, *ad hoc* wireless networks, 195–196
  concurrent packet drop probability, two node-disjoint paths, 200–201
  envisioned network model, 199–200
  future research, 228–229
  heuristic optimum multipath selection, 203–204
  packet drop probability link computation, 201–203
  routing protocols, 197–198
  simulation results, 204–206
Unlicensed mobile access (UMA), handoff management, 463
Unspecified bit rate (UBR), broadband wireless local area networks, 480
Upper bounds, video communications routing
  branch-and-bound techniques, 177–188
  initialization and relaxation, 183
Upstream node selection, parallel multiple nearly disjoint trees multicast routing protocol, 219–220
Usage scenarios, mobile television, 92–93
User ID (UID), peer-assisted video streaming, content discovery, 313–314
User interface
  high speed downlink packet access performance, 53–54
  interactive mobile television
    broadcast/multicasting streaming, 93
    protocols, 106–107
User-perceived video streaming quality, wireless local area networks
  bandwidth fluctuations, 379–380
  congestion, 379

User-perceived video (*contd.*)
delay, 380
heterogeneity, 379
jitter, 380–381
loss, 381
mathematical metrics, 382–383
model-based metrics, 383–384
multimedia clips, 397
noise, 381
objective video quality metrics, 382–384
optimum adaptation
trajectory-based adaptation scheme, 384–395
overview, 363–365
quality of service factors, 379–381
quality-oriented adaptation scheme principle, 384–386
quality-oriented multimedia streaming evaluation, 396–401
simulation models, 397
testing scenarios and results, 397–401
User service announcements, mobile TV systems, 104

V

Variable bit rate (VBR), prefix caching algorithm, peer-assisted video streaming quality enhancement, 320
Vertical handover, middleware-based handoff management, 456–457
Video bit rates, *ad hoc* wireless networks, 164–166
Video communications systems
ad hoc networks
application-layer performance, 162–164
bit rates and success probabilities, 164–166
branch-and-bound approach, 175–188
ε-optimal algorithm, 183–185
heuristic algorithm, 182–183
linearization, 181–182
numerical results, 185–188
distributed implementation, 188–189
metaheuristic approach, 168–175
genetic algorithms, 168–172
performance comparisons, 172–175
optimal double-path routing, 166–168
overview, 158–161, 194–195
wireless Internet video delivery, end-to-end quality of service (QoS)
end-system-centric support, 419–426
future research issues, 427–428
network-centric cross-layer support, 414–419
overview, 410–413
wireless local area networks, streaming protocols, 352–354
wireless sensor networks
future research issues, 254–255
multipath-based real-time video, 241–247
overview, 236–238
performance evaluations, 249–254
performance metrics, 248–249
QoS provisioning, time-constrained traffic, 239–240
simulation model, 247–248
WLAN video transmissions, 239
Video file name (VFN), peer-assisted video streaming, content discovery, 313–314
Video on demand, digital television technology and, 89
Video quality metric (VQM), multimedia streaming, 381–384

Video sensor networks (VSNs)
multipath-based real-time video communications, architecture, 241–243
overview, 236–237
Video Transfer Protocol (VTP), data-link/network/transport layers, 452
Voice activity detection (VAD)
multiparty audioconferencing, traffic reduction, 124–125
multiparty audioconferencing architecture, bandwidth consumption reduction, 147–148
Voice-over-IP (VoIP)
high speed downlink packet access, 48–49
multiparty audioconferencing architecture, 139–140
wireless local area networks, 352

W

Weighted cumulative expected transmission time (WCETT), *ad hoc* wireless networks, IWM protocol, 206–207
Weighted fair queuing (WFQ), broadband wireless LANs, wireline networks, 484
Weighted fair queuing squared (WFQ^2), broadband wireless LANs, wireline networks, 484–485
Weighted preference path, optimal adaptation trajectory, 393–394
Weighted round robin (WRR), broadband wireless LANs, wireline networks, 483–484
Weighted signal to noise ratio (WSNR), multimedia streaming, 382–383
WFS algorithm, Gilbert–Elliot channel model, 486–487
Wideband Code Division Multiple Access (WCDMA)-based systems, high speed downlink packet access, 49
WiMax system, peak-to-average power ratio reduction, 526–527
Window-based control, adaptive multimedia streaming, 375
Wireless channel capacity, *ad hoc* wireless networks, packet drop probability, 202–203
Wireless Internet video delivery, end-to-end quality of service (QoS)
end-system-centric support, 419–426
future research issues, 427–428
network-centric cross-layer support, 414–419
overview, 410–413
Wireless local area networks (WLANs)
adaptive multimedia streaming
algorithm control parameters, 376–377
algorithm responsiveness, 377–378
codec constraints, 378
enhanced loss delay adjustment algorithm, 376
feedback-based over-reaction, 377
overview, 371–372
quality control, 372–373
real video encoding parameters, 378
solutions, 373–376
TCP-friendly rate control and TFRC protocol, 377
TCP throughput model, 377
user perception, 378
broadband multimedia services, 336–338
IEEE 802.11 standards, 365–371
802.11a, 367
802.11b, 366–367
distance range/transmission rate comparisons, 368

Wireless local area networks (*contd.*)
802.11e MAC enhancement for quality of service, 368–369
enhanced distributed coordinated access, 369–371
802.11g, 367–368
multimedia, 264–265
evolution of, 306–308
peer-assisted video streaming
caching for quality enhancement, 319–322
content challenges, 308
content discovery, 312–314
dynamic wireless channel quality, 327
energy efficiency and awareness, 322–326
mobility tracking, 314–315
node cooperation incentives, 327–328
peer selection, 315–317
prior protection-based techniques, 308–309
quality-of-service requirements, 327
rate-distortion optimization, 317–319
resource exploitation and enhancement, 309–310
security issues, 328
streaming protocols, 311–319
user-perceived video streaming quality
bandwidth fluctuations, 379–380
congestion, 379
delay, 380
heterogeneity, 379
jitter, 380–381
loss, 381
mathematical metrics, 382–383
model-based metrics, 383–384
multimedia clips, 397
noise, 381
objective video quality metrics, 382–384
optimum adaptation
trajectory-based adaptation scheme, 384–395
overview, 363–365
quality of service factors, 379–381
quality-oriented adaptation scheme principle, 384–386
quality-oriented multimedia streaming evaluation, 396–401
simulation models, 397
testing scenarios and results, 397–401
video transmissions, 239
Wireless mesh networks, multimedia content provisioning, 465–466
Wireless multimedia sensor networks (WMSNs)
application layer, 14–23
collaborative in-network processing, 22
multimedia encoding, 16–20
open research issues, 23
pixel-domain Wyner–Ziv encoder, 18–20
software and middleware, 20–22
traffic classes, 15–16
transform-domain Wyner–Ziv encoder, 20
applications, 6–7
architecture, 10–14
coverage, 13
internal multimedia sensor organization, 13–14
reference architecture, 10–12
single- *vs.* multitier sensor development, 12–13
defined, 4
design and characteristics, 7–10
evolution of, 4–6
physical layer, 36–40
protocols, 23–36
application-specific and nonstandard protocols, 26–27
channel access policies, 29

contention-based protocols, 29–30
contention-free multi-channel protocols, 32–33
contention-free single-channel protocols, 30–31
cross-layer protocols, 35–36
link-layer error control, 34–35
MAC layer, 29–35
network-based QoS routing, 27–28
network layer, 27–29
scheduling, 33–34
streaming routing protocols, 28–29
TCP/UCP and TCP-friendly schemes, 24–26
traffic classes QoS routing, 28
transport-layer protocols, 24–27
ultra wideband communications, 37–40
Wireless multimedia services
end-to-end quality of service support, network-centric techniques, wireless Internet, wireless multimedia enhancements (WMEs), 415–416
middleware handoff management approach
adaptability, 445–446
application layer, 452–459
content provision, wireless mesh networks, 465–466
data-link/network/transport layers, 451–452
full context awareness, 439–445
future development issues, 462–464
handoff awareness, 441–442
IEEE 802.21 media-independent handover, 462–463
interactive mobile application session handoff, 453–454
Internet protocol multimedia subsystem, 463–464
Internet protocol multimedia system open handoff, 466
location awareness, 442–443
media-independent preauthentication, 452–453
mobile agent-based ubiquitous middleware, 458–459
mobile Internet protocol-enabled WLAN seamless streaming, 454–455
mobile network support, 464–465
Mobiware, 455–456
overview, 436–439
proactive solutions, 445–446
quality of service awareness, 443–445
reconfigurability, 445–446
service continuity design guidelines, 446–450
session initiation protocol-based handoff solutions, 457
solution evaluations, 459–462
state-of-the-art solutions, 450–462
unlicensed mobile access, 463
vertical handover, 456–457
Wireless network interface cards (WNICs), peer-assisted video streaming, energy efficiency and awareness, 322–326
Wireless networks
interactive broadcasting constraints, 125–129
customized mixing, 127
interactivity, 126
multicasting availability, 128
quality controls, 129
scalability, 127–128
signaling, 127
traffic reduction, 128–129
multiparty audioconferencing, limitations, 122–123
Wireless sensor networks (WSNs), video communications systems
future research issues, 254–255
multipath-based real-time video, 241–247
overview, 236–238

Wireless sensor networks (*contd.*)
performance evaluations, 249–254
performance metrics, 248–249
QoS provisioning, time-constrained traffic, 239–240
simulation model, 247–248
WLAN video transmissions, 239
Wireless technology, wireless multimedia sensor network integration, 10
Wireline networks, broadband wireless LANs, 482–485
Wyner–Ziv encoders, wireless multimedia sensor network, 18–20

RECEIVED
OCT - 3 2008